NATO ASI Series

Advanced Science Institutes Series

A Series presenting the results of activities sponsored by the NATO Science Committee, which aims at the dissemination of advanced scientific and technological knowledge, with a view to strengthening links between scientific communities.

The Series is published by an international board of publishers in conjunction with the NATO Scientific Affairs Division

A Life Sciences **B Physics**	Plenum Publishing Corporation London and New York
C Mathematical **and Physical Sciences** **D Behavioural and Social Sciences** **E Applied Sciences**	Kluwer Academic Publishers Dordrecht, Boston and London
F Computer and Systems Sciences **G Ecological Sciences** **H Cell Biology**	Springer-Verlag Berlin, Heidelberg, New York, London, Paris and Tokyo

Series C: Mathematical and Physical Sciences - Vol. 281

Fluid Movements – Element Transport and the Composition of the Deep Crust

Fluid Movements —
Element Transport and the
Composition of the Deep Crust

edited by

David Bridgwater

Geological Museum,
Copenhagen, Denmark

Kluwer Academic Publishers

Dordrecht / Boston / London

Published in cooperation with NATO Scientific Affairs Division

Proceedings of the NATO Advanced Research Workshop on
Fluid Movements – Element Transport and the Composition of the Deep Crust
Lindås, Norway
May 18–24, 1987

Library of Congress Cataloging in Publication Data

```
Fluid movements : element transport and the composition of the deep
   crust : proceedings of the NATO advanced research workshop held in
   Lindås, Norway, May 18-24, 1987 / edited by David Bridgwater.
        p.    cm. -- (NATO ASI series. Series C, Mathematical and
   physical sciences ; vol. 281)
     Includes index.

      1. Earth--Crust--Congresses.   2. Geothermal deposits--Congresses.
   I. Bridgwater, David.   II. Series: NATO ASI series.   Series C,
   Mathematical and physical sciences ; no. 281.
   QE511.F625   1989
   551.1'3--dc20                                              89-34355
```

ISBN-13: 978-94-010-6935-9   e-ISBN-13: 978-94-009-0991-5
DOI: 10.1007/978-94-009-0991-5

Published by Kluwer Academic Publishers,
P.O. Box 17, 3300 AA Dordrecht, The Netherlands.

Kluwer Academic Publishers incorporates the publishing programmes of
D. Reidel, Martinus Nijhoff, Dr W. Junk and MTP Press.

Sold and distributed in the U.S.A. and Canada
by Kluwer Academic Publishers,
101 Philip Drive, Norwell, MA 02061, U.S.A.

In all other countries, sold and distributed
by Kluwer Academic Publishers Group,
P.O. Box 322, 3300 AH Dordrecht, The Netherlands.

Printed on acid free paper

This book contains the proceedings of a NATO Advanced Research Workshop held within the programme of activities of the NATO Special Programme on Global Transport Mechanisms in the Geo-Sciences running from 1983 to 1988 as part of the activities of the NATO Science Committee.

Other books previously published as a result of the activities of the Special Programme are:

BUAT-MENARD, P. (Ed.) – *The Role of Air-Sea Exchange in Geochemical Cycling* (C185) 1986

CAZENAVE, A. (Ed.) – *Earth Rotation: Solved and Unsolved Problems* (C187) 1986

WILLEBRAND, J. and ANDERSON, D.L.T. (Eds.) – *Large-Scale Transport Processes in Oceans and Atmosphere* (C190) 1986

NICOLIS, C. and NICOLIS, G. (Eds.) – *Irreversible Phenomena and Dynamical Systems Analysis in Geosciences* (C192) 1986

PARSONS, I. (Ed.) – *Origins of Igneous Layering* (C196) 1987

LOPER, E. (Ed.) – *Structure and Dynamics of Partially Solidified Systems* (E125) 1987

VAUGHAN, R. A. (Ed.) – *Remote Sensing Applications in Meteorology and Climatology* (C201) 1987

BERGER, W. H. and LABEYRIE, L. D. (Eds.) – *Abrupt Climatic Change – Evidence and Implications* (C216) 1987

VISCONTI, G. and GARCIA, R. (Eds.) – *Transport Processes in the Middle Atmosphere* (C213) 1987

SIMMERS, I. (Ed.) – *Estimation of Natural Recharge of Groundwater* (C222) 1987

HELGESON, H. C. (Ed.) – *Chemical Transport in Metasomatic Processes* (C218) 1987

CUSTODIO, E., GURGUI, A. and LOBO FERREIRA, J. P. (Eds.) – *Groundwater Flow and Quality Modelling* (C224) 1987

ISAKSEN, I. S. A. (Ed.) – *Tropospheric Ozone* (C227) 1988

SCHLESINGER, M.E. (Ed.) – *Physically-Based Modelling and Simulation of Climate and Climatic Change* 2 vols. (C243) 1988

UNSWORTH, M. H. and FOWLER, D. (Eds.) – *Acid Deposition at High Elevation Sites* (C252) 1988

KISSEL, C. and LAY, C. (Eds.) – *Paleomagnetic Rotations and Continental Deformation* (C254) 1988

HART, S. R. and GULEN, L. (Eds.) – *Crust/Mantle Recycling at Subduction Zones* (C258) 1989

GREGERSEN, S. and BASHAM, P. (Eds.) – *Earthquakes at North-Atlantic Passive Margins: Neotectonics and Postglacial Rebound* (C266) 1989

MOREL-SEYTOUX, H. J. (Ed.) – *Unsaturated Flow in Hydrologic Modeling* (C275) 1989

ANDERSON, D.L.T. and WILLEBRAND, J. (Eds.) – *Ocean Circulation Models: Combining Data and Dynamics* (C284) 1989

TABLE OF CONTENTS

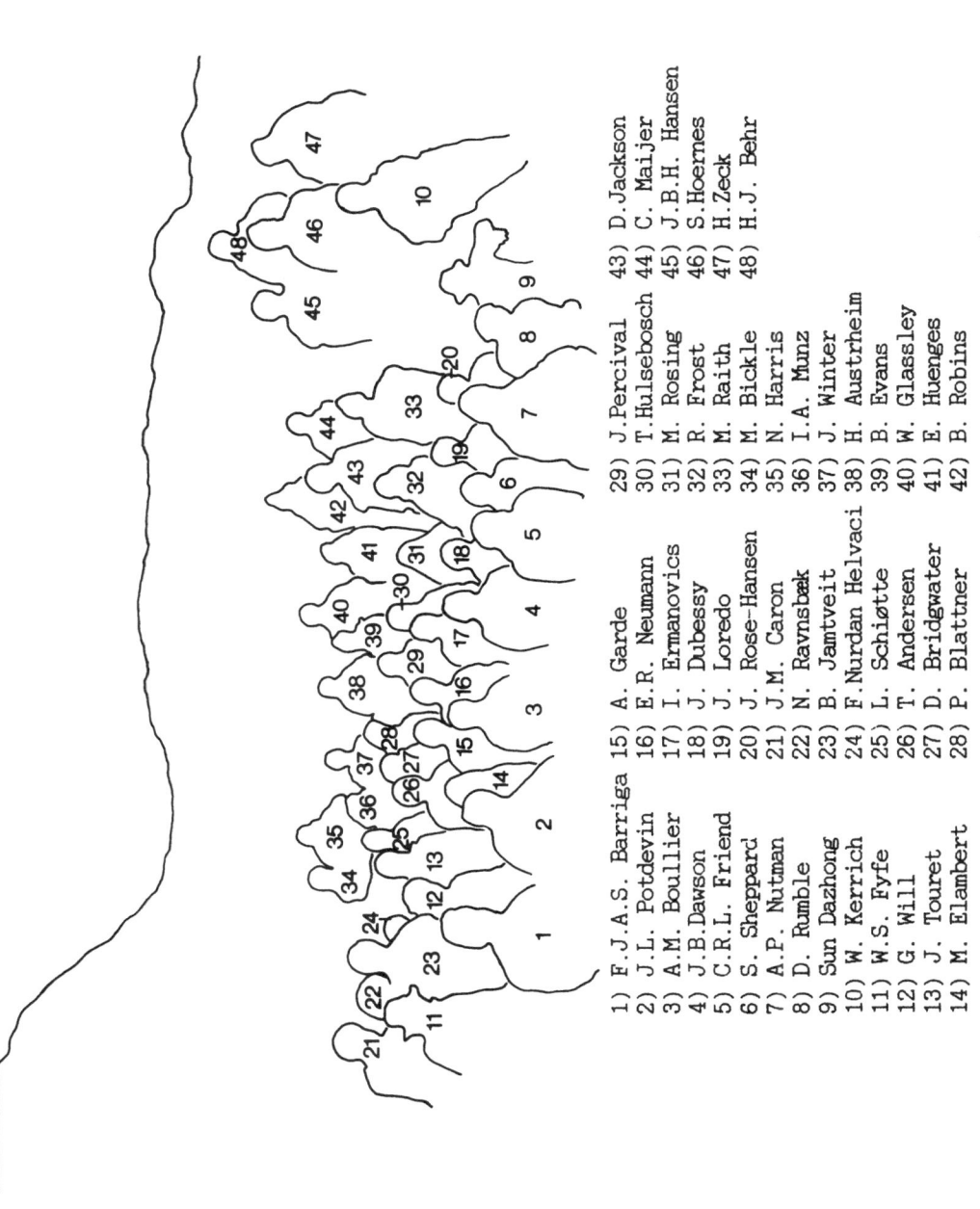

1) F.J.A.S. Barriga
2) J.L. Potdevin
3) A.M. Boullier
4) J.B.Dawson
5) C.R.L. Friend
6) S. Sheppard
7) A.P. Nutman
8) D. Rumble
9) Sun Dazhong
10) W. Kerrich
11) W.S. Fyfe
12) G. Will
13) J. Touret
14) M. Elambert

15) A. Garde
16) E.R. Neumann
17) I. Ermanovics
18) J. Dubessy
19) J. Loredo
20) J. Rose-Hansen
21) J.M. Caron
22) N. Ravnsbæk
23) B. Jamtveit
24) F. Nurdan Helvaci
25) L. Schiøtte
26) T. Andersen
27) D. Bridgwater
28) P. Blattner

29) J.Percival
30) T.Hulsebosch
31) M. Rosing
32) R. Frost
33) M. Raith
34) M. Bickle
35) N. Harris
36) I.A. Munz
37) J. Winter
38) H. Austrheim
39) B. Evans
40) W. Glassley
41) E. Huenges
42) B. Robins

43) D.Jackson
44) C. Maijer
45) J.B.H. Hansen
46) S.Hoernes
47) H.Zeck
48) H.J. Behr

PREFACE

Many geologists have an equivocal attitude to fluid movements within the crust and the associated changes in the chemical and physical properties of crustal rocks. The controversies earlier this century between the "soaks" and the "pontiffs" memorably summarised by H.H. Read (1957) in **The Granite Controversy** have largely been resolved. Few would now advocate the formation of large granitic bodies by in situ transformation of pre-existing crust as the result of the passage of ichors without the formation of a granitic melt. To many geochemists fluid transport and metasomatism have become slightly suspect processes which at the most locally disturb the primary geochemical and isotopic signatures. While there is common agreement that there are marked differences in the composition of the lower and upper crust, the role of fluid movement as one of the controls of this differentiation is often neglected in favour of suggested primary differences in the composition of igneous rocks emplaced at different depths. Selective fluid transport however provides many geologists with their livelihood. Without the secondary concentration of commercially important elements by fluids within the crust the mining industry, geological science and human activities based on their products would be very different.

In this situation it perhaps not surprising there is still a polarisation between the more theoretical geochemists, often with a training in physics or chemistry and interests in processes on a global scale or even astronomical scale, and those with a more local field oriented approach who have made the attempt to apply standard geochemical models to rocks from high grade terranes and found them wanting. A workshop on fluid movements and element transport was timely even if by its title it was likely to attract those who were at least sympathetic to the concept.

The individual papers show there is wide range of opinion about the degree in which different elements can be transported during fluid movements. This illustrates one of the fundemental difficulties of working with fluid transport within the crust. There can be a general agreement that at the temperatures and pressures found more than few kilometers down, fluids must be capable of dissolving and redepositing large amounts of material, but virtually everyone who has worked on the subject has different experiences and different views of just what is taken into solution. We are not dealing with a single fluid in a single set of conditions, we are rarely dealing with processes which go to completion. An immobile element used as a reference of original composition in one set of conditions can become the most mobile constituent in other conditions (see for example Al in papers by Rosing, Zeck and Toft, and Potdevin et al., in this volume).

Lindås, just north of Bergen, Norway, was a highly appropriate location to discuss fluid movements and element transport in the deep crust. Scandinavia has provided some of the main litigants in the debate between those who advocated regional transformation of earlier crustal rocks to give rise to granites and gneisses and those who advocated formation from melts. The Bergen Arc area contains some of the most spectacular exposures of high grade gneisses anywhere. Coarse-grained anorthosites and associated igneous rocks which were affected by granulite facies metamorphism during the Grenville Orogeny were piled up as a series of thrust slices during the Caledonian. During the thrusting pressures approached 15-17 kilobars and the very dry Grenville granulite facies assemblages in the basic rocks recrystallised to eclogite along discordant fracture zones, the centers of which contain hydrated minerals. The importance of fluid movement in controlling the formation of new mineral assemblages (and thereby element redistribution) in this transition zone between the crust and the mantle is seen by the development of eclogite minerals for tens of meters on either side the shear zones.

Workshops such as that at Lindås which have a slightly informal character and in which the emphasis is on where should we go, rather where have we got, grow from the ideas of a small group of people with common interests. The Lindås meeting started as a result of conversations between myself, W.S. Griffin, and H. Austrheim (Oslo), W. Fyfe (University of Western Ontario) J. Winter (Whitman College), W. Glassley (Livermore Laboratories), R. Frost (University of Wyoming), and the research group with whom I have worked for several years in Greenland and Labrador centered in the Geological Surveys of Canada and Greenland. We found that this interest was shared by other research groups particularly in France and Germany. Professor H. Sørensen, (Institute of Petrology, University of Copenhagen) suggested that a meeting on fluid transport in the deep crust was an ideal subject as part of the NATO Science committee special programme on global transport mechanisms. Without both the considerable financial support from the NATO science division, and administrative advice from the staff at Brussels there would have been no meeting. Special thanks are due to Dr Brian Robins at the University of Bergen and his department for the very effective practical organisation of both the meeting and considerable help with the accounts. H. Austrheim ran two excellent excursions during an unexpected heat wave on the Norwegian coast. The series of papers in the present volume reflect the wide range of subject discussed at the workshop Although there have been considerable delays in publication for which I take most of the responsibility - the overall objectives of the volume are those of a workshop - to present the ongoing state of research in different aspects of fluid transport in the deep crust rather than finished and polished accounts. Finally I would like to thank the members of the workshop -it was a good meeting with an excellent atmosphere - you made it so.

D. Bridgwater, The Geological Museum, Copenhagen, March 1989.

Reference: Read, H.H. (1957) The Granite Controversy. T. Murby & Co. London.

LIST OF PARTICIPANTS

Dr. T. Andersen
Mineralogisk-Geologisk Museum
Sars' Gate 1
N-0562   Oslo 5
NORGE

Dr. H. Austrheim
Mineralogisk-Geologisk Museum
Sars' Gate 1
N-0562   Oslo 5
NORGE

Dr. F.J.A.S. Barriga
Departamento de Geologia
Faculdade de Ciências
Rua da Escola Politécnica, 58
P-1200   Lisboa
PORTUGAL

Dr. H.J. Behr
Inst. für Geologie und
Dynamik der Lithosphäre
Universität Göttingen
Goldschmidt-Str. 3
D-3400   Göttingen
WEST GERMANY

Dr. M. Bickle
Dept. of Earth Sciences
University of Cambridge
Downing Street
Cambridge   CBQ 3EQ
U.K.

Dr. P. Blattner
New Zealand Geological Survey
P.O. Box 30 368
Lower Hutt
NEW ZEALAND

Dr. A. M. Boullier
C.N.R.S., C.R.P.G.
Boîte Postale 20
F-54501 Vandoeuvre-les-Nancy Cedex
FRANCE

Dr. D. Bridgwater
Geologisk Museum
Øster Voldgade 5-7
DK-1350   Copenhagen K
DANMARK

Dr. J.M. Caron
Department des Sciences de la Terre
Université Claude-Bernard
27-43 bd du 11 Novembre
F-69622 Villeurbanne, Cedex
FRANCE

Dr. I. Cartwright
Department of Geology and Geophysics
Lewis G Weeks Hall for Geol. Sci.
1215 W Drayton St.
Madison, Wisconsin 53706
U.S.A.

Dr. J.B. Dawson
Dept. of Geology
University of Sheffield
Mappin Street
Sheffield S1 3JD
U.K.

Dr. J. Dubessy
Centre de Recherches sur la Géologie
 de l'Uranium
3 rue du Bois-de-la-Champelle
F-54500 Vandoeuvre-les-Nancy, Cedex
FRANCE

M. Elambert
Mineralogisk-Geologisk Museum
Sars' Gate 1
N-0562   Oslo 5
NORGE

Dr. I. Ermanovics
Geological Survey of Canada
588 Booth Street
Ottawa K1A OE4
CANADA

Dr. B. Evans
Geological Sciences, AJ 20.
University of Washington
98195
U.S.A.

Dr. C.R.L. Friend
Dept. of Geology and Physic. Sci.
Oxford Polytechnic
Oxford OX3 0B0
U.K.

Dr. R. Frost
Dept. of Geology and Geophysics
The University of Wyoming
P.O. Box 3006
Laramie, Wyoming 82071
U.S.A.

Dr. H. Fossen.
Geologisk Inst, Avd. A
Realfagsbygget
Bergen Universitet
N-5014 Bergen
NORGE

Dr. W.S. Fyfe
Department of Geology
University of Western Ontario
London, Ontario
CANADA N6A 5B7

A.A. Garde
Geological Survey of Greenland
Øster Voldgade 10,
DK-1350 Copenhagen K.
DANMARK

Dr. W. Glassley
Lawrence Livermore National Lab.
University of California
P.O. Box 808
Livermore, California 94550
U.S.A.

Dr. N. Harris
Department of Geology
Open University
Milton Keynes
U.K.

T. Hansteen
Mineralogisk-Geologisk Museum
Sars Gate 1 Seattle, WA
N-0562, Oslo 5
NORGE

F.Nurdan Helvaci
MTA Genel Mudurlugu
Maden Etud ve Arama Dairesesi
Ankara
TURKEY

Dr. S. Hoernes
Mineralogisches-
Petrologisches inst.
und Museum der Universitat
Poppelsdorfer Schloss
D-5300 Bonn 1
WEST GERMANY

Dr. E. Huenges
Mineralogisches-
Petrologisches inst.
und Museum der Universitat
Poppelsdorfer Schloss
D.5300 Bonn
WEST GERMANY

Dr. T. Hulsebosch
Department of Geology and Geophysics
The University of Wyoming
P.O. Box 3006
Laramie, Wyoming 82071
U.S.A.

Dr. D. Jackson
Department of Geology
Open University
Milton Keynes
U.K.

Dr. J.B.H. Jansen
P M K G B, afd. Petrologie
Rijksuniversiteit Utrecht
Instituut voor aardwetenschappen
Postbus 80.021
NL-3508 TA Utrecht
HOLLAND

B. Jamtveit
Mineralogisk-Geologisk Museum
Sars'Gaten 1
N-0562 Oslo
NORGE

Dr. R. Kerrich
Department of Geological Sciences
University of Saskatchewan
S7N OWO Saskatoon,
CANADA

Dr. E. Klatt
Mineralogisch-
Petrologisches inst.
und Museum der Universitat
Poppelsdorfer Schloss
5300 Bonn
WEST GERMANY

L. Kullerud
Mineralogisk-Geologisk Museum
Sars'Gate 1
N 0562, Oslo 5.
NORGE

Dr. Jorge Loredo
Dept. of Geology
University of Oviedo
Oviedo
SPANIEN

Dr. C. Maijer
P M K G B, afd. Petrologie
Rijksuniversiteit Utrecht
Instituut voor aardwetenschappen
Postbus 80.021
NL-3508 TA Utrecht
Nederlands

I.A. Munz
Mineralogisk-Geologisk Museum
Sars' Gate 1
N-0562 Oslo
NORGE.

Dr. E. R. Neumann
Mineralogisk-Geologisk Museum
Sars' Gate 1
N-0562 Oslo 5
NORGE.

Dr. D.I. Norman,
Department of Geoscience
New Mexico Inst. of Mining and Tech
Socorro, N M 87801
U.S.A.

Dr. G. Nover
Mineralogisches-
Petrologisches inst.
und Museum der Universitat
Poppelsdorfer Schloss
D.5300 Bonn 1
WEST GERMANY

Dr. A.P. Nutman
Dept. of Geology
Memorial University
St. John's, Newfoundland
CANADA   A1B 3X5

Dr. J.A. Percival
Geological Survey of Canada
588 Booth Street
Ottawa K1A 0E4
CANADA

Dr. J.L. Potdevin
Département des Sciences de la Terre
Université Claude-Bernard
27-43 bd du 11 Novembre
F-69622 Villeurbanne, Cedex
FRANCE

N. Ravnsbæk
Geologisk Museum
Østervoldgade 5-7
DK 1350, Copenhagen
DANMARK

Dr. M. Raith
Mineralogisches-
Petrologisches inst.
und Museum der Universitet
5300 Bonn
WEST GERMANY

Dr. B. Robins
Geological Institute, Avd A
Realfagbyget
Bergen Universitet
N-5014 Bergen
NORGE

Dr. J. Rose-Hansen
Institute of Petrology
University of Copenhagen
Øster Voldgade 10
DK-1350 Copenhagen K
DANMARK

Dr. M. Rosing
Geologisk Museum
University of Copenhagen
Øster Voldgade 5-7
DK-1350 Copenhagen K
DANMARK

Dr. D. Rumble III
Geophysical Laboratory
2801 Upton Street N W
Washington, DC 20008
U.S.A.

Dr. L. Schiøtte
Geologisk Museum
University of Copenhagen
Øster Voldgade 5-7
DK-1350 Copenhagen K
DANMARK

Dr. S.M.F. Sheppard
C.N.R.S.
C.R.P.G.
Boîte Postale 20
F-54501 Vandoeuvre-les-Nancy Cedex
FRANCE

Dr. Sun Dazhong
Tianjin institute of Geology and
Mineral resources
Chinese Academy of Geological Sciences
No.4, 8th Road
Dazhigu
Tinajin 300170
THE PEOPLE'S REPUBLIC OF CHINA.

Dr. J. Touret
Instituut voor Aardwetenschappen
Vrije Universiteit
Postbus 7161
NL-1007 MC Amsterdam
HOLLAND

Dr. G. Will
Mineralogisches-Petrologisches
 Inst. und Museum der Universität
Poppelsdorfer Schloss
D-5300 Bonn 1
WEST GERMANY

Dr. J. Winter
Division III, Basic Sciences
Whitman College
Walla Walla, Washington 99362
U.S.A.

Dr. H. Zeck
Institute of Petrology.
University of Copenhagen
Øster Voldgade 10
DK-1350 Copenhagen K
DANMARK

MAGMAS AS A SOURCE OF HEAT AND FLUIDS IN GRANULITE METAMORPHISM

B. Ronald Frost, Carol D. Frost
Department of Geology and Geophysics
University of Wyoming
Laramie, Wyoming 82071
U.S.A.

Jacques L. R. Touret
Instituut voor Aardwetenschappen
Virje Universiteit
N-1007 MC Amsterdam
Netherlands

ABSTRACT. Three diverse modes of granulite formation, CO_2-streaming, partial melting, and recrystallization of originally anhydrous rocks, can be aspects of the same process: movement of magmas through the lower crust. CO_2-saturated silicic and mafic magmas can exsolve enough CO_2 to dehydrate a volume of country rock approximately equal to 20%that of the magma itself. Consequently, movement of magmas through the crust can provide both the heat and the CO_2 necessary for granulite metamorphism. Furthermore, silicic magmas emplaced into the deep crust are likely to produce anhydrous pyroxene-bearing cumulates (ie. charnockites) while more hydrous portions of the magma would be forced to migrateto shallower, cooler levels before they could crystallize to theH2O-saturated liquidus. Thus, magmas may form conduits by which CO_2 of mantle origin is transported into the lower crust while H_2O is extracted from the lower crust and moved to shallower levels. Evidencesupporting this hypothesis lies in the abundance of CO_2 fluid inclusionsin clearly igneous charnockitic rocks, in the elevated geotherms suggested by P-T conditions of some granulites, and in the relict igneous features found in the highest grade areas of some granulite terranes. This theory implies that some felsic rocks with high K/Rb ratios may be cumulates, and that such K/Rb ratios are not diagnostic of CO_2-fluxing.

INTRODUCTION

The problem of granulite genesis and its geochemical signature has received considerable attention in the past decade. One theory maintains that granulites form as the result of massive influx of CO_2 from lower crustal or mantle sources (Heier, 1973; Collerson and Fryer, 1978; Newton and Hansen, 1983). Such an influx could be responsible for the marked depletion many granulite terranes show in large ion lithophile elements

1

D. Bridgwater (ed.), Fluid Movements – Element Transport and the Composition of the Deep Crust, 1–18.
© *1989 by Kluwer Academic Publishers.*

(LILE) (Heier, 1973; Tarney and Windley, 1977). Others, however, maintain that granulite metamorphism is a necessary complement of crustal melting (Brown and Fyfe, 1970, 1972; Fyfe, 1973; Nesbitt, 1980). Recently Lamb and Valley (1984,1985) noted that evidence for pervasive CO_2 streaming is absent in the relatively reducing, graphite-free metamorphosed plutonic rocks in some terranes, because influx of CO_2 into such rocks would have deposited significant amounts of graphite and homogenized oxygen fugacities to those of the graphite-saturation surface. They argue that granulite metamorphism may be a product of one of three processes: CO_2-streaming, partial melting, or recrystallization of originally dry rocks. It is our contention that these processes need not be independent and that they all may be related to the passage of melts through the crust (Frost and Frost, 1987). Melts moving through the lower crust will provide heat for melting of fusible lithologies, they may exsolve a CO_2-rich vapor that can dehydrate surrounding crustal rocks (see Wells, 1979), and they may themselves crystallize to anhydrous rocks, which, when deformed, will be examples of the "metamorphosed dry rocks" mentioned by Lamb and Valley (1984).

CO_2 IN MELTS

Numerous experiments show that CO_2 has a finite, albeit rather limited, solubility in silicate melts and that this solubility is enhanced at high pressures and temperatures (Kadik and Lukanin, 1973; Mysen, 1976; Mysen et al., 1975; 1976; Brey, 1976; Eggler and Rosenhauer, 1978; Eggler and Kadik, 1979). Carbon dioxide dissolves into melts according to the following equilibria:

$$CO_2 \text{ fluid} = CO_2 \text{ melt} \tag{1}$$

$$CO_2 \text{ melt} + 2\,NBO^- = CO_3^{2-} \text{ melt} + BO \tag{2}$$

where NBO = non-bridging oxygens in the melt and BO = bridging oxygens (Mysen, 1976; Fine and Stolper; 1985). In addition, data in Spera and Bergman (1980) and Fine and Stolper (1985, 1986) indicate that carbonate is not present merely as a loose radical (ie. CO_3^{2+}), but that most of it is bound to a metal ion. Thus, a third equilibrium must also be considered:

$$CO_3^{2-} \text{ melt} + 2/n\,R^{n+} \text{ melt} = R_{2/n}CO_3 \text{ melt} \tag{3}$$

where R represents a metal ion of valence n. Anything that drives equilibria (2) or (3) to the right will increase CO_2 solubility in melts. For example CO_2 is more soluble in hydrous melts than in anhydrous melts of the same composition because H_2O tends to depolymerize melts (Mysen, 1976; Eggler and Rosenhauer, 1978). Similarly, CO_2 solubility increases with increasing calcium content at constant T and P in the systems

$CaSiO_3$-$MgSiO_3$ and Ca_2SiO_4-Mg_2SiO_4 (Holloway et al., 1976), reflecting the tendency for Ca^{+2} to bond with CO_3^{2-} by equilibrium (3) more readily than does Mg^{2+}.

Unfortunately, experimental work to date is insufficient to allow for the development of thermodynamic models that will predict the CO_2 solubility in silicate melts as a function of temperature, pressure, and melt composition. Furthermore, because most experiments were done at temperatures well above those of the liquidus, it is difficult even to determine CO_2 saturation for most melts under typical lower crustal conditions. Despite these problems, some significant conclusions can be derived from the literature on CO_2 saturation of silicate melts. These data indicate that under anhydrous conditions CO_2 is more soluble in melts rich in non-bridging oxygens, such as basalt, than in highly polymerized melts, such as in melts of albite composition (Mysen et al., 1975; Mysen, 1976). However, because H_2O depolymerizes melts, the maximum solubility of CO_2 in the presence of H_2O is nearly equal in basaltic and albitic melts (Kadik et al., 1972; Mysen et al., 1975; Mysen, 1976). Experimental data indicates that H_2O-bearing granitic melts at mid-crustal levels (ca. 3 to 5 kilobars) can dissolve 0.4 weight % CO_2 (Kadik et al., 1972; Holloway and Lewis, 1974) and that dry basaltic melt at 25 kilobars can accommodate more than 1.0 weight % CO_2 (Mysen et al., 1975; Spera and Bergman, 1980).

Because CO_2 is much less soluble in silicate melts than is H_2O, melts that are saturated in CO_2 will crystallize at higher temperature than those that are H_2O-saturated. Furthermore, because the CO_2 solubility of melts increases with increasing pressure and temperature (Mysen, 1976; Mysen et al., 1975, Spera and Bergman, 1980), during emplacement into the crust most melts will become saturated in a vapor phase long before H_2O saturation is attained (Kadik and Lukanin, 1973; Holloway, 1976). Once a melt is vapor-saturated, further decrease in pressure or temperature, even without any crystallization, will lead to the evolution of a fluid phase. Initially, this fluid phase will be CO_2-enriched (see Fig. 1). If the fluid is progressively removed from the

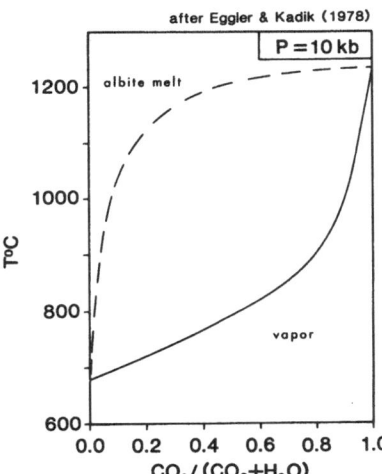

Figure 1. T – X diagram showing the partitioning of CO_2 between albite melt and a binary fluid phase. Data from Eggler and Kadik (1979).

magma body as the melt rises, then the magma will be able to remain molten down to lower temperatures and the exsolved fluid will become progressively enriched in H_2O (see Fig. 1). Along with CO_2, the early fluids will also be enriched in other components, such as NaCl, that are not strongly soluble in the silicate melt (Kadik and Lukanin, 1973). If sufficient NaCl were present in the fluid, it could lead to the unmixing of the fluid to a CO_2-rich vapor and a saline brine even at temperatures in excess of $1000^{\circ}C$ (Bowers and Helgeson, 1983; Skippen and Trommsdorff, 1986). This is important because such unmixing could further enrich the CO_2 content of the vapor phase, which, being less dense than the associated brine, will be more likely to infiltrate and dehydrate the country rocks.

PETROLOGIC EVIDENCE

Evidence for CO_2-rich Fluids from Melts

Although CO_2-streaming is not the sole process that produces granulites, the validity of any theory calling for melts as a source of heat and CO_2 during granulites metamorphism is dependent on two questions: 1) can magmas dissolve sufficient CO_2 to produce significant amounts of granulite and 2) is there evidence that melts in the lower crust are ever CO_2-saturated?

Before answering the question of how much granulite can be produced due to exsolution of CO_2 from melts, we must first determine how large a volume of rock can be dehydrated by infiltration of one mole of CO_2 into amphibolite-grade crustal material. In this calculation it is assumed that the rock has a porosity of 0.1% and that these pores contain a fluid with $X(CO_2) = 0.0$. In addition to the fluid present in the pores, the fluid initially present in the rock also includes the H_2O incorporated in the hydrous silicates. We will consider two rock types; a biotite granite with 5% biotite and amphibolite with 50% hornblende. Using expression (4) of Greenwood (1975) one can calculate how much CO_2 will be required to dehydrate the silicates and to drive the fluid in these rock to $X(CO_2) = 0.8$, a common value in granulites (see Wells, 1979). These calculations indicate that one mole of CO_2 will dehydrate 640 cm^3 of biotite granite or 130 cm^3 of amphibolite.

These crude calculations probably underestimate the volume of rock dehydrated because they assume that H_2O is passively driven off during CO_2 streaming and that no other process operates. Grant (1986) noted, however, that the first effect of CO_2-influx into a biotite-bearing granitic rock will be to produce an H_2O-bearing melt. If this melt is incorporated into the through-moving magmas, then the CO_2-streaming process will be far more efficient than is modelled here. Even so these simplified calculations indicate that a granitic melt with 0.4 weight % CO_2 at 5 kilobars can accommodate enough CO_2 to granulatize a volume of felsic gneiss equal to about 20% that of the melt itself (Frost and Frost, 1987).

These calculations indicate that if CO_2-saturated melts are emplaced into the lower crust, they could exsolve enough CO_2 to produce significant amounts of granulite. Furthermore there is ample evidence to indicate that significant amounts of CO_2 are transported through the crust in melts. For example basaltic glass from the sea floor contains 100 to 300 ppm CO_2 (Fine and Stolper, 1985). This value probably reflects substantial degassing during eruption, since melts included in olivine from Kilauea have an order of magnitude more CO_2 (Harris and Anderson, 1983). These data indicate that oceanic basalts and magmatic arc basalts contain substantial amounts of CO_2 and that major CO_2-degassing accompanies eruption (Harris and Anderson, 1984). However, it is not known whether CO_2-evolution accompanies the rise of magmas into the crust or whether CO_2-saturation is attained only at upper crustal levels. Fluid inclusions in phenocrysts and mantle nodules indicate that at least some basaltic magmas become CO_2-saturated at lower crustal levels. CO_2-rich fluid inclusions are common in mantle nodules, and those examined generally have densities of 0.8 to 1.2 g/cm^3 (Roedder, 1965, 1984), indicating trapping at magmatic temperatures and at pressures of less than 11 kilobars. These fluid inclusions were either primary inclusions that stretched during decompression or inclusions that were trapped during the uplift of the nodules. If they are primary, they indicate that the alkali basalts in which the inclusions are trapped formed in the presence of a CO_2 fluid and therefore were CO_2-saturated throughout their ascent. If the fluid inclusions were trapped during ascent, and it takes only a few hours to a few days to trap fluid inclusions in quartz during healing of fractures (Sterner and Bodner, 1984), then the basalts were CO_2-saturated from depths of 11 kilobars.

It is much more difficult to characterize the fluid composition of granitic melts because of the explosive nature of silicic volcanism. However, available evidence indicates that silicic magmas may also contain significant amounts of CO_2. Even though it must have expelled considerable amounts of CO_2-rich vapors, the Bishop tuff contains 800 +/- 400 ppm CO_2 (Sheridan and Moore, 1981). Phenocrysts from the dacitic tuffs of St. Helens apparently equilibrated with fluids that had $X(H_2O)$ = 0.60, (Rutherford and Devine, 1986) the remainder of the fluid was largely CO_2. Indeed, in four months following the 1980 eruption, 2.2 x 10^{10} moles of CO_2 were emitted from St. Helens (Harris et al., 1981).

Fluid inclusion studies of charnockitic granitoids also provide evidence of CO_2-saturated melts. These show that the fluid composition changes from carbonic to aqueous as the rock evolves from charnockite to biotite granite (Konnerup-Madsen, 1977; Hulsebosch et al., 1985). In fact, in some of the granitoids studied, the presence of a two-phase fluid is suggested by the occurrence of both carbonic and saline fluid inclusions (Konnerup-Madsen, 1977; 1984; Konnerup-Madsen and Rose-Hansen, 1982; Konnerup-Madsen et al., 1985). Admittedly, there is some question as to whether the fluid inclusions in charnockites reflect the composition of the primary fluid. However, the occurrence of narrow zones of granulite metamorphism around thin charnockite dikes in the Wind River Mountains of Wyoming (see Frost and Frost, 1987, Fig. 1) is persuasive evidence that charnockitic melts exsolve fluids that dehydrate the country rock (see also Koesterer et al., in press).

Finally, the magnitude of the present-day carbon flux from the mantle is evidence that large volumes of CO_2 are being carried into and through the crust by magmatic processes. By considering the ^{3}He flux from the mantle and the mantle ^{3}He/C ratio, Des Marais (1985) estimated the present-day carbon flux from the mantle to be on the order of 1 to 8 x 10^{12} moles C/yr. Although most of this carbon will be exhaled at mid-ocean spreading centers, some fraction will be emplaced into the crust. Let us assume that because submarine volcanism accounts for more than 90% of the mantle ^{3}He exhaled (O'Nions and Oxburgh, 1983), such volcanism also delivers 90% of mantle carbon (Des Marais, 1985). Ten percent of the mantle carbon flux, about 1 to 8 x 10^{11} moles of carbon/year, may be delivered to the base of the continents. As with the mantle flux in general, this CO_2 will be concentrated in areas of active magmatism, and if it interacted with crustal rocks during ascent, it could dehydrate 0.1 to 0.5 km^3 crust/year.

Thermobarometric Evidence

Pressure and temperature conditions recorded by the mineral assemblages in granulite terranes also suggest that granulite metamorphism is directly related to an igneous event. Many granulite facies assemblages require elevated geothermal gradients (Fig. 2). Bohlen (1987) reached a similar conclusion based on evidence that many granulite terranes show evidence for isobaric cooling, a phenomenon that is more consistent with igneous heating than with tectonic loading. Extensional regimes, such as the Basin and Range, have higher geothermal gradients that encompass some granulite conditions, but this does not necessarily imply that all granulites formed in such settings. They could equally well be produced in areas of lower regional geothermal gradient where the geotherms have been locally elevated by the passage of melt.

Figure 2 also documents that there is a complete gradation from low-pressure contact environments to high pressure granulite terranes. In the low-pressure terranes, the contact aureoles of the Kiglapait intrusion and Laramie Anorthosite Complex, the participation of a melt as a heat is source easily recognized, both by the high temperatures encountered and the isobaric nature of the transition from amphibole-bearing to pyroxene-bearing assemblages. In a slightly higher pressure terrane, the Rogaland, the isograd pattern and extreme temperatures of metamorphism (Tobi et al., 1985) leave little doubt that the heat source was the Rogaland Anorthosite massif. In other low-pressure granulite terranes, the Wind River (Koesterer et al., in press) and Williyama (Phillips and Wall, 1981), the role of igneous activity as a heat source is equally clear.

It is only in medium and high pressure granulite terranes that the role of igneous processes becomes more problematic. This problem is partly caused by the fact that an obvious igneous source is missing from many terranes and partly by the fact that the P-T conditions for higher pressure regimes lie within the range of the thermal gradient expected in orogenic regimes. The lack of obvious igneous heat sources in many high-pressure terranes may be an artifact of preservation; igneous cumulates emplaced at great depth shoould be expected to undergo deformation as the remaining melt is squeezed out to higher levels (Hopson and Dellinger, 1987). The common occurrence of high-pressure terranes recording

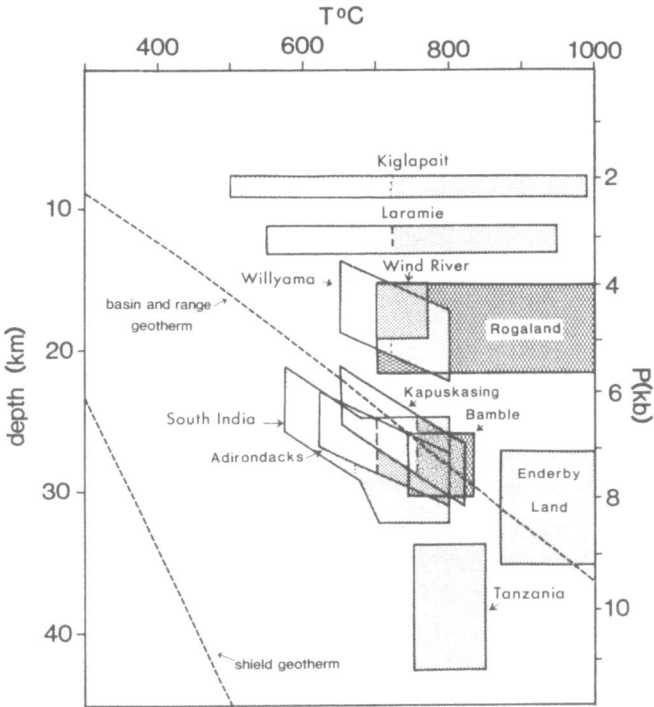

Figure 2. Depth versus temperature for some granulite and amphibolite terranes compared to Shield and Basin and Range geotherms· Heavy symbols = granulite facies, light symbols = amphibolite facies. Transition is shown in solid lines if temperature is well known, dotted line if it is poorly constrained. Data from Sclater et al., (1980) (geotherms); Berg and Docka (1983) (Kiglapait), Grant and Frost (1986) (Laramie); Koesterer et al. (in press) (Wind River);Janson et al. (1985) (Rogaland); Phillips and Wall (1981) (Willyama); Bohlen et al. (1985) (Adirondacks); Percival (1983) (Kapuskasing); Raase et al. (1986) (S. India); Lamb et al. (1986) (Bamble); Ellis (1980) and Harley (1986) (Enderby Land); Newton (1983) (Tanzania).

temperatures of 700 - 800 °C may be a reflection of the fact that ion-exchange geothermometers are also geospeedometers. The kinetics of ion-exchange thermometers are such that even in a quickly cooled igneous body, such as the Sybille Monzosyenite, such thermometers do not yield temperatures much above 800°C (Fuhrman et al., in press). Higher temperatures can be obtained only through reintegration of "fossil" thermometers: ie. exsolution lamallae in pyroxenes, feldspars, or oxides. These fossil thermometers, however, are delicate features that are easily destroyed by deformation. More importantly, data from fossil thermometry

may also be lost in a slowly cooled pluton through extensive granular
exsolution, as demonstrated by Lindsley and Andersen (1983).Therefore,
ion-exchange thermometers may not record the maximum temperature of a
metamorphic terrane. Indeed, because most granulite terraneshave had a
complex geologic history, it may be exceedingly difficult to prove
whether the rocks in a given terrane had experienced temperatures or
pressures higher than those recorded by ion-exchange thermobarometry (cf.
O'Hara, 1975). In this regard, Enderby Land (Fig. 2) may beunique, not
so much because it attained extreme temperatures, but becauseevidence
for these temperatures has been preserved.

Evidence from the Bamble Area

Because the Bamble area of southern Norway was the first areawhere
granulite metamorphism was shown to be associated with CO_2 rich fluids
(Touret, 1971), it serves as an appropriate location to examine in more
detail the possibility that the CO_2 is of magmatic origin. These rocks,
collectively described as the charnockitic gneisses of Tromoy(near
Arendal, Southern Norway) (Field et al., 1980), constitute a complete
igneous suite ranging from basic to acidic quartz-bearing varieties. All
display a conspicuous LILE-depleted character, formerly attributed to
element scavenging during high-grade metamorphism, but it hasbeen
convincingly demonstrated (Field et al., 1980, Smalley et al., 1983;
Field et al., 1985) that most geochemical features are related to
synmetamorphic igneous fractionation. These magmas, intrudedat a deep
crustal level, crystallized directly to their present metamorphic
assemblages (Lamb et al., 1986).
    Carbonic inclusions are present in all rock types but they are
distinctly more abundant in the basic varieties. They concentrate
especially in coarse-grained segregations that probably represent the
last, fluid-enriched parts of the crystallizing magmas (Touret, 1985).
In addition to trace element evidence, an igneous origin of these rocks
is established by the persistance of typical magmatic features in some
plagioclase. This includes remnants of igneous textures not completely
obliterated by metamorphic recrystallization and typical igneous-looking
solid inclusions of apatite and zircon (cf. Touret, 1985). These
magmatic inclusions are also found in some metamorphic minerals, notably
garnet, suggesting igneous crystallization at peak metamorphic
conditions.
    A magmatic origin of the CO_2 inclusions is supported by a number of
arguments:
    1) Earliest inclusions have densities consistent with the P-T
conditions that correspond to peak metamorphism and, therefore, from the
above-mentioned considerations, to the crystallization of the deep-
seated magmas (Touret, 1981; 1985).
    2) Early CO_2 inclusions are rare or absent in metasediments
associated with the orthogneisses. Most metaquartzites and metapelites,
which are well-represented in the granulite facies domain, contain NaCl-
rich brines as the dominant and earliest fluids. The same istrue for
skarns, in which concentrated brines with up to 50% NaCl are far more
abundant than CO_2 inclusions.

3) Stable isotope data are inconclusive, but they rule out large-scale derivation from sediments. Most $\delta^{13}C$ measured in inclusions are very low ($\delta^{13}C$ between -10 and -20 ‰ relative to PDB (Hoefs and Touret, 1975; Pineau et al., 1981)). They correspond to a mixture of different generations which may have different origins, with the latest fluids showing evidence of the presence of reduced species (CH_4 and graphite). Significant quantities of mantle-derived carbonates ($\delta^{13}C = -7$ ‰) have been observed in some samples (Pineau et al., 1981), but the positive $\delta^{13}C$ values one would expect if CO_2 was derived directly from breakdown of sedimentary carbonates have not been observed.

The field relations in the Bamble area can be interpreted as a high-pressure equivalent of the Rogaland Complex, with the highest grade zones representing a suite of synmetamorphic igneous plutons and the lower-grade zones being a large-scale contact aureole (Lamb et al., 1986). Unlike the Rogaland, the synmetamorphic plutons are not anorthositic, but are compositionally similar to the country rock which they intrude. Thus, distinguishing between the intrusions and the country rock requires detailed chemical comparisons (Lamb et al., 1986).

A MODEL FOR GRANULITE FORMATION

The arguments above emphasize that a substantial amount of CO_2 moves through the earth's crust entrapped in melts. Although there is no absolute estimate of the amount of CO_2 that would be expelled by these melts at lower crustal depths where it would be available to produce granulites, we contend that the total amount of CO_2 transported thorough the continents in melts is large enough that CO_2 exsolved from melts must be considered as a possible cause of granulite metamorphism. The manner in which this may happen is illustrated in Figure 3. A rifting environment of relatively high heat flow is diagrammed (cf. Sandiford and Powell, 1986), although granulites are almost certainly associated with the deeper portions of magmatic arcs as well, for example, with the Sierra Nevada (Ross, 1985) and Coast Range batholiths (Hollister, 1975) (cf. Bohlen, 1987). In this model basaltic magma of mantle origin rises into the crust. Some portions of this magma move through the crust to be erupted on the surface. During their passage they will become CO_2-saturated and will evolve a CO_2-rich fluid phase. This fluid may move upward in association with the magma but, given conduits, such as deformation zones or dikes, it may be bleed off into the adjacent country rocks where it may produce granulite metamorphism. In addition, heat from this magma may melt the more fusible portions of the country rock with the resulting relatively hydrous melt being incorporated into the magma. The end result of these processes will be dehydration in the vicinity of the feeder.

A large portion of the basaltic magma, however, is likely to pond in the lower crust, where it will produce melting of the adjacent crustal material. Because the temperatures involved lie well above the H_2O-saturated granite minimum, crustal melts generated in this environment will be either vapor undersaturated or will coexist with a CO_2-rich fluid. Like the mafic magmas, movement of these granitic magmas through the crust will tend to produce CO_2-rich fluids, while H_2O released by metamorphic reactions in the country rocks is absorbed by the magma, either in the form of a hydrous crustal melt or directly as fluids. It is important to note that granitic magmas emplaced into granulite environments cannot crystallize to the H_2O-saturated solidus because thislies below the ambient temperature. Any crystallization that takes place will leave a more hydrous magma that must move to higher crustal levels before solidifying. The lower portions of batholiths formed from such magmas may be hot enough to crystallize as charnockite, but at upper

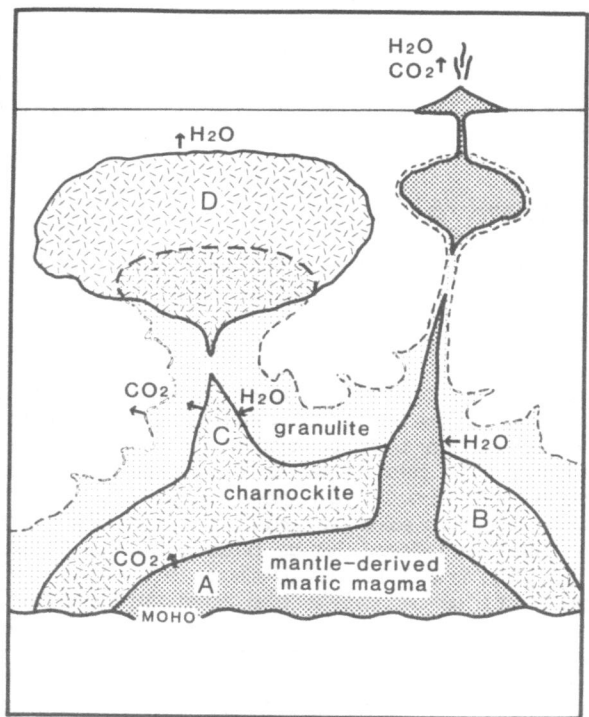

Figure 3. Diagram showing the role of melts in dehydration of the lower crust. A. Basaltic magma is emplaced into the crust. As it ascends it evolves a CO_2 rich fluid and incorporates H_2O from the country rock. B. Basaltic magma ponded at the base of the crust will induce melting in the country rock. C. Felsic melts thus produced will rise into the crust, evolving CO_2 and absorbing H_2O. Portions of these magmas that crystallize at depth will form charnockites or other pyroxene–bearing rocks. D. Felsic magmas ascending to shallow levels will balloon out to form batholiths, which will be dominated by hydrous granitoids. Modified after Frost and Frost (1987).

levels the magma will become progressively more hydrous and will crystallize to form typical hornblende or biotite-bearing granites. In such a model the granitic magma becomes both a conduit for moving CO_2 into the lower crust and a means for moving H_2O to higher crustal levels.

GEOCHEMICAL CONSEQUENCES

An important inference gained from the above model is that pyroxene-bearing felsic igneous rocks, which are generally broadly termed "charnockite", are actually cumulates. Therefore the geochemical characteristics of pyroxene-bearing granitoids were never those of a magma; that is, they are not direct representations of the melt from which they crystallized, but rather they are the product of igneous differentiation. As such "charnockites", being composed of early crystallizing phases, are likely to be depleted in incompatible elements. They may have geochemical signatures that are more similar to other LILE-depleted granulites than to most granitic rocks.

This argument was presented by Hubbard and Whitley (1979) and Field et al. (1980), who on the basis of REE or REE and other LILE abundance data demonstrated that charnockites and granites may be produced from a common parent magma by crystal fractionation (see also Drury, 1980 and Lamb et al., 1986). Geochemical modelling by Field et al. (1980) showed that a cumulate consisting of plagioclase, pyroxene, and quartz extracted from a melt with normal igneous K/Rb ratio will have an elevated K/Rb ratio similar to that found in granulites. Their argument is reinforced by a comparison of K/Rb ratios of granulites with those of charnockites and pyroxene-plagioclase cumulate rocks, including anorthosite, diorite, and monzonite, from the unmetamorphosed Nain and Laramie Anorthosite Complexes (Fig. 4). Both the granulites and the cumulates show a strong increase in K/Rb at low K abundances. In granulites, this trend has been attributed to a secondary process in which LILEs are flushed out of the rock during granulite metamorphism. The preferential removal of Rb over K may be related to its larger ionic radius (Tarney and Windley, 1977). In pyroxene-plagioclase cumulates, the trend probably results from the fact that in K-poor rocks, which are not likely to contain K-feldspar, there is no favorable site for the inclusion of Rb. Therefore the simple presence of high K/Rb ratios in a suite of felsic granulitic gneisses does not necessarily imply that the gneisses were subjected to element depletion during granulite metamorphism. The ratios may be a primary feature of the rocks if they arose as deep-level cumulates.

Previous workers have argued that granulite terranes may be formed by pyroxene-bearing granitic magmas (Holland and Lambert, 1975; Field et al., 1980; Lamb et al., 1986). Our purpose here is to emphasize that the geochemical signature of cumulates formed from such magmas may be indistinguishable from that of granulite terranes in general. Therefore, although the direct participation of synmetamorphic igneous injection, both as a fluid and a heat source, has been documented in only a few granulite terranes, it is quite possible that such bodies are present in many other terranes where their identity has been obscured by later deformation.

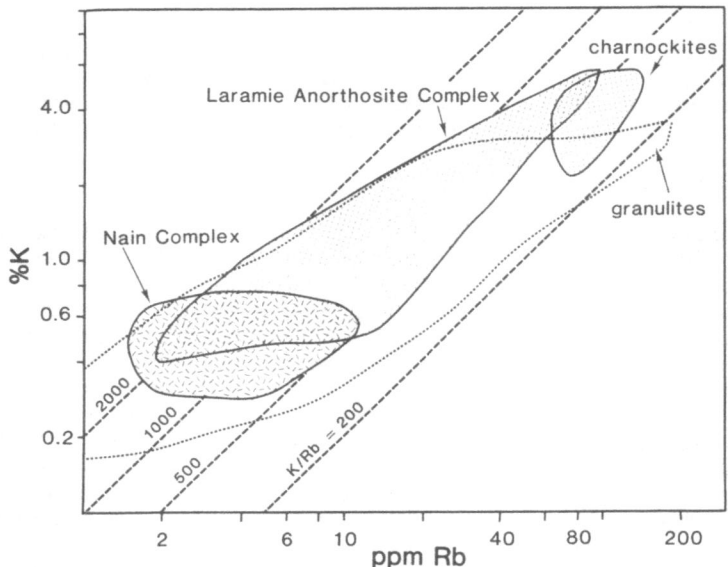

Figure 4. Diagram comparing the K/Rb ratio of granulites (field outlined by dashed line) with charnockites and cumulate pyroxene-plagioclase rock of the Nain and Laramie Anorthosite Complexes. Data from Goldberg (1984); Malm and Ormaasen (1978); Olarewaju (1987); Petersen (1980); Tarney and Windley (1977); and Wiebe (1979).

CONCLUSIONS

The model presented above contends that a single process involving the passage of melts through the crust can account for most modes of granulite formation. Experimental studies on H_2O and CO_2 solubility in silicate melts indicate that CO_2-rich fluids are likely to evolve from melts as they pass through the crust. This provides a source for the CO that is postulated to have streamed through many lower crustal terranes. It also involves large volumes of "dry" magma, much of which may crystallize to form charnockite or other pyroxene-bearing igneous rocks. The presence of these dry magmas has been documented in the Rogaland and Bamble areas of Norway and in the Wind River Mountains of Wyoming and we contend that they may be present in other terranes where later deformation has eliminated any igneous textures from rocks crystallized from these melts.

In conclusion, we reiterate that granulites which formed by CO_2 streaming, melting, or metamorphism of synmetamorphic dry igneous rocks may be different facets of a single process involving passage of melts through the crust. Each sub-process may not necessarily produce a characteristic geochemical signature. Many anhydrous cumulates, for example, have K/Rb ratios also typical of other granulites (Fig. 4). Therefore, before we can identify the sub-process involved in any given instance we must investigate further the geochemical changes that each of these sub-processes may produce.

ACKNOWLEDGEMENTS

This research was supported by NSF grant EAR-8408357 and EAR-8707296. We gratefully acknowledge John Valley for helpful discussions on this topic and Robert Newton and Dennis Geist for critical reviews of this manuscript. We also thank Eric Nye for carrying rocks and for other technical assistance.

REFERENCES

Berg, J. H., and Docka, J. A.,1983, 'Geothermometry in the Kiglapait contact aureole, Labrador' Am. J. Sci. 283, 414-434.

Bohlen, S. R.,1987, 'Pressure-temperature-time paths and a tectonic model for the evolution of granulites' J. Geol. 95, 617-632.

Bohlen, S.R., Valley, J.W., and Essene, E.J.,1985, 'Metamorphism in the Adirondacks. I. Petrology, pressure and temperature' J. Petrol. 26, 971-992.

Bowers, T. S., and Helgeson, H. C.,1983, 'Calculation of the thermodynamic and geochemical consequences of nonideal mixing in the system H_2O-CO_2-NaCl on phase relations in geologic systems: Equation of state for H_2O-CO_2-NaCl fluids at high pressures and temperatures' Geochim. Cosmochim. Acta. 47, 1247-1275.

Brey, G.,1976, 'CO_2 solubility and solubility mechanisms in silicate melts at high pressures' Contrib. Mineral. Petrol. 57, 215-221.

Brown, G.C., and Fyfe, W.S.,1970, 'The production of granitic melts during ultra-metamorphism' Contrib. Mineral. Petrol. 28, 310-318.

Brown, G. C., and Fyfe, W. S.,1972, 'The transition from metamorphism to melting status of granulite and eclogite facies' Int. Geol. Cong. 24, Section 2, 27-34.

Collerson, K. D., and Fryer, B. S.,1978, 'The role of fluids in the formation and subsequent development of the early continental crust' Contrib. Mineral. Petrol. 67, 151-167.

Des Marais, D.J.,1985, 'Carbon exchange between the mantle and the crust, and its effect upon the atmosphere: today compared to Archean time' In E.T. Sundquist and W.S. Broecker, eds., The carbon cycle and atmospheric CO_2: natural variations Archean to Present. AGU Geophysical monograph 32, 602-611.

Drury, S. A.,1980, 'Lewisian pyroxene gneiss from Barra and the geochemistry of the Archean lower crust' Scott. J. Geol. 16, 199-207.

Eggler, D. H., and Kadik, A. A.,1979, 'The system $NaAlSi_3O_8-H_2O-CO_2$ to 20 kbar pressure: I. Compositional and thermodynamic relations of liquids and vapors coexisting with albite' Am. Mineral. 64, 1036-1048.

Eggler, D. H., and Rosenhauer, M.,1978, 'Carbon dioxide in silicate melts: II. Solubilities of CO_2 and H_2O in $CaMgSi_2O_6$ (diopside) liquids and vapors at pressures to 40kb' Am. J. Sci. 278, 64-94.

14

Ellis, D. J.,1980, 'Osumilite-sapphirine-quartz granulites from Enderby Land, Antarctica: P-T conditions of metamorphism, implications for garnet-cordierite equilibria and the evolution of the deep crust' Contrib. Mineral. Petrol. 74, 201-210.

Field, D., Drury, A., and Cooper, D. C.,1980, 'Rare earth and LILE element fractionation in high grade charnockitic gneisses, South Norway' Lithos 13, 281-289.

Field, D., Smalley, P. C., Lamb, R. C., and Raheim, A.,1985, 'The 1.6 - 1.5 Ga old amphibolite-granulite terrain, Bamble sector, Norway: dispelling the myth of regional Grenvillian orogenic reworking' in A.C. Tobi and J.L.R. Touret, eds., The deep Proterozoic crust in the North Atlantic Province, NATO ASI C158, pp. 567-578.

Fine, G., and Stolper, E.,1985, 'The speciation of carbon dioxide in sodium aluminosilicate glasses' Contrib. Mineral. Petrol. 91, 105-121.

Fine, G. and Stolper, E.,1986, 'Dissolved carbon dioxide in basaltic glasses: concentrations and speciation' Earth and Planet. Sci. Lett. 76, 263-278.

Frost, B. R. and Frost, C. D.,1987, 'CO_2, melts, and granulite metamorphism' Nature 327, 503-506.

Fuhrman, M.L., Frost, B.R., and Lindsley, D.H.,in press, 'The petrology of the Sybille Monzosyenite, Laramie Anorthosite Complex, Wyoming' J. Petrol.

Fyfe, W. S.,1973, 'The granulite facies, partial melting and the Archean crust' Phil. Trans. Royal Soc. Lond. A273, 457-461.

Goldberg, S.A.,1984, 'Geochemical relationships between anorthosite and associated iron-rich rocks, Laramie Range, Wyoming' Contrib. Mineral. Petrol. 87, 376-387.

Grant, J. A.,1986, 'Quartz-phlogopite-liquid equilibria and the origins of charnockites' Am. Mineral. 71, 1071-1075.

Grant, J.A. and Frost, B.R.,1986, 'Decompression, metamorphism and melting in the aureole of the Laramie Anorthosite Complex' Geol. Soc. Amer. Abstracts w. Programs 18, 620.

Greenwood, H.J.,1975, 'Buffering of pore fluids by metamorphic reactions' Am. J. Sci. 273, 561-571.

Harley, S. L.,1986, 'A sapphirine-cordierite-garnet-sillimanite granulite from Enderby Land Antarctica: Implications for FMAS petrogenetic grids in the granulite facies' Contrib. Mineral. Petrol. 94, 452-460.

Harris, D. M., and Anderson, A. T.,1983, 'Concentrations, sources, and losses of H_2O, CO_2, and S in Kilauean basalt' Geochim. Cosmochim. Acta. 47, 1139-1150.

Harris, D. M., and Anderson, A. T.,1984, 'Volatiles H_2O, CO_2, and Cl in a subduction related basalt' Contrib. Mineral. Petrol. 87, 120-128.

Harris, D. M., Sato, M., Casadevall, T. J., Rose, W. I., and Bornhorst, T. J.,1981, 'Emission rates of CO_2 from plume measurements' U.S. Geol. Surv. Prof. Pap. 1250, 201-207.

Heier, K. S.,1973, 'Geochemistry of granulite facies rocks and problems of their origin' Phil. Trans. Royal Soc. Lond. A273, 429-442.

Hoefs, J. and Touret, J. L. R.,1975, 'Fluid inclusion and carbon
    isotopic study from Bamble granulites (south Norway): A
    preliminary investigation' Contrib. Mineral. Petrol. 52, 165-174.
Holland, J. G., and Lambert, R.StJ.,1975, 'The chemistry and origin of
    the Lewisian gneisses of the Scottish mainland: the Scourie and
    Inver assemblages and sub-crustal accretion' Precambrian Res. 2,
    161-188.
Hollister, L. S.,1975, 'Granulite facies metamorphism in the Coast
    Range crystalline belt' Can. J. Earth Sci. 12, 1953-1955.
Holloway, J. R.,1976, 'Fluids in the evolution of granitic magmas:
    Consequences of finite CO_2 solubility' Geol. Soc. Amer. Bull. 87,
    1513-1518.
Holloway, J.R., and Lewis, C. F.,1974, 'CO_2 solubility in hydrous albite
    liquid at 5Kbar' Trans. Amer. Geophys. Un. 55, 483.
Holloway,. J. R., Mysen, B. O., and Eggler, D. H.,1976, 'The
    solubility of CO_2 in liquids on the join $CaO-MgO-SiO_2-CO_2$'
    Carnegie Inst. Washington Yearbook 75, 626-631.
Hopson, C.A., and Dellinger, D.A.,1987, 'Evolution of four-dimensional
    compositional zoning, illustrated by the diapiric Duncan Hill
    Pluton, North Cascades, Washington' Geol. Soc. Amer. Abstracts w.
    Programs 19, 707.
Hubbard, F. H., and Whitley, J. E.,1979, 'REE in charnockite and
    associated rocks, southwest Sweden' Lithos, 12, 1-11.
Hulsebosch, T. P., Koesterer, M. E., and Frost, B. R.,1985, 'Late
    Archean intrusive charnockites from the west-central Wind River
    Mountains, Wyoming' Geol. Soc. Amer. Abstracts w. Programs 17,
    616.
Jansen, J.B.H., Blok, R.J.P., Bos, A., and Scheelings, M.,1985,
    'Geothermometry and geobarometry in Rogaland and preliminary
    results from the Bamble area, S. Norway' In A.C. Tobi and J.L.R.
    Touret, eds., The deep Proterozoic crust in the North Atlantic
    Provinces, NATO ASI C158, pp. 499-516.
Kadik, A. A., and Lukanin, O. A., Lebedev, Ye. B., Korovushkina, E.
    Ye., 1972, 'Solubility of H_2O and CO_2 in granite and basalt melts
    at high pressures' Geochemistry Internat. 9, 1041,1050.
Kadik, A. A., and Lukanin, O. A.,1973, 'The solubility-dependent behavior
    of water and carbon dioxide in magmatic processes' Geochemistry
    Internat. 10, 115-129.
Koesterer, M. E., Frost, C. D., Frost, B. R., Hulsebosch, T. P.,
    Bridgwater, D., and Worl, R. C.,in press, 'Development of the
    Archean crust in the Medina Mountain area, Wind River Mountains,
    Wyoming (U.S.A.)' Precambrian Res.
Konnerup-Madsen, J.,1977, 'Composition and microthermometry of fluid
    inclusions in the Kleivan Granite, South Norway' Am. J. Sci. 277,
    637-696.
Konnerup-Madsen, J., 1984, 'Compositions of fluid inclusions in granites
    and quartz syenites from the Gadar continental rift province (South
    Greenland)' Bull. Mineral. 107, 327-340.

16

Konnerup-Madsen, J., Debussy, J., and Rose-Hansen, J.,1985, 'Combined Raman microprobe spectrometry and microthermometry of fluid inclusions in minerals from Gadar province (south Greenland)' Lithos 18, 271-280.

Konnerup-Madsen, J., and Rose-Hansen, J.,1982, 'Volatiles associated with alkaline igneous rift activity: Fluid inclusions in the Ilimaussaq intrusion and the Gadar granitic complexes (South Greenland)' Chem. Geol. 37, 79-93.

Lamb, R. C., Smalley, P. C., and Field, D.,1986, 'P-T conditions for the Arendal granulites, southern Norway: implications for the roles of P,T and CO_2 in deep crustal LILE-depletion' J. Metamorphic Geol. 4, 143-160.

Lamb, W. and Valley, J.,1984, 'Metamorphism of reduced granulites in low-CO_2 vapour-free environment' Nature 312, 56-58.

Lamb, W. and Valley, J.,1985, 'C-O-H fluid calculations and granulite genesis' In A.C. Tobi and J. L. R. Touret, eds., The deep Proterozoic crust in the North Atlantic Provinces, NATO ASI, C158, pp.119-131.

Lindsley, D.H., and Andersen, D.J.,1983, 'A two-pyroxene thermometer' J. Geophys. Res. 88, A887-A906.

Malm, O.A., and Ormaasen, D. E.,1978, 'Mangerite-charnockite intrusives in the Lofoten-Vesteralen area, North Norway: petrography, chemistry and petrology' Norges Geol. Undersok. 338, 38-114.

Mysen, B. O.,1976, 'The role of volatiles in silicate melts: Solubility of carbon dioxide and water in feldspar, pyroxene, and feldspathoid melts to 30 kb and $1625^{O}C$' Am. J. Sci. 276, 969-996.

Mysen, B. O., Arculus, R. J., and Eggler, D. H.,1975, 'Solubility of carbon dioxide in melts of andesite, tholeiite, and olivine nephelinite compo- sition' Contrib. Mineral. Petrol. 53, 227-239.

Mysen, B. O., Eggler, D. H., Seitz, M. G., and Holloway, J. R.,1976, 'Carbon dioxide in silicate melts and crystals. Part I. Solubility measurements' Am. J. Sci. 276, 455-479.

Nesbitt, H. W.,1980, 'Genesis of the New Quebec and Adirondack granulites: Evidence for their production by partial melting' Contrib. Mineral. Petrol. 72, 303-310.

Newton, R.C.,1983, 'Geobarometry of high-grade metamorphic rocks' Amer. J. Sci. 283-A, 1-28.

Newton, R. C., and Hansen, E. C.,1983, 'The origin of Proterozoic and late Archean charnockites - evidence from field relations and experimental petrology' Geol. Soc. Amer. Mem. 161, 167-178.

O'Hara, M.J.,1975, 'Great thickness and high geothermal gradient of Archean crust: the Lewisian of Scotland' Inter. Conf. Geotherm. Geobarom., Abstr., Penn. St. University, 127-128.

Olarewaju, V. O.,1987, 'Charnockite-granite association in SW Nigeria: rapakivi granite type and charnockitic plutonism in Nigeria?' J. African Earth Sci. 6, 67-77.

O'Nions, R.K., and Oxburgh, E.R.,1983, 'Heat and helium in the Earth' Nature 306, 429-431.

Percival, J. A.,1983, 'High-grade metamorphism in the Chapleau-Foleyet Area, Ontario' Amer. Mineral. 68, 667-686.

Petersen, J. S.,1980, 'The zoned Kleivan granite - an end member of the anorthosite suite in southwest Norway' Lithos 13, 79-95.

Phillips, G.N., and Wall, V.J.,1981, 'Evaluation of prograde regional metamorphic conditions: their implications for the heat source and water activity during metamorphism in the Willyama Complex, Broken Hill, Australia' Bull. Mineral. 104, 801-810.

Pineau, F., Javoy, M., Behar, F., and Touret, J.,1981, 'La geochimie isotopique du facies granulite du Bamble (Norvege) et l'origine des fluides carbones dans la croute profonde' Bull. Mineral. 104, 630-641.

Raase, P., Raith, M., Ackermand, D., and Lal, R.K.,1986, 'Progressive metamorphism of mafic rocks from greenschist to granulite facies in the Dharwar craton of South India' J. Geol. 94, 261-282.

Roedder, E.,1965, 'Liquid CO_2 inclusions in olivine bearing nodules and phenocrysts from basalts' Am. Mineral. 50, 1746-1782.

Roedder, E.,1984, Fluid Inclusions Rev. in Mineral. 12, 503-532.

Ross, D. C.,1985, 'Mafic gneissic complex (batholithic root?) in the southernmost Sierra Nevada, Calif.' Geol. 13, 288-291.

Rutherford, M. J., and Devine, J.,1986, 'Experimental petrology of recent Mount St. Helens Dacites: Amphibole, Fe-Ti oxides and magma chamber conditions' Geol. Soc. Amer. Abstracts w. Programs 18, 736.

Sandiford, M. and Powell,R., 1986, 'Deep crustal metamorphism during continental extension: Modern and ancient examples' Earth and Planet. Sci. Lett. 79, 151-158.

Sclater, J.G., Jaupart, C., and Galson, D.,1980, 'The heat flow through oceanic and continental crust and the heat loss of the earth' Rev. Geophys. 18, 269-311.

Sheridan, M. F. and Moore, C. B.,1981, 'Carbon, nitrogen, and sulfur variations in the Bishop Tuff, California' Lithos 14, 23-27.

Skippen, G., and Trommsdorff, V.,1986, 'The influence of NaCl and KCl on phase relations in metamorphosed carbonate rocks' Am. J. Sci. 286, 81-104.

Smalley, P. C., Field, D., Lamb, R. C., and Clough, P. W. L.,1983, 'Rare earth, Th-Hf-Ta and large-ion lithophile element variations in metabasites from the Proterozoic amphibolite-granulite transition zone at Arendal, South Norway' Earth and Planet. Sci. Lett. 63, 446-458.

Spera, F. J. and Bergman, S. C.,1980, 'Carbon dioxide in igneous petrogenesis: I Aspects of the dissolution of CO_2 in silicate liquids' Contrib. Mineral. Petrol. 74, 55-66.

Sterner, S. M., and Bodner, R. J.,1984, 'Synthetic fluid inclusions in natural quartz I. Compositional types synthesized and applications to experimental petrology' Geochim. Cosmochim. Acta. 48, 2659-2668.

Tarney, J. and Windley, B. F.,1977, 'Chemistry, thermal gradients and the evolution of the lower continental crust' J. Geol. Soc. Lond. 134,153-172.

Tobi, A.C., Hermans, G.A.E.M., Maijer, C., and Jansen, J.B.H.,1985, 'Metamorphic zoning in the high-grade Proterozoic of Rogaland-Vest Agder, SW Norway' In A.C. Tobi and J.L.R. Touret, eds., The deep Proterozoic crust in the North Atlantic Provinces, NATO ASI C158, 477-497.

Touret, J.,1971, 'Le facies granulite en Norvege meridionale. II. Les inclusions fluides' Lithos 4, 423-436.

Touret, J.,1981, 'Fluid inclusions in high grade metamorphic rocks' In L. S. Hollister and M. C. Crawford, eds., Mineral. Assoc. Canada Short Course 6, 182-208.

Touret, J. L. R.,1985, 'Fluid regime in Southern Norway: The record of fluid inclusions' In A.C. Tobi and J. L. R. Touret, eds., The deep Proterozoic crust in the North Atlantic Province, NATO ASI C158, 517-549.

Wells, P. R. A.,1979, 'Chemical and thermal evolution of Archaean sialic crust, Southern West Greenland' J. Petrol. 20, 187-226.

Wiebe, R.A.,1979, 'Fractionation and liquid immiscibility in an anorthositic pluton of the Nain Complex, Labrador' J. Petrol. 20, 239-269.

MORPHOLOGY OF GRANULITE - AMPHIBOLITE FACIES TRANSITIONS: THE IMPORTANCE OF FLUID MOVEMENTS

C.R.L. Friend,
Department of Geology,
Oxford Polytechnic,
Oxford OX3 0BP, UK.

ABSTRACT. Information regarding the processes whereby granulite facies rocks were formed or destroyed may be gained from a study of the morphology of transitional zones. Both transitions have a complex 3D structure which may include anatectic rocks suggesting that melting can play an important role. Because of the intimate relationship of granulite and amphibolite facies assemblages such transitions appear to have been largely fluid controlled. The nature of this control could be a simple compositional change or a more complicated mechanism. Some transitions seem to have had a flux of dehydrating fluids which lead to melting in amphibolite facies rocks, whilst in others melting in the granulite facies rocks occurred.

Retrogressive transitions, which often involve some decompression, appear to be largely controlled by hydration. The morphological features preserved are in reverse to the prograde types, but usually accompanying deformation obscures the details.

Conclusions are that fluids at the transition are channelled, but as the transition advances the channels are overtaken such that the impression is that the fluids were pervasive. It is emphasised that these processes may only take place at the transition and cannot yet be applied to the bulk of granulite facies rocks. The complexities of both types of transition can be adequetely explained by the channelled fluid flow and different partial melting regimes.

1. INTRODUCTION

Granulite facies rocks formed under low a_{H2O} conditions comprise an important constituent of Precambrian continental crust. The origin of these high-grade rocks has been explained by recourse to many different hypotheses (e.g. Fyfe, 1973; Grant, 1973; Janardhan et al., 1979), some of which have more validity than others (Powell, 1983). However, many problems such as the depleted nature of some granulites (Tarney & Weaver, 1987) still exist. Transition zones into amphibolite facies rocks are of particular importance since it is from these zones that we may be able to gain knowledge of both the prograde and retrograde processes which take place and thereby clarify the origin of at least some granulites. In many cases granulite - amphibolite facies terrains have undergone a polyphase evolution and superimposed deformation and retrogression have obliterated prograde details (eg. the Lewisian). To gain some knowledge of the processes operating at both prograde and retrograde transitions it is of considerable use to examine this 3D morphology. A study of undeformed transition zones can be used to test some of the hypotheses

19

D. Bridgwater (ed.), Fluid Movements – Element Transport and the Composition of the Deep Crust, 19–28.
© 1989 by Kluwer Academic Publishers.

and help clarify some of the models proposed for the production of granulite facies rocks.

The discovery in southern India of the arrested formation of charnockite (Pichamuthu, 1960) was the first descriptive contribution to the modern study of the morphology of granulite - amphibolite transitions. This work thus stands as a landmark in our understanding of such features. Subsequently contributions particularly concerning compositions of fluid phases found in granulites (e.g. Touret, 1977) and mechanisms of formation (e.g. Janardhan *et al.*, 1979) have been made.

2. THE SOUTH INDIAN TRANSITION

Throughout southern India (Fig. 1) and Sri Lanka more examples of 3D charnockite networks have been found (e.g. Kumar, *et al.*, 1985; Hansen, *et al.*, 1987; Glassley; Raith & Hoernes, this vol.). The ages are poorly constrained (Crawford, 1969; Spooner & Fairburn, 1970), but it is apparent that there are several ages at which charnockite formation took place. Some must be synchronous with the Madras charnockites dated at c. 2700 Ma (Bhattacharaya & Sen, 1986). Others formed at the time of the Kabbaldurga event at c. 2500 Ma (Grew & Manton, 1984), whilst examples from Kerala are Proterozoic (Raith & Hoernes, this vol.). It is now clear that these transitions are of different types, having formed by different mineral reactions (Hansen *et al.*, 1987) and that the transitions represented have formed by different processes.

2.1 Prograde relationships at Kabbaldurga, Karnataka

The 3D network described by Pichamuthu (1960) was explained by the hypothesis of CO_2-flushing (Janardhan *et al.*, 1979). Subsequent fluid inclusion studies confirmed that such a process causing amphibole, biotite and quartz to yield orthopyroxene could have operated (Janardhan, *et al.*, 1982; Hansen *et al.*, 1984). The hypothesis of CO_2 influx along micro-fissures was linked to the formation of granite by partial melting of the amphibolite facies country rocks. Influx of fluids caused a migration of H_2O-dominated volatiles, which ponded in an area of crust at amphibolite facies conditions and resulted in partial melting of the quartzo-feldspathic gneisses (Friend, 1981, 1983). Even on small scales it has been shown that fluid ponding in the crust can produce significant effects (Spear & Selverstone, 1983). Additionally, it was possible to track the change in the composition of the fluid phase by the visible effects upon the rocks (Friend, 1985). The chemistry of parts of the transition zone, which runs through Karnataka and northern Tamil Nadu, have been examined in an attempt to understand the mechanism of elemental depletion that appears to be a characteristic of many granulite areas (Condie *et al.*, 1982; Weaver & Tarney, 1983). The Indian transition does not appear to show such elemetal depletion, so that it is posible that this feature is related to processes operating within rather than at the transitional boundary of granulite faices conditions.

2.2 Prograde relations in Kerala

Throughout southern Kerala granulite - amphibolite facies transitions appear to be common (Hansen *et al.*, 1987). The rocks at Ponmudi (Srikantappa *et al.*, 1985) comprise a variety of garnet-biotite gneisses and garnet-sillimanite-biotite gneisses. In places rocks interpreted as metatexites have the assemblage garnet-sillimanite-cordierite-biotite. Granulite facies assemblages have developed in a sub-rectangular,

21

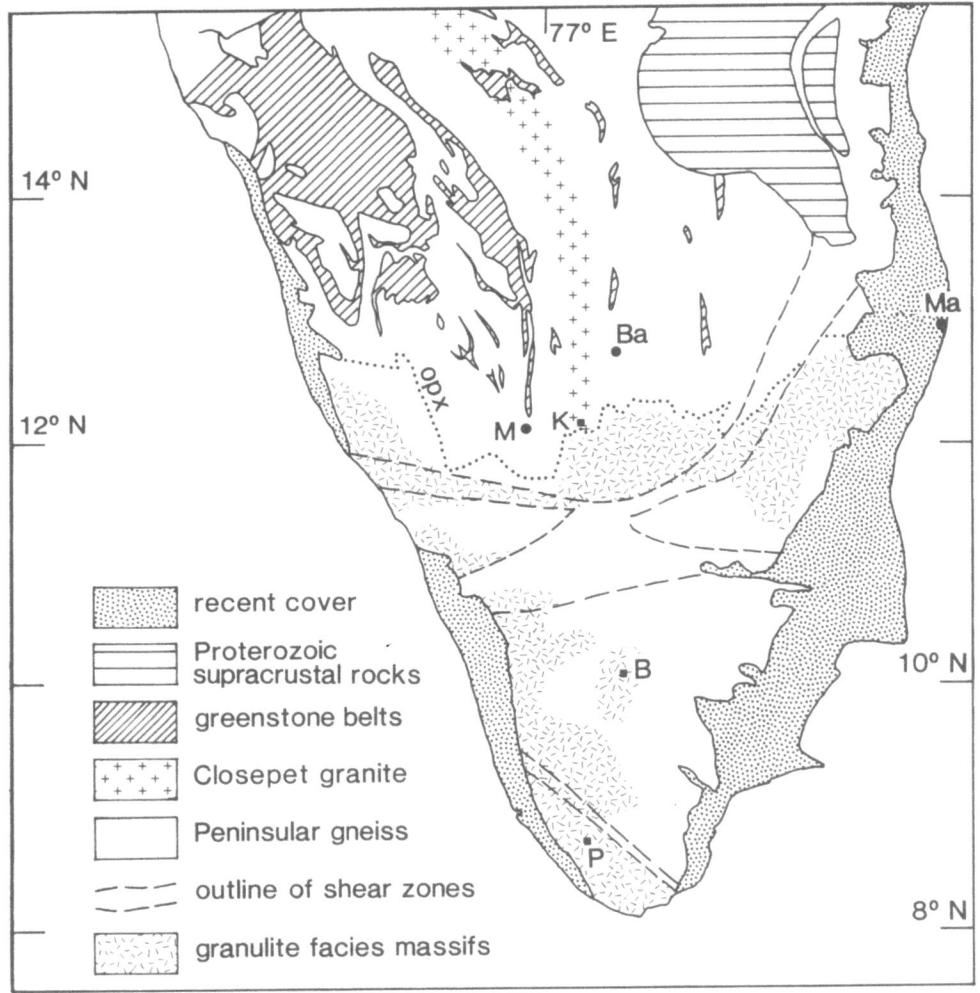

Fig. 1. Sketch map of southern India showing the position of the main granulite facies massifs and the northernmost extent of orthopyroxene.
B = Bangalore; M = Mysore; Ma = Madras. B = Bodinayakkanur, with Theni and Karumpari to the east; K = Kabbaldurga; P = Ponmudi.

3D network similar to that at Kabbaldurga, interpreted to represent pathways along fractures and the gneissosity. The fluid phase reported consists of four different assemblages; H_2O, CO_2, H_2O+CO_2, and importantly, CH_4-N_2. In comparison, the Kabbaldurga inclusions are CO_2-dominated with minor H_2O (Hansen et al., 1984). At Ponmudi the fluids in both the amphibolite and granulite facies portions are very similar and it was concluded that both had evolved under the same fluid conditions (Srikantappa et al., 1985). The formation of charnockitic assemblages was thus not related to an influx of dehydrating fluids and a new mechanism, isothermal fluid pressure release, was invoked. More detailed studies have corroborated the lack of an invasive fluid phase and have concluded that orthopyroxene formed from reaction of biotite+quartz±garnet (Hansen et al., 1987).

2.3 Retrograde relations in southern Tamil Nadu

Throughout Tamil Nadu the same complex association of granulite and amphbiolite facies rocks, some Archaean, whilst others are Proterozoic, (probably similar to those in Kerala) is found. Within this region retrograde relations, largely dominated by hydrous mineral formation occur. The morphology of these retrograde areas are very similar to that of the prograde areas.

In the vicinity of Bodinayakkanur (Holt & Wightman, 1983) the rocks are generally at granulite facies and have been described as charnockites. The rocks often have copious orthopyroxene porphyroblasts and are frequently unfoliated, massive, brownish-green acid rocks in which it is dificult to establish any lithological variation. However, the transitional relations preserved, are retrograde, *not* prograde as reported. 20 km to the east in a small quarry near Theni (Fig. 1), both granulite and amphibolite facies rocks are present. The granulite facies rocks are typical coarse-grained, greenish-grey or brownish orthopyroxene-bearing, acid gneisses. These may be variably foliated with both biotite and orthopyroxene participating. Traversing the granulite facies rocks are creamy-white, felsic patches and vein-like areas, which contain no orthopyroxene and virtually no biotite. Within these white patches small areas of greenish granulite facies rocks outlined by a pronounced brown-stained rim occur. In many cases these small areas have the same coarse-grained texture as the whitish patches. In other examples foliated granulite facies rocks are contained within non-foliated, white rocks. The field relationships are clear that there has been no introduction of intrusive granitic material and the interpretation preferred is that the granulite facies rocks have undergone static recrystallisation under hydrous, retrogressive, amphibolite facies conditions. The rocks have become bleached and have lost their foliation-forming granulite facies minerals, except in the remnant patches. In some remnants, particularly the smaller examples, there are concentrations of randomly oriented amphibole porphyroblasts which are absent from both the bleached patches and larger unretrogressed areas. The amount of orthopyroxene present in the granulite is such that to become reduced to the small amount of biotite present in the white retrogressed rocks some Fe+Mg must have left the system. A possible stage in this process is the production of the large amphibole porphyroblasts in the small granulite facies remnants.

Another example is well-displayed in a quarry at Karumpari (Fig. 1). Here the main rock type is foliated, brown, garnet+orthopyroxene-bearing gneiss which has whitish zones of diferent types within it. Some of these are linear zones about 20-30 cm wide in which the foliation and mineralogy are essentially the same as the granulite facies rocks traversed. This is most clearly displayed where the zones run obliquely to the foliation. Orthopyroxene is altered to biotite and with the bleaching is simply attributed to hydration. Irregular bleached areas, sometimes associated with pink feldspar, have a much weaker foliation which is continuous with that in the granulite facies rocks. Orthopyroxene has been virtually completely replaced by biotite. Lastly, there is the development of linear patches in which there is a loss of structure in the centre and the development of biotite clots. In these zones hydration has led to the recrystallisation of the felsic phases and the replacement of orthopyroxene by randomly oriented biotite causing the obliteration of the foliation. Essentially these structures represent the reverse of the prograde structures at Kabbaldurga.

The morphology of both prograde examples, at Kabbaldurga and Ponmudi, and the two retrograde examples from Tamil Nadu comprises a 3D network of fluid pathways which probably have been controlled by micro-fissures or the gross structure of the rock. It is, therefore, of interest to investigate whether other such transitions exist elsewhere and, if undeformed, whether their morphology is similar to the Indian examples.

3. SOUTHERN WEST GREENLAND

Between Nuuk and Fiskenaesset (Fig. 2) amphibolite - granulite transitions were known to exist and were interpreted to be essentially prograde boundaries with only relatively minor modification due to retrogression (Kalsbeek, 1976; Chadwick & Coe, 1983). These boundaries have been re-investigated and some modification of the previous hypotheses is necessary.

3.1 Retrograde relations beween Fiskenaesset and Ameralik

The Fiskenaesset region has recently been shown to comprise a portion of an excavated thrust block, the sole of which now outcrops between Tre Brødre, Ameralik and the Inland Ice (Friend et al., 1987a,b). In this block metamorphism culminated at granulite facies conditions in the late Archaean c. 2850 Ma (Black et al., 1973; Pidgeon & Kalsbeek, 1978; Taylor et al., 1980). As a result of post-thrusting crustal re-equilibration, deformation took place at c. 2650-2700 Ma under amphibolite facies conditions (Friend et al., 1987a,b). This event caused extensive retrogression of the granulite facies assemblages forming 'blebby' texture due to hornblende+biotite pseudomorphs after orthopyroxene (McGregor, et al., 1986). This type of texture is is frequently found in retrogressed granulite facies rocks (eg. Garde, this vol.) and may be used as an indicator of the former extent of granulite facies conditions.

On the basis of the mineralogies formed the retrogression was maninly caused by influx of hydrous fluids, though decompression did play a part. In particular, it may have provided open pathways for the access of fluids (Friend et al., 1987b). In the north, along the sole of the thrust (Fig. 2), retrogression is very extensive and virtually complete. Moving structurally upwards evidence of channelled fluid movement is found. These fluids are considered to have been released from the underlying amphbiolite facies rocks during the thrusting and subsequent deformation.

3.2. Prograde relations in Bjørnesund

Despite extensive retrogression throughout most of the northern part of the block (McGregor et al., 1986) the southern boundary is relatively well-preserved, cropping out in the vicinity of Bjørnesund (Fig. 2). The boundary was first investigated by Walton (1971, 1972) and interpreted as prograde. The rocks to the south have never been to granulite facies, a point which was supported by the regional metamorphic studies of Kalsbeek (1976). Northwards the rocks become affected by late Archaean

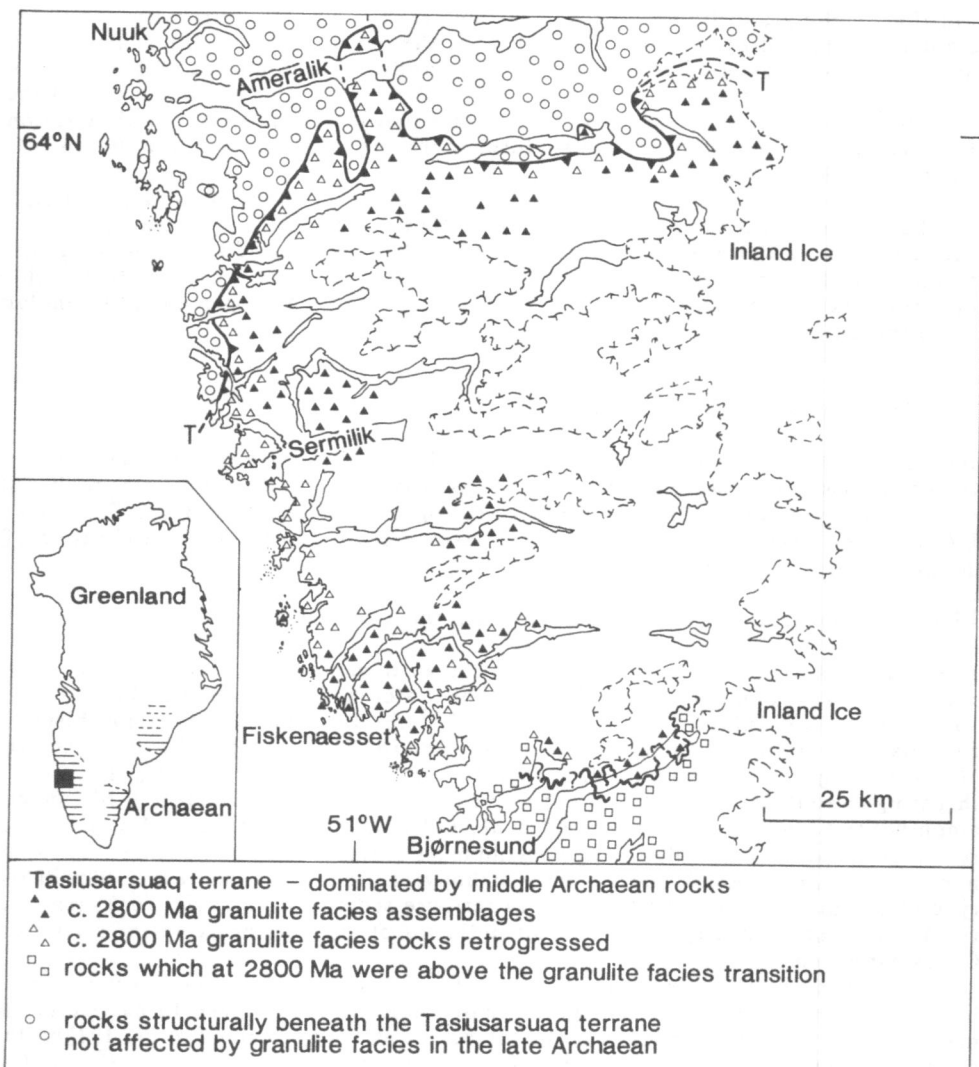

Tasiusarsuaq terrane – dominated by middle Archaean rocks
▲ c. 2800 Ma granulite facies assemblages
△ c. 2800 Ma granulite facies rocks retrogressed
□ rocks which at 2800 Ma were above the granulite facies transition

○ rocks structurally beneath the Tasiusarsuaq terrane
○ not affected by granulite facies in the late Archaean

Fig. 2. Sketch of the late Archaean granulite facies block of southern West Greenland. The position of the northern thrust (T-T) and the location of the prograde transition in the south at Bjørnesund are indicated.

deformation and retrogression and there are numerous local retrograde transitions into ampihibolite facies conditions. The true extent of granulite facies conditions in the Fiskenaesset region has only recently begun to emerge (McGregor et al., 1986; Friend et al., 1987a,b).

The country rocks in the Bjørnesund area comprise typical multiphase grey gneisses which show slightly different evolutionary histories. Approaching the transition the gneisses become very nebulitic and rather pink, but do not seem to have reached the stage where partial melting began and melt started to segregate. The transition zone itself has a resemblance to that described at Kabbaldurga (Janardhan et al., 1979, 1982) and in places comprises a three-dimensional network of brownish-stained, orthopyroxene-bearing veins and patches passing through the amphibolite facies, grey gneisses. Over a large area at the head of the fjord there is a layering of orthopyroxene-bearing brownish gneisses and grey amphbiolite facies gneisses, a development which is controlled by the banding and foliation. Fabrics in the grey gneisses can be found undergoing obliteration by the growth of orthopyroxene porphyroblasts and the recrystallization of the felsic phases to a coarser-grained, sub-equigranular, granoblastic, non-foliated groundmass. There has been little or no subsequent deformation, though some retrogression due to hydration has taken place with both orthopyroxene and garnet altering to aggregates involving cummingtonite. Because of the intricacy of the relationships and the structure of the amphibolite facies rocks, the prograde nature of this boundary is not in doubt. One significant difference to the other boundaries described is that there are cross-cutting sheets and pipes of coarse-grained, leucocratic, orthopyroxene-bearing trondhjemitic gneiss. These are thus related to the granulite facies rocks and suggest that in addition to dehydration, some melting in the granulite facies areas below the present outcrop was taking place.

4. DISCUSSION

The evidence which is now emerging suggests that granulite-amphibolite facies transitions are an integral part of many crustal blocks but are formed by a number of different processes. Because relatively young examples of granulites have not been excavated except after involvement in active orogenic episodes (eg. the Alps) the only opportunity of examining such transitions currently appears to occur in old basement gneiss terrains which have been elevated over long periods of time, essentially by erosion or block faulting, thus avoiding dynamic deformation. It seems clear that in areas of quite different age and geographical distribution, remarkably similar structures may be found in both prograde and retrograde transitions.

The 3D morphology of a prograde transition is considered to be controlled essentially by fluid fluxes, which in the immediate transition area, were channelled. This is in marked contrast to the areas of solid granulite facies at lower crustal levels which thus appear to have had pervasive fluid flow (Valley & O'Neil, 1984). It is not yet possible to know whether these also formed by the same processes as occcur at the transition zone. Depending upon the height in the crust reached by the transition, the lower portions could be overtaken by pervasive granulite facies conditions, thus obliterating sites of the main channels (cf. Friend, 1985). Because of the intricacy of the fluid flux, involving the movement of both hydrating and dehydrating volatile phases, partial melting may also take place, in either the amphibolite facies or granulite facies areas, as exemplified by the Kabbaldurga and Bjørnesund localities respectively. The products of such processes are quite different and may lead to the recognition of different types of transition zone. Thus, despite morphological similarities, the Kerala/Sri Lanka prograde relations are not the same as those at Kabbaldurga. Whilst the study of such transitions is still relatively new, it may be that on any one transition, different processes may operate in different places. Additionally, preservation (see below) may not give a totally representative picture. The highly irregular and delicate structures produced by fissure-controlled fluid flux, coupled with different partial melting regimes and products, quite adequetely explain the widespread inhomogeneities encountered in such transition zones.

Retrogression is clearly not so limited in its crustal position. Hydrous volatiles may be derived from many lower crustal or mantle sources. This implies that, given the correct conditions, granulite facies rocks deep in the crust can be retrogressed without ingress of water from above, perhaps the example from southern Tamil Nadu. More complicated scenarios have been constructed for the more generally retrogressed West Greenland gneisses where fluids are derived from underlying hydrous crustal rocks (Friend et al., 1987b; Nutman et al., this vol.). The channelled nature of this fluid movement is clear at specific structural positions, otherwise there appears to have been a partially pervasive movement. This is probably because a degree of decompression was involved thus allowing micro-fractures to develop which facilitated fluid penetration.

Preservation of the delicate and intricate structures is clearly dependent upon the degree of involvement of that particular crustal segment in later deformational and metamorphic events. Once deformation has taken place to any significant degree the information that a transition can yield would appear to be lost. Because of the extreme age and probability of later crustal reworking during excavation, these transiton zones are likely to be modified and we may never be able to find a complete record of the processes occurring in them.

Acknowledgements: Financial assistance for parts of the various aspects of this work have come from the British Council (India) and the Royal Society (Greenland). Further support came from Oxford Polytechnic and all these sources are gratefully acknowledged. The work was greatly helped by assistance from many friends and colleagues in variuos capacities. Technical assistance from D.J. Allen and C.G. Jones is much appreciated.

REFERENCES

Bhattacharaya, A. & Sen, S.K. 1986. Granulite metamorphism, fluid buffering and dehydration melting in the Madras charnockites and metapelites. *J. Petrol.* 27, 1119-1141.

Black, L.P., Moorbath, S., Pankhurst, R.J. & Windley, B.F. 1973. $^{207}Pb/^{206}Pb$ whole rock age of the Archaean granulite facies metamorphic event in West Greenland. *Nature Phys. Sci.* 244, 50-53.

Chadwick, B. & Coe, K. 1983. Descriptive text to 1:100,000 sheet Buksefjorden 63V1 Nord. *Grønlands Geologiske Undersøgelse.*

Condie, K.C., Allen, P. & Narayana, B.L. 1982. Geochemistry of the Archaean low- to high-grade transition zone, southern India. *Contrib. Mineral. Petrol.* 81, 157-167.

Crawford, A.R. 1969. Reconnaissance Rb-Sr dating of Precambrian rocks of southern Peninsular India. *J. Geol. Soc. India* 10, 117-166.

Friend, C.R.L. 1981. Charnockite and granite formation and the influx of CO_2 at Kabbaldurga. *Nature* 294, 550-551.

Friend, C.R.L. 1983. The link between charnockite formation and granite production: evidence from Kabbaldurga, Karnataka, southern India. In: *Migmatites. Melting and Metamorphism* (Atherton, M.P. & Gribble, C.D. eds.) pp.264- 276. Shiva,

Nantwich.
Friend, C.R.L. 1985. Evidence for fluid pathways through Archaean crust and the generation of the Closepet granite, Karnataka, south India. *Precambrian Res.* 27, 239-250.
Friend, C.R.L., Nutman, A.P. & McGregor, V.R. 1987a. Late-Archaean tectonics in the Faeringehavn-Tre Brødre area, south of Buksefjorden, southern West Greenland. *J. Geol. Soc. London* 144, 369-376.
Friend, C.R.L., Nutman, A.P. & McGregor, V.R. 1987b. Significance of the late Archaean granulite facies terrain boundaries, southern West Greenland (abstract). In *Workshop on the Deep Continental Crust of South India*, Lunar and Planetary Institute, Houston, in press.
Fyfe, W.S. 1973. The granulite facies, partial melting and the Archaen crust. *Phil. Trans. Roy. Soc. London* A273, 457- 462.
Grant, J.A. 1973. Phase equlibria in high grade metamorphism and partial melting of pelitic rocks. *Am. J. Sci.* 270, 281- 296.
Grew, E.S. & Manton, W.I. 1984. Age of allanite from Kabbaldurga quarry, Karnataka. *J. Geol. Soc. India* 25, 193- 195.
Hansen, E.C., Newton, R.C. & Janardhan, A.S. 1984. Fluid inclusions in rocks from the amphbiolite-facies gneiss to charnockite progression in southern Karnataka, India: direct evidence concerning the fluids of granulite facies metamorphism. *J. Metamorphic Geol.* 2, 249-264.
Hansen, E.C., Janardhan, A.S., Newton, R.C., Prame, W.K.B.N. & Kumar, G.R.R. 1987. Arrested charnockite formation in southern India and Sri Lanka. *Contrib. Mineral. Petrol.* 96, 225-244.
Holt, R.W. & Wightman, R.T. 1983. The role of fluids in the development of a granulite facies transition zone in S. India. *J. Geol. Soc. London* 140, 651-655.
Janardhan, A.S., Newton, R.C. & Smith, J.V. 1979. Ancient crustal metamorphism at low P_{H2O}: charnockite formation at Kabbaldurga. *Nature* 277, 511-514.
Janardhan, A.S., Newton, R.C. & Hansen, E.C. 1982. The transformation of amphibolite facies gneiss to charnockite in southern Karnataka and northern Tamil Nadu, India. *Contrib. Mineral. Petrol.* 79, 130-149.
Kalsbeek, F. 1976. Metamorphism of Archaean rocks of West Greenland. In *Early History of the Earth* (Windley, B.F. ed.) pp.225-235. Wiley, Chichester.
Kumar, G.R., Srikantappa, C. & Hansen, E. 1985. Charnockite formation at Ponmudi in southern India. Nature 313, 207- 209.
McGregor, V.R., Nutman, A.P. & Friend, C.R.L. 1986. The Archaean geology of the Godthåbsfjord region, southern West Greenland. In Early Crustal Genesis: The Worlds Oldest Rocks. (Ashwal, L.D. ed.) 113-170 Lunar & Palnetary Institute Tech Rept. 86-04.
Pichamuthu, C.S. 1960. Charnockite in the making. *Nature* 188, 135.
Pidgeon, R.T. & Kalsbeek, F. 1978. Dating of igneous and metamorphic events in the Fiskenaesset region of southern West Greenland. *Can. J. Earth Sci.* 15, 2021-2025.
Powell, R. 1983. Processes in granulite facies metamorphism. In *Migmatites, Melting and Metamorphism* (Atherton, M.P. & Gribble, C.D., eds.) pp.127-139. Shiva, Nantwich.
Spear, F.S. & Selverstone, J. 1983. Water exsolution from quartz: Implications for the generation of retrograde metamorphic fluids. *Geology* 11, 82-85.
Spooner, C.M. & Fairburn, H.W. 1970. Strontium 87/strontium 86 initial ratios in pyroxene granulite terrains. *Jour. Geophys. Res.* 75, 6706-6713.
Srikantappa, C., Raith, M. & Spiering, B. 1985. Progressive charnockitization of a leptynite-khondalite suite in southern Kerala, India - Evidence for formation of charnockites through decrease in fluid pressure? *J. Geol. Soc. India* 26, 849-872.
Tarney, J. & Weaver, B.L. 1987. Geochemistry of the Scourian complex: petrogenesis

and tectonic models. *Sp. Publ. Geol. Soc. London* 27, 45-56.

Taylor, P.N., Moorbath, S., Goodwin, R. & Petrykowski, A.C. 1980. Crustal contamination as an indicator of the extent of early Archaean continental crust: Pb isotopic evidence from the late Archaen gneisses of West Grenland. *Geochim. Cosmochim Acta* 44, 1427-1453.

Touret, J. 1977. The significance of fluid inclusions in metamorphic rocks. In: *Thermodynamics in Geology* (Fraser, D.G. ed.), pp. 203-227. D. Reidel Publishing, Dordrecht.

Valley, J.W. & O'Neil, J.R. 1984. Fluid heterogeneity during granulite facies metamorphism in the Adirondaks: stable isotopic evidence. *Contrib. Mineral. Petrol.* 85, 158-173.

Walton, B.J. 1971. The area north of Bjørnesund, near Fiskenaeset, southern West Greenland. Unpubl. Int. GGU Rep. 11pp.

Walton, B.J. 1972. The area north of Bjørnesund, continued. Unpubl. Int. GGU Rep., 8pp.

Weaver, B. & Tarney, J. 1983. Elemental depletion in Archaean granulite facies rocks. In: *Migmatites, Melting and Metamorphism* (Atherton, M.P. & Gribble, C.D. eds.) pp.250 -263 Shiva, Nantwich.

CONTRASTING MECHANISMS OF CHARNOCKITE FORMATION IN THE AMPHIBOLITE TO GRANULITE GRADE TRANSITION ZONES OF SOUTHERN INDIA

M. Raith, S. Hoernes, E. Klatt and H.J. Stähle
Mineralogisch-Petrologisches Institut
Universität Bonn
Poppelsdorfer Schloß
5300 Bonn, Federal Republic of Germany

ABSTRACT. Progressive development of charnockite through dehydration of amphibolite grade gneisses in southern India occurred exclusively in zones transitional to the deeper and pervasively granulitized crust. Isotope data document two distinct events of "arrested" charnockite formation: at ~2.5 b.y. in the Archaean gneiss terrane of southern Karnataka and at ~ 550 m.y. in the Proterozoic crustal segment of southern Kerala. A detailed study of field relations, petrological and geochemical characteristics in two selected exposures constrain possible mechanisms of arrested charnockitization.

Charnockitization in both terranes postdates the major event of penetrative deformation, high-grade metamorphism and migmatization. It occurred during uplift and slow cooling when the rheologic properties of the gneiss complexes changed from ductile to brittle. The dehydration process proceeded along late ductile shears (Kabbaldurga, Karnataka) or conjugate fractures (Kottavattam, Kerala) as well as the foliation planes of the gneisses. It was evidently controlled by fluid-rock interaction in the tectonically generated fluid-pathways. The final product is a coarse-grained massive hypersthene-bearing rock of granitic composition (charnockite s. str.). While charnockitization of granitic biotite-garnet gneisses (Kottavattam) was nearly isochemical, marked changes in mineralogy and bulk chemistry occurred where charnockite developed in place of granodioritic lithologies (Kabbaldurga).

The results of the present investigation indicate that arrested charnockite formation in southern India was controlled by two different mechanisms of fluid-rock interaction: At Kabbaldurga, dehydration of the hornblende-biotite assemblages was caused by influx of external carbonic fluids along ductile shears and the concommitant decrease of water activity (at ~ 700 °C and 6 kb). It is assumed that the invading fluids were released by shear deformation from a "fossil" reservoir of carbonic fluid inclusions trapped in the deeper crustal granulites underlying the gneiss terrane. At Kottavattam, dehydration of the biotite-garnet gneiss (at 750 °C and 5-6 kb) was possibly triggered by a drop in fluid pressure resulting from the escape of internally derived and buffered CO_2-N_2-rich pore fluids into a system of conjugate fractures. It appears unlikely that these locally restricted and tectonically controlled processes caused the pervasive granulitization of extensive parts of Precambrian lower crust in southern India.

D. Bridgwater (ed.), Fluid Movements – Element Transport and the Composition of the Deep Crust, 29–38.
© *1989 by Kluwer Academic Publishers.*

INTRODUCTION

In the deeply eroded Precambrian crust of southern India and Sri Lanka, a series of spectacular exposures shows progressive dehydration of amphibolite grade gneisses in different arrested stages. Their regional distribution indicates that this process was restricted to a zone transitional to the deeper and pervasively granulitized crust. In the initial stage of the process, dark coarse-grained hypersthene-bearing charnockite was formed in an irregular pattern of isolated patches. During the more advanced stages, charnockitization then continued along the foliation planes of the gneisses and a system of fractures or ductile shears. The final stage, though rarely observed, led to a complete replacement of the gneisses by massive charnockite. Though known since the pioneering work of Pichamuthu (1960), it is only recently this phenomenon has received considerable interest. This process offers a possible model for granulite formation in the deep continental crust (Janardan et al., 1982; Hansen et al., 1984, 1987; Srikantappa et al., 1985; Stähle et al., 1987).

There is a general agreement that this type of "in-situ" charnockitization was controlled by fluid-rock interaction in a system of tectonically generated fluid-pathways. Based on their studies on the well-known exposure at Kabbaldurga (southern Karnataka), Janardhan et al. (1982) and Hansen et al. (1984, 1987) have argued that charnockitization was triggered by massive influx of carbonic fluids of deep-seated origin and the concommitant reduction of water activity. It is thought that the pervasively and completely granulitized crustal domains to the south of Kabbaldurga initially have passed through a Kabbaldurga-type stage and represent the most advanced and highest-grade stage of one prograde metamorphic event. On the other hand, recent investigations in southern Kerala (Srikantappa et al., 1985; Raith et al., 1988) and Sri Lanka (Baur and Kröner, 1987) give evidence that arrested charnockitization was a late post-kinematic process which occurred subsequent to high-grade regional metamorphism. Furthermore, though influx of external carbonic fluids as envisaged in the case of Kabbaldurga, may be decisive for the generation of granulites in the lower continental crust, it is not the only possible mechanism for "in-situ"-type charnockitization (cf. Srikantappa et al., 1985; Stähle et al. 1987; Raith et al., 1988; Waters, 1988). Srikantappa et al. (1985) have suggested that arrested charnockitization in southern Kerala could also arise from a decrease of fluid pressure due to the release of trapped pore fluids into a system of developing fractures. In this model, the fluids are generated in the gneiss complex itself and their composition is internally buffered.

During the last years we have studied the field relations, petrological and geochemical characteristics of "in-situ" charnockitization in selected exposures in great detail. It is the aim of this paper to summarize the essential observations and data obtained for Kabbaldurga (southern Karnataka) and Kottavattam (southern Kerala) and to substantiate the contrasting mechanisms envisaged for incipient charnockitization (Fig. 1).

CHARNOCKITE FORMATION AT KABBALDURGA

The gneiss complex at Kabbaldurga represents an intensely deformed and migmatized unit of homogeneous grey granodioritic hbl-bio gneisses and mafic monzodioritic to monzonitic bio-hbl gneisses, interbanded with various types of metatexitic to diatexitic gneisses, transsected by strongly fragmented and boudinaged amphibolites, and intruded by anatectic granites preferentially along the foliation of the gneisses (cf. Friend, 1984). The age of the dominant grey hornblende-biotite gneisses is

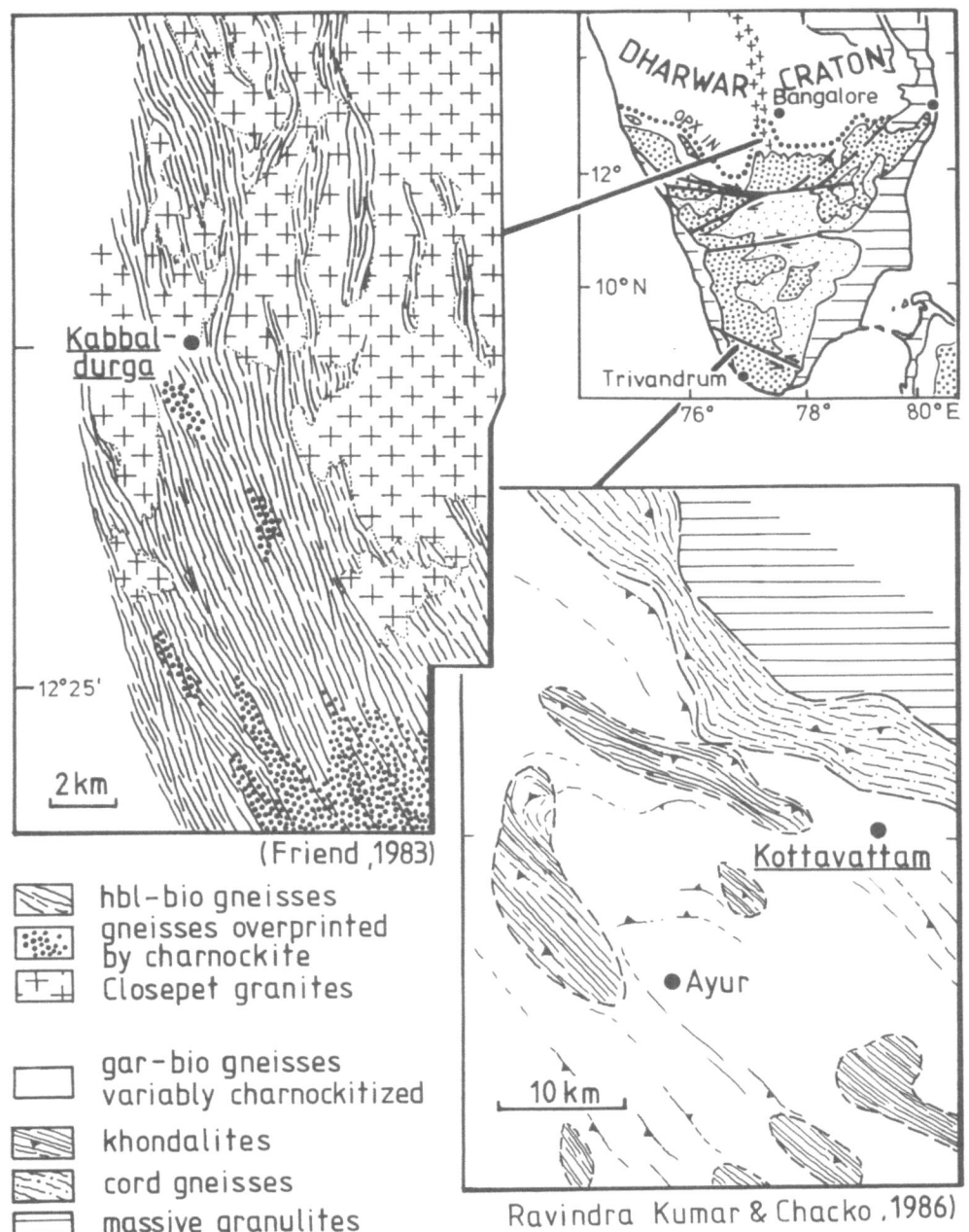

FIGURE 1. Geological sketch maps of the Kabbaldurga and Kottavattam areas

Archaean, i.e. ~ 3.4 b.y. (U-Pb zircon upper intercept data; Buhl, 1987). High-grade metamorphism and partial anatexis occurred much later, i.e. ~ 2.5 b.y. ago (U-Pb zircon and Rb-Sr whole rock data; Buhl et al., 1983; Buhl, 1987). A set of E-W trending ductile shears transsects the gneiss and migmatite structures and it is predominantly along these late shears that coarse-grained massive hypersthene-bearing charnockite of granitic composition has developed over the quartzo-feldspathic lithologies. Isotope data (Grew and Manton, 1984; Buhl, 1987) indicate that this process of dehydration closely followed regional high-grade metamorphism.

The charnockitized zones have a high concentration of carbonic fluid inclusions (T_h -25.6 to +31.3°C, peak at +22 °C), whereas the grey gneisses are essentially free of pore fluids (cf. Hansen et al., 1984; Stähle et al., 1987). It is apparent that charnockitization was controlled by influx of external carbonic fluids along tectonically generated pathways of increased permeability.

Recent petrographic and geochemical studies on large rock slabs across gneiss-charnockite transitions (Stähle et al., 1987) revealed that charnockitization was accompanied by substantial changes in mineralogy as well as bulk chemistry from granodioritic to granitic (Table 1). The metasomatic changes show a strong mineralogical control: Increased modal amounts of K-feldspar (and quartz) at the expense of plagioclase are coupled with marked gains in K, Rb, Ba and Si, whereas decreased modal amounts of hornblende, biotite and magnetite (+ ilmenite) correlate with the losses of Fe, Mg, (Ca), Ti, Zn, and V. Similarly, the lower abundances of P and Zr in the charnockites point to lower modal amounts of apatite, zircon and monazite. The modal abundances of these accessory phases (and the amphibole) obviously control the REE systematics. While the grey gneisses are characterized by moderately fractionated spectra (La_N = 270; La_N/Yb_N = 5-20; Eu_N ~ 27; Eu_N/Eu_N^* = 0.6-0.3), the charnockite spectra, depending on the intensity of metasomatism, are moderately to strongly fractionated (La_N = 200; La_N/Yb_N = 20-80; Eu_N ~ 22; Eu_N/Eu_N^* = 0.6-1.8). The systematic depletion especially in the HREE correlates with decreasing Zr contents and thus indicates progressive dissolution of zircon, a phase which strongly concentrates the HREE. In addition, the loss in apatite, monazite and amphibole might have contributed to the general depletion in the REE.

Gneisses and charnockite have similar oxygen isotopic compositions ($\delta^{18}O$ = 6.9 - 8.0 °/oo). The $\delta^{18}O$ values are typical for igneous rocks. The carbon isotopic composition of late carbonate in two charnockite samples displays typical magmatic values ($\delta^{13}C$ = -8.5 and -6.5 °/oo). These findings militate against intensive interaction with fluids derived from a metasedimentary source.

The metamorphic temperatures prevailing during charnockitization at Kabbaldurga have been constrained by oxygen isotope thermometry. The oxygen isotope fractionations between quartz and magnetite (Δ_{qz-mt} = 8.4 - 7.3 °/oo) indicate isotopic equilibration in the temperature range between 650 and 720 °C in both the gneiss and charnockite. Mineralogical thermometry and barometry on suitable granulite assemblages in areas adjacent to Kabbaldurga yielded model temperatures of 700 to 750 °C and lithostatic pressures between 5 and 7 kb (Raith et al., 1983; Mazumdar, 1987). These estimates, however, reflect the conditions of upper amphibolite to granulite facies regional metamorphism and not necessarily the P-T regime of charnockitization. The fluids involved in charnockitization were strongly water-deficient (X_{H2O} < 0.3) and moderately oxidized (fO_2 higher than defined by the QFM-buffer and controlled by bio-qtz-Kfsp-mt and bio-qtz-Kfsp-opx-mt equilibria).

TABLE 1. Modal analyses and geochemical data of gneisses and charnockites from Kabbaldurga and Kottavattam, southern India.

KABBALDURGA QUARRY (KARNATAKA) *

	plag	qtz	kfsp	myrm	hbl	bio	mt	opx	
grey gneiss:	46	26	12	3	7	4	2	---	vol.%
charnockite:	36	31	24	4	1	2	1	<1	

accessories: magnetite, ilmenite, apatite, zircon, monazite

	SiO_2	Al_2O_3	Fe_2O_3	MnO	MgO	CaO	Na_2O	K_2O	
grey gneiss:	69.5	13.8	5.0	0.08	1.10	3.07	3.67	2.68	wt.%
charnockite:	74.1	13.0	2.3	0.07	0.56	1.74	3.19	4.22	

	Rb	Sr	Ba	Zr	V	Zn	Y		TiO_2	P_2O_5		$\partial^{18}O$	
gn:	64	141	445	300	78	74	52	ppm	0.77	0.20	wt.%	7.2	º/oo
ch:	102	149	722	205	32	32	10		0.42	0.10		7.5	

	La	Ce	Nd	Sm	Eu	Tb	Dy	Ho	Er	Yb	Lu	
gn:	64	115	50	10.2	1.45	1.93	10.50	1.88	6.15	4.32	0.70	ppm
ch:	59	92	30	4.3	1.26	0.66	2.34	0.46	1.28	1.00	0.23	

KOTTAVATTAM QUARRY (KERALA) **

	plag	qtz	kfsp	gar	bio	opx	
gneiss:	22-27	28-30	26-30	6-10	6-10	---	vol.%
charnockite:	26-29	24-28	27-30	6-10	2-4	~5	

accessories: ilmenite, pyrrhotite, graphite, zircon, monazite, apatite (rutile, magnetite).

	SiO_2	Al_2O_3	Fe_2O_3	MnO	MgO	CaO	Na_2O	K_2O	
gneiss:	68.1	13.6	5.6	0.08	1.1	2.4	2.5	4.4	wt.%
charnockite:	67.9	14.0	4.7	0.04	0.9	2.3	2.7	5.3	

	Rb	Sr	Ba	Zr	V	Zn		TiO_2	P_2O_5		$\partial^{18}O$	
gn:	220	130	1055	344	105	65	ppm	0.90	0.38	wt.%	10.3	º/oo
ch:	216	141	1032	349	70	63		0.87	0.36		10.3	

	La	Ce	Nd	Sm	Eu	Tb	Dy	Er	Yb	Lu	
gn:	78	164	74	13.1	0.86	2.04	12.3	6.5	5.80	0.82	ppm
ch:	78	123	63	10.7	1.07	1.30	7.6	3.9	3.64	0.51	

* Average modal analyses and geochemical data for 10 gneiss and 10 charnockite specimens from a 2m rock profile (Stähle et al., 1987; Tables 1 and 2)

** Average modal analyses and geochemical data for 4 gneiss and 3 charnockite specimens form a 40cm rock profile (Raith et al., 1988)

CHARNOCKITIZATION AT KOTTAVATTAM

"In-situ" charnockitization is a wide-spread phenomenon in the Proterozoic crustal segment south of the Achankovil shear belt in southern Kerala and excellent examples can be studied in several active quarries around Punalur (Srikantappa et al., 1985). The Kottavattam quarry exposes fairly homogeneous light grey biotite-garnet gneisses of granitic composition (age < 2 b.y.; Buhl, 1987) (Table 1). The streaky foliation of the gneisses (N 60-80 W; dip 80-90 SW) is largely obliterated by a diffuse network of garnetiferous leucosomes which were possibly generated by biotite dehydration-melting reactions. The P-T conditions of high-grade regional metamorphism were estimated from mineralogical geothermo-barometry to be 700-800 °C and 5-7 kb.

The transformation of these gneisses into coarse-grained charnockite occurred along a system of conjugate fractures (N 70 E and N 20 W) but also followed the foliation planes about 550 m.y. ago (U-Pb zircon and monazite data; Buhl, 1987). In the charnockitized zones, the migmatitic structure of the precursor gneisses is completely extinguished due to thorough recrystallization and considerable coarsening. The hypersthene-bearing charnockite zones are commonly separated from the unaltered gneiss by a narrow transition zone almost devoid of biotite. This zonation suggests considerable lateral mass transfer via a fluid phase from an outer zone of biotite breakdown to an inner zone of hypersthene neoblastesis. The overall changes in mineralogy and modal composition, however, are small (Table 1).

Detailed microprobe work on larger rock specimens across gneiss-charnockite transitions revealed almost identical mineral compositions in both the gneiss and charnockite assemblages: garnets (alm 75-76, pyr 13-15, gro 7-9, spe ~ 2 mol%); biotites (X_{Mg} 0.47-0.53, Ti 0.28-0.32 atoms p.f.u.); plagioclases (An 32-36, Or 1-2); K-feldspars (Or 78-84, Ab 15-20, An 1-2); ilmenites (>98 $FeTiO_3$). Hypersthenes could not be analysed due to complete retrograde alteration to hydrosilicates (smectite, chlorite, talc), carbonate and opaque minerals. The absence of changes in mineral chemistry across the gneiss-charnockite transitions indicates isothermal-isobaric equilibration of the gneiss and charnockite assemblages. This is confirmed by P-T estimates obtained from up-dated calibrations of garnet-biotite Fe-Mg exchange thermometry and the garnet-plagioclase-quartz-ilmenite-rutile geobarometer (Bhattacharya, pers. comm.), i.e. 750 ± 10 °C and 5.6 ± 0.2 kbars lithostatic pressure. Furthermore, a comparable internal buffering of pore fluids to low fugacities of water and oxygen, but to high fugacities of carbon dioxide is obvious from the presence of identical opaque mineral assemblages (ilmenite, pyrrhotite, graphite ± rutile) in both the gneiss and charnockite.

Fluid inclusions are abundant in the charnockite and the associated gneiss. They occur in planar arrays, mainly in quartz but also in feldspars and garnet. In addition to these fluid- pathways, cathodo-luminescence images showed numerous systems of healed micro-fractures. A high permeability and fluid activity during charnockitization can be inferred from these observations. The properties of entrapped fluids were studied by microthermometry, Raman-Laser-Probe analysis and mass spectrometry (Klatt et al., 1987, 1988). The similarity in the fluid inclusion characteristics of both rock types testifies a comparable and complex evolution of the pore fluids. Early metamorphic fluids are entrapped in rare briny inclusions (+salt).

The dominant fluids are represented by medium- to low-density carbonic inclusions (T_h: +6 to +27 °C; T_m: -57.7 to -56.6 °C; ρ: 0.70-0.86 g/cm^3; up to 10 mol% nitrogen, less than 2 mol% hydrocarbons). They occur in several sets of healed fractures, thus indicating partial to complete physical equilibration of these fluids by progressive leakage and repeated reentrapment as a consequence of near-isothermal uplift of the rock complex. Nitrogen inclusions (T_h: -144 to -142 °C) are less abundant. A generation of these fluids by devolatilization of the NH_4-bearing phases biotite and K-feldspar mainly during partial anatexis of the gneisses (formation of garnetiferous leucosomes) appears most likely. Medium-density watery inclusions of low salinity (ρ 0.89-0.94 g/cm^3; < 4 mol% equiv. NaCl) are the texturally latest entrapped metamorphic pore fluids. Where they cross trails of earlier carbonic inclusions, mixed H_2O-CO_2 inclusions (forming clathrate ices) developed. Bulk fluid analysis by mass spectrometry on quartz concentrates showed the charnockites to be higher in CH_4 and H_2O but lower in N_2 in comparison to the gneisses. CO_2 has a similar abundance in both rock types. An evaluation of fluid composition based on graphite-fluid equilibria in the C-O-H-N system at P-T conditions of charnockitization confirms the strongly water-deficient and reduced nature (fO_2 close or lower than defined by the QFM buffer) of the pore fluids.

δ^{13}C data on graphite and entrapped carbonic fluids have been reported for several exposures of "in-situ" charnockitization in southern Kerala (Jackson et al., 1987; Klatt et al., 1988; Hoernes and Raith, 1988). The δ^{13}C systematics indicates a derivation of graphite (δ^{13}C -14 to -22 °/oo) from the degradation of organic matter and attainment of isotopic equilibrium with the coexisting carbonic fluids (δ^{13}C -10 to -15 °/oo) near peak metamorphic conditions. Further evidence of the internal nature of carbonic fluids comes from whole rock oxygen isotope data. Gneiss and charnockite from Kottavattam exhibit identical δ^{18}O values of 10.3 °/oo which are typical for psammo-pelitic metasediments. Massive influx of external carbonic fluids with mantle isotopic signature (δ^{18}O ~ 8 °/oo) thus appears unlikely and even a limited amount of mantle derived CO_2 would have lowered and homogenized the graphite δ^{13}C values as demonstrated by Vry et al. (1988).

The chemical composition of the gneisses at Kottavattam closely resembles that of arcosic sediments (Table 1). The sedimentary nature of the protoliths is corroborated by the occurrence of graphite and nitrogen fluid inclusions. The charnockitized zones have almost identical major and trace element abundances (Table 1). These results document the isochemical character of "in-situ" charnockitization at Kottavattam.

CONTRASTING MECHANISMS OF CHARNOCKITE FORMATION

The field relations, petrological and geochemical data summarized in the previous sections clearly show that "arrested" charnockite formation in southern India was not restricted to a specific metamorphic event and terrane, but occurred at different times in high-grade gneisses of different bulk chemistry and mineralogy. Furthermore, there are also characteristic differences regarding the origin of the fluids, their control on the dehydration reactions, and the extent of fluid-rock interaction. Different mechanisms, therefore, might be envisaged to explain all the features of structurally-controlled charnockitization.

At Kabbaldurga, charnockitization was induced by the influx of CO_2-dominated fluids along a set of ductile shears and the foliation planes at P-T conditions of about 700 °C and 6 kbars. The concommitant, externally controlled decrease of water activity in the percolated zones caused the almost complete breakdown of hornblende and biotite, and the new growth of hypersthene. The marked changes in mineralogy and chemistry accompanying dehydration documents the metasomatic nature of this process. The pronounced gain in K, Rb, Ba and Si is attributed to intense replacement of plagioclase by K-feldspar through cation exchange with the passing fluids, whereas the loss of Fe, Mg, (Ca), Ti, P, Zn, V, Zr and the HREE resulted from dissolution of hornblende, biotite, magnetite, apatite, zircon and monazite. Since solubility and partitioning data for mineral- carbonic fluid systems at the relevant P-T conditions are lacking, a quantitative evaluation of mass transfer and the fluid/rock ratio is not possible.

An external source for the carbonic fluids is indicated by the fluid inclusion characteristics and stable isotope data. Most workers assumed a generation of these fluids by deep-seated processes like degassing of crystallizing underplated basaltic magmas (Touret, 1971) or decarbonation of subducted carbonate sediments or the upper mantle (Newton et al., 1980; Glassley, 1983). In contrast it is suggested here that the most likely source for the CO_2-rich fluids is the "fossil" reservoir of carbonic fluids trapped in the deeper-crustal granulites underlying the gneiss terrane at Kabbaldurga. Shear deformation has tapped this reservoir and generated the channel-ways for fluid ascent.

At Kottavattam, the precursor garnet-biotite gneisses represented a system rich in pore fluids. It is likely, therefore, that during uplift of the gneiss complex isothermal decompression created fluid overpressure which - in a regime of anisotropic stress - initiated or at least promoted the simultaneous development of a system of conjugate fractures. The release of the CO_2-N_2-rich fluids from bursting micropores into the newly opened channel-ways then resulted in a drop of fluid pressure ($P_{fluid} < P_{load}$) which ultimately triggered the dehydration reactions, i.e. the breakdown of biotite and the neoblastesis of hypersthene. This process occurred at almost isothermal-isobaric conditions, i.e. at 750 °C and 5-6 kbars P_{load} and in the presence of strongly water-deficient and reduced pore fluids in both rock types. Although charnockitization at Kottavattam took place in an open system (i.e. escape of pore fluids), this process was not accompanied by noticeable changes in bulk chemistry. This is explained by the internal generation and buffering of the fluids and their probably limited migration in an entirely granitic rock system. Recently, comparable examples of near-isochemical charnockitization have been described from the high-grade terrane of Sri Lanka (Baur and Kröner, 1987; Hansen et al., 1987; Glassley, this Volume).

ACKNOWLEDGEMENTS

We would like to thank D. Bridgwater, W.E. Glassley and N.B.W. Harris for their careful reviews of the manuscript. Financial support of the investigations by the Deutsche Forschungsgemeinschaft and the Stiftung Volkswagenwerk is gratefully acknowledged.

REFERENCES

Bauer, N. and Kröner, A., 1987. Structural control of prograde and retrograde metamorphic reactions in high-grade Proterozoic gneisses of Sri Lanka. Annual Meeting Geologische Vereinigung, Basel. Terra Cognita, v. **7**, 1, p. 48.

Buhl, D., 1987. U-Pb und Rb-Sr Altersbestimmungen und Untersuchungen zum Strontium-Isotopenaustausch an Granuliten Südindiens. (Ph.D. Thesis University of Münster (FRG)).

Buhl, D., Grauert, B. and Raith, M., 1983. U-Pb dating of Archaean rocks from the amphibolite-granulite facies transition zone at Kabbal quarry, southern Karnataka. Fortschr. Mineralogie, v. **61**, 1, p. 42-45.

Friend, C.R.L., 1983. The link between charnockite formation and granite production: evidence from Kabbaldurga, Karnataka, South India. In: Atherton, M.P. and Gribble, C.D. (Eds), Migmatites, Melting and Metamorphism. Shiva Publ. Glasgow, p. 264-276.

Friend, C.R.L., 1984. The origins of the Closepet granites and the implications for the crustal evolution of southern Karnataka. J. Geol. Soc. India, v. **25**, p. 73-84.

Glassley, W.E., 1983. The role of CO_2 in the chemical modification of deep continental crust. Geochim. Cosmochim. Acta, v. **47**, p. 597-616.

Glassley, W.E., 1988. Mineral stability and element mobility in fluid-bearing systems under deep crustal conditions. (This Volume).

Grew, E.S. and Manton, W.I., 1984. Age of allanite from Kabbaldurga quarry, Karnataka. J. Geol. Soc. India, v. **25**, p. 193-195.

Hansen, E.C., Newton, R.C. and Janardhan, A.S., 1984. Fluid inclusions in rocks from the amphibolite—facies gneiss to charnockite progression in southern Karnataka, India: Direct evidence concerning fluids of granulite metamorphism. J. metam. Geology, v. **2**, p. 249-264.

Hansen, E.C., Janardhan, A.S., Newton, R.C., Prame, W.K.B.N. and Ravindra Kumar, G.R., 1987. Arrested charnockite formation in southern India and Sri Lanka. Contrib. Mineral. Petrol., v.**96**, p. 225-244.

Hoernes, S. and Raith, M., 1988. Herkunft, Mobilisation und Fällung des Kohlenstoffs in granulitfaziellen Gesteinsserien Südindiens. Forschritte Mineral., v.**66**, p. 65.

Jackson, D.H., Mattey, D.P. and Harris, N.B.W. 1988. Carbon isotope compositions of fluid inclusions in charnockites from southern India. Nature, v. **333**, p. 167-170.

Janardhan, A.S., Newton, R.C. and Hansen, E.C., 1982. The transformation of amphibolite facies gneiss to charnockite in southern Karnataka and northern Tamil Nadu, India. Contrib. Mineral. Petrol., v. **79**, p. 130-149.

Klatt, E. and Raith, M., 1987. Fluid inclusion characteristics of the progressive gneiss-charnockite transformation in southern Kerala, India. European Current Research on Fluid Inclusions, 9th Symposium; University of Porto, Portugal (Abstracts).

Klatt, E., Hoernes, S. and Raith, M., 1988. Characterization of fluids involved in the gneiss-charnockite transformation in southern Kerala (India). J. Geol. Soc. India, v. 31, p. 57-59.

Mazumdar, A.C., 1987. Chemical petrology of the granulites around Satnur-Halaguru, Karnataka. (Ph.D. Thesis): Indian Institute of Technology, Kharagpur, India..

Newton, R.C., Smith, J.V. and Windley, B.F., 1980. Carbonic metamorphism, granulites and crustal growth. Nature, v. 288, p. 45-50.

Pichamuthu, C.S., 1960. Charnockite in the making. Nature, v. 188, p. 135-136.

Raith, M., Raase, P., Ackermand, D. and Lal, R.K., 1983. Regional geothermobarometry in the granulite terrane of South India. Trans. Roy. Soc. Edinburgh, v. 73, p. 221-244.

Raith, M., Klatt, E., Spiering, B., Srikantappa, C. and Stähle, H.J., 1988. Gneiss-charnockite transformation at Kottavattam, southern Kerala (India). J. Geol. Soc. India, v. 31, p. 114-115 .

Ravindra Kumar, G.R. and Chacko, T., 1986. Mechanisms of charnockite formation and breakdown in southern Kerala: Implications to the origin of southern India granulite terrain. J. Geol. Soc. India, v. 28, p. 277-288.

Stähle, H.J., Raith, M., Hoernes, S. and Delfs, A., 1987. Element mobility during incipient granulite formation at Kabbaldurga, southern India. J. of Petrology, v. 28, p. 803-834.

Srikantappa, C., Raith, M. and Spiering, B., 1985. Progressive charnockitization of a leptynite-khondalite suite in southern Kerala, India - evidence for formation of charnockites through a decrease in fluid pressure? J. Geol. Soc. India, v. 26, p. 849 872.

Touret, J., 1971. Le facies granulite en Norvège meridionale. Lithos, v. 4, p.239-249, p. 423-436.

Vry, J., Brown, P.E., Valley, J.W. and Morrison, J., 1988. Constraints on granulite genesis from carbon isotope compositions of cordierite and graphite. Nature, v. 332, p. 66-68.

Waters, D.J., 1988. Partial melting and the formation of granulite facies assemblages in Namaqualand, South Africa. J. metamorphic Geol., v. 6, 4, p. 387-404.

CHEMICAL CHANGES ASSOCIATED WITH FORMATION OF GRANULITE AND MIGRATION OF COMPLEX C-O-H-S FLUIDS, SRI LANKA

W. E. Glassley, F. J. Ryerson, H. Shaw, Lawrence Livermore
National Laboratory, Livermore, CA. 94550, USA; P. B. Abeysinghe
Geological Survey Department, Colombo, Sri Lanka

ABSTRACT. Granulite facies gneisses forming from amphibolite facies gneisses are well preserved and exposed in the Highland Series of Sri Lanka. Invasion of carbon- and sulfur-bearing fluids into the amphibolites resulted in breakdown of hydrous phases to generate orthopyroxene-bearing assemblages, and led to significant chemical modification, including an increase in La/Yb, Th/U and Sr/Rb ratios, and a decrease in K/Rb and total K. Granulite formation is associated with precipitation of graphite, sulfides and allanite at one site and not another, even though the chemical changes are the same at both sites. The similarities and contrasts of the sites indicate that local chemical conditions existing prior to fluid invasion influence the mineralogical response of the site. Field relationships indicate the fluid was derived from marbles in the vicinity and migrated along well defined pathways.

INTRODUCTION

In 1965, Heier and Adams described depletion of K, Th, U, and Rb in granulite facies rocks, relative to lower grade metamorphic rocks. Although similar depletions have since been reported from a number of granulite terrains by other authors (Heier, 1965, 1973; Lambert and Heier, 1967, 1968; Sighinolfi, 1969, 1971; Heier and Thoresen, 1971; Condie et al., 1982; Okeke et al., 1983; Smalley et al., 1983; Sheraton and Collerson, 1984) variations in most other major elements have also been reported (Ramberg, 1951; Tarney et al., 1972; Drury, 1973; Kalsbeek, 1976a,b; Wells, 1979; Clough and Field, 1980; Hansen et al., 1987). However, examples of terrains that show enrichment in so-called incompatible elements have also been reported (Weaver, 1980; Janardhan et al., 1983; Weaver and Tarney, 1983), as have terrains that exhibit no apparent change in bulk composition during granulite facies metamorphism (Gray, 1977; Sighnolfi et al., 1981; Barbey and Cuney, 1982; Iyer et al., 1984).

Recently, Rudnick et al. (1985) surveyed the literature describing variations in Th, U, K, Rb and La, and demonstrated that covariation in K/Rb ratios suggest that protolith composition may have an important influence on the bulk composition of the metamorphic rocks. Nevertheless, they present strong evidence for varying degrees of Rb depletion and enhanced Th/U ratios in most granulite facies gneisses. The extent to which potassium is depleted in these same rocks cannot be clearly established. Complete overlap in K/La values between igneous and metamorphic rocks does not allow

39

D. Bridgwater (ed.), Fluid Movements – Element Transport and the Composition of the Deep Crust, 39–49.
© *1989 by Kluwer Academic Publishers.*

conclusions to be drawn regarding the extent to which changes in this ratio reflect metamorphic mass transport. Thus, despite concerted study, the nature and extent of chemical modification of rocks during granulite facies metamorphism remains somewhat ambiguous.

To explain some of the apparent changes in bulk composition associated with granulite formation Tarney et al. (1972), Heier (1973), Drury (1973), Bridgwater (1979), Glassley (1983), and Hansen et al. (1987) have suggested that migration of a CO_2-rich fluid through amphibolite facies rocks can lead to development of granulite facies parageneses, and may result in changes in bulk composition via preferential partitioning of certain elements into the fluid phase. Because speciation and thermodynamic properties of complex C-O-H fluids remains unknown at high temperatures and pressures, the chemical mechanisms that would accomplish the partitioning and mass transport have not been outlined or described.

We describe here results from Sri Lanka in which more precise definition of the chemical changes associated with granulite formation is possible through detailed sampling of arrested granulite formation in amphibolite facies gneisses. The locations studied are areas where precursor amphibolites and later granulites are intimately associated on the outcrop scale, allowing unequivocal identification of the chemical characteristics of the protolith to the granulite.

REGIONAL GEOLOGY

Detailed descriptions of the rocks of Sri Lanka can be found in Cooray (1961, 1962), and Hapuarachi (1968, 1975); only a sketch of the geological relationships significant for the geochemical studies is presented here.

Upper amphibolite and granulite facies gneisses occur within the Highland Series that traverses the island from southwest to northeast. Flanking the Highland Series are amphibolite facies gneisses of the Vijayan Complex. The Highland Series rocks are composed of a complexly folded, polymetamorphic terrain of tonalitic and metasedimentary rocks that were intruded by the Arena granites. The Vijayan Complex rocks are predominately migmatitic gneisses which generally lack the marbles and quartzites common to the Highland Series. The relationship between the Highland Series and the Vijayan Complex is a matter of debate. Arguments have been presented that the Vijayan Complex is a retrogressed equivalent of the Highland Series (Cooray, 1961, 1962), and, conversely, that the Vijayan is basement upon which the Highland Series rocks were deposited (Katz, 1971).

Crawford and Oliver (1969) established that three clusters of ages could be obtained from the high grade metamorphic rocks. Rb/Sr model ages in excess of 2.0 Ga were obtained from some rocks within the Highland Series. Ages in the range of 1.1 to 1.3 Ga were obtained from rocks within the Highland Series and the Vijayan Complex, and were interpreted as representing the main metamorphism during which the granulite facies parageneses developed. An age of 400 to 600 Ma was obtained for some samples, but was not correlated with any particular event.

More recent Sm/Nd, Rb/Sr, and U-Th-Pb data (Shaw et al., 1987; Kroner et al., 1987) demonstrate that the granulite facies parageneses at the sites described in this paper developed during the ca. 500 Ma. metamorphic event.

STUDY LOCATIONS

Within the Highland Series there exist a number of recent quarries in which occur well exposed examples of arrested granulite formation. We describe in this paper the results of detailed studies carried out on samples from two of these sites, Kurunegala and Digana.

At the Kurunegala site (Fig. 1), orthopyroxene-bearing parageneses have developed as highly elongate fingers that occur superimposed on quartzo-feldspathic garnet-biotite amphibolite facies gneisses. These granulite fingers lie within the foliation plane and are parallel to the foliation dip. In cross section the fingers have irregular forms and vary in size from several cms. to ca. 1 meter in maximum width. The length of these fingers exceeds 10 meters; exposures were insufficient to determine their maximum length. There was no systematic correlation between the location of the granulite parageneses and deflection of the preexisting gneissic fabric; features that define the gneissic fabric in the amphibolitic gneisses (quartz lenses and rods, and

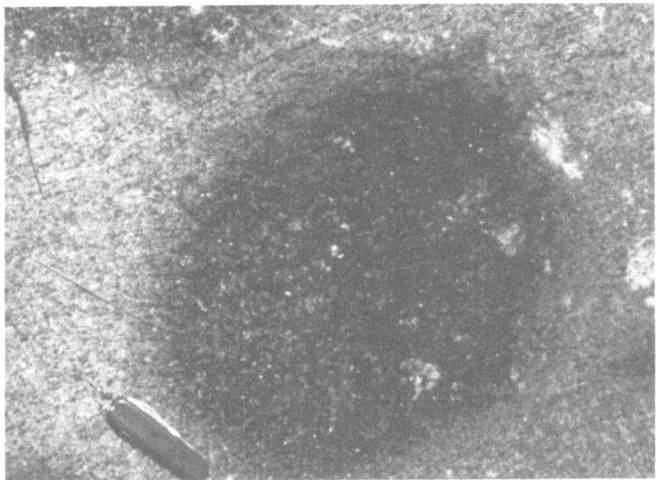

Figure 1. Arrested granulite in amphibolite facies gneiss at Kurunegala.

concentrations of mafic minerals) persist in the granulite parageneses without detectable deformation. Plagioclase and potassium feldspar maximum grainsize (ca. 1 cm) in the core of the granulite fingers is approximately 20-50% greater than in the adjacent amphibolite facies gneisses.

At the Digana site, which is more than 30 km northeast of Kurunegala, a greater abundance of quarries allowed a more thorough characterization of the local geology. At the highest structural level, quarries expose quartzo-feldspathic hornblende-garnet amphibolite facies gneisses that had imposed on them orthopyroxene-bearing granulite facies parageneses that generally occurred as 5 cm to ca. 5 m wide irregularly shaped bodies. Associated with these bodies are significant concentrations of graphite, sulfides, and allanite. In one location an occurrence of granulite was also noted in which the pyroxene-bearing assemblages had formed along the margins of a post-kinematic, ca. 10 cm wide, quartzo-feldspathic dike. Exposure was insufficient to accurately establish the three dimensional form of these bodies, with the exception of the dike occurrence where the granulite clearly formed parallel-sided sheets adjacent to the planar dike. As at Kurunegala, the granulite facies paragenesis had been superimposed on the amphibolite facies gneiss without evidence of deflection or deformation of the preexisting fabric. The preservation of highly elongate quartz rods within the granulites emphasized the static nature of the recrystallization process. With the exception of the dike, no correlation between structural elements and granulite occurrence could be found.

In a road cut structurally below the quarries occur granulite facies gneisses that contain ubiquitous orthopyroxene, clinopyroxene, and graphite, and lack evidence of preservation of the amphibolite facies gneisses. These 'massive' granulites possess the same fabric features and orientations seen within the overlying quarries. Because of the preservation of the delicate quartz textures, the structural proximity of these gneisses to the overlying mixed amphibolite-granulite quarries, and the similarity in mineralogy of the arrested granulite and the 'massive' granulite, we interpret these 'massive' gneisses to be thoroughly recrystallized equivalents of the overlying amphibolite facies gneisses.

Structurally below the 'massive' granulite occurs a very massive dolomitic marble. The marble is in excess of twenty meters thick, and extends for a minimum distance of .5 km along strike. Forsterite, diopside, phlogopite, and spinel are dispersed throughout the carbonate. Blocks of massive calcsilicates are scattered throughout the marble, and are mineralogically zoned. The margins of the calcsilicates are composed of phlogopite and spinel, while the interiors are composed of pargasitic amphibole and phlogopite. Silicates occupy approximately 20-40% of the total volume of the marble bodies.

METHODS

Large blocks were collected from the Kurunegala and Digana quarries. Individual blocks were selected that contained both unmodified amphibolite facies gneiss and associated arrested granulite. A sequence of slabs was then cut from each sample, such that the plane defined by each slab was oriented parallel to the margin of the arrested granulite. Polished thin sections were made of each slab for microscopic and electron microprobe analysis, and each slab was chemically analyzed for major and trace elements (SiO_2, TiO_2, Al_2O_3, FeO*, MgO, MnO, CaO, Na_2O, K_2O, P_2O_5, SO_3, Cu, Zn, Nb, Cr, V, Ni, Ba, Zr, Y, Sr, Rb, Pb, Th by standard X-ray fluorescence techniques; and La, Ce, Nd, Sm, Eu, Tb, Yb, Lu, Ba, Na_2O, Hf, Sc, Ta, Cr, Th, U, Ni, Cs, using INAA). For selected samples Rb, Sr, Sm, and Nd were also determined in the course of isotopic studies (Shaw et al., 1986).

RESULTS

Petrography

Kurunegala: The amphibolite facies mineralogy of the country rock at this location consists of biotite-garnet-perthitic microcline-plagioclase-quartz-zircon-ilmenite. The gneissic fabric is defined by quartz rods (elongation ratios of >40:1) and concentrations of biotite, garnet and ilmenite. Moving from the country rock into the dark colored rods of granulite, the first mineralogical change that is noted is the appearance of iron oxide and chlorite(?) staining along the grain boundaries, accompanied by an increase in the irregularity of grain boundaries. Approximately half way into the dark colored granulite rods the mineralogy abruptly changes. Biotite virtually disappears, garnet decreases in abundance, and large orthopyroxene and clinopyroxene grains appear. An increase in grain size is also noteable.

Digana: At this location the amphibolite facies country rock consists of hornblende-garnet-perthitic microcline-plagioclase-zircon-sphene-il-menite. Lepidoblastic hornblende, quartz rods (elongation ratios > 35:1), and elongated feldspars define a strong gneissic fabric. As at Kurunegala, the first mineralogical change noted in the dark colored pods is the appearance of iron oxide and chlorite (?) staining on the mineral grains. In addition, feldspars become somewhat turbid, sulfides, allanite and graphite become common, carbonate occurs interstitially, and simplectites of orthopyroxene-clinopyroxene-plagioclase-quartz-magnetite-ilmenite form around hornblende grains (Fig. 2). Thorite also occurs associated with the simplectites and with the allanite.

Chemical Changes

Chemical changes (Fig. 3) accompany the transformation from amphibolite to granulite facies at both locations. The largest absolute change is in K_2O, which decreases by 1.2 wt. percent in the granulites at Kurunegala and by 1.0 wt. percent at Digana. Rb and Sr are positively correlated with K_2O, and CaO is negatively correlated with K_2O. At both locations K/Rb is lower and Sr/Rb is higher in the granulites, relative to the corresponding amphibolite facies gneisses. The greatest relative changes occur with the light rare earth elements (LREEs) and volatiles lost on ignition (LOI). Relative to the amphibolite facies country rocks, the La/Yb ratio increases by 80% to 100% at both sites, which primarily reflects an enrichment in the LREEs La to Eu and little change in the heavy rare earth elements (HREEs). This enrichment at Digana is accompanied by a ca. 40%-50% increase in the Th/U ratio; at Kurunegala no noticeable change in the Th/U ratio was noted.
    Although there is significant scatter in the LOI values, there is an apparent 200% increase in the granulites at both sites, relative to the respective amphibolite facies country rocks.

INTERPRETATION

    The mineralogical changes at both sites reflect a complex dehydration process in which pyroxenes and feldspars are generated from the breakdown of biotite (Kurunegala) and amphibole (Digana). The development of sulfides

44

and graphite as reaction products at Digana suggests that dehydration was driven by introduction of a fluid rich in oxidized, transportable forms of carbon and sulfur which were reduced during the dehydration process. This introduced fluid presumably resulted in the destablization of the hydrous phases by decreasing the water activity, either by dilution or displacement.

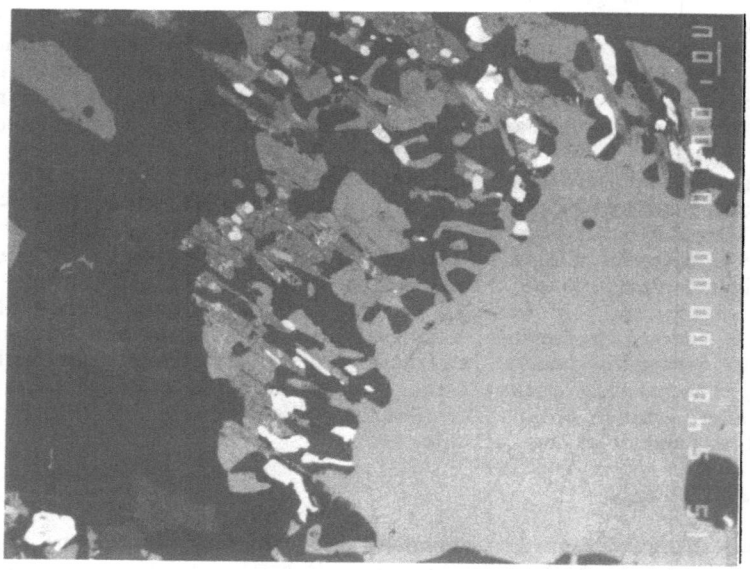

Figure 2. Back scattered electron image of pyroxene-feldspar-Fe/Ti oxide simplectite around hornblende, Digana.

The field relationships and the geometry of the granulite rods and fingers suggests that, in some cases, this dehydrating fluid migrated along pathways that were structurally controlled (e.g. dike margins at Digana and foliation surfaces at Kurunegala). Irregular granulite fingers that crosscut the foliation at Digana appear to reflect a different, and as yet unknown, form of control on the migration pathway.

The consistent chemical changes at the two widely separated sites suggests that the chemical processes responsible for the compositional changes are similar. The precipitation of allanite, thorite, graphite and sulfides at Digana and the absence of these phases at Kurunegala, suggests that the chemical environment in which these changes occurred was different; a relatively higher initial activity of Ca and Fe^{2+}, and lower fO_2, at Digana would be sufficient to explain the contrasts in reaction product mineralogy at the two sites.

Mass transport of LREEs can be accomplished by ligands of the

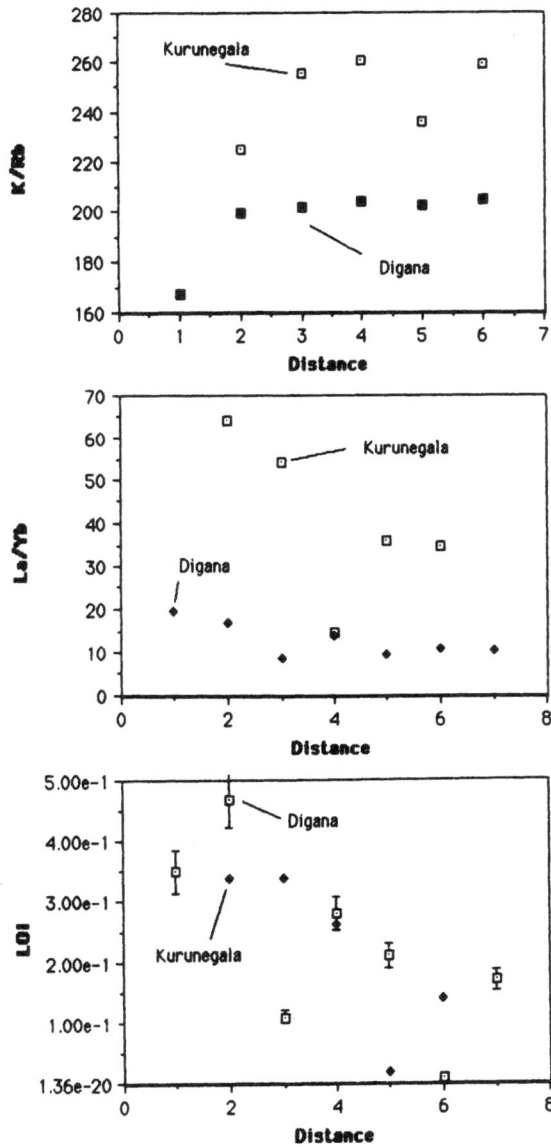

Figure 3. Chemical variations associated with arrested granulite formation. At Digana, "massive" granulite is shown at 1 on the Distance scale; points at distances 2 (core of arrested granulite) through 6 or 7 (country rock amphibolite) are from a sequence of 2 cm. thick slabs from a single sample. At Kurunegala, points 2 (core of arrested granulite) through 6 (country rock amphibolite) are from a sequence of 2.5 cm. thick slabs from a single sample.

form $(CO_3)_x(OH)_y$, which complex with the cations (cf. Sherry and Marinsky, 1964; Sklyarenko and Ruzaikina, 1970; Rard, 1985; Firsching and Mohammadzadel, 1986; Cantrell and Byrne, 1987); similar transport mechanisms have been suggested for Th (Simpson et al., 1982). The transport process is schematically represented by the equilibria

$$(LREE) + .5yH_2O + xCO_2 = (LREE)(CO_3)_x(OH)_y{}^Z$$

$$Th^{4+} + .5yH_2O + xCO_2 = Th(CO_3)_x(OH)_y{}^{4+}$$

Many such equilibria, some of which will involve sulfur and other carbon-bearing species, will govern the solution chemistry involving these constituents. Nevertheless, at the site where the fluid is generated equilibrium will be approached such that, for the CO_2 and H_2O activities at the site, various concentrations of a range of complexes will be formed. The concentrations of the cations in solution will then be a function of the pressure, temperature, fluid composition, and mineralogy of the rock at the fluid generation site; sites rich in phosphates (e.g. apatite), zircon, fluorides, and certain carbonates are potentially rich sources for the LREEs and Th. Decarbonation reactions in marbles, for example, will favor formation of such a mixed C-O-H-S fluid, with associated high concentrations of LREEs and Th if appropriate phases are present.

Advective transport of such a fluid along pathways that transect different rocks will result in development of gradients in activities of the components transported in the fluid. These gradients are equivalent to mixing fronts, and may result in 1) development of dehydration reactions, 2) precipitation of minerals as the activities of fluid constituents change during mixing, and 3) non-stoichiometric dissolution of phases as the original host rock mineralogy adjusts to a new chemical environment. These processes alone are sufficient to explain the similarities and differences observed at Digana and Kurunegala. Thus, the development of pyroxene-feldspar simplectites around hornblende would represent the products of fluid-induced dehydration, the formation of thorite, graphite, sulfides and allanite would represent the precipitation of minerals as the fluid composition changes during mixing, and depletion of the rock in K, Rb, and Sr would represent the consequences of nonstoichiometric dissolution of some mineral phase or phases, possibly the feldspars.

CONCLUSIONS

At the Digana site, decarbonation reactions that took place in marbles resulted in the generation of a C-O-H-S fluid that entered overlying amphibolite facies gneisses. Similarities between the sites studied suggest that Kurunegala experienced a similar history. Mixing of the advecting fluid with the in situ fluid resulted in changes in the activities of fluid components, leading to the development of dehydration reactions involving biotites and hornblendes, chemical modification of the rocks and, where chemical conditions were appropriate, precipitation of secondary minerals such as graphite, sulfides, thorite and allanite.

The similarities and contrasts at the sites we studied emphasize the following points regarding the generation of granulites from amphibolites in deep crustal environments: 1) The chemical features of the granulites may reflect complex chemical processes involving mixing of fluids of contrasting compositions. 2) The mineralogy and chemistry of the source region for the advecting fluid will strongly influence the nature of the chemical and mineralogical changes that take place during the transformation process. The granulites that form during the transformation may be depleted or enriched in a range of elements, depending upon the composition of the fluid that results from mixing of the advecting and in situ fluids. 3) Fluid migration pathways may be strongly influenced by structural features, although in many cases no clear relationship with structural elements can be identified.

REFERENCES

Barbey, and Cuney, 1982. 'K, Rb, Sr, Ba, U and Th geochemistry of the Lapland granulites (Fennoscandia). LILE fractionation controlling factors'. Contrib. Mineral. Petrol., 81, 304-316.
Bridgwater, D. 1979. 'Chemical and isotopic redistribution in zones of ductile deformation in a deeply eroded mobile belt'.U.S. Geol. Surv. Open File Report 79-1239, 505-512.
Cantrell, K.J. and Byrne, R.H., 1987. 'Rare earth element complexation by carbonate and oxalate ions'. Geochim. Cosmochim. Acta, 51, 597-605.
Clough, P.W.L. and Field, D. 1980. 'Chemical variation in metabasites from a Proterozoic amphibolite-granulite transition zone, South Norway'. Contrib. Mineral. Petrol., 73, 277-286.
Condie, K., Allen, P. and Narayana, B.L. 1982. 'Geochemistry of the Archean low- to high-grade transition zone, southern India. Contrib. Mineral. Petrol., 81, 157-167.
Cooray, P.G. 1961. 'The geology of the country around Rangala'. Mem. Geol. Surv. Ceylon, 3.
Cooray, P.G. 1962. 'Charnockites and their associated gneisses in the Pre-Cambrian of Ceylon'. Q. Jour. Geol. Soc. London, 118, 239-273.
Crawford, A.R. and Oliver, R.L. 1969. 'The Precambrian geochronology of Ceylon'. Spec. Publ. Geol. Soc. Aust., 2, 283-306.
Drury, S.A. 1973. 'The geochemistry of Precambrian granulite facies rocks from the Lewisian complex of Tiree, Inner Hebrides'. Chem. Geol., 11, 167-188.
Glassley, W.E. 1983. 'The role of CO_2 in the chemical modification of deep continetal crust'. Geochim. Cosmochim. Acta, 47, 597-616.
Gray, C.M. 1977. 'The geochemistry of central Australian granulites in relation to the chemical and isotopic effects of granulite facies metamorphism'. Contrib. Mineral. Petrol., 65, 79-89.
Firsching, F.H. and Mohammadzadel, J. 1986. 'Solubility products of the rare earth carbonates'. J. Chem. Eng. Data, 31, 40-42.
Hansen, E.C, Janardhan, A.S., Newton, R.C., Prame, W.K.B.N, and Ravindra Kumar, G.R. 1987. 'Arrested charnockite formation in southern India and Sri Lanka'. Contrib. Mineral. Petrol., 96, 225-244.

Hapuarachi, D.J.A.C 1968. 'Cordierite and wollastonite-bearing rocks of Southwestern Ceylon'. Geol. Mag., 105, 317-324.

Hapuarachi, D.J.A.C. 1975. 'Granulite facies in Sri Lanka'. Geol. Surv. Dept. Sri Lanka, Prof. Pap. 4, 29pp.

Heier, K. 1965. 'Radioactive elements in the continental crust'. Nature, 208, 479-480.

Heier, K. 1973. 'Geochemistry of granulite facies rocks and problems of their origin'. Phil. Trans. Roy. Soc. London, A 273, 429-442.

Heier, K. and Adams, J.A.S. 1965. 'Concentration of radioactive elements in deep crustal material'. Geochim. Cosmochim. Acta, 29, 53-61.

Heier, K. and Thoresen, K., 1971. 'Geochemistry of high grade metamorphic rocks, Lofoten Vesteralen, North Norway'. Geochim. Cosmochim. Acta, 35, 89-99.

Iyer, S.S., Chouduri, a., Vasconcellos, M.B.A., and Cordani, U.G. 1984. 'Radioactive element distribution in the granulite terrane of Jequie-Bahia, Brazil'. Contrib. Mineral. Petrol., 85, 95-101.

Janardhan, A.S., Newton, R.C., and Hansen, E.C. 1983. 'Transformation of Peninsular gneiss to charnockite in southern Karnataka.' In *Precambrian of South India* (eds. S.M. Naqvi and J.J.W. Rogers), Geol. Soc. India. Mem. 4, 417-435.

Kalsbeek, F. 1976a. 'Metamorphism of the Fiskanaesset region'. Gronlands Geol. Unders. Rapp., 73, 34-41.

Kalsbeek, F. 1976b. 'Metamorphism of Archaean rocks of West Greenland'. In The Early History of the Earth (ed. B. Windley), J. Wiley and Sons, London.

Katz, M.B. 1971. 'The Precambrian metamorphic rocks of Ceylon'. Geol. Rundsch., 60, 1523-1549.

Kroner, A., Williams, I.S., Compston, W., Vitanage, P.W., and Perera, L.R.K. 1987. 'Zircon ion microprobe dating of granulites in Sri Lanka'. Jour. Geol.,

Lambert I. and Heier, K. 1967. 'The vertical distribution of uranium, thorium and potassium in the continental crust'. Geochim. Cosmochim. Acta, 31, 377-390.

Lambert I. and Heier, K. 1968. 'Chemical investigation of deep seated rocks in the Australian shield'. Lithos, 1, 30-53.

Okeke, P.O., Borley, G.D., and Watson, J. 1983. 'A geochemical study of Lewisian metasedimentary granulites and gneisses in the Scourie-Laxford area of the northwest Scotland'. Min. Mag., 47, 1-9.

Ramberg, H. 1951. 'Remarks on the average chemical composition of granulite facies and amphibolite to epidote amphibolite facies in West Greenland'. Medd. Dansk Geol. Foren., 12, 27-34.

Rard, J.A. 1985. 'Chemistry and thermodynamics of Europium and some of its simpler inorganic compounds and aqueous species'. Chem. Rev., 85, 555-582.

Rudnick, R.L., McLennan, S.N., and Taylor, S.R. 1985. 'Large ion lithophile elements from high pressure granulite facies terrains'. Geochim. Cosmochim. Acta, 49, 1645-1655.

Shaw H., Niemeyer, S., Glassley, W.E., Ryerson, F.J., and Abeysinghe, P.B. 1987. 'Isotopic and trace element systematics of the amphibolite to granulite facies transition in the Highland Series of Sri Lanka'. EOS, 68, 464.

Sheraton, J.W. and Collerson, K. D. 1984. 'Geochemical evolution of Archaean granulite facies gneisses in the Vestfold Block and comparisons with other

Archaean gneiss complexes in the East Antarctic Shield'. Contrib. Mineral. Petrol., **87**, 51-64.

Sherry, H.S. and Marinsky, J.A. 1964. 'Carbonate and bicarbonate complexes of neodymium and europium'. Inorganic Chem., 3, 330-334.

Sighinolfi, G.P, 1969. 'K-Rb ratio in high grade metamorphism: A confirmation of a hypothesis of continual crustal evolution'. Contrib. Mineral. Petrol., **21**, 346-356.

Sighinolfi, G.P, 1971. 'Investigations into deep crustal levels: fractionating effects and geochemical trends related to high grade metamorphism'. Geochim. Cosmochim. Acta, 35, 1005-1021.

Sighinolfi, G.P., Figueredo, M.C.H., Fyfe, W.S., Kronberg, B.I., and Oliveira, M.A.F.T. 1981. 'Geochemistry and petrology of the Jeqioe granulitic complex (Brazil): an Archean basement complex.' Contrib. Mineral. Petrol.,**78**, 263-271.

Simpson, H.J., Trier, R.M., Toggweiler, J.R., Mathieu, G., Deck, B.L., Olsen, C.R., Hammond, D.e., Fuller, C. and Ku, T.L. 1982. 'Radionuclides in Mono Lake, California'. Science, **216**, 512-514.

Sklyarenko, Yu.S. and Ruzaikina, L.V. 1970. 'Formation of Nd, Eu, and Yb carbonates and their behavior in aqueous K2CO3 solutions'. Russian J. of Inorganic Chem., **15**, 399-402.

Smalley, P.C., Field, D. and Raheim, A. 1983. 'Resetting of Rb-Sr whole rock isochrons during Sveconorwegian low gr de events in the Gjerstad augen gneiss, Telemark, Southern Norway'. Isotope Geosci., **1**, 269-282.

Tarney, J., Skinner, A.C., and Sheraton, J.W. 1972. 'A geochemical comparison of major Archean gneiss units from northwestern Scotland and East Greenland'. 24th Intl. Geol. Cong., Montreal, Sect. 1, 162-174.

Weaver, G.L. 1980. 'Rare earth geochemistry of Madras granulites'. Contrib. Mineral. Petrol., **71**, 271-279.

Weaver, G.L. and Tarney, J. 1983. 'Elemental depletion in Archaean granulite-facies rocks'. In Migmatites. Melting and Metamorphism (eds. M.P. Atherton and C.D. Gribble), 250-263.

Wells, P.R.A. 1979. 'Chemical and thermal evolution of Archaean sialic crust, Southern West Greenland'. J. Petrol., **20**, 187-226.

METAMORPHISM AND MELTING AT AN EXPOSED EXAMPLE OF THE CONRAD
DISCONTINUITY, KAPUSKASING UPLIFT, CANADA

John A. Percival and David M. Fountain
Geological Survey of Canada, 588 Booth St.,
Ottawa, Ontario, Canada K1A OE4 and
Dept. of Geology and Geophysics,
University of Wyoming,
Laramie, Wyoming, U.S.A. 82071

ABSTRACT. The nature of the mid-crustal velocity discontinuity,
identified in seismic refraction experiments in widely-spaced probes of
the continents, has been debated for many years; recent reflection
profiles have added new constraints. In central Ontario, Canada, the
thrustbounded Kapuskasing uplift exposes a continuous oblique
cross-section of Archean crust to a depth of some 25 km, including a
Conrad-like discontinuity, allowing direct observation of the nature and
origin of this complex feature. Lithologically, it represents a
gradational change from higher level, homogeneous tonalitic gneiss in
the amphibolite facies (Wawa gneiss terrane) to a deeper-level, layered
heterogeneous sequence in the upper amphibolite and granulite facies
(Kapuskasing structural zone). Accompanying lithological changes are
increases in density of the order of 0.1 g.cm^{-3} and aggregate P-wave
velocity of approximately 0.35 km.sec^{-1}. Migmatitic structures in the
high-grade Kapuskasing rocks suggest that partial melting during
metamorphism was responsible for production of tonalite, which could
have coalesced, risen and ponded above the discontinuity in the Wawa
terrane. Other effects that contributed to the gross crustal density
stratification include the pre-metamorphic emplacement of more mafic
intrusions and anorthosites at depth.

INTRODUCTION

Fluids attending metamorphism in the deep crust are not directly
observable, but their composition is a critical parameter in the
interpretation of processes forming and modifying the continental crust.
For example, if the process of streaming by mantle-derived CO_2 is the
driving mechanism for dehydration and granulite production (e.g. Newton
et al., 1980), then this has profound implications for the nature of
regional metamorphism, which generally is temporally related to
deformation and thought to be caused by large scale tectonic processes.
Thus it is important to document fluid composition and behaviour at any
exposed crustal level.
    Fluid fluxes were at a maximum during regional metamorphism, when

51

D. Bridgwater (ed.), Fluid Movements – Element Transport and the Composition of the Deep Crust, 51–60.
© *1989 by Kluwer Academic Publishers.*

heat was available to drive devolatilization and anatectic reactions. Mineral assemblages, textures and fluid inclusions frozen into the rock at this time provide the basis for interpretation of fluid composition and behaviour. Such observations do not represent a snap-shot in time, but rather, present blurred records of prolonged heating and cooling.

The present work deals with the behaviour of H_2O-rich fluids and melts in the vicinity of a widely-recognized feature in the crust, the Conrad seismic discontinuity. First we examine the physical nature of this boundary as exposed in the Kapuskasing uplift of Ontario, then focus on the inferred processes of tonalitic melt generation, segregation and crystallization, and finally, consider the significance of a mid-crustal discontinuity.

1. THE KAPUSKASING UPLIFT

The Kapuskasing uplift occurs within the Superior Province, an Archean terrane of diverse lithological character. The southern part of the province consists of east-trending, alternating granite-greenstone and metasedimentary gneiss belts, whereas the north appears to be uniformly high-grade gneiss (Card and Ciesielski, 1986; Fig. 1). Seismic refraction studies over granite-greenstone terranes detect a mid-crustal discontinuity with an increase of P-wave velocity (Vp) on the order of 0.3 km.sec^{-1} (Berry and Fuchs, 1973; Green et al., 1979; Hall and Brisbin, 1982).

In the south-central part of the Province, the Kapuskasing uplift exposes rocks that show continuous transitions from greenstones, to amphibolite-facies gneisses, into granulites (Fig. 1). The lithological transition coincides with an increase in metamorphic grade and pressure, from 200-300 MPa (2-3 kb) greenschists, through amphibolites, to 700-900 MPa (7-9 kb) granulites (Percival, 1983; Percival and McGrath, 1986). This transect has been interpreted as an uplifted, oblique cross-section through the upper two-thirds of the crust (Percival and Card, 1983; 1985), based on the criteria of Fountain and Salisbury (1981). The amphibolite-granulite boundary zone corresponds to an increase in average density (Percival, 1986) and seismic velocity (Fountain and Salisbury, 1986) that makes it analogous to the Conrad discontinuity observed at depth elsewhere. Anomalously high near-surface velocities were recorded over the Kapuskasing structural zone in recent seismic reflection (Cook, 1985) and refraction (Northey and West, 1986) experiments.

The region affected by the Kapuskasing uplift can be divided into four distinct lithological domains, from west to east (1) the Michipicoten greenstone belt; (2) the Wawa gneiss terrane; (3) the Kapuskasing structural zone; and (4) the Abitibi greenstone belt. Only domains 2 and 3 are relevant to this discussion; more background information is available in Percival and Card (1985).

The Wawa gneiss terrane comprises predominantly amphibolite facies biotite ± hornblende tonalitic rocks, with variable amounts of clinopyroxene-bearing mafic xenoliths, intruded by later granodiorite and granite. Several large-scale domal structures have been identified

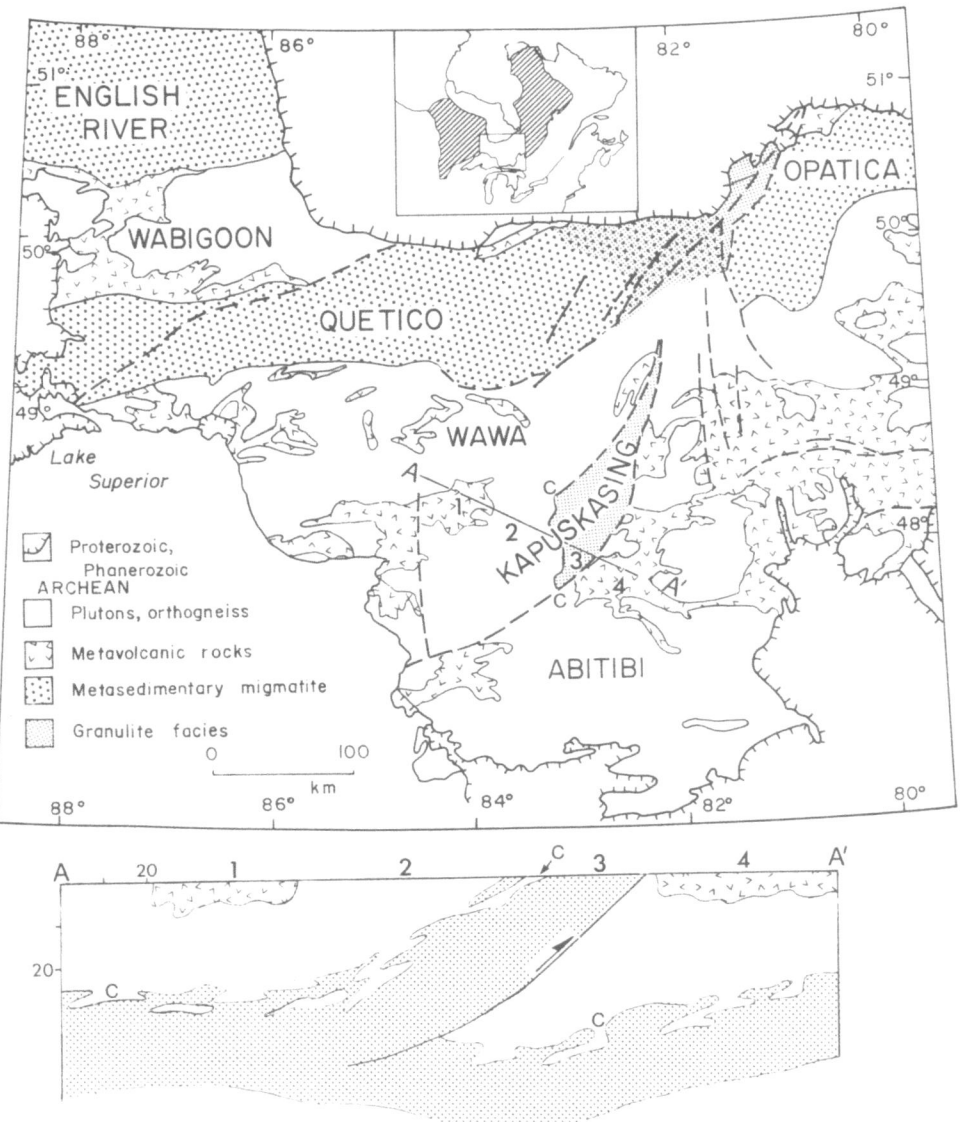

Figure 1: a) Generalized geological map of central Superior Province, showing location of Michipicoten greenstone belt (1), Wawa gneiss terrane (2), Kapuskasing structural zone (3) and Abitibi greenstone belt (4). Line c-c, separating the Wawa and Kapuskasing zones, corresponds to the exposed Conrad seismic discontinuity; b) Cross-section A-A' showing depth relationships of domains 1 through 4 and inferred position of Conrad discontinuity (C). Moho topography in section after Northey and West (1986); dip of uppermost 10 km of basal thrust of Kapuskasing zone constrained by seismic reflection data (Cook, 1985).

in these complexly deformed gneisses.

In the southern part of the uplift, tonalites of the Wawa terrane can be traced into the Kapuskasing structural zone, where they take on a strong northeast foliation and are interlayered with 500-8000 m thick, NE-striking belts of paragneiss, mafic and dioritic gneiss and the Shawmere anorthosite complex. Metamorphic grade is uniformly in upper amphibolite to granulite facies, characterized by orthopyroxene in paragneiss and garnet-clinopyroxene-plagioclase in mafic gneiss (Percival, 1983). The high-grade rocks are thrust onto the low-grade rocks of the Abitibi belt to the east along the northwest-dipping Ivanhoe Lake fault (Percival and Card, 1983; Cook, 1985).

2. SEISMIC VELOCITIES AND THE CONRAD DISCONTINUITY

Many continental regions exhibit a mid-crustal discontinuity in seismic velocity, commonly referred to as the Conrad discontinuity and thought to result from crustal density/velocity stratification. The existence of such a discrete mid-crustal discontinuity was questioned recently, however, owing to the lack of reflectors at the level identified by refraction studies (Oliver et al., 1983) and to reinterpretation of refraction data (Mueller, 1987). Beneath Superior Province, however, the discontinuity is distinct and is marked by a velocity increase of approximately 0.3 km.sec^{-1} at depths of 18-22 km (Berry and Fuchs, 1973; Green et al., 1979). A preliminary analysis of density from rocks of the Kapuskasing uplift (Percival, 1986) showed an increase in density of about 0.1 g.cm^{-3} across the Wawa-Kapuskasing boundary, that, when converted to seismic velocity, is of the appropriate magnitude for that expected at a mid-crustal discontinuity.

To test this possibility, compressional wave velocities at varying confining pressures up to 600 MPa (6 kb) were measured in a suite of representative samples from the Wawa gneiss terrane and Kapuskasing structural zone. Cores of about 2.65 cm diameter and 3.8 cm in length were cut from samples in three orthogonal directions with respect to fabric elements evident in the samples. The cores were placed in a pressure vessel that was loaded to a maximum pressure of 600 MPa, monitored through a manganin coil. Compressional wave velocities were determined using the pulse-transmission technique described by Birch (1960). Travel times of ultrasonic waves were measured on core samples of known length to calculate the P-wave velocity. Corrections for sample length change at high pressures are negligible.

The detailed velocity data will be reported elsewhere and here we summarize the key results. The average velocities at 600 MPa for the major rock units are given in Table 1 where we tabulate the mean Vp and ranges in Vp for the bulk velocity and for velocities propagated parallel to foliation. Velocities measured parallel to foliation correspond to horizontal propagation directions in the crust when the section is restored to its pre-emplacement geometry (Percival and Card, 1985). Also tabulated are mean densities, ranges of densities and relative abundance of the respective lithologic units from Percival (1986).

TABLE 1: AVERAGE DENSITY AND P-WAVE VELOCITY FOR ROCK SAMPLES FROM THE
WAWA GNEISS TERRANE AND KAPUSKASING STRUCTURAL ZONE

Lithologic Unit	Area (%)	Density (g.cm-1) Mean	Range	Bulk Vp (km.sec-1) Mean	Range	Foliation Vp Mean	Range
WAWA GNEISS TERRANE							
Tonalite gneiss	100	2.70	2.66-2.75	6.55	6.51-6.64	-	-
KAPUSKASING STRUCTURAL ZONE							
Tonalite gneiss	20	2.71	2.65-2.79	6.71	6.55-6.87	6.83	6.62-7.04
Paragneiss	20	2.77	2.70-2.88	6.59	6.58-6.60	6.74	6.68-6.79
Diorite	25	2.80	2.76-2.83	6.64	-	6.79	-
Anorthosite	20	2.82	2.72-2.92	7.26	7.12-7.39	-	-
Mafic gneiss	15	3.10	3.01-3.31	7.33	7.18-7.44	7.46	7.34-7.60
Aggregate	100	2.82	-	6.87	-	6.98	-

It is evident from Table 1 that the variation in seismic velocity
in the Wawa and Kapuskasing zones reflects the lithological diversity,
or lack thereof, of each zone. The tonalitic gneisses of the Wawa gneiss
terrane have relatively uniform mean bulk velocities between 6.5 and
6.65 km.s^{-1} and exhibit little seismic anisotropy. In marked contrast,
the rocks of the Kapuskasing zone exhibit a wide range of velocities, as
rock types range from low-velocity paragneiss to high-velocity mafic
gneiss. The high velocities in anorthosite are a consequence of the high
percentage of calcic plagioclase, garnet, hornblende and clinopyroxene.
Orthopyroxene, clinopyroxene, hornblende, garnet and plagioclase all
contribute to the high velocities in mafic gneiss.

Seismic anisotropy can be related to preferred orientation of
hornblende in some mafic gneiss samples and to alignment of biotite in
paragneiss. In general, seismic velocities parallel to foliation are
higher than mean velocity or velocity normal to foliation. This suggests
that seismic waves that travel along horizontal paths in the crust
(refraction surveys) may be faster than those propagated along vertical
paths (reflection surveys). When compared to the Wawa tonalite gneisses,
the Kapuskasing tonalite gneisses exhibit higher velocities because
velocities for several anisotropic, highly strained samples are included
in the mean.

Table 1 also shows the weighted average of the velocities and
densities of the Wawa and Kapuskasing zones based on the distribution of
the major lithologic units. Aggregate velocities of the respective
terranes are important as seismic waves have large wavelengths and
sample large volumes of rock over long travel paths in the crust. The
aggregate bulk velocity of the Wawa zone is 6.55 km.s^{-1} in contrast to
the 6.9 km.s^{-1} aggregate velocity for the Kapuskasing terrane. Thus,
aggregate compressional wave velocities increase by 0.35 km.s^{-1} in the
transition zone between the Wawa gneiss terrane and Kapuskasing
structural zone, an increase close to that observed at mid-crustal
levels in seismic surveys elsewhere in the Superior Province. We
therefore regard the transition from the Wawa to Kapuskasing zone as an

exposed portion of this discontinuity, generally observed only at mid-crustal levels.

3. MELTING, DEHYDRATION AND MATERIAL TRANSPORT

The density/velocity discontinuity in the Kapuskasing uplift is not a sharp geological feature but rather, a complex zone of lithologic change over several kilometres. The lithological transition from homogeneous tonalitic composition of the Wawa gneiss terrane to heterogeneous layered gneiss of the Kapuskasing structural zone is complicated by locally intense deformation, both prograde and retrograde amphibolite-granulite reactions and late pegmatitic injections.

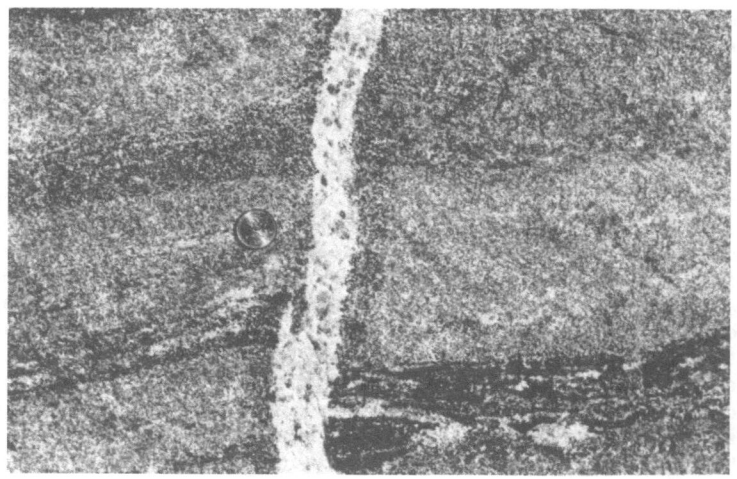

Figure 2: Outcrop photograph of mafic gneiss in Kapuskasing zone. Light grey layers are garnet-, clinopyroxene-rich and dark grey layers are hornblende-rich. Note white, concordant, discontinuous pods of in-situ tonalitic leucosome, showing no adjacent hydration selvedge. In contrast, the discordant 2 cm tonalite dyke is bordered by a dark grey zone 2 cm thick, in which garnet-, clinopyroxene-rich layers are hydrated to hornblende-plagioclase assemblages. Coin is 1.8 cm in diameter.

Petrological observations suggest that Kapuskasing gneisses may be the source of some tonalite bodies within the Wawa gneiss terrane. Gneisses in the Kapuskasing structural zone are characterized by migmatitic structures. Paragneiss has up to 20% white leucosome of tonalitic composition, in melanosome of garnet-biotite-plagioclase-quartz ± orthopyroxene schist. Minor biotite and/or orthopyroxene occur locally in leucosome. Rocks without orthopyroxene probably melted in the presence of an aqueous vapour:

$$H_2O + NaCa\ plg + Qtz = Ca\text{-}richer\ plg + tonalitic\ melt \qquad (1)$$

The presence of orthopyroxene in some rocks suggests biotite instability and vapour-absent melting according to reaction 2 (Percival, 1983):

$$Bio + Plg + Qtz = Opx + granodioritic\ melt \qquad (2)$$

K-feldspar components and water released by biotite breakdown would be incorporated into the melt.

Mafic gneiss is compositionally heterogeneous, with hornblende or garnet, clinopyroxene-rich layers and up to 20% tonalitic leucosome in pods and layers (Fig. 2). The discontinuous nature of the leucosome pods, in conjunction with their near-eutectic composition (Percival, 1983), suggest the operation of a continuous vapour-absent melting reaction such as (3):

$$Hbl + Plg + Qtz = Grt + Cpx + tonalitic\ melt \qquad (3)$$

Calculations based on observed mineral compositions suggest conditions of 700-800°C, 700-900 MPa (7-9 kb) at water activity of 0.1-0.6 (Percival, 1983).

The above reactions are inferred to have produced the in-situ leucosome observed in migmatitic gneisses. It is possible that some melt migrated away from the melting site and rose to form intrusions at higher structural levels. Evidence for the operation of this process is observed locally, where sub-metre-scale discordant tonalite dykes cut mafic gneiss (Fig. 2). Adjacent to the contacts, mafic gneiss is hydrated from garnet-clinopyroxene to hornblende-plagioclase assemblages, suggesting that water infiltrated from crystallizing hydrous tonalitic magma, driving reaction (3) from right to left.

The general pattern implied by these observations is that of net upward movement of water, felsic components and heat in the form of hydrous tonalitic melts, leaving a denser, large-ion-lithophile-element (LILE)-depleted residue. This inference is supported by geochemical data which show complementary LILE depletion in granulites and enrichment in tonalite (Rudnick and Taylor, 1986) and dacite (Sylvester et al., 1986) in the Kapuskasing-Michipicoten area. Similar hornblende or garnet, clinopyroxene-bearing mafic sources for Archean tonalite have been inferred by Arth and Hanson (1972) and Gower et al.(1982).

The concept of upward transport of lithophile components in hydrous melts is well established (e.g. Quensel, 1951; Taylor and McLennan, 1985; Crawford and Hollister, 1986). However, the relationships observed in the Kapuskasing uplift suggest that this process has contributed to large-scale lithological, density and seismic-velocity zonation in the crust.

4. DISCUSSION

The Conrad-like discontinuity in the Kapuskasing uplift is an irregular,

58

gradational lithological, and prograde/retrograde metamorphic boundary
several km wide (Fig.3). There is an intimate relationship between
lithology and metamorphism, as tonalite melts, produced at depth in the
Kapuskasing structural zone through dehydration melting, rose and
crystallized, releasing entrained water which hydrated granulites in the
vicinity of the prograde amphibolite-granulite boundary. Extraction and
upward migration of hydrous felsic magma is one process contributing to
the density and metamorphic stratification of the Superior Province
crust.

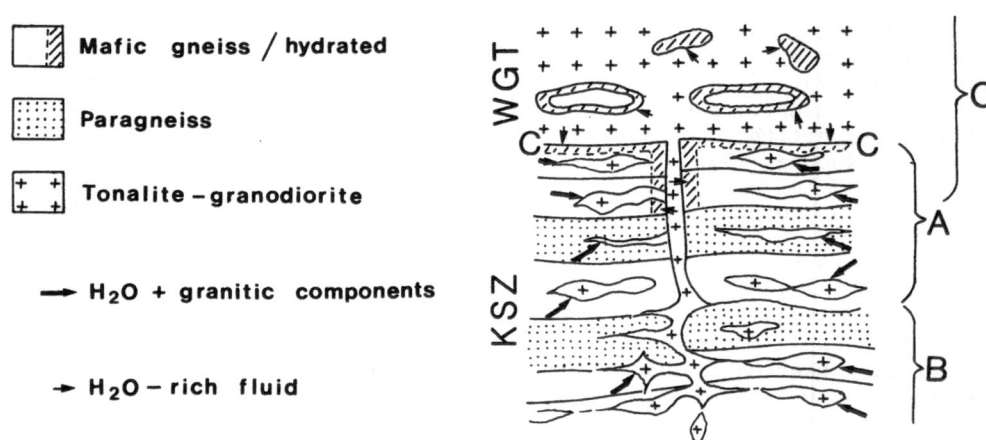

*Figure 3: Schematic representation of metamorphic relationships adjacent
to the exposed Conrad discontinuity (C-C) in the Kapuskasing uplift. In
zone A, exposed in the Kapuskasing structural zone (KSZ), migmatitic
paragneiss and mafic gneiss produced leucosome in reactions 1 through 3
(see text), involving transfer of tonalite-granodiorite melt and hydrous
fluid into pods. At deeper levels (zone B; not exposed) melt pods are
inferred to have coalesced to form discordant bodies which rose
gravitationally and ponded in the Wawa gneiss terrane (WGT). Tonalite
that crystallized in zone C hydrated granulitic passageways and
inclusions.*

A second process is equally important, if not dominant in
accounting for the higher densities and velocities at depth. Intrusions
in the deep parts of the section are mafic with respect to the mid-
crustal tonalites. Some of these, for example the Shawmere anorthosite,
have not suffered extensive migmatization and hence the high density and
velocity cannot be related to melt extraction. Similarly, mafic and
dioritic rocks, although migmatitic, had original compositions more
mafic than tonalite. Thus the changes induced by metamorphism and
melting appear to have been superimposed on an already densitystratified
sequence. The geometry and distribution of intrusions is consistent with
pre-metamorphic, gravity-controlled magma ponding.
Mid-crustal seismic velocity discontinuities have been observed in
terranes which are at granulite grade at surface, and presumably also at

depth, such as the Grenville Province of Canada (Berry and Fuchs, 1973) and the Baltic Shield (Luosto and Korhonen, 1986). In these terranes the velocity discontinuity cannot be related to the amphibolite-granulite boundary, which has been eroded, but may be caused by gravity-controlled emplacement of magmas of different composition and density, metamorphically-induced intracrustal segregation, or a combination of both.

ACKNOWLEDGEMENTS

I. Ermanovics and A. Davidson are thanked for constructive comments. This is Geological Survey of Canada contribution 29787.

REFERENCES

Arth JG, Hanson GN (1972) Quartz diorites derived by partial melting of eclogite or amphibolite at mantle depths. *Contrib Mineral Petrol* 37:161-174

Berry MJ, Fuchs K (1973) Crustal structure of the Superior and Grenville Provinces of the northeastern Canadian Shield. *Seis Soc Amer Bull* 63: 1393-1432

Birch F (1960) The velocity of compressional waves in rocks to 10 kilobars,1. *J Geophys Res* 65:1083-1102

Card KD, Ciesielski A (1986) Subdivisions of the Superior Province of the Canadian Shield. *Geosci Can* 13:5-13

Cook FA (1985) Geometry of the Kapuskasing structure from a Lithoprobe pilot reflection survey. *Geology* 13:368-371

Crawford ML, Hollister LS (1986) Metamorphic fluids: The evidence from fluid inclusions. In: Walther JV, Wood BJ (eds) *Fluid-rock interactions during metamorphism*. Springer Berlin 1-35

Fountain DM, Salisbury MH (1981) Exposed cross-sections through the continental crust: Implications for crustal structure, petrology and evolution. *Earth Planet Sci Lett* 56:263-277

Fountain DM, Salisbury MH (1986) Seismic properties of the Superior Province crust based on seismic velocity measurements on rocks from the Michipicoten, Wawa and Kapuskasing terranes, Ontario. *Geol Assoc Can* 11:69

Gower CF, Paul DK, Crocket JH (1982) Protoliths and petrogenesis of Archean gneisses from the Kenora area, English River subprovince, northwest Ontario. *Precam Res* 17:245-274

Green AG, Anderson NL, Stephenson OG (1979) An expanding spread seismic reflection survey across the Snake Bay- Kakagi Lake greenstone belt, northwestern Ontario. *Can J Earth Sci* 16:1599-1612

Hall DH, Brisbin WC (1982) Overview of regional geophysical studies in Manitoba and northwestern Ontario. *Can J Earth Sci* 19:2049-2059

Luosto U, Korhonen H (1986) Crustal structure of the Baltic Shield based on off-Fennolora refraction data. *Tectonophysics* 128:183-208

Mueller S (1987) The Conrad discontinuity- what is it? *Int Un Geod Geophys* 19,1:68

Newton RC, Smith JV, Windley BF (1980) Carbonic metamorphism, granulites and crustal growth. *Nature* 288:45-50

Northey DJ, West GF (1986) A crustal scale seismic refraction experiment over the Kapuskasing structural zone. *Geol Assoc Can* 11:108

Oliver J, Cook F, Brown L (1983) COCORP and the continental crust. *J Geophys Res* 88:3329-3347

Percival JA (1983) High-grade metamorphism in the Chapleau-Foleyet area, Ontario. *Amer Mineral* 68:667-686

Percival JA (1986) A possible exposed Conrad discontinuity in the Kapuskasing uplift, Ontario. In: Barazangi M, Brown L (eds) Reflection seismology: The continental crust. *Amer Geophys Un Geodynam Ser* 14:135-141

Percival JA, Card KD (1983) Archean crust as revealed in the Kapuskasing uplift, Superior Province, Canada. *Geology* 11:323-326

Percival JA, Card KD (1985) Structure and evolution of Archean crust in central Superior Province, Canada. In: Ayres LD, Thurston PC, Card KC, Weber W (eds) Evolution of Archean supracrustal sequences. *Geol Assoc Can Spec Pap* 28:179-192

Percival JA, McGrath PH (1986) Deep crustal structure and tectonic history of the northern Kapuskasing uplift of Ontario: An integrated petrological-geophysical study. *Tectonics* 5:553-572

Quensel P (1951) The charnockite series of the Varberg district on the southwest coast of Sweden. *Arkiv Min Geol* 1:229-332

Rudnick RL, Taylor SR (1986) Geochemical constraints on the origin of Archean tonalitic-trondhjemitic rocks and implications for lower crustal composition. In: Dawson JB, Carswell DA, Hall J, Wedepohl KH (eds) The nature of the lower continental crust. *Geol Soc Spec Publ* 24:179-191

Sylvester PJ, Attoh K, Schulz KJ (1986) Did anatexis in the Kapuskasing structural zone produce the HREE-depleted dacites of the Michipicoten greenstone belt? *Geol Assoc Can* 11:133

Taylor SR, McLennan SM (1985) *The continental crust: Its composition and evolution*. Blackwell Oxford 312 p

MELT-INDUCED FLUID PUMPING AND THE SOURCE OF CO₂ IN GRANULITES

J.A. Percival
Geological Survey of Canada
588 Booth St. Ottawa, Ontario, Canada, K1A 0E4

ABSTRACT. Observations of fluid composition and behaviour at various structural levels in metamorphic terranes have led to an integrated crustal model for metamorphism with $H_2O >> CO_2$ fluids and high fluxes at shallow levels and less abundant $CO_2 >> H_2O$ compositions at depth. Anatectic removal of granitic components and H_2O from the lower crust accounts for LILE-depleted granulite compositions and CO_2-rich inclusions, but not for the abundance of CO_2, which many workers have attributed to external (mantle) sources. A crustal-scale, closed-system circulation model is proposed, in which $H_2O > CO_2$ fluids released in low temperature devolatilization reactions are pumped downward by volume-reducing anatectic reactions. The fluid is filtered by removal of much of the H_2O component in an amphibolite-facies migmatite zone; the remaining CO_2-rich portion is pumped down into granulite-facies migmatites where it is trapped as inclusions or absorbed to a limited extent by magmas.

INTRODUCTION

Recent advances in the understanding of phase relations, fluid inclusions and stable isotopic characteristics of igneous and metamorphic rocks have provided a new perspective on fluid composition, distribution and circulation throughout the earth's crust (see Touret, 1987 for a brief review). In general, fluids in the upper crust are dominated by H_2O, with some CO_2 (e.g. Graham et al., 1983), and those in the lower crust by CO_2 (Newton et al., 1980; Touret and Dietvorst, 1983; Newton, 1986). Large fluid fluxes have been inferred during metamorphic events in the upper 10 km of crust in various case studies (e.g. Ferry, 1980; Wickham and Taylor, 1985) whereas low fluid volumes are documented for high-grade regions representing deep-crustal metamorphism (e.g. Valley and O'Neil, 1984; Wickham and Taylor, 1987).

Many workers agree that the major agent of dehydration at deep crustal levels is partial melting, which fractionates water (Taylor, 1977) and large-ion-lithophile elements (LILE) (Fyfe, 1973a) into leucosome, which then rises as granite (Fyfe, 1973b), leaving LILE-depleted restite (e.g. Rudnick et al., 1985) and CO_2-rich fluid (Touret and Dietvorst, 1983; Crawford and Hollister, 1986). Others, noting the

61

nature of LILE depletion (Collerson and Fryer, 1978; Weaver, 1980; Weaver and Tarney, 1983; Hansen et al., 1987) and high volume of CO_2rich fluid inclusions (e.g. Touret, 1971; Janardhan et al., 1982), postulate that streaming of externally-derived CO_2-rich fluids was the fundamental control on granulite metamorphism (e.g. Newton et al., 1980). In attempting to reconcile these diverse observations, Frost and Frost (1987) showed that CO_2-rich fluids, emanating from deep-crystallizing, crustally-derived granite magmas, could have been the source of streaming CO_2 in granulite terranes.

This contribution attempts to reconcile two apparently unrelated phenomena in high-grade metamorphic terranes: the apparent abundance of water available to flux amphibolite-facies migmatites, and the apparently dry (CO_2-rich) nature of fluids in migmatitic rocks (Touret and Dietvorst, 1983; Olsen, 1987; Percival, 1983; Rudnick et al., 1984). A model is proposed in which $H_2O>>CO_2$ fluids released in dehydration reactions at high structural levels are drawn down into the melt zone by volume-reducing melt reactions. Fluids are purged of most of their water as they are sucked through the amphibolite-facies migmatite zone, but a small volume of CO_2-rich fluid is pumped into the granulites below. Voluminous melts, generated at high temperature in the granulite facies, rise gravitationally, leaving a LILE-depleted solid and the CO_2-rich residue of the imported fluid.

1. WATER AND MIGMATITES

Migmatites form in both amphibolite and granulite facies and an orthopyroxene isograd has been recognized in several migmatite terranes (e.g. Percival, 1983; Perkins and Chipera, 1985; Percival and McGrath, 1986). Three hypotheses are commonly invoked to explain migmatitic veins of granitic composition: 1) segregation of granitic components in a melt phase; 2) solid-state segregation of granitic components; and 3) lit-par-lit injection of granitic melt from an external source. A combination of processes has been documented in several terranes (e.g. Olsen, 1984; Sawyer and Robin, 1986). However, each of these processes is variably problematic. For example, the third does not address the key question of granite genesis. The second is not likely at super-solidus conditions, where in the presence of an aqueous vapour needed to transport granitic components, the stable assemblage would be a melt. The most common problem with the first hypothesis is accounting for the abundance of water. In the 650-750°C, 500-700 MPa (5-7 kbar) range, some 10-14 wt% water is required to saturate granitic melt, depending on composition (Burnham, 1975; Clemens and Wall, 1981; Clemens, 1984). This amount of water could be produced by the breakdown of hydrous minerals (Brown and Fyfe, 1970), however, many amphibolite-facies migmatites retain these phases without solid dehydration products, precluding dehydration melting as the water source. Alternatively, ambient pore fluid could be the source. This is possible for the first increment of melting, but it is unlikely that the 2-3 wt% free water necessary to saturate the 20-30 % granitic leucosome present in many migmatites could be retained at conditions near the solidus. Nevertheless, melting remains a widely-invoked hypothesis, even if externally-derived water is an adjunct requirement.

2. CO_2 AND GRANULITES

Several lines of evidence point to relatively dry conditions in granulites. First, calculations of orthopyroxene stability indicate that PH_2O must have been much less than Ptotal in most granulites, leading to the conclusion that the fluid phase consisted of species other than H_2O (Phillips, 1980) or that fluid was virtually absent (Valley et al., 1983; Valley,1985). Second, syn-metamorphic fluid inclusions in granulites are almost invariably high-density CO_2 (Touret, 1971; 1987; Rudnick et al., 1984; Crawford and Hollister, 1986), with minor components such as CH_4, N_2, H_2O, etc. Aqueous inclusions may contain abundant dissolved salts (Touret, 1987). Not only does the ratio $CO_2:H_2O$ increase in the migmatitic amphibolite and granulite facies, but so does the absolute abundance of CO_2 (Touret and Dietvorst, 1983; Olsen, 1987; Yardley et al., 1983).

Several theories have been proposed to explain the origin of CO_2-rich fluids in granulites. Foremost among these are: 1)"indigenous" CO_2, remaining in the fluid phase after fractionation of H_2O into a granitic melt (Touret, 1971; Crawford and Hollister, 1986); 2) "para-indigenous" CO_2, introduced through decarbonation of adjacent carbonates (Glassley, 1983; Newton, 1987) or crystallization of CO_2-bearing granites (Frost and Frost, 1987); and 3) juvenile CO_2 introduced from the mantle directly (Schuiling and Kreulen, 1979; Newton et al., 1980; Harris et al., 1982) or through degassing mafic intrusions (Selverstone, 1982).

Water and carbon dioxide are miscible species at high-grade metamorphic conditions (see review in Crawford and Hollister, 1986). Therefore the effect of introducing a CO_2-rich fluid is to dilute an ambient hydrous fluid, promoting dehydration as well as possible redox reactions (Lamb and Valley, 1984). Passage of a CO_2-rich fluid along discrete channelways is thought to have been the cause of charnockite veins at several localities in southern India (Friend, 1981; Raase et al., 1986; Hansen et al., 1987).

3. THE GRANITE SOLIDUS

Hydrous melting reduces volume relative to solid + vapour reactants, accounting for the familiar negative slope of the granite solidus on P-T projections. The magnitude of volume change is proportional to the amount of vapour involved, up to saturation level, which increases with pressure (Burnham, 1975; Clemens and Wall, 1981). Provided that a small amount of hydrous pore fluid is present initially, some volume reduction accompanies the onset of melting.

The most readily diffuseable material to fill the "space" created by anatexis is fluid, which if hydrous, would also act as a flux to hydrate the melt. Sibson (1987) postulated that a similar pumping mechanism draws fluids into dilational zones in faults. Metamorphic fluids would

probably not be escaping from the deeper crust at or near the
metamorphic peak, because the hotter, more molten migmatites below would
also be attempting to scavenge fluids. Water released in dehydration
melting reactions in these rocks would be absorbed by local H_2O-
undersaturated leucosome. A source with more available fluids is the
rock sequence above, devolatilizing during prograde metamorphism. Under
the downward hydraulic gradient created by anatexis, fluids would
descend into the developing melt zone and the hydrous component become
incorporated by melts. Other fluid species less soluble in granitic melt
would fractionate into the fluid phase, where they could be trapped as
inclusions or pumped down to deeper levels. Such a mechanism could lead
to catastrophic crustal weakening of the type envisioned by Hollister
and Crawford (1986).

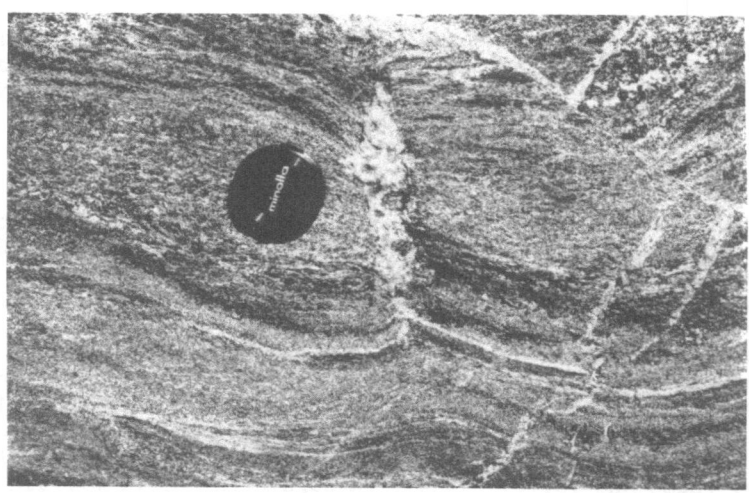

*Figure 1: Boudinaged layer in tonalitic gneiss, Wawa gneiss terrane
(Percival and Fountain, this volume). Note ductile deformation of
gneissic layering into boudin neck and granitic leucosome in pressure
shadow. The photograph suggests the simultaneous operation of ductile
flow and melt pumping.*

Ductile flow could also act to fill the "space" created by hydrous
melting. Many theoretical and experimental studies show that rocks at
migmatitic conditions have very little strength. Although the relative
magnitude of the two processes is not intuitively obvious, field
evidence suggests that both operate simultaneously. In migmatite
terranes, one commonly observes evidence of ductile flow of gneissic
layering into boudin necks, as well as granitic leucosome in pressure
shadows (Fig. 1; Hanmer, 1986, Figs. 3,4). The presence of leucosome
indicates that not only fluids, but also "solid" granitic components
have been transported into regions of low pressure. As fluid components
are more mobile and easily transported than solids (Bickle and McKenzie,
1987), there may be some steady-state proportion of strain:fluid

65

compensation for the melt-pumping effect. Several factors may enter into
calculations of this ratio: 1) the relative rates of melting and ductile
flow; if melting is relatively slow, volume changes will be accomodated
by deformation; 2) the relative strength of rocks; although viscosity is
generally low in migmatites, competency contrasts between different rock
types could produce large-scale effects similar to that illustrated in
Fig. 1; and 3) the tectonic environment; the melt-pumping process would
be more likely to occur in extensional than in compressional settings.

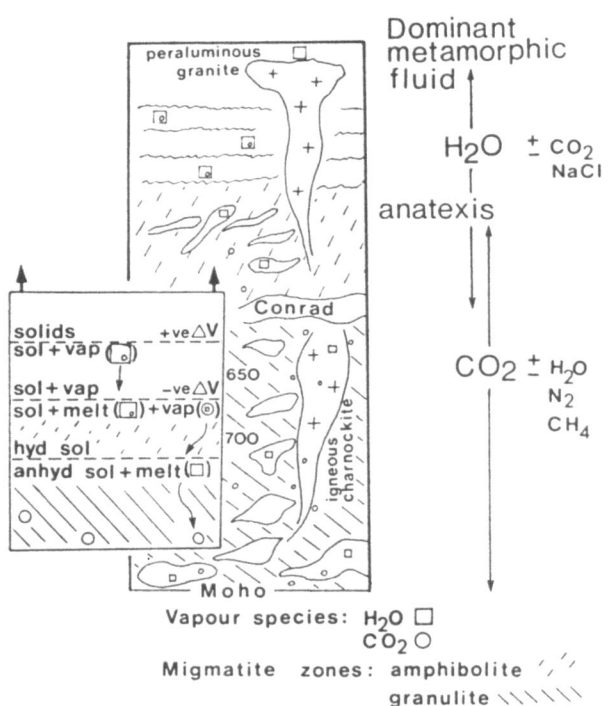

*Figure 2: Distribution (after Touret and Dietvorst, 1983) and behaviour
of fluids in the crust.In the inset box, water-rich fluids released at
relatively low temperature are being pumped downward by anatectic
reactions which are reducing volume at depth. Water is being filtered by
an amphibolite-facies migmatite zone and the CO2-rich residue is
descending into the granulite-facies migmatite zone. During progressive
metamorphism the reference frame moves upward in the crust.*

It is common to assume that fluids produced in devolatilization
reactions ascend gravitationally (Wood and Walther, 1986). Recently
Wickham and Taylor (1985; 1987) presented oxygen isotopic evidence to
suggest downward sea-water infiltration to depths of 10 km. At deeper
levels, high lithostatic pressure may prevent fluid escape, as suggested

66

by the presence of excess argon in some granulites (Foland, 1979).

4. A DYNAMIC MODEL OF FLUID CIRCULATION DURING METAMORPHISM

In the most general terms, without specifying causative tectonic process or mode of heat transfer, the process of progressive metamorphism takes place by upward movement of isotherms. The position in the crust of specific dehydration reactions also moves upward, followed from below by melting reactions (Fig. 2). Anatexis produces a negative hydraulic gradient, pumping fluids released at higher crustal levels into the melt zone. Migmatites at the top of the zone filter fluid destined for drier, more voluminous granite at deeper levels by incorporating water, enriching the descending fluid in CO_2 and other non-hydrous components. Deeper migmatites may incorporate some CO_2 into melts (Frost and Frost, 1987) and hydrous minerals may be destabilized by the changing fluid composition (e.g. Grant, 1986).

At some critical volume of melt, leucosome will begin to coalesce and rise gravitationally (Fyfe, 1973a;b; Wickham, 1987) to form plutons and extrusive equivalents. The magma may fractionate CO_2-rich fluids and dry igneous charnockites (Frost and Frost, 1987) during ascent, leaving water-rich peraluminous liquids (Todd and Shaw, 1985; White et al., 1986). Water released during crystallization at depth may have hydrating effects if the surroundings are dry (Percival and Fountain, this volume).

The melt filter effect could explain relationships between the Closepet granite, in the amphibolite facies, and charnockite veins at Kabbaldurga quarry in southern India (Friend, 1983; 1985). Descending $H_2O>CO_2$ fluid could have provided the flux to produce the Closepet granite; passage of the CO_2-rich residue through discrete passageways in the subjacent terrane could have produced the dehydration selvedges.

The crustal-scale closed-system model for introduction of fluids into the lower crust predicts the low volumes of CO_2-rich fluids necessary and eliminates the requirement of an externally-derived streaming fluid as a cause of regional metamorphism, which generally is intimately linked to deformation and large-scale tectonic processes. It is consistent with observations which point to fluid-absent (neither melting nor CO_2 flushing) metamorphism in the Adirondacks (Valley and O'Neil, 1984; Valley, 1985; Lamb et al., 1987): without some initial pore fluid to prime the system, the melt-driven pump would not function.

ACKNOWLEDGEMENTS
A question by J. Touret on the writer's presentation at the NATO "Fluids" workshop led to the thought train summarized here. J.W. Valley is thanked for an objective and constructive review. Geological Survey of Canada contribution 29887.

REFERENCES

Bickle MJ, McKenzie D (1987) The transport of heat and matter by fluids during metamorphism; *Contrib Mineral Petrol* 95:384-392

Brown GC, Fyfe WS (1970) The production of granitic melts during ultrametamorphism; *Contrib Mineral Petrol* 28:310-318
Burnham CW (1975) Water and magmas: A mixing model; *Geochim Cosmochim Acta* 39:1077-1084
Clemens JD (1984) Water contents of silicic to intermediate magmas; *Lithos* 17:273-287
Clemens JD, Wall VJ (1981) Origin and crystallization of some peraluminous (S-type) granitic magmas; *Can Mineral* 19:111-131
Collerson KD, Fryer BJ (1978) The role of fluids in the formation and subsequent development of early continental crust; *Contrib Mineral Petrol* 67:151-167
Crawford ML, Hollister LS (1986) Metamorphic fluids: The evidence from fluid inclusions; In Walther JV, Wood, BJ (eds) *Fluid-rock interactions during metamorphism*; Springer New York, 1-35
Ferry JM (1980) A case study of the amount and distribution of heat and fluid during metamorphism; *Contrib Mineral Petrol* 77:373-385
Foland KA (1979) Limited mobility of argon in a metamorphic terrane; *Geochim Cosmochim Acta* 43: 793-801
Friend CRL (1981) Charnockite and granite formation and influx of CO2 at Kabbaldurga; *Nature* 294:550-552
Friend CRL (1983) The link between charnockite formation and granite production: evidence from Kabbaldurga, Karnataka, southern India; In Atherton MP, Gribble CD (eds) *Migmatites, melting and metamorphism*; Shiva, Nantwich, 264-276
Friend CRL (1985) Evidence for fluid pathways through Archaean crust and the generation of the Closepet granite, Karnataka, South India; *Precamb Res* 27:239-250
Frost BR, Frost CD (1987) CO2, melts and granulite metamorphism; *Nature* 327:503-506
Fyfe WS (1973a) The granulite facies, partial melting and the Archaean crust; *Phil Trans Roy Soc Lond* A273:457-461
Fyfe WS (1973b) The generation of batholiths; *Tectonophysics* 17:273-283
Glassley WE (1983) Deep crustal carbonates as CO2 fluid sources: Evidence from metasomatic reaction zones; *Contrib Mineral Petrol* 84:15-24
Graham CM, Greig KM, Sheppard SMF, Turi B (1983) Genesis and mobility of the H2O-CO2 fluid phase during regional greenschist and epidote amphibolite facies metamorphism: A petrological and stable isotope study in the Scottish Dalradian; *J Geol Soc Lond* 140:577-599
Grant JA (1986) Quartz-phlogopite-liquid equilibria and the origins of charnockites; *Amer Mineral* 71:1071-1075
Hanmer S (1986) Asymmetrical pull-aparts and foliation fish as kinematic indicators; *J Struct Geol* 8:111-122
Hansen EC, Janardhan AS, Newton RC, Prame WKBN, Ravindra Kumar GR (1987) Arrested charnockite formation in southern India and Sri Lanka; *Contrib Mineral Petrol* 96:225-244
Harris NBW, Holt RW, Drury SA (1982) Geobarometry, geothermometry and late Archean geotherms from the granulite facies terrane of South India; *J Geol* 90:509-528
Hollister LS, Crawford ML (1986) Melt-enhanced deformation: A major tectonic process; *Geology* 14:558-561

Lamb WM, Valley JW (1984) Metamorphism of reduced granulites in low-CO_2 vapour-free environment; *Nature* 312:56-58

Lamb WM, Valley JW, Brown PE (1987) Post-metamorphic CO_2-rich fluid inclusions in granulites; *Contrib Mineral Petrol* 96:485-495

Janardhan AS, Newton RC, Hansen EC (1982) The transformation of amphibolite facies gneiss to charnockite in southern Karnataka and northern Tamil Nadu India; *Contrib Mineral Petrol* 79:130-149

Newton RC (1986) Fluids of granulite facies metamorphism; In Walther JV, Wood BJ (eds) *Fluid-rock interactions during metamorphism*, Springer, New York, 36-59

Newton RC (1987) Petrologic aspects of Precambrian granulite facies terrains bearing on their origins; In Kroner A (ed) Proterozoic lithosphere evolution; *Amer Geophys Un Geodyn Ser* 17:11-26

Newton RC, Smith JV, Windley BF (1980) Carbonic metamorphism, granulites and crustal growth; *Nature* 288:45-50

Olsen SN (1984) Open- and closed-system migmatites in the Front Ranges, Colorado; *Amer J Sci* 282:1596-1622

Olsen SN (1987) The composition and role of fluids in migmatites: A fluid inclusion study of the Front Range rocks; *Contrib Mineral Petrol* 96:104-120

Percival JA (1983) High-grade metamorphism in the Chapleau-Foleyet area, Ontario; *Amer Mineral* 68:667-686

Percival JA, Fountain DM (1988) Metamorphism and melting at an exposed example of the Conrad discontinuity, Kapuskasing uplift, Canada (this volume)

Percival JA, McGrath PH (1986) Deepcrustal structure and tectonic history of the northern Kapuskasing uplift of Ontario: An integrated petrological-geophysical study; *Tectonics* 5:553-572

Perkins D, Chipera SJ (1985) Garnet-orthopyroxene-plagioclase-quartz barometry: Refinement and application to the English River subprovince and the Minnesota River Valley; *Contrib Mineral Petrol* 89:69-80

Phillips GN (1980) Water activity changes across an amphibolitegranulite transition, Broken Hill, Australia; *Contrib Mineral Petrol* 75:377-386

Raase P, Raith M, Ackermand D, Lal RK (1986) progressive metamorphism of mafic rocks from greenschist facies to granulite facies in the Dharwar craton of South India; *J Geol* 94:261-282

Rudnick RL, Ashwal LD, Henry DJ (1984) Fluid inclusions in high grade gneisses of the Kapuskasing structural zone, Ontario: Metamorphic fluids and uplift/erosion path; *Contrib Mineral Petrol* 87:399-406

Rudnick RL, McLennan SM, Taylor SR (1985) Large ion lithophile elements in rocks from high-pressure granulite facies terrains; *Geochim Cosmochim Acta* 49:1645-1655

Sawyer EW, Robin PYF (1986) The subsolidus segregation of layer-parallel quartz-feldspar veins in greenschist to upper amphibolite facies metasediments; *J Metamorphic Geol* 4:237-260

Schuiling RD, Kreulen R (1979) Are thermal domes heated by CO_2-rich fluids from the mantle? *Earth Planet Sci Lett* 43:298-302

Selverstone J (1982) Fluid inclusions as petrogenetic indicators in granulite xenoliths, Pali-Aike volcanic field; *Contrib Mineral*

Petrol 79:28-36

Sibson RH (1987) Earthquake rupturing as a mineralizing agent in hydrothermal systems; *Geology* 15:701-704

Todd VR, Shaw SE (1985) S-type granitoids and an I-S line in the Peninsular Ranges batholith, southern California; *Geology* 13: 231-233

Touret J (1971) Le facies granulite en Norvege meridionale. II Les inclusions fluides; *Lithos* 4:423-436

Touret J (1987) Fluid distribution in the continental lithosphere; In Kroner A (ed) Proterozoic lithosphere evolution, *Amer Geophys Un Geodynam Ser* 17:27-33

Touret J, Dietvorst P (1983) Fluid inclusions in high-grade anatectic metamorphites; *J Geol Soc Lond* 140:635-649

Taylor HP (1977) Water/rock interactions and the origin of H2O in granitic batholiths; *J Geol Soc Lond* 133:509-558

Valley JW (1985) Polymetamorphism in the Adirondacks: Wollastonite at contacts of shallowly intruded anorthosite; In Tobi AC, Touret J (eds) *The deep Proterozoic crust in the North Atlantic provinces.* Reidel, Dordrecht, 217-236

Valley JW, O'Neil JR (1984) Fluid heterogeneity during granulite facies metamorphism in the Adirondacks: Stable isotope evidence; *Contrib Mineral Petrol* 85:158-173

Valley JW, McLelland J, Essene EJ, Lamb WM (1983) Metamorphic fluids in the deep crust: Evidence from the Adirondacks; *Nature* 301:226-228

Weaver BL (1980) Rare earth element geochemistry of Madras granulites; *Contrib Mineral Petrol* 71:271-279

Weaver BL, Tarney J (1983) Elemental depletion in Archaean granulite facies rocks; In Atherton MP, Gribble, CD (eds) *Migmatites, melting and metamorphism*, Shiva, Nantwich, 250-263

White AJR, Clemens JD, Holloway JR, Silver LT, Chappell BW, Wall VJ (1986) S-type granites and their probable absence in southwestern North America; *Geology* 14:115-118

Wickham SM (1987) The segregation and emplacement of granitic magmas; J *Geol Soc Lond* 144:281-297

Wickham SM, Taylor HP (1985) Stable isotope evidence for large-scale seawater infiltration in a regional metamorphic terrane: The Trois Seigneurs massif, Pyrenees, France; *Contrib Mineral Petrol* 91: 122-137

Wickham SM, Taylor HP (1987) Stable isotopic constraints on the origin and depth of penetration of hydrothermal fluids associated with Hercynian regional metamorphism and crustal anatexis in the Pyrenees; *Contrib Mineral Petrol* 95:255-268

Wood BJ Walther JV (1986) Fluid flow during metamorphism and its implications for fluid-rock ratios; In Walther JV, Wood BJ (eds) *Fluid-rock interactions during metamorphism*, Springer, New York, 89-108

Yardley BWD, Shepperd TJ, Barber JP (1983) Fluid inclusion studies of high grade rocks from Connemara, Ireland; In Atherton MP, Gribble CD (eds) *Migmatites, melting and metamorphism*, Shiva, Nantwich, 110-126

P-T AND FLUID EVOLUTION OF THE ANGMAGSSALIK "CHARNOCKITE" COMPLEX, SE GREENLAND

T. Andersen[1], H. Austrheim[1] and D.Bridgwater[2]
1: Mineralogisk-Geologisk Museum, Sarsgate 1,
   N-0562 Oslo 5, Norway
2: Geologisk Museum, Øster Voldgade 5-7,
   DK-1350 København K, Denmark

ABSTRACT   The Angmagssalik "charnockite" complex is a belt of Proterozoic orthopyroxene bearing metamorphic and igneous rocks situated within the central parts of the Nagssugtoqidian mobile belt of East Greenland. The oldest intrusive rock unit in the complex is a melagabbro, which may represent a pyroxene cumulate from a mafic melt. This unit is cross-cut by leuconorite (the most abundant unit in the complex) and by minor anorthosite and hypersthene-veins. The country rock of the intrusives is an amphibolite facies quartzofeldspathic garnet gneiss, which has developed opx-bearing (granulite facies) mineral assemblages near the intrusive contact, and which has been physically mobilized and intermixed with the intrusives.

Quartz and plagioclase in melagabbro contain primary and secondary pure CO_2 inclusions, with accidentally trapped solid silicate and carbonate crystals. These inclusions have liquid homogenization temperatures in the range 0 to -20 °C. Pure CO_2 is also characteristic for inclusions in plagioclase in the leuconorite and in the anorthosite veins. The inclusions in leuconorite are distinctly secondary, and have higher densities (homogenization temperatures ranging down to -30 °C). Secondary carbonic fluid inclusions in the country rock gneisses contain minor methane, and range in homogenization temperatures from +10 °C (garnet) to -10 °C (quartz, plagioclase).

Combined with mineral thermobarometry on cpx-opx, gnt-opx and gnt-bi assemblages, the fluid inclusion data suggest an interpretation of the cooling history and fluid evolution of the complex: Primary magmatic rocks crystallized in the middle-lower crust (P: 6-8 kbar, T: 1000-1100 °C). The higher density, secondary inclusions reflect isobaric cooling in the presence of CO_2 as a free fluid phase. This fluid phase migrated into the country-rock gneisses, where interaction between rocks and fluid led to the evolution of dry, granulite facies mineral assemblages. In this process, the CO_2 mixed with CH_4-bearing fluids, or reacted with hydrous minerals to yield fluids with minor methane contents.

D. Bridgwater (ed.), Fluid Movements – Element Transport and the Composition of the Deep Crust, 71–94.
© 1989 by Kluwer Academic Publishers.

INTRODUCTION

Processes which generate or consume volatile components are important
for the evolution of metamorphic terranes (e.g. Ferry & Burt 1982).
Numerous studies of fluid inclusions in minerals from high-grade
metamorphic rocks have shown that there is a general correllation
between "dry" (granulite facies) mineralogies and the presence of a
CO_2-dominated fluid phase (e.g. Touret 1971, Bilal & Touret 1976, Olsen
1978, see also reviews by Touret 1977, 1981, 1986). Other fluid species
that may be present (as minor components) in the deep crustal environment include H_2O, CH_4 and N_2 (e.g. Hollister & Burrus 1976, Swanenberg
1980, Glassley et al. 1984, Touret & v.d.Kerkhof 1986). "Flushing" of a
rock volume by CO_2 at high metamorphic grade may lead to stabilization
of anhydrous mineral assemblages and to rock dehydration (Touret 1971),
but CO_2 is also a geochemically active substance, which can remobilize
major oxide components from the protolith (Glassley 1983).
    Different sources for the deep crustal CO_2 have been suggested:
(i) Hoefs & Touret (1975) suggested that the CO_2 in the Bamble
granulites (S. Norway) originated from the mantle, because of their
mantle-like $\delta^{13}C$ values. This would require an agent transporting the
fluid into the deep crust. The discovery of deep crustal syn-metamorphic charnockite intrusions in the Bamble region (Field et al.1980,
Smalley et al. 1983) may suggest that mantle-derived magmas acted as a
carrier for the carbonic fluids, although the actual mechanism of the
process is not fully understood (Touret 1986, 1987, Touret & Hansteen
1987). (ii) In other terranes, e.g. in West Greenland, an origin of
metamorphic CO_2-fluid by devolatilization reactions in local
metasediments has been suggested (Glassley 1983). The presence of N_2
and/or CH_4 in the metamorphic fluid may be an indication that at least
a significant proportion of the fluid has been derived by intracrustal
processes involving metasediments (Kreulen & Schuiling 1982).
    Although several studies have indicated deep crustal intrusions as
the sources of heat and fluids in granulite terranes (e.g. Frost &
Frost 1987), it has generally not been possible to trace out the fluid
evolution from magmatic PT conditions through the cooling history of
the intrusive rocks and the metamorphic aureole. We have, however, been
able to locate such an example in the Proterozoic Angmagssalik
"charnockite" complex of South East Greenland. In the present paper,
preliminary results of mineral thermometry and fluid inclusion studies
of these rocks will be presented.

GEOLOGY

The Angmagssalik "charnockite" complex (Wager 1934) consists of a
series of dominantly basic rocks (melagabbro, leuconorite, plagioclase
veins (anorthosite)), together with some intermediate to acid
components (charnockite ss) emplaced into leococratic garnet gneisses
which are mainly at amphibolite-facies, but which locally develop
orthopyroxene (Wager 1934, Bridgwater 1976, Bridgwater et al. 1987).
This "charnockite" - garnet gneiss association outcrops in the center
of the Nagssugtoqidian mobile belt in southern East Greenland, a circa

270 km wide structure[1] trending approximately E-W across high grade
Archaean gneisses. Field evidence and available U-Pb data from zircons
(B.T. Hansen, personal communication, 1986) suggest that a major part
of the mobile belt is made up of thrust sheets of tectonically reworked
Archaean gneisses with a minimum age of 2800 Ma, intercalated with
supracrustal rocks some of which give a Proterozoic metamorphic age and
no Sr isotope evidence of an extended crustal residence (F. Kalsbeek
personal communication, 1987). Proterozoic igneous and metamorphic
activity was widespread in the belt in the period 1550-1900 Ma
(Pedersen & Bridgwater 1979, Austrheim et al. 1987).

In the center of the mobile belt, Proterozoic metamorphism reached
high-pressure amphibolite conditions, with copious kyanite in pelitic
units and garnet-clinopyroxene assemblages in the basic rocks, prior to
the emplacement of the "charnockite" complex. The high pressure
assemblage in the regional gneiss complex are partly replaced by lower
pressure, higher temperature assemblages with sillimanite replacing
kyanite and plagioclase-hornblende symplectites replacing garnet-
pyroxene assemblages during the later stages of the Proterozoic
activity.

The garnet gneisses and the charnockitic rocks outcrop in a zone
extending for 30 x 70 km from the outer coast of Kulusuk island,
through Angmagssalik island to the inland ice (Fig. 1). Contacts
between the gneisses and the igneous rocks are irregular and in places
it is impossible to be certain in the field whether a particular
outcrop consists of igneous rock or partially remobilized country rock.
This is interpreted on a regional scale as the result of emplacement
of high temperature igneous rocks into rocks which were at amphibolite
facies metamorphic conditions at the time of igneous intrusion. The
garnet gneisses extend for up to 10 km away from the main body of
"charnockitic" rocks as exposed on the present surface. On a scale of
several hundered metres the garnet gneisses preserve a gross earlier
lithological layering corresponding to compositional layering in the
dominently metasedimentary protoliths. Both siliceous and calc-silicate
horizons occur locally within the garnet and biotite metasedimentary
gneisses. Concordant units of garnet-poor quartzofeldspathic gneiss,
some of which contain basic dyke remnants and fragments of gabbro-
anorthositic material, are interpreted as slices of the Archaean gneiss
complex intercalated with the metasedimentary garnet gneiss prior to
the intrusion of the igneous complex. Similar orthogneiss units are
preserved as rafts within the igneous complex. On outcrop scale the
garnet gneisses show considerable anatexis and mobility. Intrusive
relations are seen in which one phase of garnet gneiss with disoriented
fragments of earlier garnet gneiss cuts discordantly through earlier

------
1) The term **Ammassalik mobile belt** is now used by geologists working
with the Geological Survey of Greenland (F. Kalsbeek, personal
communication, 1988).

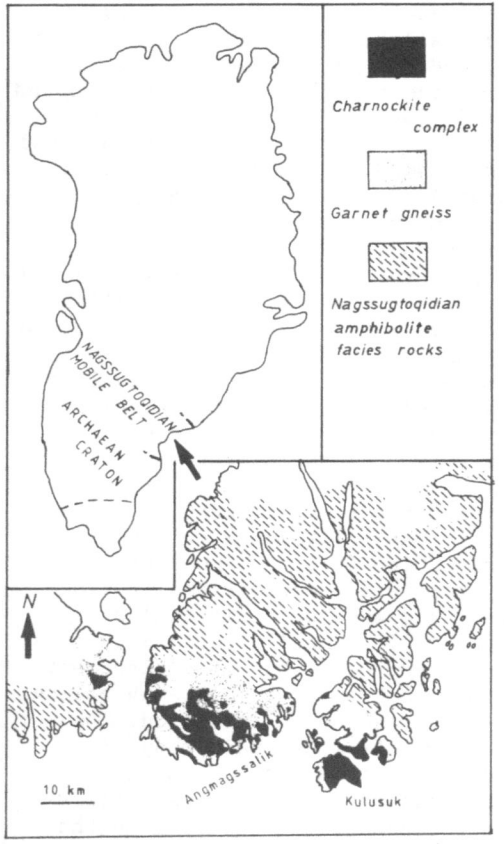

Fig. 1
Much simplified
geological map of the
Kulusuk- Angmagssalik
area, from
Bridgwater (1976).

structures in an adjacent unit. The main mineral phases in the garnet
gneisses are plagioclase, alkali feldspar, quartz, garnet and biotite.
Orthopyroxene occurs locally. The calc-silicate units are generally
broken up into pods which frequently show a zoned structure with
garnet-rich centers and diopside rims suggesting a reaction with
adjacent silica-rich units.

Several different varieties of "charnockitic" rocks, with
intrusive relations to the garnet gneiss can be recognized. Quartz-
bearing **melagabbro** (andesine-labradorite, hypersthene, clinopyroxene,
hornblende, biotite and minor quartz) is the oldest intrusive. It
locally shows small scale rhythmic layering defined by variations in
the pyroxene/plagioclase ratio, suggesting a magmatic origin for the
rock (Fig. 2).

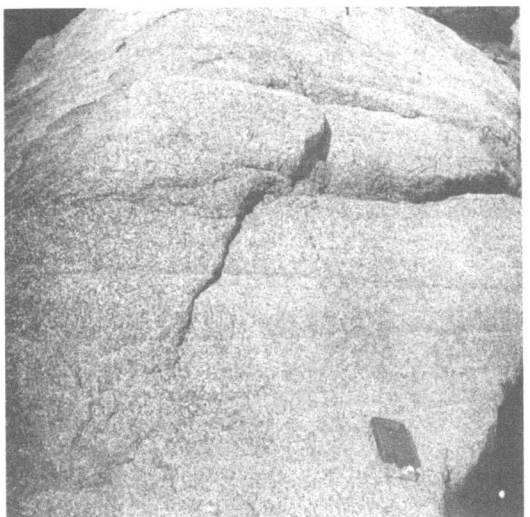

Fig. 2
Rhythmic layering in
melagabbro.

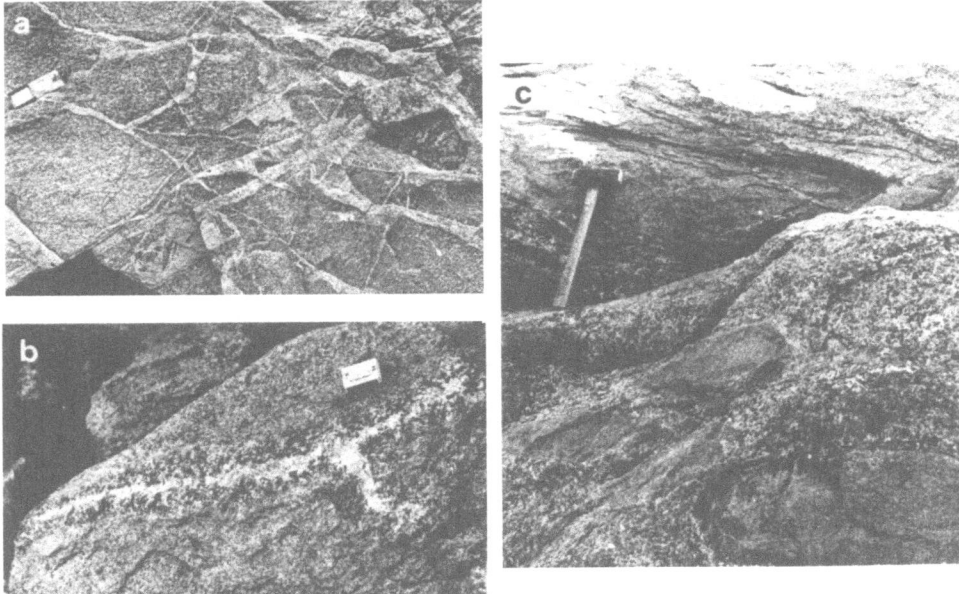

Fig. 3
a: Veins of leuconorite cutting melagabbro.
b: Leuconorite cut by an anorthosite vein. Next to the vein,
   feldspar and biotite have grown within the leuconorite.
c: Rounded fragments of "charnockitic" rocks (leuconorite) embeded in
   mobilized, hypersthene-bearing garnet gneiss.

Leuconorite (andesine, hypersthene, biotite, minor quartz)
dominates the complex in terms of area of outcrop. The rock shows a
foliation defined by parallel orientation of pyroxene and biotite. This
foliation is parallel to the regional structures, and is regarded as
developed during tectonism at the time of intrusion. In places, the
leuconorite is banded with orientated inclusions of mafic material
(melagabbro ?). Together with dykes and veins of **anorthosite** and
hypersthene rock, the leuconorite cross-cuts the melagabbro,
demonstrating the younger relative age and the intrusive nature of
these rock-units (Fig. 3a).

Undeformed irregular anorthosite dykes and veins cut foliated
leuconorite which show extensive feldspathisation for tens of
centimetres away from the vein margin (Fig. 3b). The feldspathisation
is frequently accompanied by growth of biotite concentrated in a
distinct zone cross-cutting the original foliation in the leuconorite a
few centimetres from the vein margins. Many of the anorthosite veins
show a change in composition, with plagioclase - orthopyroxene bearing
rocks where the veins are widest to quartz-biotite-feldspar bearing
rocks where the veins thin. Much of the plagioclase in the altered
gabbros close to the veins is coloured by sub-microscopic inclusions
(presumably haematite). Where the anorthosite dykes cut melagabbro,
extensive feldspathisation of the wall-rock melagabbro has taken place.
Acid intrusive rocks (**charnockite proper**) are scarce in the
Angmagssalik complex, but occur around Kulusuk (Fig. 1).

Contact relations to the garnet gneiss.

Approaching the contacts to the "charnockitic" rocks, the garnet gneiss
looses structural coherence, and develops a granulite-facies
mineralogy (opx + gt + plg + qz + bi $\pm$ cpx), at the expense of the
earlier amphibolite facies assemblages. Near the contact itself, the
modified gneiss becomes strongly mixed with the "charnockitic" rocks
(Fig. 3c). The resulting, unfoliated hypersthene-bearing rock with
relict garnet is referred to as **mobilized garnet gneiss**. Hypersthene-
bearing assemblages also develop along minor fractures and shear-zones
extending out into the non-mobilized garnet gneiss and orthogneiss
screens, resulting in the formation of garnet charnockite and acid
charnockite. The in situ formation of charnockite can be proven on the
scale of a few centimetres to several metres. It is uncertain how much
of the larger acid charnockite bodies are formed by this process.

AGE

A 6-point whole-rock Sm-Nd isochron defined by samples from all
intrusive units yields an age of 1907 $\pm$ 36 (1σ) Ma (Austrheim et al.
1987). This age agrees within error of a U-Pb concordia intercept age
of 1886 $\pm$ 2 Ma (B.T. Hansen, personal communication 1987) and a Rb-Sr
errorchron on the garnet gneisses (Pedersen & Bridgwater 1979 and

unpublished), and is interpreted as the age of emplacement of the
"charnockite" complex, and of development of the granulite facies
paragenesis in the mobilized garnet gneiss. The initial Nd isotopic
composition indicated by the isochron ($\epsilon_{Nd}t=-4$) and the presence of
zircon fractions lying on a discordia between 1850 and 2700 Ma both
point to a major component of older material in the complex.

PETROGENESIS

The origin of the Angmagssalik complex has been somewhat controversial.
Wright et al. (1973) argued a metasedimentary origin for some of the
hypersthene-bearing rocks, whereas other units, including cross-cutting
anorthosites, were assumed to originate by local anatexis. In order to
explain the origin of anorthosite by partial melting, a high water-
pressure (i.e. the presence of a water-dominated fluid phase) seems
indicated (Moorlock et al. 1972). The intrusive nature of parts of the
body and the existence of primary magmatic structures in the rocks,
suggest an igneous intrusive origin for at least the basic parts of
the "charnockite" complex. Rhythmic layering, as observed in the
melagabbro are commonly interpreted as formed by accumulation of the
phases making up the contrasting layers (Wager et al. 1960). The mafic
members of the complex are assumed to be cumulates (cpx, opx, etc.),
whereas the more feldspar-rich cross-cutting units may represent
liquids emplaced at a somewhat later stage of magmatic evolution,
possibly during actual deformation.
    Structural evidence (Bridgwater 1976, Bridgwater et al. 1987)
suggests that the emplacement was contemporaneous with regional
amphibolite facies metamorphism and deformation. The country rocks were
locally upgraded to granulite facies close to the heat source provided
by the intrusive body. The Angmagssalik "charnockite" complex thus is
an analogy to the synmetamorphic intrusive charnockites of the Bamble
region in S. Norway (Smalley et al. 1983, Touret 1986) and the rocks of
the rapakivi granite suite of S. Greenland (Bridgwater & Windley 1973).
From the point of view of fluid evolution, it is of importance to trace
out the variations of fluid composition and molar volume (density)
from the stage of magmatic crystallization _via_ the mobilized garnet
gneiss, into the unmodified garnet gneiss country rock. In order to be
able to interpret the fluid inclusion data, it is first necessary to
outline the thermal history of the rocks, from mineral thermobarometry.

MINERAL CHEMISTRY AND PT ESTIMATES

Microprobe analyses of orthopyroxene, clinopyroxene, garnet and
biotite, selected from a total of 15 samples from all units, are
presented in Table 1, with pertinent PT estimates listed in Table 2.
    The orthopyroxene cores range in composition between En_{75} and
En_{50} and have a low wollastonite content (less than Wo_2), typical for
granulite facies terraines. Application of the Lindsley (1983) one-

78

Table 1   Selected mineral compositions from the Angmagssalik complex

Sample: Min:	250345 opx rim	250345 cpx	250345 gt	250872 gt rim	250872 gt core	250272 bi	250872 bi incl	250862 opx	250862 cpx	250862 cpx core
SiO_2	51.41	51.24	37.73	38.06	38.53	38.76	35.92	52.00	51.05	50.95
TiO_2	0.05	0.05	0.00	0.00	0.06	3.96	5.03	0.10	0.34	0.30
Al_2O_3	1.90	1.61	21.76	21.55	21.80	15.89	14.92	1.99	2.98	3.04
FeO	25.32	8.73	26.95	28.29	27.09	15.20	15.85	22.56	8.86	10.35
MnO	0.09	0.00	0.65	0.78	0.70	0.09	0.03	0.55	0.27	0.28
MgO	20.52	13.03	6.70	4.68	6.09	13.79	12.99	22.36	13.10	13.67
CaO	0.41	22.81	5.30	6.59	6.56	0.19	0.07	0.55	21.89	18.81
Na_2O	0.00	0.54	0.00	0.00	0.00	0.31	0.21	0.00	0.82	0.65
K_2O	0	0	0	0	0	9.91	9.25	0	9.25	0
Sum	99.70	98.01	99.09	99.95	100.83	98.10	94.27	100.11	99.31	98.05

Cation-based structural formulae

Si	1.941	1.945	2.955	2.991	2.975	5.782	5.623	1.932	1.718	1.933
Al(IV)	0.059	0.055	0.045	0.009	0.025	2.218	2.377	0.068	0.282	0.067
Al(VI)	0.025	0.017	1.964	1.986	1.958	0.576	0.375	0.019	-0.163	0.069
Ti	0.001	0.001	0.000	0.000	0.003	0.444	0.592	0.003	0.009	0.009
Fe2+	0.799	0.277	1.765	1.859	1.749	1.896	2.075	0.701	0.249	0.328
Mn	0.003	0.000	0.043	0.052	0.046	0.011	0.004	0.017	0.008	0.009
Mg	1.155	0.737	0.782	0.548	0.701	3.066	3.031	1.238	0.657	0.773
Ca	0.017	0.928	0.445	0.555	0.543	0.030	0.012	0.022	0.790	0.765
Na	0.000	0.040	0.000	0.000	0.000	0.090	0.064	0.000	0.054	0.048
K	0.000	0.000	0.000	0.000	0.000	1.886	1.847	0.000	0.397	0.000
SUM:	4.00	4.00	8.00	8.00	8.00	16.00	16.00	4.00	4.00	4.00

Samples: 250345 and 250872 = mobilized garnet gneiss, 250862 = melagabbro.

Table 2   PT-estimates from the Angmagssalik complex.

Sample GGU No.	Rock type	H/W Core		H/W Rim		WB/W Core	
		T °C	P kbar	T °C	P kbar	T °C	P kbar
250345	Mo.Gnt.Gn.	750	10	680	5.5	720	7.2
250872	Mo.Gnt.Gn	780	5.3	560	-0.2		
268360	Mo.Gnt.Gn	719	5.6	605	3.2		
EG 401	Mo.Gnt.Gn	836	5.2	710	5.6		
250862	Melagabbro					790	
250873	Melagabbro					890	
268360	Melagabbro					871	

Sample GGU No.	Rock type	WE	FS Core	FS Rim	L_o Core	L_c Core
		T °C	T °C	T °C	T °C	T °C
250345	Mo.Gnt.Gn.			670	700	
250872	Mo.Gnt.Gn		760	587	600	
268360	Mo.Gnt.Gn			611		
EG 401	Mo.Gnt.Gn					
250862	Melagabbro	780				1085
250873	Melagabbro	870				
268360	Melagabbro	840				

H/W:	Temperature: Harley (1984), pressure: Wood (1974).
WB/W:	Temperature: Wood & Banno (1973), pressure: Wood (1974).
WE:	Temperature: Wells (1977)
FS:	Temperature: Ferry & Spear (1978)
L_o:	Orthopyroxene temperature from Lindsley (1983)
L_c:	Clinopyroxene temperature from Lindsley (1983)

pyroxene thermometer indicate temperatures in the range of 550 - 700°C for all rock units. In the norite and melagabbro the orthopyroxene and most clinopyroxenes have uniform compositions, and are typically unzoned. Lindsley's (1983) one-pyroxene thermometer give temperatures in the range of 550-750 °C also for the clinopyroxenes. The Wood and Banno (1973) and Wells (1977) thermometers, on the other hand, give markedly higher temperatures for ortho and- clinopyroxene pairs, in the range 870 - 720 °C.

The magma forming the main series of intrusive rocks (melagabbro, leuconorite, anorthosite) in the Angmagssalik complex must have been emplaced at a temperature at least as high as the solidus of comparable basic silicate melts. By analogy with experimental data on tholeitic and high-alumina basaltic systems (Yoder & Tilley 1962, Green & Ringwood 1967), this limit can be estimated to at least 1100 °C at crustal pressures. In melagabbro sample GGU 250762, clinopyroxenes preserve cores with extensive exsolution. Composition obtained by averaging 5 spot analyses show that the core analysed has lower wollastonite and higher ferrosilite content than the rest of the clinopyroxenes. A temperature of 1085 °C is obtained by the Lindsley thermometer, suggesting that this clinopyroxene core crystallized from a magma.

The garnets in the mobilized garnet gneiss have core compositions ranging between $Alm_{50-60}, Pyr_{25-50}, Gross_{5-15}$. In the mobilized garnet gneiss the garnets are zoned with increasing FeO and decreasing MgO outwards. The orthopyroxenes of the garnet gneiss are also zoned with increasing En content outward and a decrease in Al_2O_3 content from core to rim. Assuming that the orthopyroxene and garnet cores were once in equilibrium this type of zoning suggests decreasing temperature. By applying the Harley (1984) thermometer and Wood (1974) barometer to core and rim compositions of garnet and orthopyroxene, two points on the cooling PT path of the Angmagssalik complex can be determined. Temperatures ranging between 840 and 720 °C at pressures between 5 and 10 kbar were obtained for core values, while the corresponding rim values varied between 710 and 570 °C at pressures between 5 and 0 kbar.

Ti-rich biotites occur as inclusions in garnets as well as in contact with garnet rims. The Ferry and Spear (1978) garnet-biotite thermometer gives temperatures in the range of 750-650 °C for biotite inclusions in garnet cores, whereas biotites in equilibrium with garnet rims suggest lower temperatures (550-600 °C).

FLUID INCLUSIONS

Occurrence and characteristics

Fluid inclusions of a size and quality suitable for microthermometric analysis are found in all rock types studied. Depending upon the rock type, the host phases are quartz, feldspar or garnet. The inclusions can be classified as primary (trapped during crystal growth) or secondary (trapped along healed fracture planes after primary crystal growth) by criteria discussed by Roedder (1981, 1984). Inclusions related to growth zones or occurring isolated or scattered without any relationship to secondary healed fracture trails are interpreted as of primary origin. From their inclusion textures and assemblages, it is convenient to discuss the different rock types in terms of three sub-groups: (i) Melagabbro, (ii) Leuconorite and anorthosite veins (red plagioclase) and (iii) Country-rocks (garnet gneiss, mobilized garnet gneiss).

In the melagabbro, euhedral, negative crystal-shaped inclusions occur isolated or scattered at random within some quartz and plagioclase grains, with no relation to secondary healed fracture trails (Fig. 4a). These inclusions are interpreted as primary, trapped during crystal growth; their host crystals being relics from the magmatic stage of evolution. The inclusions can attain a size of ≈30 μm (largest inclusions of all the present material), and invariably contain a homogeneous fluid phase + one or more birefringent crystals (Fig. 4a,b). Strongly birefringent crystals are identified as calcite from their optical properties and Raman spectra (Fig. 4c), whereas the more moderately birefringent crystals have not been positively identified (plagioclase is, however, a possible mineral in inclusions in quartz). The solid assemblage varies both with respect to the identity, number and relative size of the crystals between neighbouring fluid inclusions.

Both quartz and plagioclase grains in the melagabbro are cut by healed fracture trails with small, secondary (5-10 μm) fluid inclusions of approximate negative crystal shapes, and irregular, somewhat larger, highly birefringent (calcite ?) single crystal inclusions. The fluid inclusions contain one fluid phase; some, but not all, also contain free, highly birefringent crystals (calcite ?), others are attached to the crystal inclusions. Again, the relative volume of the solid varies between neighbouring inclusions.

In leuconorite and anorthosite veins (red plagioclase) fluid inclusions are invariably found along secondary trails in plagioclase. In the leuconorite, the abundance of inclusion trails can be very high (Fig. 4e), whereas the distribution of inclusion trails in the red plagioclase is more irregular, and their abundance is lower (Fig. 4f). In both rock types, secondary inclusion trails resemble those in melagabbro, with fluid inclusions and inclusions of strongly bi-refringent minerals occurring along the same healed fractures.

In the country-rock garnet gneiss and the mobilized garnet gneiss, fluid inclusions have been observed in quartz, plagioclase and garnet. The inclusions are always secondary, defining healed fractures in the minerals. The inclusion trails in plagioclase strongly resemble those in leuconorite/red plagioclase, and are associated with "clouding" of the plagioclase; in the mobilized garnet gneiss, their abundance is very high (Fig. 4g). In contrast to this, the fluid inclusion trails in quartz do not contain solid inclusions. These fluid inclusions are equant negative crystals (≈10 μm). The garnet contains moderately abundant trails of inclusions of homogeneous fluid and anhedral single birefringent crystals, probably carbonates (Fig. 4h). The pseudo-irregular inclusion shapes appear to be governed by negative crystal

82

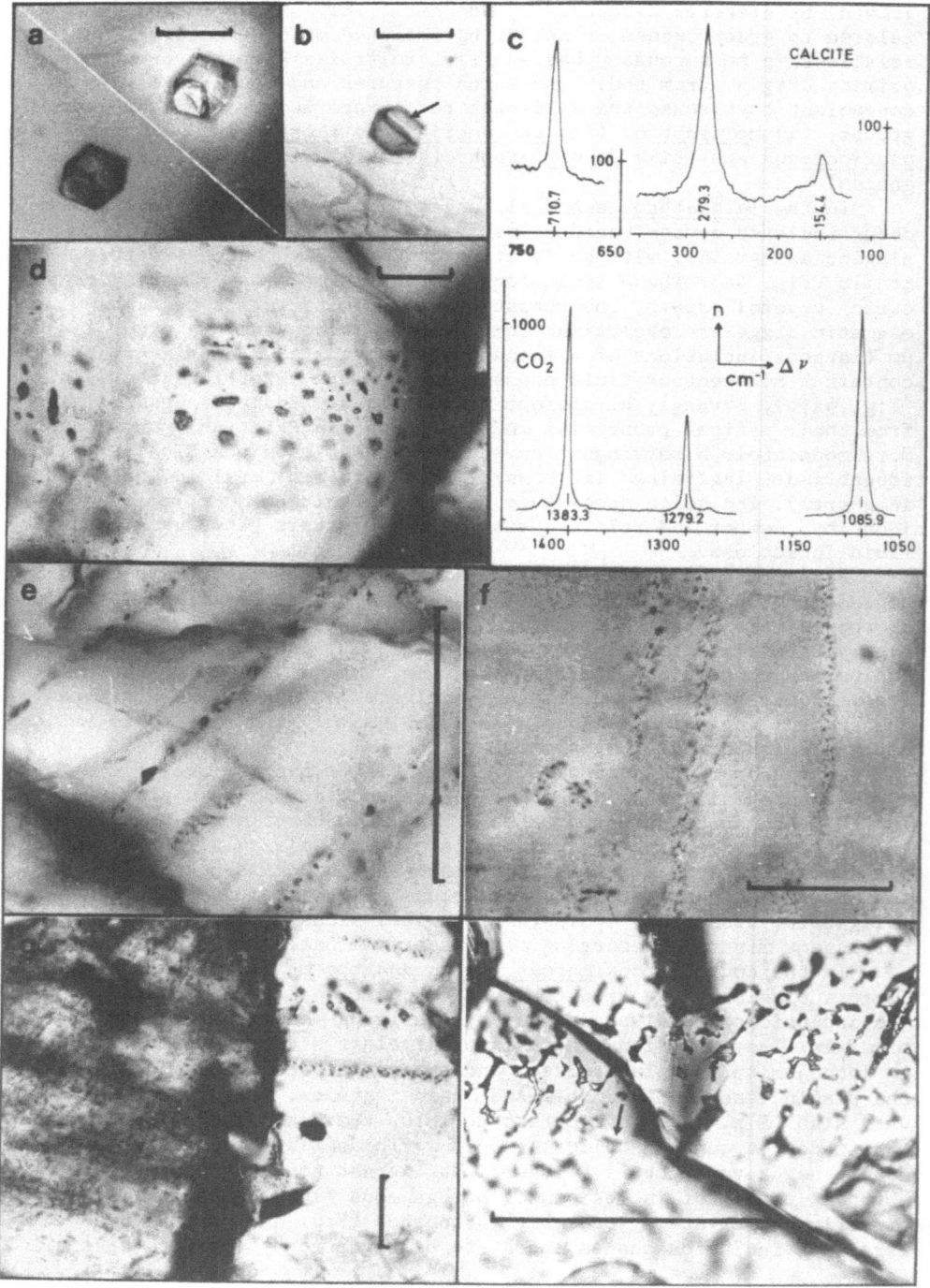

Fig. 4
Fluid inclusion assemblages in the Angmagssalik rocks. Length of
scale bars: 0.1 mm.

a: Primary fluid inclusions in plagioclase from melagabbro (sample
GGU 250919). Viewed in polarized light (lower left half) and with
crossed polarizers (upper right half). Note birefringent crystal
within the inclusion cavity.

b: An inclusion similar to **a**, which has been analysed by laser Raman
Microprobe. Arrow indicates large, birefringent crystal.

c: Raman spectra of crystal and fluid in the inclusion shown in **a**,
identifying the mineral as calcite, and the fluid as pure CO_2. The
CO_2 is characterized by T_m=-56.6 $^\circ$C, Th=-10.6 $^\circ$C.

d: Secondary inclusions in quartz from melagabbro (sample GGU
250919), postdating the inclusions in **a** and **b**. The inclusions
contain a homogeneous fluid phase and one to several crystals.

e: Trails with small, secondary fluid inclusions in plagioclase from
leuconorite (sample GGU 250864).

f: Secondary fluid inclusions in red anorthosite plagioclase (sample
GGU 250910). Away from the inclusion trails, the feldspar is
stained with submicroscopic, opaque solid inclusions (hematite ?).

g: Secondary fluid inclusion trails in feldspar and quartz from
garnet gneiss. Note clouding of the feldspar adjacent to the
inclusion trails.

h: Secondary inclusions in garnet from mobilized garnet gneiss.
The inclusions comprise carbonate (**c**) and fluid (**f**). The Raman
spectrum in Fig. 6 was recorded from the inclusion somewhat below
the present plane of focus, at the arrowhead. This inclusion is
characterized by T_e=-59.9 $^\circ$C, T_m=-58.3 $^\circ$C, T_h=+5.3 $^\circ$C.

faces of the garnet. Commonly the two types of inclusions are closely
associated with each other within healed fractures, sometimes in direct
contact (Fig. 4h).

Microthermometry

The fluid inclusions were studied from low temperature (-190 $^\circ$C) to the
critical point of CO_2 (+31.4 $^\circ$C), using CHAIXMECA and LINKAM THM 600
heating-freezing stages (e.g. Shepherd et al. 1985), with precooled,
gaseous N_2 as a cooling agent. The calibration and measurement
procedures for carbonic fluid inclusions were as described by Andersen
et al. (1984). After careful microscopy, a selection of ca. 200 fluid
inclusions was studied. The results are summarized in Fig. 5, showing
the temperatures of final melting and homogenization of CO_2.

When heated from low temperature after solidification of the fluid
contents, CO_2 melting was the first phase-transition observed. In one
group of samples, consisting of melagabbro (primary and secondary
inclusions), leuconorite and red plagioclase, melting took place

84

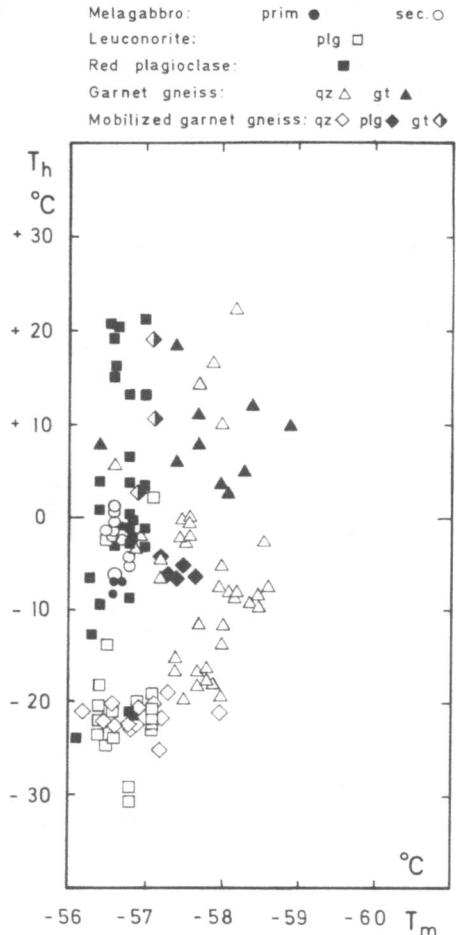

Melagabbro: prim ●   sec. ○
Leuconorite: plg □
Red plagioclase: ■
Garnet gneiss: qz △ gt ▲
Mobilized garnet gneiss: qz◇ plg◆ gt◈

Fig. 5
Final melting of CO_2
(T_m) vs. homo-
genization tempera-
ture (T_h) for the
inclusions studied.

abruptly close to the triple point of pure CO_2 (T_T=-56.6 °C). Fluid
inclusions in garnet gneiss and its mobilized equivalent showed a
well-defined melting interval, with first melting at a eutectic between
-59 and -60 °C. The melting temperature (T_m, i.e. the temperature of
disappearance of the last CO_2 crystal), as recorded in Fig. 5,
was significantly lower than T_T. In the garnet gneiss, T_m typically
ranged from -57.5 °C to - 59 °C, whereas the mobilized garnet gneiss
had T_m between -56.6 and -57.5 °C (Fig. 5).
    On further heating from the melting point, the fluid inclusions
homogenized by vapour-bubble disappearance (or, rather, partly

homogenized, as the solid phases within the inclusion cavities remained unaffected). Homogenization temperatures (T_h) are recorded in Fig. 5, all temperatures refer to liquid homogenization. As can be seen from the figure, there is a considerable range of T_h, from $\approx$-30 to $\approx$+20 °C. Fluid inclusions in garnet generally homogenize in the higher part of the total range. The majority of T_h, including all observations from melagabbro, fall between -10 and +10 °C. Another cluster of points, mainly from secondary inclusions in plagioclase in leuconorite and mobilized garnet gneiss fall between -20 and -30 °C, whereas the inclusions in red plagioclase from anorthosite veins cover the entire T_h range observed.

Raman spectroscopy

A selection of fluid inclusions have been analysed with Laser Raman microprobe, in order to determine the existence and nature of minor components in the CO_2-dominated fluid inclusions.

The primary inclusion in plagioclase from melagabbro illustrated in Fig. 4b,c shows a Raman signal from CO_2, but no response from either CH_4, N_2 or CO, effectively confirming the purity of the CO_2. The secondary inclusion in garnet (Fig. 4h, arrowhead), on the other hand, shows peaks from CO_2 and CH_4, on top of a fluorescence continuum from the garnet matrix (Fig. 6).

Given cross-sections for Raman exitation for CO_2 and CH_4, the molar composition of the inclusion illustrated in Fig. 6 can be determined by integration of the Raman peaks (Burke & Lustenhouwer 1987). For this particular inclusion, the Raman analysis indicates a molar composition of 91 % CO_2, 9 % CH_4.

Density and composition of the fluid inclusions

Among the present material, the fluid inclusions in melagabbro, leuconorite and red plagioclase melt within experimental error of the triple point of CO_2, and thus consist of pure CO_2, as also indicated by the Raman spectrum in Fig 4c. For a pure CO_2 inclusion, the homogenization temperature is a direct function of molar volume (density) (Burrus 1981). It is therefore possible to recalculate the T_h in terms of molar volume (Fig. 7), the results are given in histogram form for each of the geological units.

In binary CO_2-CH_4 fluids, the molar volume and composition can be determined from measurements of T_m and T_h (Burrus 1981, Heyen et al. 1982). In Fig. 7, the composition and molar volume of inclusions from garnet gneiss and mobilized garnet gneiss are plotted, determined from the T_h and T_m contoured diagrams of Heyen et al. (1982), as replotted by Shepherd et al. (1985). The thermal data on the secondary inclusion in garnet which has been analysed by Raman spectroscopy (Fig. 6) give a molar CO_2/(CO_2+CH_4)-ratio of 0.9, which agrees very well with the

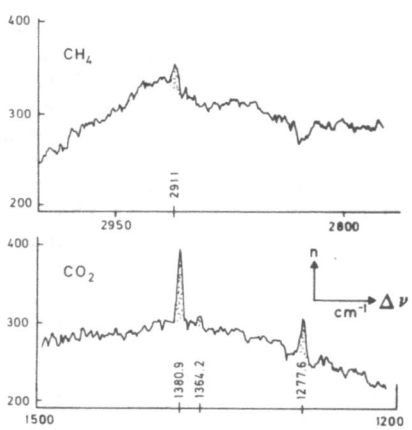

Fig. 6
Raman spectrum of a
fluid inclusion in
garnet from mobilized
garnet gneiss (Fig.4h).
The inclusion was
analysed using a
multichannel Dilor
Microdil 28 (Burke &
Lustenhouwer 1987), at
the Instituut voor
Aardwetenschappen,
Vrije Universiteit,
Amsterdam (E.A.J.
Burke, analyst).

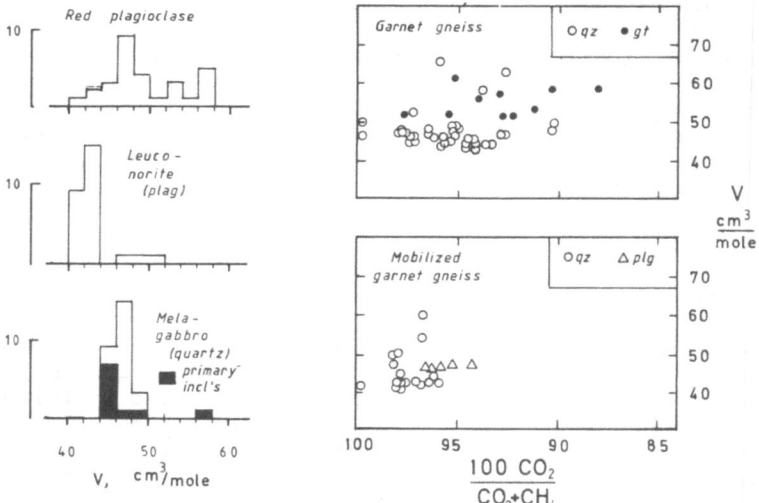

Fig. 7
Composition and molar volume of CO_2 inclusions, as derived from the
microthermometric measurements in Fig. 5. The principles of the
interpretation is discussed in the text.

analytical results.
    Two different trends of fluid evolution are suggested by Fig. 7:
(i): Among the intrusive rocks, there is a shift towards decreasing CO_2
molar volume, from inclusions in melagabbro, to secondary inclusions in
plagioclase from leuconorite.
(ii): Among the country rocks, molar volume varies (V_m in quartz < V_m
in garnet), covering approximately the same range as the total spread
of the intrusive units. More important, the total compositional range,
deduced for the unaltered garnet gneiss (87-100 mole% CO_2) is larger
than that of the mobilized rock (93-100 mole% CO_2).

THERMAL HISTORY

Mineral thermobarometry

In Fig. 8, the pertinent temperature or temperature-pressure intervals
for equilibration of the mineral assemblages of the Angmagssalik
complex is illustrated, using data from Table 2 as discussed above. Of
the mafic silicates, only the high temperature clinopyroxene-cores from
melagabbro retain evidence of the magmatic or early sub-solidus cooling
history of the intrusive rocks. Although emplacement at some depth in
the crust can be assumed from the relationship to the country rocks, no
pressure estimate for this stage of evolution can be made from mineral
thermobarometry.
    The PT estimates from coexisting minerals in the intrusive rocks
and the mobilized garnet gneiss reflect a common part of the PT path of
these units, cooling to ambient temperatures, which may be represented
by 550-700 °C at ca. 5 kbar, i.e. upper amphibolite facies conditions.

P-T interpretation of the fluid inclusion data

Unless disturbed by decrepitation or leakage processes after trapping
(e.g. Roedder 1981), a fluid inclusion behaves as a closed, isochoric
system, which allows the calculation of a PT isochore intersecting the
P and T of trapping for the inclusion, given an appropriate equation of
state for the fluid contained in the inclusion (e.g. Holloway 1981a).
The slope and level of an isochore in a PT diagram is controlled by the
composition and density of the fluid, which can be determined from
microthermometry and Raman spectroscopy. In the present case, isochores
have been calculated for selected combinations of composition and
density derived from the data in Figs. 5, 6 and 7, using the modified
Redlich-Kwong equation of state of Holloway (1977, 1981a), with the
help of a corrected version of Holloway's (1981a) computer program.
    Although the number of fluid inclusions measured is modest, all
units except the anorthosite yield clearly defined fluid density
distributions (Fig. 7), which allows a relatively simple interpretation
of the history of fluid evolution.
    Melagabbro The primary fluid inclusions in melagabbro are
adequately represented by pure CO_2, with molar volume between 44 and 47

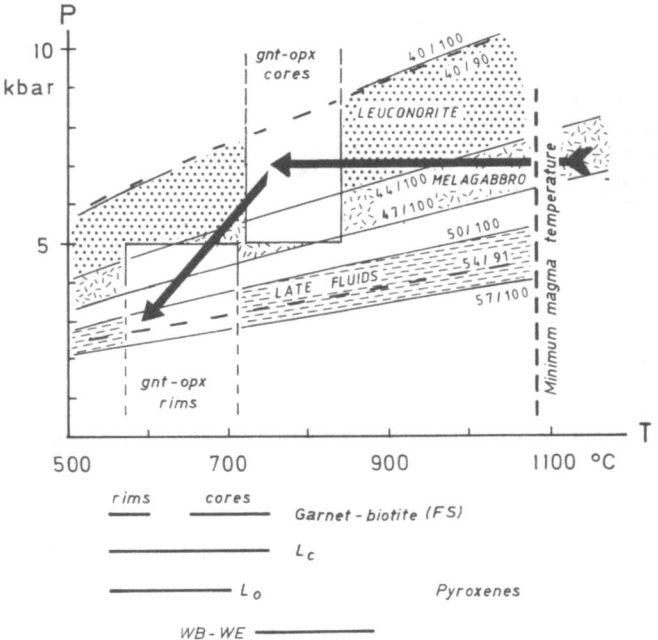

Fig. 8
PT interpretation of the mineralogical and fluid inclusion results. The
PT-"boxes" represent garnet-orthopyroxene equilibria in the mobilized
garnet gneiss, as discussed in the text. Temperature ranges obtained
from garnet-biotite and pyroxene geothermometry (see Table 2 and
discussion in the text) are shown below the temperature axis. The
minimum magma temperature (1085 °C) has been derived from analyses of
clinopyroxene cores in melagabbro. Isochores for CO_2 and pertinent CO_2-
CH_4 mixtures have been calculated by the program of Holloway (1981a).
Molar volumes (cm^3/mole) and composition (mole percent CO_2) are
indicated at the isochores (e.g. 40/100, i.e. 100 % CO_2, 40 cm^3/mole).
The shaded fields represent fluid inclusions in different rock types as
indicated. The mobilized garnet gneiss cover the entire area limited by
the heavy, broken isochores (40/90 and 54/91). A cooling path for the
system is indicated by arrows, as discussed in the text.

cm^3/mole. The corresponding isochores are indicated in Fig. 8. At temperatures above the inferred minimum for the primary crystallization of the melagabbro (1085 ^oC), the isochore sector suggests pressures from 6 to 8 kbar for trapping of fluid inclusions (Fig. 8). The fluid inclusion data thus indicate an essentially isobaric cooling history from magmatic to ambient temperatures.

Leuconorite The denser CO_2 inclusions in leuconorite plagioclase (40-44 cm^3/mole) define an isochore sector adjoining that of melagabbro on its high P - low T side. This sector overlaps with the estimated PT field of reequilibrated opx-cpx pairs in the melagabbro, and with the field of mobilized garnet gneiss.
　　The low-density fluids, (V_m>50 cm^3/mole), mainly observed in the red plagioclase, define a lower-pressure isochore sector than any of the other inclusions. It should, however, be noted that the red plagioclase contains inclusions with a considerable range in CO_2 density; the isochores of the more dense (V_m≤48 cm^3/mole) overlap with the isochore sector of the melagabbro.

Country rocks For two-components CO_2-CH_4 fluids, the isochore slope is a function of both composition and molar volume. For the present range of compositions and densities, the molar volume is, however, the controlling factor. The fluid inclusions in the garnet gneiss and the mobilized garnet gneiss define a broad isochore sector, limited by the fat, broken curves in Fig. 8. This sector spans the range of the CO_2 isochores from the intrusive rocks, and includes the PT-field of equilibration of the mobilized garnet gneiss.
　　The isochore sector of the secondary inclusions in garnet is not separately indicated on the figure, but overlaps roughly with the "late fluids" sector in Fig. 8. This may be interpreted as evidence that the garnet trapped secondary fluid inclusions during a late, low pressure stage of its evolutionary history. Tentatively, this may be equated with the retrograde event recorded by the garnet rim compositions (Fig. 8).

DISCUSSION

Composition and evolution of the fluid phase

The presence of separate solid inclusions of carbonate in the intrusive rocks and the variable abundance of calcite crystals in the fluid inclusions, suggest that carbonate crystals were trapped accidentally, together with the fluid phase. This implies that calcite belongs to the magmatic (and early postmagmatic) mineral assemblage of the mafic melt + fluid system. A similar mode of occurrence for carbonates has been described from the Bamble granulite terrane in S. Norway, where their presence has also been attributed to primary magmatic processes (Touret 1986, 1987).
　　The association of pure CO_2 fluid to the intrusive rocks suggests that the source of the CO_2 was the magmatic system itself. The carbonic

material may thus have been transported from a deeper source region, dissolved in the mafic melt, to exsolve during cooling and crystallization (Holloway 1981b, Frost & Frost 1987).

The systematic decrease in the CO_2/CO_2+CH_4 ratio of the fluid inclusions from the intrusive rocks through the mobilized garnet gneiss into the country rock garnet gneiss, can be an effect of mixing between magmatic CO_2 and more reduced crustal fluids of meta-sedimentary origin, or of reaction between magmatic fluids and more reduced country rocks. The reaction hypothesis requires the presence of free water or hydrous minerals in the country rocks, with which the magmatic CO_2 could interact by the reaction

$$CO_2 + 2 H_2O = CH_4 + O_2$$

The CO_2/CO_2+CH_4 ratio observed would then be a function of oxygen fugacity, water activity and temperature, a high water activity, low oxygen fugacity and/or low temperature favouring a CH_4-rich fluid composition (Holloway 1981a). Which of these mechanisms actually control the fluid composition can only be determined after a more comprehensive study of fluid inclusions and oxygen fugacity conditions in the country rocks during emplacement of the Angmagssalik complex, which is currently in progress.

Anorthosite by anatexis ?

Indirectly, the fluid inclusion data throw some light on the petrogenesis of the anorthositic members of the intrusive series, which were interpreted as the products of local anatexis by Moorlock et al. (1972) and Wright et al. (1973). In order to explain the major and trace element characteristics of the supposed anatectic melts, a water-dominated fluid regime had to be postulated. Although the scatter of the fluid inclusion homogenization temperatures (Figs. 5, 7) precludes any PT interpretation of this unit, the composition of the fluid phase is very well constrained by the present data, as essentially pure CO_2. Although the melt forming the anorthosite was not anhydrous, as shown by the presence of biotite in the veins and in the feldspathic alteration zones in the leuconorite, the absence of water inclusions from this unit apparently contradicts the anatectic hypothesis of Moorlock et al. (1972). More probably, the anorthosite represents the crystallization products of residual liquids or fluids derived from the same magmatic source as the melagabbro and leuconorite.

CONCLUSIONS

The intrusive members of the Angmagssalik complex (melagabbro, leuconorite, anorthosite) penetrated garnet gneiss at amphibolite facies conditions in the middle crust. The mafic intrusions represent a series of compositions including cumulates (melagabbro) and possible melts (leuconorite, anorthosite), probably derived from a common source. As a contact effect induced by the intrusions, opx-bearing ("granulite facies") mineral assemblages developed at the expense of amphibolite parageneses in the garnet gneiss, which was also physically mobilized and intermixed with the intrusive rocks at the contact. The intrusives and the mobilized garnet gneiss record a common thermal history at 600-750 $^\circ$C, 6-8 kbar, with possible retrogression to T<600 $^\circ$C.

Emplacement (at T>1085 $^\circ$C) and crystallization of the intrusives took place in the presence of a CO_2 fluid phase, which coexisited with solid carbonate (calcite) in addition to melt and the major mineral phases of the intrusive rocks. The early cooling history of the intrusives from magmatic temperatures to the 600-750 $^\circ$C sub-solidus equilibration event was reflected by increasing density of the CO_2 fluid, as recorded by fluid inclusions ranging from primary magmatic to secondary. The later, lower T reequilibration of garnet may be reflected in late, secondary fluid inclusions, trapped at lower pressures (e.g. ca. 2-3 kbar, 550 $^\circ$C). No other fluid species than CO_2 have been identified in inclusions in the intrusive rocks.

CO_2 lost from the intrusion during cooling and crystallization, flooded the garnet gneiss, even beyond the zone of mobilization. In this process, the fluid composition changed from pure CO_2 towards increasing CH_4-contents away from the intrusive contact. The maximum CH_4-concentration encountered amounts to ca. 13 mole %. The mechanism responsible for the change in fluid composition could be mixing between compositionally distinct fluids of magmatic and metasedimentary origin, or reaction between magmatic CO_2 and hydrous minerals in the country rock.

Acknowledgements

The field-work in Greenland (Austrheim and Bridgwater) and laboratory work was funded by the Danish Natural Science Research Council, and by NAVF. Permission to publish results on material collected during earlier fieldwork as part of reconnaissance mapping by the Geological Survey of Greenland is acknowledged. Facilities for Raman microprobe analysis were provided by the Free University, Amsterdam, and by WACOM, a working group for the chemical analysis of minerals and rocks, subsidied by the Netherlands Organization for Pure Research (ZWO), through the courtesy of prof. J.L.R Touret. Special thanks are due to E.A.J. Burke, who did the Raman microprobe analyses, and computed the results. Critical comments from K.I. Olsen, A.P. Nutman and F. Mengel are gratefully acknowledged. This publication is contribution No. 28 to the Norwegian program of the International Lithosphere Project.

REFERENCES

Andersen T, O'Reilly SY, Griffin WL (1984) The trapped fluid phase in
upper mantle xenoliths from Victoria, Australia: implications for
mantle metasomatism. Contrib Mineral Petrol 88:72-85
Austrheim H, Bridgwater D, Hansen BT, Pedersen S (1987) Proterozoic
magmatism and mantle-crust interaction in Archaean high grade
gneisses from the Nagssugtoqidian mobile belt of SE Greenland.
Abstr Symp on Proterozoic Geochemistry, IGCP 217 in Lund, June 1987
Bilal A, Touret JLR (1976) Les inclusions fluïdes des encaves cata-
zonals de Bournac (Massif Central). Bull Soc Minéral Cristallogr
99:134-139
Bridgwater D (1976) Nagssugtoqidian mobile belt in East Greenland.
In: Escher A, Watt WS (eds) Geology of Greenland. The Geological
Survey of Greenland, Copenhagen pp 97-103
Bridgwater D, Gormsen K (1968) Precambrian rocks of the Angmagssalik
area, East Greenland Rapp Grønlands geol Unders 15:61-67
Bridgwater D, Windley BF (1973) Anorthosites, post-orogenic granites,
acid volcanic rocks and crustal development in the North Atlantic
shield during the Mid-Proterozoic. Geol Soc S Africa Spec Publ 3:307-
317.
Bridgwater D, Austrheim H, Pedersen S, Winter J (1987) Element
mobility, and isotopic changes during retrogression and prograde
metamorphism of Archean gneisses in the Naqssuqtoquidian mobile belt
of southern East Greenland. Abstracts, Fluid movements, element
transport, and the composition of the deep crust, Nato ARW, Lindås,
Norway.
Burke EAJ, Lustenhouwer WL (1987) The application of multichannel
laser Raman microprobe (Microdil-28) to the analysis of fluid
inclusions. Chem Geol 61:11-17
Burrus RC (1981) Analysis of phase equilibria in C-O-H-S fluid
inclusions. Min Assoc Canada Short Course Handbook 6:39-74
Ferry JM, Burt DM (1982) Characterization of metamorphic fluid
composition through mineral équilibria. Reviews in Mineralogy 10:207-
262
Ferry JM, Spear FS (1978) Experimental calibration of the partitioning
of Fe and Mg between biotite and garnet. Contrib Mineral Petrol
66:113-117.
Field D, Drury SA, Cooper DC (1980) Rear-earth and LIL element
fractionation in high-grade charnockitic gneisses, southern Norway.
Lithos 13:281-289
Frost BR, Frost CD (1987) CO_2, melts and granulite metamorphism.
Nature 327:503-506
Glassley WE (1983) The role of CO_2 in the chemical modification of the
deep continental crust. Geochim Cosmochim Acta 47:597-616
Glassley WE, Bridgwater D, Konnerup-Madsen J (1984) Nitrogen in fluids
effecting regression of granulite facies gneisses: a debatable mantle
connection. Earth Planet Sci Lett 70:417-425
Green DH, Ringwood AE (1967) The genesis of basaltic magmas. Contrib
Mineral Petrol 15:103-190
Harley LS (1984) An experimental study of the partitioning of Fe and

Mg between garnet and orthopyroxene. Contrib Mineral Petrol 86:359-373

Heyen G, Ramboz C, Dubessy J (1982) Simulation des équilibres de phases dans le système CO_2-CH_4 en dessous de 50 °C et de 100 bar. C R Acad Sc Paris Ser II 294:203-206

Hoefs J, Touret JLR (1975) Fluid inclusion and carbon isotope studies from Bamble granulite, southern Norway. A preliminary investigation. Contrib Mineral Petrol 52:165-174

Holloway JR (1977) Fugacity and activity of molecular species in supercritical fluids. In Frazer DG (ed) Thermodynamics in Geology. D Reidel Publ, Dordrecht, pp 161-181

Holloway JR (1981a) Compositions and volumes of supercritical fluids in the earth's crust. Min Assoc Canada Short Course Handbook 6:13-38

Holloway JR (1981b) Volatile interaction in magmas. In Newton RC, Navrotsky A, Wood BJ (eds) Thermodynamcis of minerals and melts. Springer Verlag, New York, pp 273-293

Hollister LS, Burrus RC (1976) Phase equilibria in fluid inclusions from the Khtada Lake metamorphic complex. Geochim Cosmochim Acta 40:163-175

Kreulen R, Schuiling RD (1982) N_2-CH_4-CO_2 fluids during formation of the Dôme de l'Agout, France. Geochim Cosmochim Acta 46:193-203

Lindsley DH (1983) Pyroxene thermometry. American Mineralogist 68:477-493

Moorlock BSP, Tarney J, Wright AE (1972) K-Rb ratios of intrusive anorthosite veins from Angmagssalik, East Greenland. Earth Planet Sci Lett 14:39-46

Olsen KI (1978) Metamorphic petrology and fluid inclusion studies of granulites and amphibolite facies gneisses on Langøy and W-Hinnøy, Vesterålen, North Norway. Unpublished thesis, University of Oslo.

Pedersen S, Bridgwater D (1979) Isotopic re-equilibration of Rb-Sr whole rock systems during reworking of Archaean gneisses in the Nagssugtoqidian mobile belt, East Greenland. Rapp Grønlands geol Unders 89:133-146

Roedder E (1981) Origin of fluid inclusions and changes that occur after trapping. Min Assoc Canada Short Course Handbook 6:101-137

Roedder E (1984) Fluid inclusions. Reviews in Mineralogy 12:644pp

Shepherd TJ, Rankin AH, Alderton DHM (1985) A practical guide to fluid inclusions studies. Blackie, Glasgow, 239pp

Smalley PC, Field D, Lamb RC, Clough PWL (1983) Rare earth, Th-Hf-Ta and large ion lithophile element variations in metabasites from the Proterozoic amphibolite-granulite transition zone at Arendal, South Norway. Earth Planet Sci Lett 63:446-458

Swanenberg H (1980) Fluid inclusions in high-grade metamorphic rocks from S.W. Norway. Geologica Ultraiectina, University of Utrecht 25:146 pp

Touret JLR (1971) Le facies granulite en Norvège méridionale, II: les inclusions fluides. Lithos 4:423-436

Touret JLR (1977) The significance of fluid inclusions in metamorphic rocks. In Frazer DG (ed) Thermodynamics in Geology. D Reidel Publ, Dordrecht, pp 203-227

Touret JLR (1981) Fluid inclusions in high grade metamorphic rocks.

Min Assoc Canada Short Course Handbook 6:182-208
Touret JLR (1986) Fluid inclusions in rocks from the lower continental
crust. In Dawson JB, Carswell DA, Hall J, Wedepohl KH (eds) The
nature of the lower continental crust. Geol Soc Spec Pub 24:161-172
Touret JLR (1987) Metamorphic fluids: data from fluid inclusions. In
Helgeson HC (ed) Proceedings of a NATO Advanced Study Institute on
Chemical Transport in Metasomatic Processes. In press.
Touret JLR, Hansteen TH (1987) Geothermobarometry and fluid inclusions
in a rock from the Doddabetta charnockite, Southwest India. Rend Soc
It Min Pet, in press
Touret JLR, v.d.Kerkhof AM (1986) High density fluids in the lower
crust and upper mantle. Physica 139&140:834-840
Wager LR (1934) Geological investigations in East Greenland. Part I.
General geology from Angmagsalik to Kap Dalton. Meddr Grønland
105,2:44 pp
Wager LR, Brown GM, Wadsworth WJ (1960) Types of igneous cumulates. J
Petrol 1:73-86
Wells PRA (1977) Pyroxene thermometry in simple and complex systems.
Contrib Mineral Petrol 62:129-139
Wood BJ (1974) Solubility of alumina in orthopyroxene coexisting with
garnet. Contrib Mineral Petrol 46:1-15
Wood BJ, Banno S (1973) Garnet-orthopyroxene and orthopyroxene-
clinopyroxene relationships in simple and complex systems. Contrib
Mineral Petrol 42:109-124
Wright AE, Tarney J, Palmer KF, Moorlock BSP, Skinner AC (1973) The
geology of the Angmagssalik area, East Greenland and possible
relationships with the Lewisian of Scotland. In Park RG, Tarney J
(eds) The early Precambrian of Scotland and related rocks of
Greenland. University of Keele, pp 157-177
Yoder HS, Tilley CE (1962) Origin of basalt magmas: An experimental
study of natural and synthetic rock systems. J Petrol 3:342-532.

PROCESSES OF FORMATION AND RETROGRESSION OF SCOURIAN GRANULITES

I. Cartwright
Department of Geology and Geophysics
Weeks Hall, University of Wisconsin
Madison WI 53706
USA.

ABSTRACT.   Partial melting of Scourian gneisses during the 2.7Ga
Badcallian granulite facies metamorphism produced H_2O undersaturated
granitic to trondhjemitic melts. During cooling, melts which remained in
the complex crystallized and the residual liquids became richer in H_2O
until water-rich fluids were exsolved at 620-700°C. These fluids
probably caused the Inverian retrogression which commenced at around
650°C. This proposition is supported by the correlation of areas which
contain abundant neosomes and regions which underwent the highest
degrees of retrogression. Pegmatites representing H_2O-rich residual
liquids were intruded during the later stages of cooling as predicted by
viscosity and separation velocity calculations. During prograde
metamorphism, most volatiles probably were located in the mineral
phases. Anatexis at the peak of metamorphism transferred volatiles
(especially H_2O) into the melts. On cooling, fluids were formed which
rehydrated the minerals.

INTRODUCTION

The mainland Lewisian of NW Scotland comprises two distinct metamorphic
terrains. The rocks of the Scourian complex preserve mainly Archaean
metamorphic parageneses and tectonic fabrics; by contrast, the
Laxfordian areas underwent extensive reworking in the Proterozoic. The
Scourian complex is dominated by tonalitic gneisses (75-80% of the total
outcrop) with subordinate volumes of basic to ultramafic orthogneisses
and metasediments. Many of these lithologies are cross-cut by
leucocratic granitic to trondhjemitic sheets and veins.

Abbreviations used in text and figures:
ab = $NaAlSi_3O_8$; or = $KAlSi_3O_8$.
ksp = or-rich alkali feldspar; liq = silicate liquid; plg = plagioclase
feldspar; qtz = quartz; vap = H_2O-dominated fluid.

D. Bridgwater (ed.), Fluid Movements – Element Transport and the Composition of the Deep Crust, 95–110.
© 1989 by Kluwer Academic Publishers.

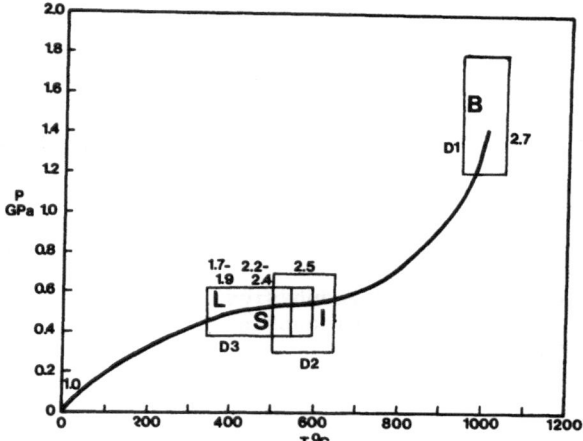

Fig. 1. P-T-time path for the Scourian complex constructed using the data in table 1. B = Badcallian metamorphism, I = Inverian retrogression, S = intrusion of Scourie dykes, L = Laxfordian retrogression; D1-D3 = major deformation events. Data from table 1.

Table 1. Geological history of the Scourian complex

	Ga	Event	T (°C)	P(GPa)
(a)	2.95	Formation of protoliths gneisses		
(b)	2.7	Badcallian granulite facies metamorphism, deformation and anatexis	1000±50	>1.2
(c)	2.55-2.45	Inverian retregrossion, shear zones and intrusion of Scourian pegmatites	650-500	0.3-0.7
(d)	2.39-2.20	Intrusion of the Scourie dyke swarm	450-550	0.4-0.6
(e)	1.9 -1.7	Laxfordian deformation and retrogression	350-600	0.4-0.6
(f)	1.0	Exposiure of complex at surface		

Geochronology: (a) Hamilton et al. (1979), Whitehouse and Moorbath(1986); (b) Pidgeon and Bowes (1972), Chapman and Moorbath (1977), Humphries and Cliff (1982); (c) Giletti et al. (1961), Evans(1965), Evans and Lambert (1974); (d) Evans and Tarney (1964), Chapman (1979); (e) Moorbath and Park (1972); (f) Moorbath (1969) P-T estimates: from summaries in O'Hara (1977), Barnicoat (1983, table 1), Barnicoat et al. (in press, table 1), Cartwright and Barnicoat (1987, table 3), Sills and Rollinson (1987, table 1, fig. 4).

GEOLOGICAL HISTORY OF THE SCOURIAN COMPLEX

The geological history of the Scourian complex is summarised in
table 1 and fig. 1. The protoliths to the gneisses were formed at 2.95Ga
and underwent granulite facies metamorphism accompanied by anatexis and
intense subhorizontal deformation in the 2.7Ga Badcallian tectono-
metamorphic episode. Peak metamorphic conditions throughout the Scourian
complex are estimated at 1000±50°C and >1.2GPa (table 1). In the 2.5Ga
Inverian event the Scourian complex underwent locally intense
retrogression penecontemporaneously with the formation of numerous
dipslip shear zones and the intrusion of a suite of pegmatites (Sills,
1983; Cartwright, in press). Inverian conditions are estimated as being
500-650°C and 0.3-0.7GPa; this implies extremely slow cooling and
excavation rates for this period of metamorphism of 2-3°C Ma^{-1} and
0.1-0.2km Ma^{-1} (Barnicoat, 1983, 1987; Cartwright in press; Cartwright
and Barnicoat, 1987 and in prep; Sills and Rollinson, 1987). The Sm-Nd
study of Humphries and Cliff (1982) indicates that, between 2.7 and
2.5Ga, the complex cooled from 1000°C to 600°C without intervening
metamorphic events.

The Scourian complex was intruded by the Scourie dykes at 2.4-2.2Ga
and underwent minor retrogression and shearing in the 1.9-1.7Ga
Laxfordian episode. A period of steady recovery followed the Laxfordian
and the complex was exposed at the subTorridonian land surface at around
1.0Ga. This paper discusses the formation of granulite facies
assemblages during the Badcallian event and their subsequent
retrogression in the Inverian episode.

PROGRADE METAMORPHISM - FORMATION OF GRANULITES AND ANATEXIS

At the peak of Badcallian metamorphism, most of the Scourian gneisses
developed granulite-facies mineralogies containing few hydrous minerals
(Rollinson and Windley, 1980, table 1; Barnicoat, 1983, table 3).
However, from geochemical constraints and comparison with similar
lithologies in analogous amphibolite facies terrains, the precursors to
the majority of the gneisses almost certainly contained hydrous minerals
(Sheraton et al., 1973; Rollinson and Windley, 1980). The Scourian
complex also shows extreme depletion in K, Rb, Th and U with respect to
comparable lower grade Archaean terrains (Weaver and Tarney, 1981).
Tarney and Windley (1977) proposed that the formation of the granulite
parageneses and the elemental depletion was brought about by the passage
of CO_2-rich fluids through the crust. However, several Scourian
lithologies contain fO_2-buffering granulite facies assemblages
(Rollinson, 1980). If a significant volume of CO_2-rich fluid had passed
through the complex, it would have overcome the fO_2 buffers and
precipitated graphite, which is not observed (Lamb and Valley, 1984).
Additionally, certain Badcallian metabasic, ultramafic and supracrustal
lithologies contain aH_2O-buffering assemblages (Barnicoat, 1983;
Cartwright and Barnicoat, 1986, 1987) which would also have been
overcome if large volumes of externally buffered fluid passed through
the crust at that time. Hence, it is considered more plausible that

these granulites are the restites of anatexis (c.f. Fyfe, 1973).

Anatexis in the Scourian complex

At the conditions of Badcallian metamorphism (1000°C and >1.2GPa), many
lithologies would have undergone anatexis at moderate aH_2O values even
in the absence of a fluid phase. For example, a fluid-absent
metatonalite which contains amphibole and biotite will commence melting
at approximately 800°C, while an amphibole-bearing fluid-absent
metabasic gneiss will start to melt at around 1000°C (Wyllie, 1983, fig.
7).

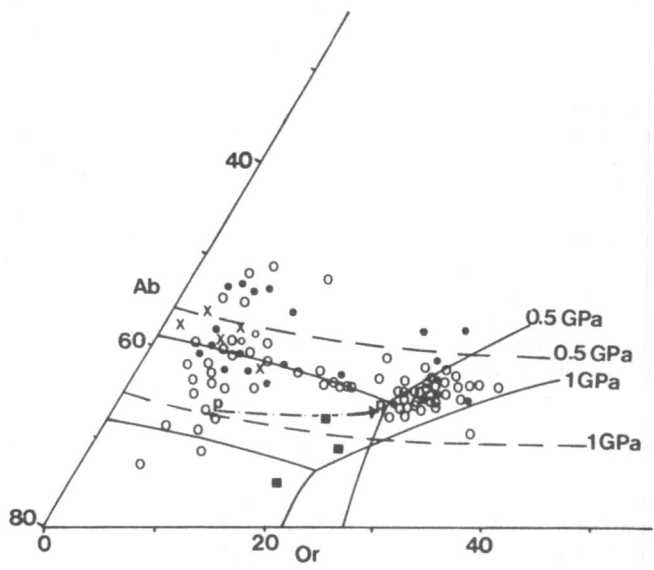

Fig. 2. Composition of veins and segregations from Scourian
para and orthogneisses projected into the qtz-ab-or system.
Composition of leucogneiss sheets in tonalites from Rollinson
and Windley (1980 open circles) and Cartwright and Barnicoat
(1987 - filled circles); Composition of veins in the basic
gneisses (crosses) and metasediments (squares) from XRF
analyses. P-P is the crystallization path discussed in the
text. Solid lines are the water-saturated cotectics, broken
lines are the anhydrous equivalents; data from Luth (1969) and
Huang and Wyllie (1975). The composition of the veins is
consistent with them being the products of local anatexis in
the mid crust.

Granitic to trondhjemitic veins and segregations which vary from a
few millimetres to several centimetres in width are present in Scourian
metabasic gneisses (Barnicoat, 1983; Cartwright and Barnicoat, 1987) and
metasediments (Cartwright and Barnicoat, 1986, 1987). The tonalitic

gneisses also contain granitic to trondhjemitic sheets and veins, the acid leucogneisses, which range up to hundreds of metres in thickness (Rollinson and Windley, 1980; Cartwright et al., 1985).

Rollinson and Windley (1980) and O'Hara and Yarwood (1978) proposed that the acid leucogneisses were emplaced close to the peak of the Badcallian metamorphism. The leucogneisses and the acidic veins in the metasediments and basic gneisses locally cut the layering of the host lithologies (Cartwright and Barnicoat, 1986, fig. 2, 1987, fig. 5) but are folded by flat-lying Badcallian folds. This is consistent with all of these sets of veins being formed during the Badcallian metamorphism. The compositions of the acidic veins lie close to the cotectic surfaces in the qtz-ab-or system (fig. 2). The veins within individual lithologies are largely confined to that lithology, with the exception of a small number of veins which transect the boundaries between rock types. These data are most consistent with the suites of granitic and trondhjemitic veins being neosomes formed by local anatexis (Sheraton et al., 1973; O'Hara and Yarwood, 1978; Pride and Meucke, 1982; Barnicoat, 1983; Cartwright et al., 1985; Cartwright and Barnicoat, 1986, 1987; Cartwright, in press).

The remaining discussion will focus on the leucogneisses, as these are volumetrically the most important of all the suites of veins.

Initial water content of the leucogneisses

The minimum initial water content of the leucogneiss melts can be constrained by the available experimental data. At the peak of metamorphism ($1000 \pm 50 °C$ and >1.2GPa) granitic melts must have contained at least 2wt% H_2O (Johannes, 1985, fig. 2.3). The maximum water content of a granitic melt under these conditions is 13wt% (Whitney, 1975; Luth, 1976). However, lithologies in the lower crust probably have low fluid:rock ratios, or may be fluid absent; anatexis of such rocks will almost certainly form waterundersaturated liquids (Powell, 1983b). It is likely that most melts in the Lewisian complex had low initial H_2O contents (O'Hara and Yarwood, 1978; Rollinson, 1982; Rollinson and Windley, 1980; Cartwright in press).

Even if melts with high initial H_2O contents formed, it is unlikely that they would have remained within the complex during the slow ($2-3°C$ Ma^{-1}) cooling from the peak of metamorphism. At $1000°C$, a melt with the composition of the granitic leucogneisses (Cartwright, in press, table 1) which contains 2wt% H_2O has a viscosity of $10^{6.5}-10^7$ Pa s (calculated using the method of Shaw, 1972). Similar melts with water contents of 13wt% have viscosities of $10^2-10^{2.5}$ Pa s. McKenzie (1985) shows that a melt with a viscosity of 10^2 Pa s will escape from the crust 2-3 orders of magnitude faster than melts with viscosities of around 10^7 Pa s.

INVERIAN RETROGRESSION

The Scourian complex underwent local retrogression during the 2.5Ga Inverian tectono-metamorphic event which involved the replacement of the pyroxene-bearing Badcallian granulites by amphibole-bearing assemblages and a reduction in the anorthite content of plagioclase; supracrustal lithologies also underwent mineralogical changes (Beach, 1974; Cartwright and Barnicoat, 1986). The degree of alteration of the Badcallian assemblages varies markedly; certain areas, such as around Stoer, south of Loch Laxford, and near Gruinard Bay (see fig. 4), show extensive replacement of the granulite parageneses. By contrast, the Scourie area and the region south of Lochinver contain virtually

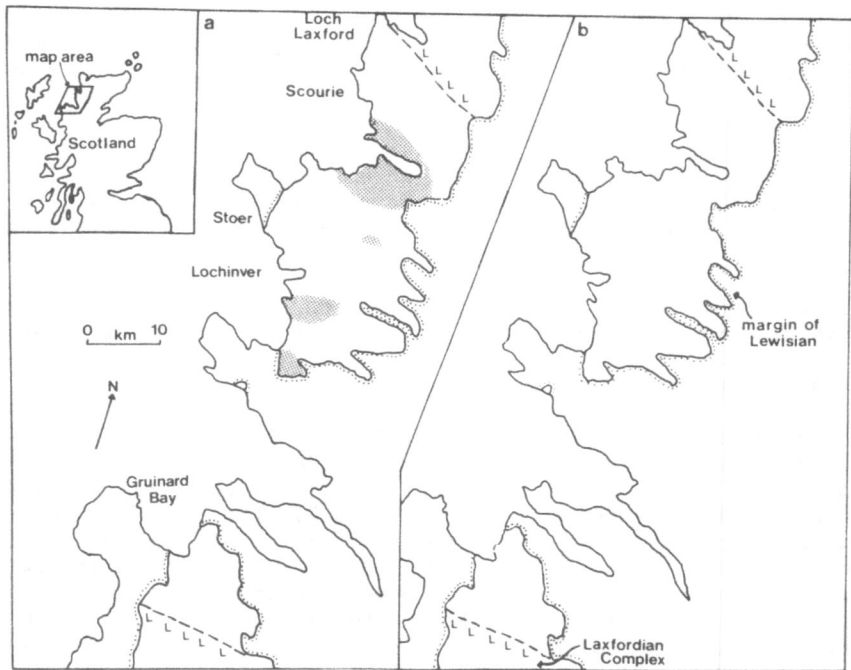

Fig. 3. Comparative maps of the Scourian complex showing: a) The areas which preserve largely unaltered granulite parageneses (heavy stipple) and regions which underwent intense Inverian retrogression (light stipple) where relic granulites are only rarely observed; the unornamented Scourian areas show partial alteration of the granulite facies assemblages. b) The location of areas rich in acid leucogneisses (light stipple). The correlation between the areas which underwent intense Inverian alteration and the regions which contain abundant leucogneisses supports the proposition that the fluids exsolved from the leucogneiss melts during cooling caused the retrogression. Data from mapping in the Stoer, Scourie-Laxford and Gruinard Bay areas and Peach et al. (1907); Bowes (1969); Holland and Lambert (1973); Park (1973); Sheraton et al. (1973); Beach (1974); Evans and Lambert (1974); Davies (1977); Sills (1983); Cartwright (1986); Cartwright et al. (1985).

unaltered granulite facies mineralogies with retrogression confined
to Inverian shear zones.

Crystallization of melts and the production of fluids

Figure 4 is a pseudobinary T:wt% H_2O section in the K_2O-Na_2O-CaO-Al_2O_3-
SiO_2-H_2O (KNCASH) haplogranite system at 0.8GPa (0.8GPa was chosen as
experimental data exists at this pressure). This phase diagram may be
used to describe crystallization sequences of granitic liquids. Melt I
has an initial water content of 3 wt% and, when cooled from 1000°C, will
undergo fractional crystallization. When composition I entersthe plg+
ksp+qtz+liq+vap field in fig. 3, between T_a and T_b (630-620°C), a fluid
phase is exsolved. This is due to the initially water undersaturated
residual liquid becoming progressively richer in H_2O as the anhydrous
phases in this system crystallize, until water saturation occurs. 12wt%
H_2O is needed to saturate a granitic melt at 630-620°C and 0.8GPa

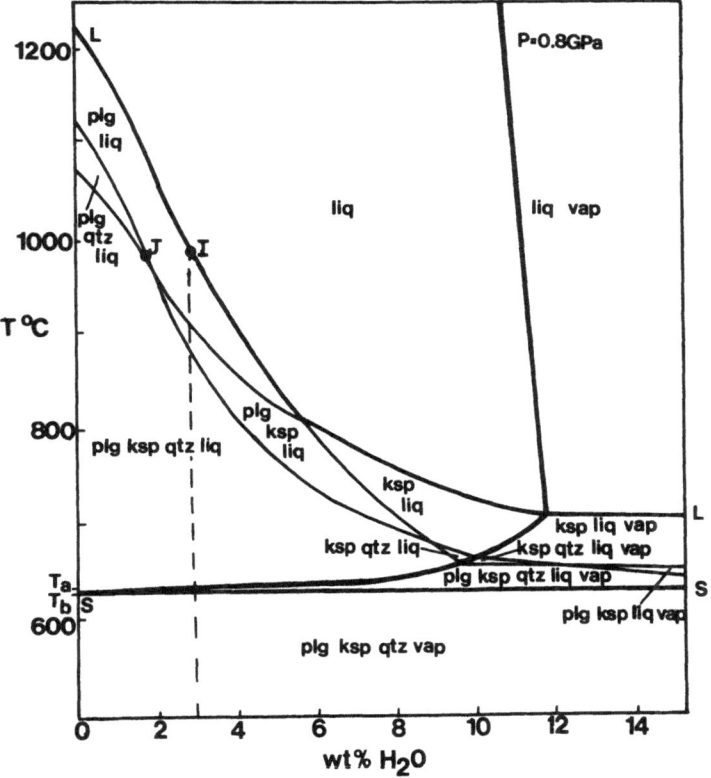

Fig. 4. Pseudobinary T-wt%H_2O diagram showing phase relations
between plg, ksp, qtz, liq and vap at 0.8GPa, data from Whitney
(1975, fig.4) and Luth (1976, fig. 23). Bold line indicates
limit of water saturation, I and J are compositions discussed
in the text.

(Whitney, 1975); to increase the water content from 3wt% to 12wt% requires that 75% of the original liquid crystallizes.

The data of fig. 4 imply that, unless the melt is originally close to being water saturated, a fluid will not be exsolved until temperatures have fallen below 700°C. This is because, although melts with higher initial H_2O contents achieve water saturation after lower degrees of crystallization, such melts will not commence crystallizing until lower temperatures are attained. Under conditions of polybaric cooling, water saturation does not occur at higher temperatures unless the melts are initially close to being saturated with H_2O (Cartwright, in press). As discussed above, it is unlikely that liquids with high water contents would have been present in the Scourian complex during cooling and melts would have probably exsolved fluids late in their crystallization history.

Crystallization of trondhjemitic melts

During crystallization, potassium-free trondhjemites achieve water saturation at comparable temperatures to granitic melts. Whitney (1975) shows that at 0.8GPa a crystallizing trondhjemite melt which initially contained <10wt% H_2O would exsolve a fluid at 650-700°C.

A melt which is intermediate in composition between a granite and trondhjemite (as many of the leucogneisses are) would be initially located along the cotectic curves in the ab-or-qtz system (fig. 2). As crystallization of quartz and plagioclase occurs, the residual liquid becomes progressively more potassic and water rich. The Scourian complex underwent cooling and uplift from 1000°C and >1.2GPa to 620°C at around 0.6GPa (fig. 1). If a leucogneiss melt initially projected as or_{10} ab_{66} qtz_{24}, the residual liquid would have reached eutectic composition (or_{22} ab_{52} qtz_{26} at 0.6GPa) after ca.45% of the melt had crystallized (path p-p in fig. 2). The maximum H_2O content of the melt at 620°C and 0.6GPa is 11-12wt%; hence, if the melt had an initial water content of <6wt%, water-saturation would have occurred after >45% of the melt had crystallized and the residual liquid was of eutectic composition.

Effects of additional phases

In the model described above, a fluid is exsolved because there are no hydrous phases crystallizing. In addition to the phases in the haplogranite system, the leucogneisses contain mafic minerals (amphibole, biotite, garnet and pyroxene) and muscovite. Crystallization of amphibole or mica would have consumed H_2O from the melt. However, the leucogneisses generally contain <10% of these additional minerals and, even if 10% mica had crystallized from a melt which initially contained 2wt% H_2O, >1.6wt% H_2O would have been exsolved. Crystallization of amphibole, pyroxene, garnet or H_2O-undersaturated micas (such as are found in high grade terrains) will result in proportionally more fluid being released.

Melts as a source of fluids for the Inverian retrogression

The data discussed above predict that the melts formed in the Scourian complex at the peak of metamorphism became progressively more water rich during cooling until they exsolved in excess of 2wt% fluid at around 620-700°C. Due to the limited solubility of CO_2 in crustal acidic melts, these fluids would be H_2O rich. From the P-T-time path of fig. 1, it is evident that the Scourian complex reached temperatures of 620-700°C at 2.55-2.45Ga. As shown in table 1, the Inverian retrogression took place at this time, and it is proposed that the fluids exsolved from melts during cooling were the cause of this retrogression.

This proposition is supported by the correlation of areas which contain abundant leucogneisses and the regions which underwent the highest degrees of Inverian retrogression. Figure 3 shows that the areas that display the strongest Inverian overprint (notably near Stoer, south of Loch Laxford, and near Gruinard Bay) are those which contain abundant leucogneisses. By contrast, the gneisses in the Scourie and Lochinver areas contain relatively few leucogneiss sheets and preserve extensive tracts of granulite facies lithologies. The correlation between melts and retrogression is also observed on a small scale; granulite facies gneisses adjacent to small leucogneiss veins show replacement of clino- and orthopyroxene by amphibole (Cartwright, in press, fig. 6c).

A typical Scourian granulite facies tonalite contains around 20% pyroxene and hence requires only 0.4wt% H_2O to convert the pyroxene to amphibole. If a leucogneiss melt exsolved 2wt% water during crystallization, an area which contains >20% leucogneisses would have generated sufficient fluid to have hydrated the surrounding tonalites. More mafic rocks, such as the metabasic gneisses, contained around 50% pyroxene at the peak of metamorphism and so would have required proportionally more fluid to cause amphibolitization. However, such lithologies generally comprise <10% of the total outcrop. In the areas of intense Inverian retrogression the leucogneisses comprise at least 20% of the outcrop and in certain parts of these regions they are even more abundant. Hence, there are sufficient leucogneisses in these areas to have provided the fluids for the retrogression.

Intense retrogression is restricted to the leucogneiss-rich areas since the volumes of fluid liberated during cooling would have been insufficient to hydrate large volumes of overlying leucogneiss-poor terrain. Fluids intersecting the Inverian shears (which were active during retrogression, table 1) would have been channelled to higher crustal levels. This probably caused the limited retrogression associated with shear zones in otherwise pristine granulites (e.g. Beach, 1980).

Intrusion of pegmatites

Granitic pegmatites were intruded at 2.6-2.4Ga (table 1) penecon- temporaneously with the Inverian retrogression. These pegmatites probably represent the water-rich granitic liquids formed during the later stages of crystallization of the leucogneisses. Observations of a leucogneiss sheet at Scourie being the source of a later pegmatite

(Cartwright, in press, fig. 6) support this proposition.

In a terrain where melts are cooling and crystallizing, the timing of the escape of residual liquids is controlled by several factors. The increase in the H_2O content of the residual liquid as crystallization proceeds decreases its viscosity making it more likely to escape from

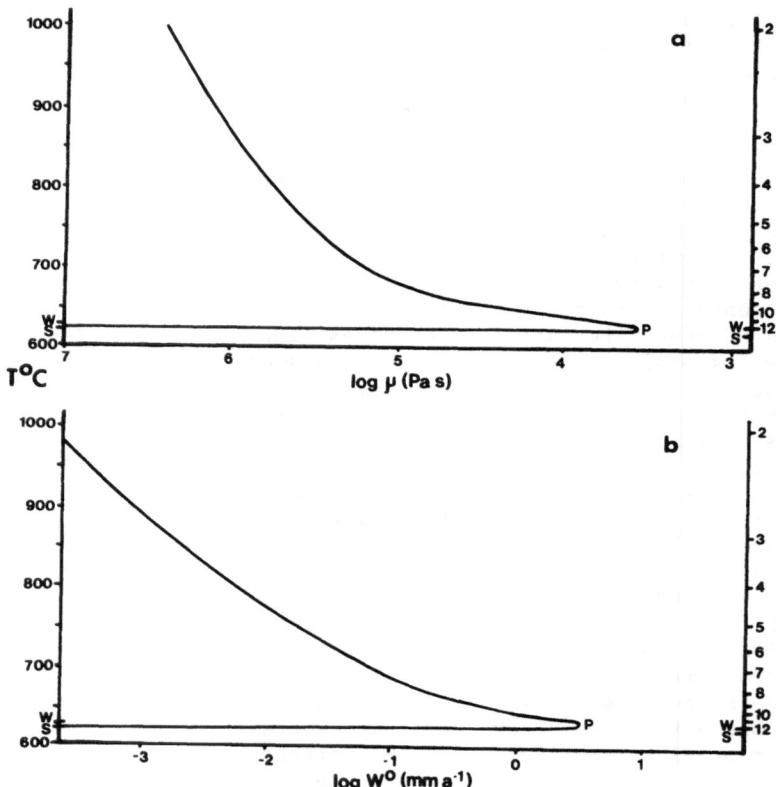

Fig. 5. Variation in: a) viscosity (m) and b) separation velocity (W^o) in a residual granitic liquid during cooling. W^o is the steady state velocity of an upwardly percolating liquid relative to its matrix which may be used as a measure of the rate of liquid extraction. Water content of the residual liquid given assuming the crystallization model outlined in the text; H_2O saturation occurs at 630°C (S) and crystallization is complete at 620°C. a) calculated using the method of Shaw (1972) and the composition of a Scourian pegmatite (Pride and Muecke, 1982, table 1, analysis 65-160). b) from the equations of McKenzie (1985) using constants listed in McKenzie (1985, table 1), a grain size of 1mm and m from (a). The conditions that residual liquid are most likely to separate and intrude as pegmatites are at P.

its matrix. However, this trend is counteracted by the falling
temperature and the decreasing volume of residual liquid (McKenzie
1985).

Figure 5 shows the changes in viscosity (μ) and separation velocity
(W^{o}) in the residual liquid as a granitic melt with an initial water
content of 2wt% crystallizes (composition J in fig. 4). In these
calculations it has been assumed that crystallization commences at
1000°C, water saturation occurs at 12wt% H_2O, and crystallization
terminates at the solidus at 620°C. Melt J undergoes initial
crystallization in the plg+ksp+qtz+liq field in fig. 4, and the water
content of the residual liquid in this composition is given by the line
bounding this field.

The minimum viscosity and maximum separation velocity occur just as
the crystallizing melt becomes water saturated at around 630°C when
16-17% of residual liquid containing 12wt% H_2O remains (P in fig. 5).
The maximum separation velocity of this model composition is around
4mm a^{-1} (4000m Ma^{-1}); if the melt had a higher initial water content,
the maximum separation velocity would occur at the same temperature but
would be greater due to a larger proportion of residual liquid remaining
when water saturation occurs. The most likely time for residual liquids
to separate and intrude as pegmatites is at temperatures below 700°C
when the separation velocity increases rapidly to >1mm a^{-1}. It would
take 10^4-10^6 years for almost all of a batch of residual loquid with
separation velocities of 0.1-4mm a^{-1} to escape from a crystallizing
sheet of 10-100m thickness (c.f. McKenzie, 1985). At a cooling rate of
2-3-C Ma^{-1}, it would have taken the Scourian complex >10^6 years to cool
from 700 to 630°C which is ample time for the liquids to separate and
form pegmatites.

These calculations assume that the liquid percolates upward through
a sheet which is undergoing compaction. Deformation may cause the melts
to aggregate and separate at an earlier stage. However, as shown in
table 1 and fig. 1, no major deformational episodes occurred between the
Badcallian and Inverian events and it is considered likely that the
residual liquids escaped from their matrices during the later stages of
crystallization. As discussed above the residual liquids at these
conditions would be granitic even if the original melts were potassium
poor. The Scourian pegmatites were intruded at 2.55-2.45Ga (i.e. close
to the time that the residual liquids in the crystallizing leucogneisses
became water saturated) and are granitic which is consistent with these
data.

DISCUSSION

Partial melting in the Scourian complex occurred during prograde
metamorphism and the subsequent crystallization of melts resulted in the
production of fluids. The precursors to the majority of the Scourian
gneisses are likely to have been plutonic igneous rocks which contained
hydrous minerals but no carbonates (Holland and Lambert, 1975; Windley
and Smith, 1976). These lithologies may have also contained a small
amount of a fluid phase or may have been locally fluid absent. In such

rocks, the majority of volatiles are located in the mineral phases and the activity of volatile components, in both fluidpresent and fluid-absent assemblages, will be internally buffered throughout metamorphism (Greenwood, 1975; Powell, 1983a,b; Cartwright and Barnicoat, 1987). If aH_2O remains below unity, most reaction in fluid-present assemblages, and all finite amounts of reaction in fluid-absent parageneses, will occur at H_2O-conserving equilibria. Under amphibolite and granulite facies conditions, the majority of H_2O-conserving equilibria are melting reactions which transfer H_2O directly from the mineral phases into the melts (Powell, 1983a,b; Cartwright and Barnicoat, 1987). By contrast, the amount of reaction that occurs at fluid-present dehydration equilibria is small and it is unlikely that prograde metamorphism of the Scourian gneisses resulted in the production of large volumes of fluid.

During the Inverian retrogression, a free fluid phase was exsolved and fluid-present reactions were probably more important. The Inverian fluids would eventually have been consumed by the formation of hydrous phases in the rocks through which they passed. Thus, the location of the bulk of the volatile components changed during metamorphism from the mineral phases to the melts and eventually back to the minerals. If lithologies were fluid absent at the onset of metamorphism, then they probably remained fluid absent at least until the Inverian retrogression when they may have transiently contained a fluid phase. Certain lithologies which preserve granulite facies mineralogies may never have been fluid bearing.

It is proposed that the Scourian complex did not experience the introduction of large volumes of fluids from external sources (e.g the mantle or underlying crust) at any stage of metamorphism. The formation and retrogression of granulites is envisaged in terms of changing the location of volatile components. The total amount of volatiles in the complex at any stage of metamorphism is likely to have been little more than could be accommodated in the mineral phases.

The preservation of granulite facies parageneses in high grade terrains requires that H_2O is removed from the system; this is efficiently accomplished by anatexis and escape of the melts during metamorphism. Thus, regions which preserve granulite facies mineralogies will generally contain few neosomes. Many high grade terrains (e.g. the Adirondacks and Ivrea zone) have extensive migmatite regions on the low grade side of the amphibolite-granulite transition zone. The migmatites are locally intrusive and may, in part, be formed by melts escaping from the underlying granulites. In the Scourian complex, the Gruinard Bay area contains a higher volume of leucogneisses than can be accounted for by local partial melting, and possibly represents a deeper region where melts collected.

Many other granulite facies terrains may show similar stages of evolution to the Scourian complex. The crystallization of melts was proposed as the cause for the regional retrogression of the Wilyama complex, Australia by Corbett and Phillips (1981). The North Atlantic craton as a whole also contains neosomes which may have provided fluids for retrogression.

ACKNOWLEDGMENTS

I wish to thank Prof M.J. O'Hara, Dr A.C. Barnicoat and Dr W.R. Fitches
for initiating this study and their advice throughout. Dr J.W. Valley's
careful review helped improve the text. Drs P.K. Harvey and B.P. Atkin
of Nottingham University, U.K helped with the XRF analyses. T.R. Weaver
helped proof the text. I acknowledge receipt of a University of Wales
studentship and a NERC postdoctoral fellowship.

REFERENCES

Barnicoat, A.C, 1983. Metamorphism of the Scourian complex, north-west
    Scotland. *J. metamorphic Geol.*, 1, 163-183.
Barnicoat, A.C, 1987. The causes of the high-grade metamorphism of the
    Scourian complex, NW Scotland. In: Park, R.G and Tarney, J. (eds),
    Evolution of the Lewisian and comparable Precambrian high grade
    terrains, *Geol. Soc. Lond. Spec. Pub.*, 27, 73-79.
Barnicoat, A.C., Cartwright, I. and O'Hara, M.J., in press. Kyanite in
    the mainland Lewisian complex. *Scott. J. Geol.*
Beach, A., 1974. Amphibolitization of Scottish granulites. *Scott. J.
    Geol.*, 10, 35-43.
Beach, A., 1980. Retrogressive metamorphic processes in shear zones with
    special reference to the Lewisian complex. *J. Struct. Geol.* 2,
    257-263.
Bowes, D.R., 1969. The Lewisian of the NW Highlands of Scotland. In:
    Kay, M. (ed), North Atlantic - geology and continental drift - a
    symposium. *Mem. Am. Ass. Pet. Geol.* 579-594.
Cartwright, I., 1986. The geological history of the Lewisian complex at
    Stoer, NW Scotland. *Unpbl. Ph.D Thesis, Univ. of Wales.*
Cartwright, I., in press. Crystallization of melts, pegmatite intrusion
    and the Inverian retrogression of the Scourian complex, north-west
    Scotland. *J. metamorphic Geol.*
Cartwright, I., and Barnicoat, A.C., 1986. The generation of
    quartz-normative melts and corundum-bearing restites by crustal
    anatexis: petrogenetic modelling based on an example from the
    Lewisian of north-west Scotland. *J. metamorphic Geol.*, 4, 79-99.
Cartwright, I., and Barnicoat, A.C., 1987. Petrology and petrogenesis of
    Scourian supracrustals and orthogneisses from the Lewisian at Stoer,
    NW Scotland, and their bearing on the geological evolution of the
    Lewisian complex in: Park, R.G. and Tarney, J. (eds), Evolution of
    the Lewisian complex and comparable Precambrian high grade terrains,
    Geol. Soc. Lond. Spec. Pub., 27, 93-108.
Cartwright, I., and Barnicoat, A.C., in prep. Evolution of the Scourian
    complex. Submitted to Evolution of metamorphic belts, *Geol. Soc.
    Lond. Spec. Pub.*
Cartwright, I., Fitches, W.R., O'Hara, M.J., Barnicoat, A.C. and O'Hara,
    S., 1985. Archaean supracrustals from the Lewisian near Stoer,
    Sutherland. *Scott. J. Geol.*, 21, 187-196.
Chapman, H.J., 1979. 2390myr Rb-Sr whole rock age for the Scourie dykes
    of north-west Scotland. *Nature*, 277, 642-643.

Chapman, H.J. and Moorbath, S., 1977. Lead isotope measurements from the oldest recognised Lewisian gneiss of NW Scotland. *Nature*, 268, 41-42.

Corbett, G.J. and Phillips, G.N., 1981. Regional metamorphism of a high grade terrain: The Wilyama complex, Broken Hill, Australia. *Lithos*, 14, 59-79.

Davies, F.B., 1977. The Archaean evolution of the Lewisian complex of Gruinard Bay. *Scott. J. Geol.*, 5, 279-284.

Evans, C.R., 1965. Geochronology of the Lewisian basement near Lochinver, Sutherland. *Nature*, 207, 54-56.

Evans, C.R. and Lambert, R.St.J., 1974. The Lewisian of Lochinver, Sutherland: the type area for the Inverian metamorphism. *J. Geol. Soc. Lond.*, 130, 125-150.

Evans, C.R. and Tarney, J., 1964. Isotopic ages of Assynt dykes. *Nature*, 204, 638-641.

Fyfe, W.S., 1973. The granulite facies, partial melting and the Archaean crust. *Phil. Trans. R. Soc. Lond.*, A273, 457-462.

Gilletti, B.J., Moorbath, S. and Lambert, R.St.J., 1961. A geochronological study of the metamorphic complexes of the Scottish highlands. *Quat. J. Geol. Soc. Lond.*, 117, 233-272.

Greenwood, H.J., 1975, Buffering of pore fluids by metamorphic reactions. *Am. J. Sci.* 275, 573-593.

Hamilton, P.J., Evanson, N.M., O'Nions, R.K. and Tarney, J., 1979. Sm-Nd systematics of Lewisian gneisses: implications for the origins of granulites. *Nature*, 277, 25-28.

Holland, J.G. and Lambert, R.St.J., 1973. Comparative major element geochemistry of the Lewisian of the mainland of Scotland. In: Park, R.G. and Tarney, J. (eds). *The early Precambrian of Scotland and related rocks in Greenland. Univ. Keele.* 51-62.

Holland, J.G. and Lambert, R.St.J., 1975. The chemistry and origin of the Lewisian gneisses of the Scottish mainland: the Scourie and Inver assemblages and subcrustal accretion. *Precamb. Res.*, 2, 161-174.

Huang, W.L. and Wyllie, P.J., 1975. Melting relations in the system $NaAlSi_3O_8-KAlSi_3O_8-SiO_2$ to 35 kilobars, dry and with excess water. *J. Geol.*, 83, 737-748.

Humphries, F.J, and Cliff, R.A., 1982. Sm-Nd dating and cooling history of Scourian granulites. *Nature*, 295. 515-517.

Johannes, W., 1985. The significance of experimental studies in the formation of migmatites. In: Ashworth, J.R. (ed), *Migmatites*, Blackie, Glasgow. 36-85.

Lamb, W. and Valley, J.W., 1984. Metamorphism of reduced granulites in low-CO_2 vapour-free environment. *Nature*, 312, 5658.

Luth, W.C., 1969. The systems $NaAlSi_3O_8-SiO_2$ and $KAlSi_3O_8-SiO_2$ to 20kb and the relationship between H_2O content, PH_2O and P_{total} in granitic magmas. *Am. J. Sci.*, 267A, 325-341.

Luth, W.C., 1976. Granitic rocks. In : Bailey, D.K. and MacDonald, R. (eds), *The evolution of the crystalline rocks*, Academic Press, London. 335-417.

McKenzie, D., 1985, The extraction of magma from the crust and mantle. *Earth planet Sci. Lett.*, 74, 81-91.

Moorbath, S., 1969. Evidence for the age of deposition of Torridonian sediments of NW Scotland. *Scott. J. Geol.*, 5. 154170.

Moorbath, S. and Park, R.G., 1972. The Lewisian chronology of the southern region of the Scottish mainland. *Scott. J. Geol.*, 8, 5174.

O'Hara, M.J., 1977. Thermal history of excavation of Archaean gneisses from the base of the continental crust. *J. Geol. Soc. Lond.*, 134, 185-200.

O'Hara, M.J. and Yarwood, G., 1978. High pressure-temperature point on an Archaean geotherm, implied magma genesis by crustal anatexis and consequences for garnet-pyroxene thermometry and barometry. *Phil. Trans. R. Soc. Lond.*, A288, 441-456.

Park, R.G., 1973. The Laxfordian belts of the Scottish mainland. In: Park, R.G. and Tarney, J. (eds). *The early Precambrian of Scotland and related rocks in Greenland. Univ. Keele*, 65-76.

Peach, B.N., Horne, J., Gunn, W., Clough, C.T., Hinxman, L.W. and Teall, J.J.H., 1907. The geological structure of the northwest highlands of Scotland. *Mem. Geol. Soc. GB*.

Pidgeon, R.T and Bowes, D.R., 1972. Zircon U-Pb ages of granulites from the central region of the Lewisian, NW Scotland. *Geol. Mag.*, 109, 247-258.

Powell, R., 1983a. Fluids and melting under upper amphibolite facies conditions. *J. Geol. Soc. Lond.*, 140, 629-634.

Powell, R., 1983b. Processes in granulite facies metamorphism. In: Atherton, M.P. and Gribble, C.D. (eds), *Migmatites, melting and metamorphism*, Shiva, Nantwich. 127-139.

Pride, C. and Muecke, G.K., 1982, Geochemistry and origin of granitic rocks, Scourian complex, NW Scotland. *Contrib. Mineral. Petrol.*, 80, 379-385.

Rollinson, H.R., 1980. Iron-titanium oxides as an indicator of the fluid phase during cooling of granites metamorphosed to granulite grade. *Mineral. Mag.*, 43, 165-170.

Rollinson, H.R., 1982. Evidence from feldspar compositions of high temperatures in granitic sheets in the Scourie complex, NW Scotland. *Mineral. Mag.*, 46, 73-76.

Rollinson, H.R. and Windley, B.F., 1980, An Archaean granulite grade tonalite-trondhjemite-granite suite from Scourie, NW Scotland. *Contrib. Mineral. Petrol.*, 72, 257-263.

Shaw, H.R., 1972. Viscosities of magmatic silicate liquids: an empirical method of prediction. *Am. J. Sci.*, 272, 870-893.

Sheraton, J.W., Skinner, A.C. and Tarney, J., 1973. The geochemistry of the Scourian gneisses of the Assynt district. In: Park, R.G. and Tarney, J. (eds), *The early Precambrian of Scotland and related rocks of Greenland, Univ. Keele*. 13-30.

Sills, J.D., 1983. Mineralogical changes occurring during the retrogression of Archaean gneisses from the Lewisian complex of NW Scotland. *Lithos*, 16, 113-124.

Sills, J.D., and Rollinson, H.R., 1987. Metamorphic evolution of the mainland Lewisian complex. In: Park, R.G. and Tarney, J. (eds), Evolution of the Lewisian and comparable Precambrian high grade terrains, *Geol. Soc. Lond. Spec. Pub.*, 27, 81-92.

Tarney, J. and Windley, B.F., 1977. Chemistry, thermal gradients and evolution of the lower continental crust. *J. Geol. Soc. Lond.*, 134,

153-172.

Weaver, B.L. and Tarney, J., 1981. Lewisian gneiss geochemistry and Archaean crustal development models. *Earth planet Sci. Lett.*, 51, 171-180.

Whitehouse, M.J. and Moorbath, S., 1986. Pb-Pb systematics of Lewisian gneisses - implications for crustal differentiation. *Nature*, 319, 448-449.

Whitney, J.A., 1975. The effects of pressure, temperature and X(H2O) on phase assemblages in four synthetic rock compositions. *J. Geol.*, 83, 1-31.

Windley, B.F. and Smith, J.V., 1976. Archaean high grade complexes and modern continental margins. *Nature*, 319, 488-489.

Wyllie, P.J., 1983, Experimental studies on biotite- and muscovite-bearing granites and some crustal magmatic sources. In: Atherton, M.P. and Gribble, C.D. (eds), *Migmatites, melting and metamorphism*, Shiva, Nantwich. 12-26.

PROTEROZOIC METASEDIMENTS FROM THE DEEP CRUST OF ROGALAND, SW NORWAY:
Chemistry of metabasites and granofelses.

Luc C.G.M. Bol and J. Ben H. Jansen
Department of Geochemistry and Experimental Petrology,
Institute of Earth Sciences, State University of Utrecht,
Budapestlaan 4, 3584CD Utrecht, The Netherlands.

EXTENDED POSTER ABSTRACT. Metamorphic basites and alumina-rich granofelses from the supracrustal Faurefjell Formation and dark layers (amphibolites and metabasites) from the Basic-Acid Migmatite Sequence from Rogaland/Vest-Agder, SW Norway, have been studied chemically (figs. 1 and 2). Major elements, trace elements and REE confirm the magmatic

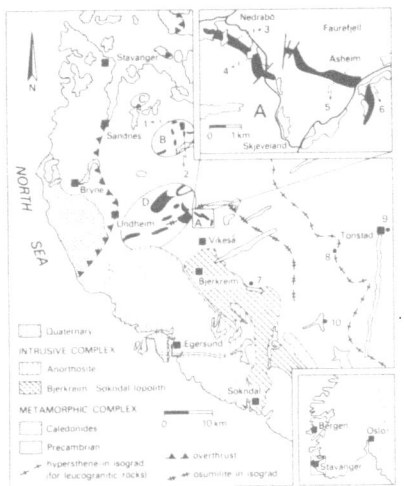

Fig.1: *Geological sketch-map of SW Norway showing outcrops of the Faurefjell formation at areas A,B,C and D (indicated in black) and sample locations (1 to 10).*

provenance of the metabasites and the continuous chemical trends strongly suggest a common progenitor for the granofelses (fig.3). When compared to metabasites, the granofelses are enriched in immobile elements and depleted in silica and calcium. Sapphirine-bearing granofelses differ chemically from the other granofelses by a higher Mg and lower P, Zr, Hf, Y and REE content. Spinel-bearing metabasites and granofelses from one specific layer show the most pronounced enrichment trends of immobile elements. During the main metamorphism most rocks were variably affected by K-metasomatism (fig.4). The chemical data have been plotted in discriminant immobile element diagrams in order to charactarize the metabasites in terms of magma series and to reveal a possible tectonic setting. Metabasites, alumina-rich granofelses and amphibolites yield sub-alkaline, tholeiitic affinities on most diagrams. This is also supported by rather flat REE-patterns. The diagrams are not conclusive for the exact tectonic setting but from the coexistence of the metabasites together with quartzites and metadolomites in the Faurefjell Formation (Hermans et al.; 1975, Sauter, 1981) a back-arc basin position within the Proterozoic orogeny (> 1200 Ma) is tentatively suggested.

111

D. Bridgwater (ed.), Fluid Movements – Element Transport and the Composition of the Deep Crust, 111–113.
© *1989 by Kluwer Academic Publishers.*

112

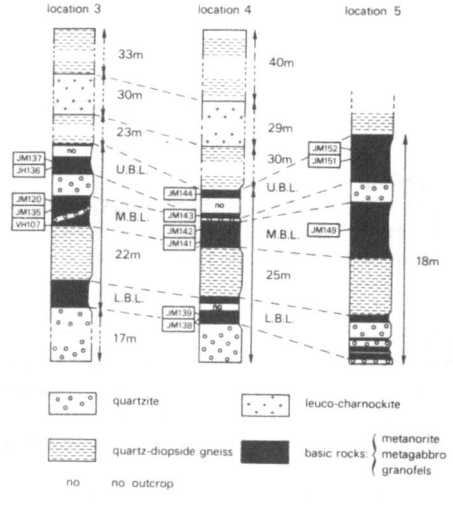

Fig.2: *Simplified profiles through the Faurefjell formation in area A at locations 3, 4 and 5 (see inset fig.1). Samples taken from the three different basic layers are shown.*
UBL = Upper Basic Layer
MBL = Middle Basic Layer
LBL = Lower Basic Layer
"Metabasites" (plag+opx+mgt+/-cpx,bio, herc,zr,ap) are found in the UBL, MBL, LBL and ELS (uncorrelated basic layers ELSwhere in the Faurefjell Formation). Granofelses (crd+plag+mgt+herc+/-opx,sil,bio,zr,sapph) are found in the UBL (spinel rich) and the MBL (sapphirine bearing).

Fig.3: *Variation diagrams of major and trace elements against SiO₂ and Zr, K₂O vs. Rb. Granofelses are systematically higher in "immobile" elements.*

Symbols:
△:UBL-metabasites
▲:UBL-granofelses
○:MBL-metabasites
●:MBL-granofelses
▽:LBL-metabasites
✳:ELS-metabasites
□:BMAS-metabasites
■:BMAS-amphibolites

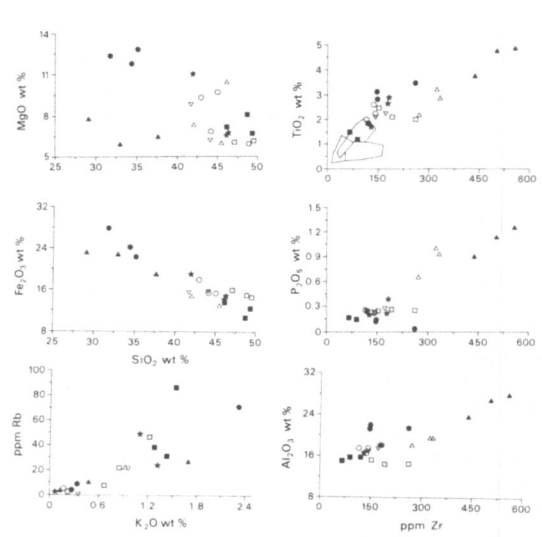

Fig.4: $TiO_2-K_2O-P_2O_5$ diagram. The heavy line discriminates between oceanic (near the TiO_2 apex) and non-oceanic basalts (Pearce et al., 1975). The light lines show metasomatic enrichment trends towards the K_2O apex.

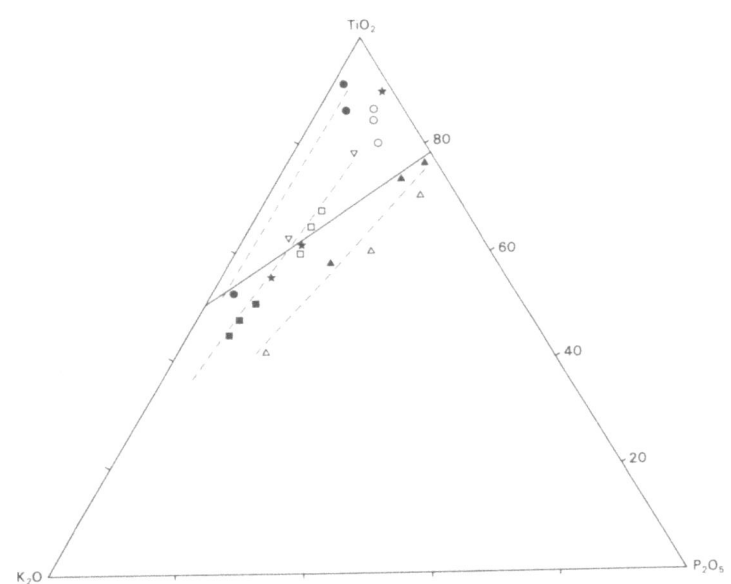

Hermans GAEM, Tobi AC, Poorter RPE, Maijer C (1975) The high-grade metamorphic Precambrium of the Sirdal-Ørsdal area, Rogaland/Vest-Agder, South-west Norway. Nor Geol Unders 318: 51-74.

Pearce TH, Gorman BE, Birkett TC (1975) The $TiO_2-K_2O-P_2O_5$ diagram: a methode of discriminating between oceanic and non-oceanic basalts. Earth Plan Sci Lett 24: 419-426.

Sauter PCC (1981) Mineral relations in siliceous dolomites and related rocks in the high-grade metamorphic Precambrium of Rogaland, SW Norway. Nor Geol Tidsskr 61: 35-45.

LIMITED FLUID TRANSPORT FROM THE GNEISS CORE TOWARD THE SCHIST ENVELOPE OF THE DOME DE L'AGOUT, FRANCE.

Ariejan Bos, Willem Duit[*] and J. Ben H. Jansen
Department of Geochemistry and Experimental Petrology,
Institute for Earth Sciences, University of Utrecht,
Budapestlaan 4, 3584 CD Utrecht, The Netherlands.

[*]Died July 26, 1985

EXTENDED POSTER ABSTRACT. The Dome de l'Agout consists mainly of gneisses and granites, mantled by low to medium grade micaschists with relatively large amounts of graphite. Kreulen and Schuiling (1982) observed high nitrogen contents in fluid inclusions, with the highest gas contents increasing with metamorphic grade. They suggested that the nitrogen was mantle derived and moved from the centre of the dome outwards. Duit et al. (1986) analyzed the ammonium- and rubidium-contents in biotite and muscovite fractions and they observed a negative correlation between the highest ammonium contents in both minerals and metamorphic grade (Figure 1). Bos et al. (in press) studied the

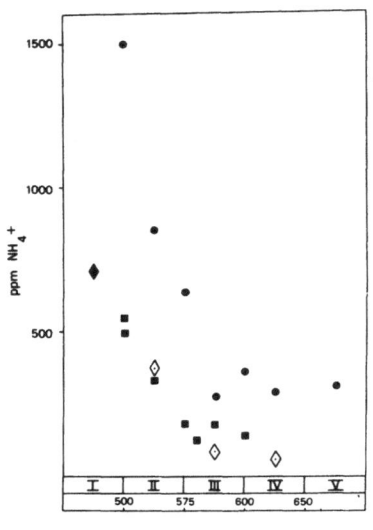

Figure 1: Average NH_4^+ concentrations in minerals and rocks in the different metamorphic zones of the Dome de l'Agout area. The estimated temperatures for the isograds are indicated at the bottom. Solid symbols are measured, open symbols are calculated values based on modal composition of the rocks and on estimated distribution coefficients. The symbols are ● for biotite, ■ for muscovite and ◆ for the whole-rock.

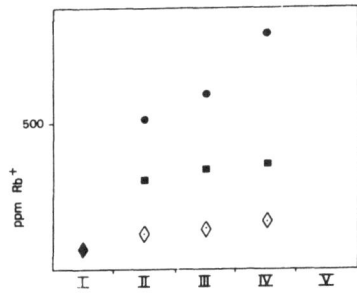

Figure 2: Average Rb^+ concentrations.

115

partitioning of ammonium and potassium between synthetic phlogopite and a chloride vapour. The distribution coefficient of ammonium was determined to be larger than 1 at the PT-conditions of 2000 bars at 550 and 650·C. The distribution coefficient increases, although slightly, with temperature, indicating that nitrogen storage in biotite is a valid source for N_2-gas in the lower crust (Bos et al., in press). In contrast to ammonium, rubidium contents increase in both micas with increasing metamorphic grade (Figure 2), pointing at a difference in behaviour and origin of the ions in the metamorphic rocks. Rubidium and ammonium both substitute for potassium in natural silicates. The ions are almost equal in size, and a similar behaviour should be expected. Experimental studies by Beswick (1973) and Volfinger (1976) on the distribution of rubidium between phlogopite and vapour yielded similar results as for the ammonium distribution experiments.

In order to explain the contrasting trends of ammonium and rubidium with the available natural and experimental data, two models were examined: model (A) in which a pervasive fluid was supplied, flowing from the centre of the dome outward, and model (B) in which only local fluid movement has taken place at the border of the gneissic and migmatitic core with the schist envelope. From calculations based on a number of assumptions regarding rock and fluid composition, and on estimated PT-conditions, the ammonium concentration pattern can be explained by model (A), only with a fluid/rock-ratio of ca 5. This amount of fluid is thought to be hardly realistic in a regionally metamorphosed area. Besides, the rubidium contents cannot be explained with model (A). Model (B), based on local fluid movements at the boundary of the two rock units, together with minor outward flow of a nitrogen-rich fluid, may better describe the now present concentration patterns of both ammonium and rubidium.

Beswick AE (1973) An experimental study of alkali metal distributions in feldspars and micas. Geochim Cosmochim Acta 37:183-208.
Bos A, Duit W, Eerden A van der, Jansen JBH (in press) Nitrogen storage in biotite: Experimental study of the ammonium and potassium partitioning between 1M-phlogopite and vapour at 2 kb. Geochim Cosmochim Acta.
Duit W, Jansen JBH, Breemen A van, Bos A (1986) Ammonium micas in pelitic rocks as exemplified by Dome de l'Agout (France). Amer J Sci 286:702-732.
Kreulen R, Schuiling RD (1982) N_2-CH_4-CO_2 fluids during the formation of the Dome de l'Agout, France. Geochim Cosmochim Acta 46:193-203.
Volfinger M (1976) Effet de la température sur les distributions de Na, Rb et Cs entre la sanidine, la muscovite, la phlogopite et une solution hydrothermale sous une pression de 1 kbar. Geochim Cosmochim Acta 40:267-282.

HYDROTHERMAL GRAPHITE VEINS AND ACADIAN GRANULITE FACIES METAMORPHISM,
NEW HAMPSHIRE, USA

D. Rumble, III[1], C. P. Chamberlain[2], P. K. Zeitler[3], and
B. Barreiro[2]
[1]Geophysical Lab, 2801 Upton St. NW, Wash., DC 20008
[2]Dept. Earth Sciences, Dartmouth College, Hanover, N.H. 03755
[3]Research School of Earth Sciences, Australian National
   University, Canberra 2600, Australia

1. HYDROTHERMAL GRAPHITE

Hydrothermal graphite has been found over an area of 3,000 Km^2 in New
Hampshire, USA. Its pervasive presence records fluid flow of regional
extent during metamorphism. Graphite precipitated from metamorphic
fluids crosscuts and replaces Silurian and Devonian metasedimentary and
metaplutonic rocks. The metamorphic grade of host rocks ranges from
kyanite + staurolite, to sillimanite, sillimanite + alkali feldspar, or
cordierite + sillimanite + alkali feldspar.
    Hydrothermal graphite is found in two characteristic forms: (1)
Spherulites and irregular clumps of graphite (10-100 μm in diameter)
are disseminated throughout pre and synmetamorphic granitic plutons.
Spherulites are locally concentrated in halos surrounding healed fractures and retrograded shear zones. (2) Veins containing graphite measure up to 1 m thick and 20 m long. The vein contacts crosscut bedding, schistosity, and pluton/wall rock contacts. Wall rock alteration
zones are present where mica schist has been replaced by graphite +
tourmaline up to 6-8 cm from the vein. Crustiform depositional structures of graphite coprecipitated with quartz, tourmaline, muscovite,
ilmenite, sillimanite, or rutile are found in the veins (Rumble, Duke,
and Hoering,]986).
    The precipitation of spherulites is believed to have been caused
by infiltration of metamorphic fluids along fractures in granitic
plutons. Reaction between the fluids and biotite and feldspar of the
plutons produced the hydrous minerals chlorite and muscovite. The
remaining fluid became increasingly depleted in H_2O but enriched in
CO_2 + CH_4 as hydration reactions proceeded. In consequence, the
composition of residual fluid was driven towards the graphite saturation surface. Graphite precipitation followed, provided that a
favorable oxygen fugacity, approximately that of quartz + fayalite
+ magnetite, was maintained (Duke and Rumble,]986).
    Graphite was deposited in veins when aqueous fluids with different
CO_2/CH_4 ratios were mixed in fractures. The saturation surface of
graphite in a fluid consisting chiefly of H_2O, CO_2, and CH_4 is
strongly curved in the region rich in H_2O. Mixing unsaturated fluids

117

D. Bridgwater (ed.), Fluid Movements – Element Transport and the Composition of the Deep Crust, 117–119.
© *1989 by Kluwer Academic Publishers.*

from the convex side of the curved phase boundary results in a meta-
stable fluid that achieves equilibrium by precipitating graphite
(Rumble and Hoering, 1986).

Vein fluids and fluids infiltrating fractures were produced by metamorphic devolatilization reactions taking place in heated sediments.
Carbon isotope compositions of both graphite veins and spherulites range
from -25 to -9°/oo $\delta^{13}C$ (PDB), intermediate between the crustal reservoirs of biogenic carbonate and reduced organic matter. The intermediate $\delta^{13}C$ values suggest mixing carbon from the two reservoirs, an
hypothesis consistent with the proposal that graphite precipitation is
caused by mixing different fluids (Rumble and Hoering, 1986).

Our current research is focused on the relationship between graphite veins and regional metamorphism. Of particular interest are the
graphite veins of Bristol, New Hampshire, USA, that lie in the center
of a granulite facies hot spot.

2. GRANULITE FACIES HOT SPOTS

Hot spots are localized, discrete areas of granulite facies metamorphism
surrounded by a regional metamorphic sillimanite zone. The hot spots
in New Hampshire measure 5-10 km across in a sillimanite zone that extends over 150 km north and south and 40 km east and west. Mineral
assemblages in pelitic schists change over a distance of a few kilometers
from sillimanite + staurolite + muscovite + garnet + biotite + quartz to
sillimanite + alkali feldspar + cordierite + garnet + biotite. Temperatures increase abruptly over the same distance from 550° in the sillimanite zone to 700°C in the granulite facies. The outcrop patterns of
the hot spots resemble those of contact metamorphism but they are not
centered on exposed plutons nor is there evidence from geophysical surveys of subjacent, buried plutons. Surprisingly, at least one of the
hot spots, near Bristol, N. H., has quartz + graphite veins in its core.

The Bristol hot spot is relatively small, about 2 to 3 km across,
as outlined by the sillimanite + alkali feldspar isograd. There is a
concentric, innermost zone of cordierite + sillimanite + alkali feldspar assemblages. Both the 600° and 650°C isotherms form closed loops
centered on the 700°C cordierite Zone. The isotherms overlap the location of the isograds. Analysis of $^{18}O/^{16}O$ in quartz from quartz +
graphite veins and wall rocks shows a halo of ^{18}O depletion circum-
scribed about the veins in the core of the hot spot. Quartz values inside the cordierite zone range from +13.9 to 14.6°/oo $\delta^{18}O$ (SMOW), but,
outside the sillimanite + alkali feldspar isograd, quartz from the same
stratigraphic units is +14.7 to 16.0°/oo $\delta^{18}O$ (Chamberlain and Rumble,
in preparation).

The question arises: Is the concentric location of isograds, isotherms, $\delta^{18}O$ depletion halo, and veins merely a coincidence of unrelated features, or is there a cause/effect relationship between heated
rocks and flowing fluids? Was the hydrothermal activity that produced
quartz-graphite veins and their ^{18}O halo responsible for heating wall
rocks into the granulite facies?

Radiometric age dating demonstrates that the hydrothermal activity
and granulite facies metamorphism are identical in age, within error

of measurement. The SHRIMP ion microprobe has been used for U-Pb dating of hydrothermal overgrowths on zircons from one of the veins in the hot spot at 390 $\pm$ 10 my (Zeitler, Barreiro, Chamberlain, and Rumble, in preparation). Monazite from cordierite gneisses in the wall of the same vein gives a conventional U-Pb age of 397 $\pm$ 10 my (Barreiro, in preparation).

It is concluded that the Bristol hot spot formed as a result of heating by flowing metamorphic fluids focussed through a fracture system now recorded by quartz + graphite veins.

REFERENCES

Barriero, B. (in preparation) Age of monazite from metasediments of the Bristol, N. H., hot spot.

Chamberlain, C. P., and Rumble, D. (in preparation) Thermal anomalies in a regional metamorphic terrane: the role of advective heat transport by fluids.

Duke, E. F., and Rumble, D. (1986) Textural and isotopic variations in graphite from plutonic rocks, south-central New Hampshire, Contrib. Mineral. Petrol. 93, 409-419.

Rumble, D. and Hoering, T. C. (1986) Carbon isotope geochemistry of graphite vein deposits from New Hampshire, USA, Geochim. Cosmochim Acta, 50, 1239-1247.

Rumble, D., Duke, E. F., and Hoering, T. C. (1986) Hydrothermal graphite in New Hampshire: evidence of carbon mobility during regional metamorphism, Geology 14, 452-455.

Zeitler, P. K., Barreiro, B., Chamberlain, C. P., and Rumble, D. (in preparation) Adacian age for zircon from quartz + graphite veins at Bristol, N. H.: evidence for synmetamorphic transfer of heat by fluids.

THE CRUSTAL ORIGIN OF ECLOGITE - STATIC OR DYNAMIC

Henning Sørensen
Institute for Petrology
Øster Voldgade 10
DK-1350 Copenhagen K
Denmark

ABSTRACT. Griffin and colleagues have presented an impressive wealth of
new information on the crustal origin of eclogites. With reference to
Backlund's 1936-paper, in which eclogites are considered to be the
products of extreme dynamical pressure or stress concentrated in layers
of amphibolite, it is recommended to combine an evaluation of the
dynamic factors of metamorphism with mineral-chemical data when studying
massive rocks as eclogite and some types of alpine peridotites.
    The eclogites of Western Norway, which were made world-famous by
Eskola's 1921-paper, have attracted renewed interest in recent years
since Bryhni et al. (1970) proposed a crustal origin for these rocks.
    In an impressive series of publications Griffin and collaborators
have presented a wealth of new information not only about the Norwegian
occurrences of eclogites but also about high pressure metamorphism and
its products. Two main types of occurrence of eclogite have been
described in these papers:
1. In the Western Gneiss Region eclogites are demonstrated to have been
formed by prograde metamorphism of gabbroic or amphibolite rocks at high
temperatures and pressures (750o C, 20 kbar) in a thickened crust. The
enclosing Pre-cambrian gneisses are thought to have been exposed to
similar high temperatures and pressures, but retrograde metamorphism has
obliterated all traces of the eclogite facies mineralogy of these rocks.
The eclogites are also extensively retrograded (Griffin et al., 1985).
2. In the granulite facies assemblage of anorthositic-noritic rocks in
the Bergen Arcs (Austrheim & Griffin, 1985; Austrheim, 1987) eclogites
are demonstrated to have formed in the vicinity of Caledonian shear
zones in the anorthosite complex at 700-750o C, 16-19 kbar. Granulite-
facies mineral assemblages are preserved a few centimetres from the
shear zones which is interpreted as an indication of the importance of
deformation in forming the eclogites. Tectonic "overpressure" is not
considered important (Austrheim & Griffin, 1985).

COMMENTS AND DISCUSSION

Two modes of formation of the eclogites of Western Norway have been
considered in the papers referred to in the above: Caledonian regional

D. Bridgwater (ed.), Fluid Movements – Element Transport and the Composition of the Deep Crust, 121–123.
© 1989 by Kluwer Academic Publishers.

metamorphism of rocks of basaltic composition deep in a continental
crust composed of Pre-cambrian gneisses; and localized Caledonian
reworking along shear zones in a Pre-cambrian granulite facies complex.
    It is evident from the descriptions that the eclogites of the
Western Gneiss Region are strongly tectonized and often occur as
boudined layers. Contact relations have in most cases been obliterated
by high T,P deformation and metamorphism. I have not found any detailed
structural analysis of the region, most data concern the
mineral-chemical relations of the rocks.
    In the Bergen Arcs the eclogites occurring in shear zones in the
anorthosite complex are evidently located in their site of origin.
    One may pose the question: have all eclogites originally been
formed in shear zones as seen in the Bergen Arcs and are the lenses of
eclogite found in the Western Gneiss Region a strongly tectonized
equivalent of a complex in which eclogite originated in shear zones? Or
may the high T,P necessary for the formation of eclogites be realized in
the deep parts of a thickened crust as well as in shear zones in higher
crustal levels.
    I would at this point like to draw the attention to Backlund's
paper: *"Zur genetischen Deutung der Eclogite"* (1936) in which eclogites
are explained as "tectonites". According to him the high pressures,
which are needed to form eclogites, are not caused by great depths in
the crust, but by "extreme dynamical pressure" or stress concentrated in
layers of amphibolite (*"ausserordentlich starke tektonische Verschuppung
der Amphibolite"*, op.cit. p. 56). This is stated in a more precise way
by Eberth (1986, p. 75) in his comments on Backlund's paper: *"Damit
diese Überdrucke tatsächlich wirksam werden können, ... müssen über
engbegrenzte Gebiete tektonische Spannungszustände herschen, die den
Druckausgleich verhindern"*.
    Along this line of arguing the role of "dynamical pressure" should
be reconsidered when studying massive high T,P rocks such as eclogites
and peridotites. I may in this connection refer to my own attempts at
explaining the formation of some occurrences of alpine-type peridotites
(Sørensen, 1953, 1955, 1967) at a time when the origin of these rocks
was an even more puzzling petrological problem than now (cf. 1st edition
of Turner-Verhoogen, 1951). The mechanism envisaged is that the
peridotitic rocks examined are the products of metamorphic
differentiation in zones of high nonhydrostatic stress in rigid bands of
amphibolite enclosed in less rigid gneisses, the strain stored in the
rocks, probably partly in dislocations, being relieved under
recrystallization (recovery) under the formation of minerals with small
mol volumes such as olivine and orthopyroxene (cf. Bennington, 1956,
Reitan & Geul, 1959). This process may result in the formation of
"polygonal" textures. White (1979) has discussed the phenomenon of
dynamic recrystallization as an integral part of deformation - a
phenomenon largely neglected by geologists. Internal strain energy is
relieved by the formation of new strain-free grains. The driving force,
according to White (op.cit.), is the reduction in strain energy achieved
by strain induced grain boundary migration. Much can be learned by
applying metallurgical principles to the deformation and
recrystallization of rocks (White, op.cit.; Sørensen, 1967).

It is my view that a structural geological analysis of the occurrences of eclogite, including an evaluation of the dynamic factors of metamorphism, may add valuable information to the already existing impressive geochemical and mineral-chemical data.

REFERENCES

Austrheim, H., 1987: Eclogitization of lower crustal granulites by fluid migration through shear Zones. *Earth Planet. Sci. Lett.*, 81, 221-232.

Austrheim, H. & Griffin, W.L., 1985: Shear deformation and eclogite formation within granulite-facies anorthosites of the Bergen Arcs, Western Norway. *Chem. Geol.*, 50, 267-281.

Backlund, H.G., 1936: Zur genetischen Deutung der Eclogite. *Geol. Rundschau*, 27, 47-61.

Bennington, K.O., 1956: Role of shearing stress and pressure in differentiation as illustrated by some mineral reactions in the system $MgO-SiO_2-H_2O$. *Journ. Geol.*, 64, 558-577.

Bryhni, I., Green, D.H., Heier, K.S. & Fyfe, W.S., 1970: On the occurrence of eclogite in Norway. *Contrib. Mineral. Petrol.*, 26, 12-19.

Ebert, H., 1936: Bemerkung zu den Vorträgen Backlund und Eskola. *Geol. Rundschau*, 27, 74-75.

Eskola, P., 1921: On the eclogites of Norway. *Skr. Norske Vidensk.- Akad.*, Oslo, Kl. I,8, 118 pp.

Griffin, W.L., Austrheim, H., Brastad, K., Bryhni, I., Krill, A.G., Krogh, E.J., Mørk M.B.E., Quale, J. & Tørudbakken, B., 1985: High pressure metamorphism in the Scandinavian Caledonides. In: Gee, D.G. & Sturt, B.A. (eds.): *The Caledonide Orogen - Scandinavia and Related Areas*. John Wiley & Sons Ltd., 783-801.

Reitan, P.H. & Geul, J.J.C., 1959: On the formation of a carbonate-bearing ultrabasic rock of Kviteberg, Lyngen, Northern Norway. *Norges Geol. Unders.*, 205, 111-127.

Sørensen, H., 1953: The ultrabassic rocks at Tovqussaq, West Greenland a contribution to the peridotite problem. *Meddr Grønland*, 136,4, 86 pp. (also *Bull. Geol. Survey Greenland* 4).

Sørensen, H., 1955: A preliminary note on some peridotites from Northern Norway. *Norsk Geol. Tidsskr.*, 35, 93-104.

Sørensen, H., 1967: Metamorphic and Metasomatic Processes in the Formation of Ultramatic Rocks. In Wyllie, P.J. (ed.) *Ultramafic and Related Rocks*. J. Wiley & Sons, Ltd., 204-212.

Turner, F.J. & Verhoogen, J., 1951: *Igneous and Metamorphic Petrology*. 1st. Edition. McGraw-Hill Book Co., Inc. 602 pp.

White, S., 1979: The effects of strain on the microstructures, fabrics, and deformation mechanisms in quartzites. *Phil. Trans. R. Soc. London*, A 283, 69-86.

RETROGRESSION AND FLUID MOVEMENT ACROSS A GRANULITE-AMPHIBOLITE FACIES
BOUNDARY IN MIDDLE ARCHAEAN NÛK GNEISSES, FISKEFJORD, SOUTHERN WEST
GREENLAND.

Adam A. Garde
Grønlands Geologiske Undersøgelse
Øster Voldgade 10
DK-1350 København K
Denmark

ABSTRACT. Middle Archaean Nûk gneisses in the Godthåbsfjord-Fiskefjord
area in southern West Greenland contain a gradational retrogressive
metamorphic boundary between granulite and amphibolite facies gneisses.
Movement of hydrous fluids during retrogression in this boundary zone
transported and homogenised mobile major and trace elements, upset Rb-Sr
and U-Pb isotope systems, contaminated the retrograded rocks with early
Archaean lead, and led to formation of anatectic granites.

1. FIELD RELATIONS AND PETROGRAPHY

1.1. Amphibolite and granulite facies gneisses with equilibrium textures

East of Qugssuk, an embayment in northern Godthåbsfjord (fig. 1), the
Nûk gneisses consist of polyphase but essentially coeval gneisses which
range from quartz-dioritic to granitic in composition. On outcrop scale,
boundaries between individual gneiss phases are sharp, and it appears
that the rocks have never been severely disturbed in spite of
deformation and amphibolite facies metamorphism. The Nûk gneisses east
of Qugssuk are biotite- and hornblende-bearing, with equilibrium mineral
textures. Biotite and hornblende are subhedral, c. 2-5 mm in size, and
evenly distributed.
    In Nordlandet west of Godthåbsfjord, brownish hornblende-granulite
facies gneisses of mafic tonalitic to dioritic composition prevail. At
most outcrops the rocks appear to be homogeneous, but here and there
also these gneisses can be seen to have a polyphase origin (on a scale
of metres). Also the granulite facies gneisses essentially have
equilibrium mineral textures. They are rarely completely dehydrated, as
both hornblende and biotite are common in equilibrium with pyroxene.
Locally a second generation of biotite is present, replacing
orthopyroxene and indicating hydration after the granulite facies
metamorphism.

125

D. Bridgwater (ed.), Fluid Movements – Element Transport and the Composition of the Deep Crust, 125–137.
© *1989 by Kluwer Academic Publishers.*

126

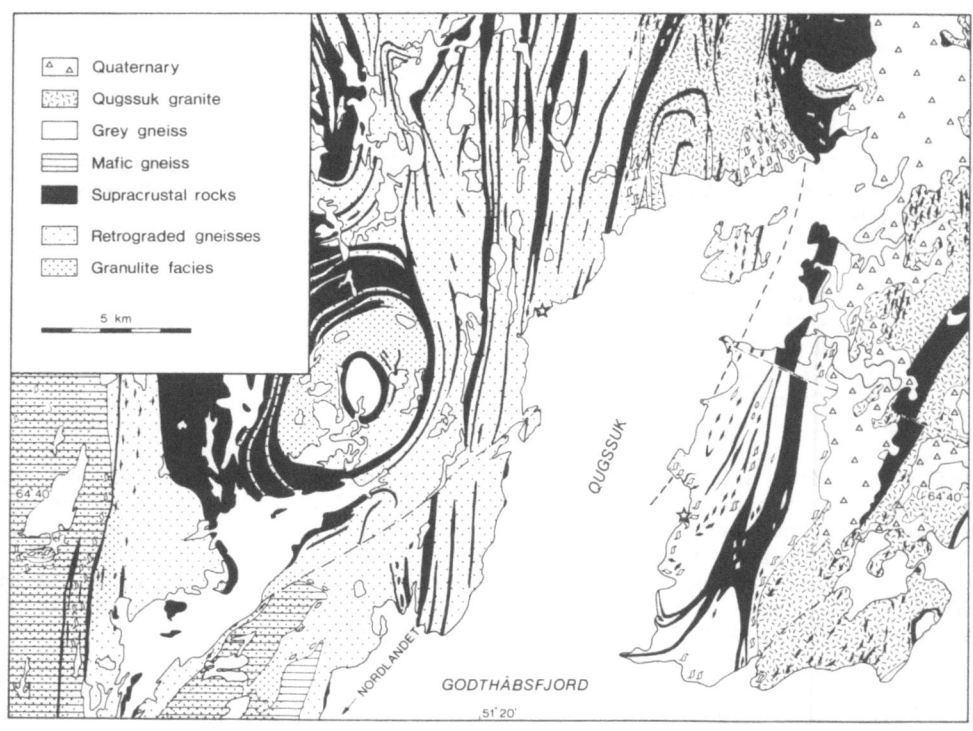

Fig. 1. Simplified and schematic map of the area around Qugssuk,
northern Godthåbsfjord. The stippled line indicates the eastern limit of
retrogression. Asterisks indicate sample localities of unaltered
amphibolite facies and retrograded gneisses (see text).

1.2. Retrograded gneisses

Around the head of Qugssuk, in a wide boundary zone between the
amphibolite and granulite facies areas, retrograded grey gneisses
prevail. Contrary to the unaltered amphibolite facies gneisses, the
contacts between gneiss phases are diffuse, so that their polyphase
origin can rarely be demonstrated. The retrograded gneisses typically
have a patchy texture of centimetre-sized clots of fine-grained mafic
minerals (blebby texture, McGregor et al., 1986). This patchy or blebby
texture may be either pervasive or concentrated in a reticular meshwork,
showing that deformation did not normally accompany the retrogression.
    Textures in thin section indicate static disequilibrium
recrystallisation under low amphibolite facies P-T conditions. Relic
orthopyroxene may be found within the mafic clots, which otherwise
consist of blue-green spongy amphibole (probably orthopyroxene
pseudomorphs, Beach, 1974) and fine-grained radial aggregates of
biotite. Ilmenite is partially decomposed and overgrown with biotite,

and secondary epidote may occur as well.

The retrograded grey gneisses crop out in a large area between Godthåbsfjord and Fiskefjord with local small patches of granulite facies gneisses. East of Qugssuk the patchy texture is gradually lost and the retrograded gneisses grade into unaltered upper amphibolite facies gneisses. The retrogression thus extends into areas that have not been in granulite facies.

1.3 Granites

Remobilised rocks with granitic composition occur both within the granulite and amphibolite facies areas. The largest of these granites, the Qugssuk granite (Garde et al., 1986) occurs at the head of Qugssuk and on the peninsula east of Qugssuk and is contemporary with or postdates the retrogression. It is a white or pink, medium- to coarse-grained rock with granitic texture and evenly dispersed, c. 1-5 mm large biotite flakes, and only minor secondary recrystallisation. The granite intrudes unaltered amphibolite facies grey gneiss east of Qugssuk, and cuts the retrograded gneiss at the head of Qugssuk. North of Qugssuk there is sporadic but clear evidence that the Qugssuk granite has been derived by partial melting of the local gneisses (Garde, 1984).

South-west of Qugssuk in the granulite facies terrain of Nordlandet, sheets of granite up to c. 100 m thick were intruded during the granulite facies metamorphism. The granite sheets have sugary, granular textures, and the feldspar is mesoperthite. Other granites in the same area postdate the granulite facies event. They resemble the Qugssuk granite and are two-feldspar rocks with granitic textures, and along their margins the host gneisses are bleached and recrystallised.

1.4 Relative timing of retrogression

Field and petrographic data show that the retrogressive metamorphic transition between granulite and unaltered amphibolite facies gneisses postdates major polyphase deformation. The retrogression extends from the granulite facies terrain into an area that has never been in granulite facies. The retrogression was accompanied or postdated by intrusion of granite s.s., the Qugssuk granite, which was partially or wholly derived by anatectic melting of the country gneisses.

2. GEOCHEMISTRY

2.1. Geochemistry of the gneisses

In the following the geochemistry of mobile elements in the three groups of gneisses is compared in order to assess the influence of granulite facies metamorphism and retrogression. Variation diagrams of major and trace elements against SiO_2 in the unaltered amphibolite facies, granulite facies and retrograded gneisses are presented in fig. 2. Table I contains analyses of representative individual samples. Most of the samples were collected c. 5 km apart at two localities at the east and

128

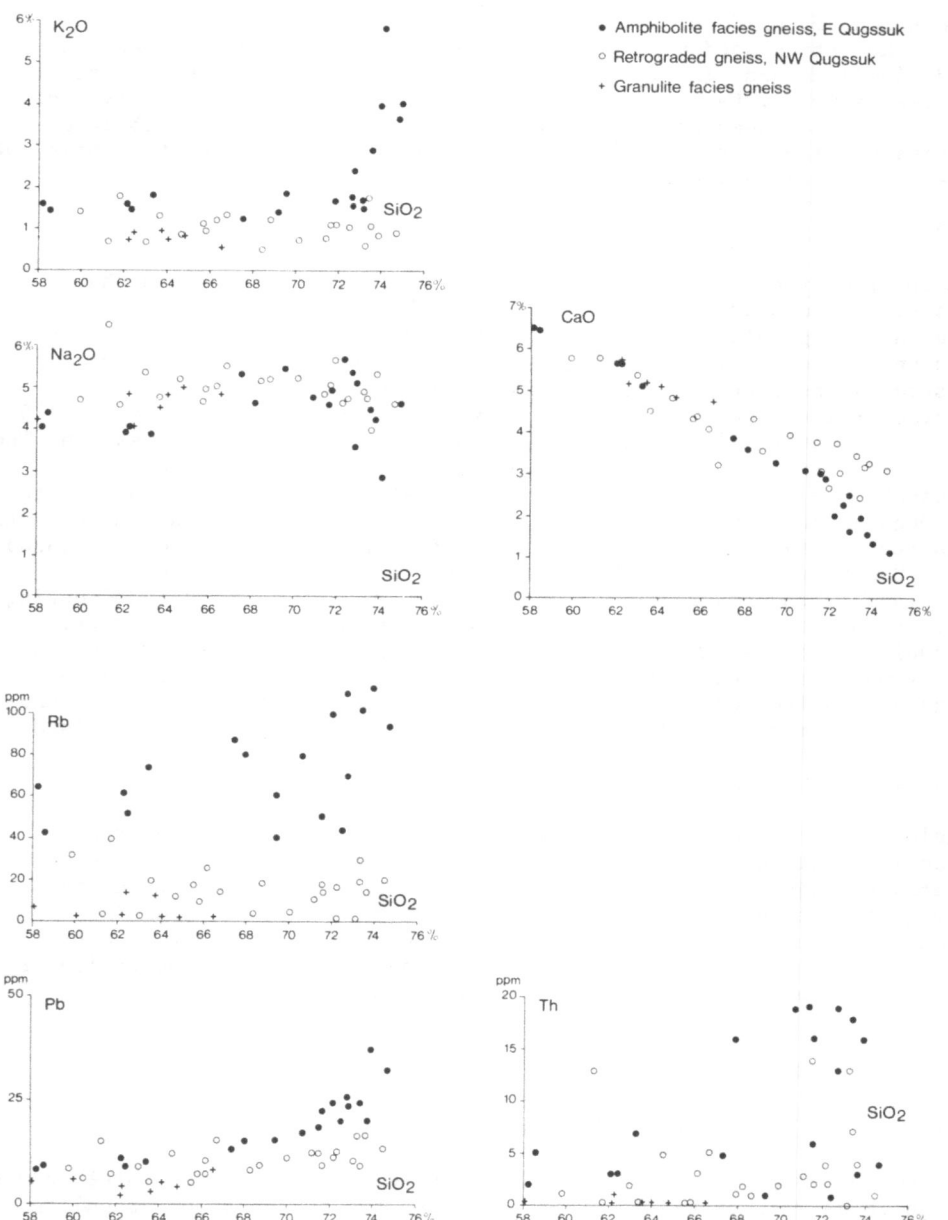

Fig. 2a. Variation diagrams of K_2O, Na_2O, CaO, Rb, Pb and Th vs. SiO_2 in gneisses around Qugssuk, northern Godthåbsfjord.

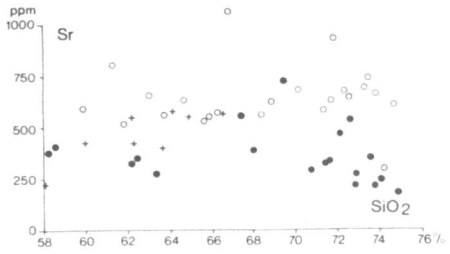

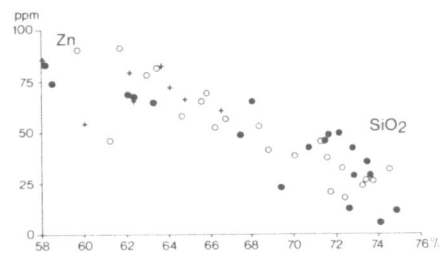
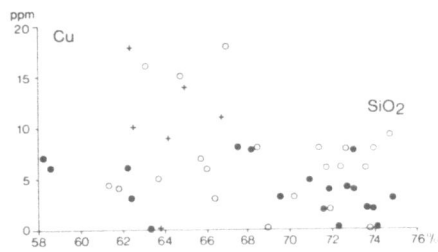

Fig. 2b. Variation diagrams of Sr, Zn and Cu vs. SiO₂ in gneisses around
Qugssuk, northern Godthåbsfjord.

north sides of Qugssuk, within amphibolite facies and retrograded
gneisses respectively. It is not known whether these particular
retrograded gneisses were once in granulite facies since no relic
orthopyroxene is preserved within the retrograded gneisses at the
locality north of Qugssuk.

It cannot be proven that the unaltered amphibolite facies and the
retrograded gneisses are cogenetic, but it is considered highly likely
and is fundamental for the conclusions presented below. The granulite
facies samples were collected in the granulite facies terrain of mafic
gneisses between Godthåbsfjord and Fiskefjord, and probably correspond
to the most mafic members of the other groups.

Fig. 2a shows that the retrograded gneisses have low K_2O, Rb, Pb
and Th concentrations relative to the amphibolite facies gneisses, but
generally have higher contents of these elements than granulite facies
gneisses with comparable bulk compositions. This trend is particularly
clear in the cases of Rb and Pb, with intermediate Rb and Pb contents in
the retrograded rocks relative to unaltered amphibolite facies and
granulite facies rocks. At the same time CaO and particularly Na_2O are
high and almost constant through a wide range of SiO_2 contents in the
retrograded gneisses. Contrary to K_2O and Rb, Sr (fig. 2b) is high in
the retrograded gneisses, which have Sr concentrations of 500-1000 ppm,
average c. 650 ppm, through a wide range of bulk compositions. In
equivalent unaltered amphibolite facies gneisses Sr shows a normal
igneous trend, with a maximum of about 500 ppm around 70% SiO_2.

In the cases of Zn and Cu, fig. 2b, no effects of retrogression can
be shown: Zn shows good negative correlation and Cu poor negative

130

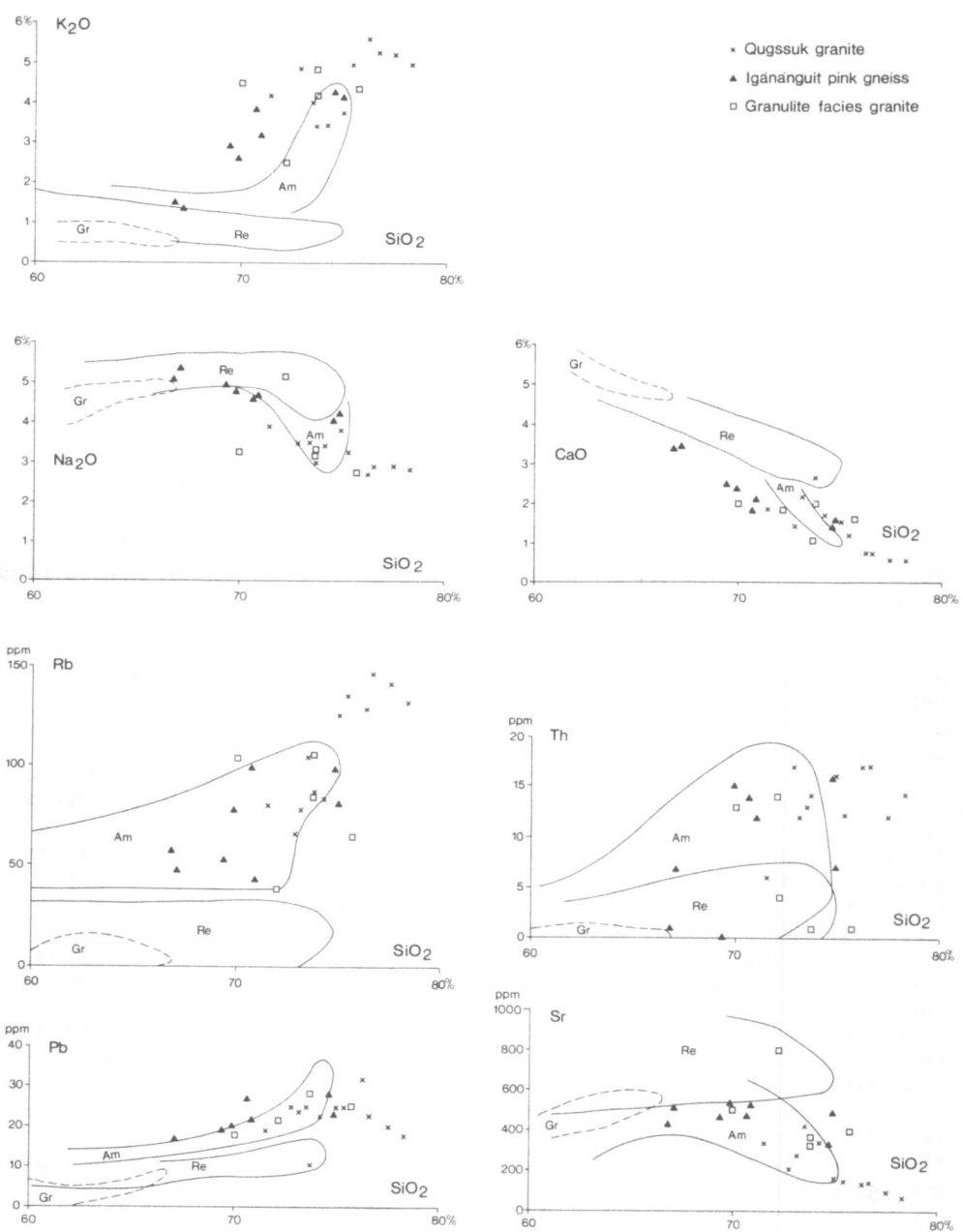

Fig. 3. Variation diagrams of K_2O, Na_2O, CaO, Rb, Pb, Th, and Sr vs.
SiO_2 in remobilised granitic rocks around Qugssuk, northern
Godthåbsfjord. Composition fields of the gneisses in fig. 2 are shown
with solid and dashed lines (Am = amphibolite facies, Gr = granulite
facies, Re = retrograded gneisses).

Table I.

No.	278738	278774	278786	278877	289156	289272
SiO_2	64.04	63.56	75.30	72.92	72.15	3.25
TiO_2	0.55	00.61	0.10	0.10	0.15	0.57
Al_2O_3	16.80	17.06	13.30	14.70	15.24	5.43
Fe_2O_3	1.31	1.85	0.23	0.12	0.48	1.33
FeO	3.40	3.17	0.88	0.54	0.48	3.99
MnO	0.06	0.05	0.01	0.01	0.01	0.10
MgO	2.00	1.79	0.05	0.24	0.37	3.23
CaO	5.13	4.57	1.22	1.61	1.83	5.12
Na_2O	4.80	4.82	3.24	4.25	5.13	3.88
K_2O	0.73	1.29	4.92	4.11	2.50	1.78
P_2O_5	0.18	0.10	0.01	0.06	0.04	0.17
l.o.i.	0.62	0.55	0.08	0.21	0.28	0.44
Sum	99.62	99.42	99.34	98.87	98.66	99.29

Trace elements in ppm

Rb	1.7	19	135	82	39	74
Pb	5	5	25	23	22	10
Th	<1	<1	12	7	4	7
Sr	568	549	146	493	800	286
Zn	72	81	16	12	32	65
Cu	9	5	3	<1	2	<1

287838: Granulite facies gneiss
278774: Grey retrograded gneiss
278786: Qugssuk granite
278877: Igánánguit pink gneiss (amphibolite facies)
289156: Granulite facies granite
289272: Unaltered amphibolite facies gneiss

Table I. Representative analyses of gneisses and remobilised granitic rocks from the northern Godthåbsfjord-Fiskefjord area. All sample numbers are GGU (Geological Survey of Greenland) numbers. Analyses by standard XRF methods.

correlation with silica, but the patterns are similar for all the three gneiss groups.

2.2. Geochemistry of the remobilised granites

The younger, remobilised granodioritic and granitic rocks are compared with the gneisses in fig 3. The trends of both the granulite facies granites and the Qugssuk granite resemble those of the most acid varieties of the amphibolite facies gneisses, and show enrichment in large-ion lithophile elements. Furthermore the granites have high contents of those elements that are depleted in the retrograded grey gneisses, whereas Sr contents are low in the granites and high in the retrograded gneisses.

2.3. Geochemical changes during retrogression

In general the retrograded gneisses have uniform concentrations of mobile elements through a wide range of chemical compositions (viz. 58-75% SiO_2). K_2O, Rb, Pb and Th contents in retrograded gneisses are intermediate compared to unaltered amphibolite and granulite facies gneisses, whereas Na_2O and particularly Sr are considerably enriched in the retrograded gneisses. Remobilised granites have high mobile element contents also in granulite facies.

3. ISOTOPE GEOCHEMISTRY

A zircon U-Pb age of 2982 +/-7 Ma has been obtained for a late phase of grey gneiss, the Taserssuaq tonalite c. 30 km north-east of Qugssuk, by R. T. Pidgeon, Western Australian Institute of Technology (Garde et al., 1986).
Fig. 4 is a Rb-Sr whole-rock isochron diagram of unaltered amphibolite facies and retrograded grey gneisses, sampled at either side of the retrogression front, c. 5 km apart at the north and east coasts of Qugssuk. The amphibolite facies gneisses plot along an errorchron of 2954 +/- 120 Ma, MSWD = 6.85, initial $^{87}Sr/^{86}Sr$ = 0.7014. The retrograded gneisses, not included in the calculation, all plot along the same line but very near its origin. This implies that the retrograded gneisses have lost their Rb not significantly later than the closure of the Rb-Sr system in their unaltered precursors.
The Qugssuk granite sampled north of the head of Qugssuk within a few kilometres of the two groups of grey gneiss, yields a Rb-Sr isochron age of 2969 +/-32 Ma (2 o), MSWD = 1.09, initial $^{87}Sr/^{86}Sr$ = 0.7020 (Garde et al., 1986).
Lead isotopic compositions of amphibolite and granulite facies gneiss, retrograded gneiss, Igánánguit pink gneiss and Qugssuk granite are shown on fig. 5 (measured at the Age and Isotope Laboratory, University of Oxford, P. N. Taylor, pers. comm. 1987). The amphibolite and granulite facies gneisses, together with the Igánánguit pink gneiss, plot on a common isochron of 3112 +40 -38 Ma. Separate calculations for each of the three groups in the isochron of fig. 5 give the same age

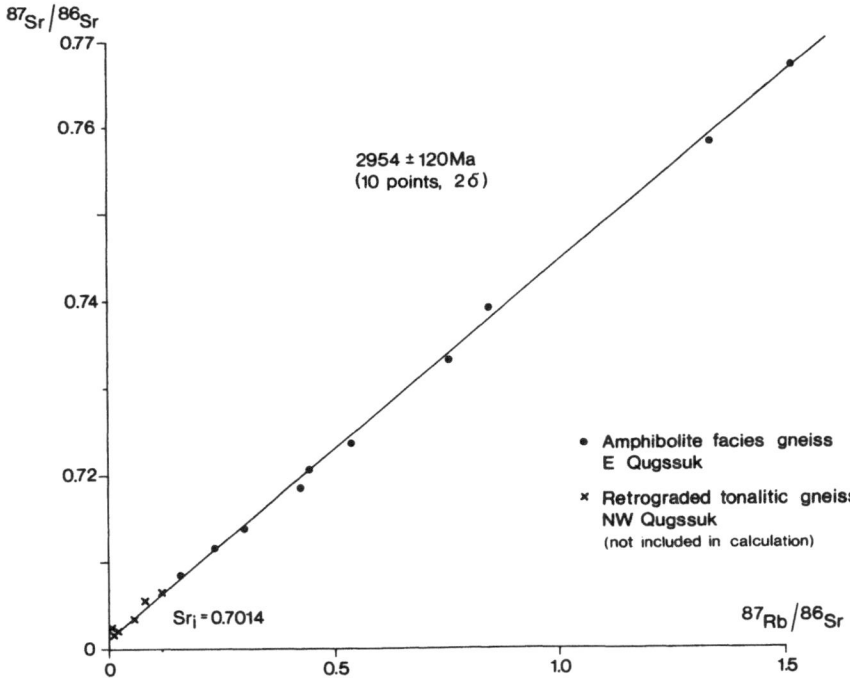

Fig. 4. Rb-Sr isochron diagram of unaltered amphibolite facies gneiss (dots), with samples of retrograded gneiss (crosses) for comparison. The decay constant 1.42×10^{-11} y^{-1} was used for $^{87}Rb/^{87}Sr$. See Garde et al. (1986) for details of analytical methods.

although the error is increased. The retrograded gneisses (open circles) plot below this line in an elongated cluster of points with less radiogenic Pb-Pb compositions than the other gneisses. The lead isotopic composition of the Qugssuk granite (crosses) resembles that of the retrograded gneiss. Lead from the Qugssuk granite also plots below the isochron, but is somewhat more radiogenic.

4. DISCUSSION

4.1. Rb-Sr and Pb-Pb isotope data

The poorly defined Rb-Sr age of 2954 +/-120 Ma of the amphibolite facies gneisses is considered to be in agreement with the Pb-Pb age obtained by Taylor et al. (1980) from granulite facies gneisses and with the zircon U-Pb age of 2982 +/-7 Ma from the Taserssuaq tonalite (see above).

The combined Pb-Pb age of 3112 +40 -38 Ma (fig. 5) is significantly older than the Pb-Pb age of 3000 +/-70 Ma obtained by Taylor et al.

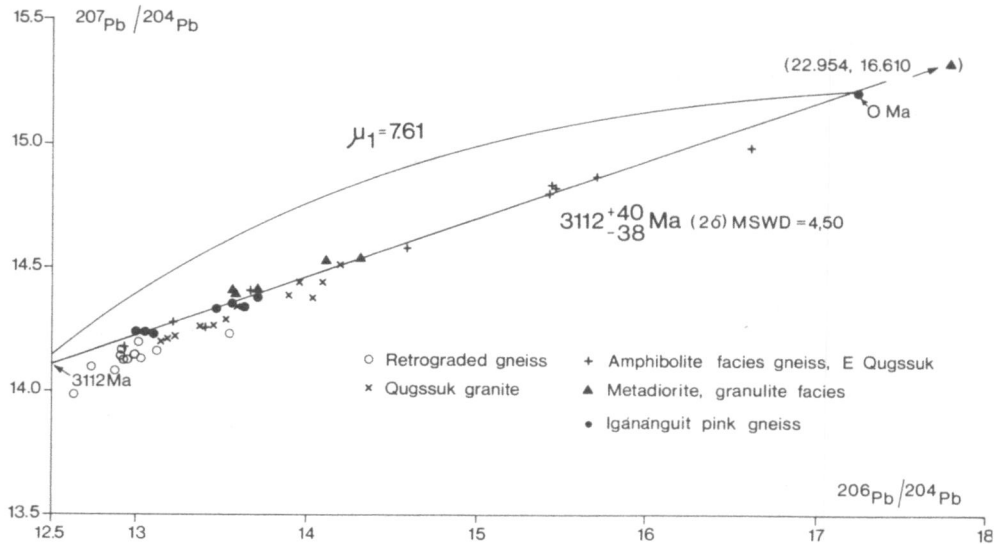

Fig. 5. ^{207}Pb/^{204}Pb vs. ^{206}Pb/^{204}Pb plot of gneisses and granites around
Qugssuk, northern Godthåbsfjord. The age calculation is based on
amphibolite facies gneiss, granulite facies metadioritic gneiss and
Igánánguit pink gneiss. Analytical methods and decay constants as in
Taylor et al. (1980).

(1980) from granulite facies rocks sampled along the outer coast of
southern West Greenland from Nordlandet to Sukkertoppen. The ul value of
7.61 from the three groups of gneisses seems to indicate a short crustal
residence time of their parent rocks. The 3112 Ma age may represent the
time of intrusion; however it is possible that there has been a slight
contamination of the some of the points to the left in the diagram with
unradiogenic lead, see also below. This would tilt the reference line a
little and increase the apparent age. Therefore the interpretation is
preferred that the three groups gneisses were formed about 3000 Ma ago,
as suggested by the Rb-Sr and zircon U-Pb ages.
    The Rb-Sr age of the Qugssuk granite, 2969 +/- 32 Ma, is considered
to be the minimum age of the retrogression event in the grey gneisses,
since the Qugssuk granite cuts retrograde textures in the gneisses.
    The retrograded gneisses do not give a Pb-Pb isochron. Their lead
isotopic compositions cannot have been derived solely from precursors
similar to the unaltered amphibolite facies gneisses at the closure of
their U-Pb systems c. 3000 Ma ago, the lead in the retrograded gneisses
being less radiogenic than that in the unaltered gneisses. The total
lead content of the retrograded gneisses is lower than that in unaltered
amphibolite facies gneisses but higher than in the granulite facies
gneisses (fig. 2a). It is believed that the retrograded gneisses lost
most of their original uranium and some of their lead during granulite
facies metamorphism (without alteration of its isotopic composition).

They were then contaminated during the retrogression with variable
amounts of less radiogenic lead, probably old lead from U-depleted lower
crustal rocks.

Early Archaean Amîtsoq gneisses with very unradiogenic lead
isotopic compositions occur in the Godthåbsfjord area, and Taylor et al.
(1980) have shown that some Nûk gneisses in Godthåbsfjord were emplaced
with significant proportions of old, unradiogenic lead. In the Qugssuk
area the gneisses initially may only have contained very small amounts
of old crustal lead.

The lead isotopic composition of the Qugssuk granite, with points
in an elongated cluster below the isochron (fig. 5), supports field
evidence that it formed by anatexis of retrograded gneisses. The Qugssuk
granite seems to have inherited isotopically heterogeneous lead with an
old component from the retrograded gneisses, but obtained a higher U/Pb
ratio and therefore plots to the right of the retrograded gneisses.

While lead with a foreign isotopic composition was brought into the
area during the retrogression and contaminated the gneisses, this cannot
be demonstrated for Sr. The retrograded gneisses plot on and close to
the origin of the Rb-Sr reference line for unaltered amphibolite gneiss,
fig. 4. Furthermore the strontium source for the Qugssuk granite had a
homogeneous isotopic composition, or strontium homogenisation took place
during its emplacement.

4.2. Retrogression

The retrograded gneisses resemble the unaltered amphibolite facies
gneisses in terms of general large-scale structure, relations to older
supracrustal rocks and younger remobilised rocks of crustal origin.
Field observations and petrography indicate that the retrogression took
place in a static environment in a large halo around preserved granulite
facies rocks, and that it penetrated into fresh amphibolite facies
areas. There is a tight age constraint of the retrogression, which took
place around 3000 Ma ago, very shortly after the formation of the
gneisses. Retrogression took place in essentially solid rocks but was
locally accompanied by anatectic melting.

The retrograded gneisses can be characterised geochemically by
their general low contents of large lithophile cations, and by their
almost constant level of these elements irrespective of SiO_2 content.
The isotope data show that U, Pb and Rb were highly mobile during the
retrogression. While lead with a foreign isotopic composition was
introduced during retrogression, this cannot be shown for Sr.

4.3. Mechanism of retrogression

It is suggested that the retrogression in the northern
Godthåbsfjord-Fiskefjord region was caused by hydrous fluids, which were
released during ongoing static granulite facies metamorphism, at and
below the present level of erosion around 3000 Ma ago, not long after
the peak of amphibolite facies metamorphism of the exposed Nûk gneisses
in northern Godthåbsfjord. It is thought (see e.g. McGregor et al.,
1986) that the granulite facies metamorphism was a result of rising

temperature accompanied by thermal dehydration (rather than CO_2 flushing), due to accumulation of heat by continuous injection of tonalitic and dioritic magmas.

Wells (1979) suggested this mechanism of prograde granulite facies metamorphism in the Buksefjorden area south of Godthåbsfjord. However there is evidence that granulite facies metamorphism south of Godthåbsfjord is considerably younger (c. 2800 Ma, Pidgeon & Kalsbeek, 1978; Taylor et al., 1980), and recent field work by Friend et al. (1987) has demonstrated that interpretation of facies boundaries in the area studied by Wells (1979) is complicated by the presence of large-scale post-2800 Ma thrusts. Hence the model of Wells (1979) seems to fit the northern Godthåbsfjord-Fiskefjord region better than the Buksefjorden area where it was proposed.

Fluid movement during retrogression enabled large-scale element transport. Furthermore, the lead isotope data presented here show that not all the rocks and fluids involved in the retrogression were of local, middle Archaean origin. Lead was also remobilised from older, U-depleted rocks. It is possible that the trace element patterns in the retrograded gneisses were directly related to recrystallisation and growth of sodic plagioclase, sometimes at the expense of K-feldspar, during retrogression of the gneisses, but further work on the mineral chemistry of the gneisses is needed to confirm this.

5. ACKNOWLEDGEMENTS

I am grateful to P. N. Taylor, University of Oxford, for the isotope analyses and discussions about their interpretation. Ib Sørensen, Geological Survey of Greenland, and J. C. Bailey, University of Copenhagen, kindly provided XRF major and trace element analyses, respectively. F. Kalsbeek, Geological Survey of Greenland, is thanked for reviewing the manuscript. The director of the Geological Survey of Greenland is thanked for permission to publish this paper.

6. REFERENCES

Beach, A. 1974: Amphibolisation of Scourian granulites. Scott. J. Geol.
10, 35-43.

Friend, C. R. L., Nutman, A. P. & McGregor, V. R. 1987: Late-Archaean
tectonics in the Færingehavn-Tre Brødre area, south of Buksefjorden,
southern West Greenland. J. Geol. Soc. London, 144, 369-376.

Garde, A. A. 1984: Field work between Fiskefjord and Godthåbsfjord,
southern West Greenland. Rapp. Grønlands geol. Unders., 120, 45-50.

Garde, A. A. 1986: Field observations around northern Godthåbsfjord,
southern West Greenland. Rapp. Grønlands geol. Unders., 130, 63-68.

Garde, A. A., Larsen, O. & Nutman, A. P. 1986: Dating of late Archaean
crustal mobilisation north of Qugssuk, Godthåbsfjord, southern West
Greenland. Rapp. Grønlands geol. Unders., 128, 23-36.

McGregor, V. R., Nutman, A. P. & Friend, C. R. L. 1986: The Archean
geology of the Godthåbsfjord region, southern West Greenland. In
Ashwal, L.D. (ed.) Workshop on early crustal genesis: the World's
oldest rocks. Tech. Rep. Lunar planet. Inst., 86-4, 113-169.

Pidgeon, R. T. & Kalsbeek, F. 1978: Dating of igneous and metamorphic
events in the Fiskenaesset region of southern West Greenland. Can.
J. Earth Sci., 15, 2021-2025.

Taylor, P. N., Moorbath, S., Goodwin, R. & Petrykowski, A. C. 1980:
Crustal contamination as an indicator of the extent of early Archaean
continental crust: Pb isotopic evidence from the late Archaean
gneisses of West Greenland. Geochim. cosmochim. Acta, 44, 1437-1453.

Wells, P. R. A. 1979: Chemical and thermal evolution of Archaean sialic
crust, southern West Greenland. J. Petrology, 20, 187-226.

MINERAL CHANGES, ELEMENT MOBILITY, AND FLUIDS ASSOCIATED WITH DEEP
SHEARING IN THE MOUNT HELEN STRUCTURAL BELT, WYOMING, U.S.A.

Thomas P. Hulsebosch, B. Ronald Frost
Department of Geology and Geophysics
University of Wyoming
Laramie, Wyoming 82071
U.S.A.

ABSTRACT. The Mount Helen Structural Belt, located in the Wind River
Mountains, Wyoming, formed during high temperature ductile deformation at
about 2.7 to 2.8 Ga. Rock types within the structural belt are dominated
by mylonitic orthogneiss which anastomoses around boudins of relatively
undeformed orthogneiss. Changes in petrology, chemistry, and fluid
inclusions were studied at orthogneiss - mylonitic gneiss transitions.
Deformation occurred at approximately 700o C and 4.5 to 5.0 kb. At these
P-T conditions the dominant petrological change was physical grain size
reduction by ductile processes. No major mineral reactions accompanied
shearing and the mineralogy of both orthogneiss and mylonitic gneiss
remained basically unchanged. Fluid inclusions within the mylonitic
gneiss contain saline brines while inclusions in the relatively
undeformed orthogneiss are CO_2 rich. No systematic major element
redistribution can be attributed to shearing in the Mt. Helen Structural
Belt. Local modification of trace element chemistry is related to minor
changes in accessory mineralogy related to the formation of sphene. High
temperature ductile deformation in the presence of aqueous brines is not
a sufficient mechanism to modify the chemistry of the deep crust when
major mineral reactions are absent.

1. INTRODUCTION

Element mobility associated with shearing has been documented in many
shear zones where major mineral reactions accompany shearing (Bridgwater,
1977; Etheridge and Cooper, 1981; Hickman and Glassley, 1984; Kerrich et
al. 1977; Sinha et al. 1986; Vocke et al. 1987). In general, moderate to
low temperature shear zones were studied where retrogressive mineral
changes have occurred. Few studies have assessed the effects of high-
temperature shearing on element mobility where mineral reactions are
largely absent. Kerrich et al. (1977,1980) have reported essentially
isochemical behavior for high-temperature shearing where mineral
reactions are limited and no significant transport of aqueous fluid
through the system occurred. This study investigates chemical
modification of the deep crust in the Mount Helen Structural Belt (MHSB)
where high temperature ductile deformation was not accompanied by major
mineral changes, but where infiltration by aqueous fluids did occur.

2. GEOLOGICAL SETTING

The MHSB represents a major ductile deformation zone in the Wind River
Mountains of Wyoming (Granger, et al., 1971) (fig. 1). The structural
belt, which is dominated by mylonitic gneiss, forms a northwest trending
zone approximately 40 km long and 1/2 to 3 km wide. An upper age limit
for deformation of 2.66 +/-.18 Ga (Koesterer et al, in press) is provided
139

D. Bridgwater (ed.), Fluid Movements – Element Transport and the Composition of the Deep Crust, 139–150.
© *1989 by Kluwer Academic Publishers.*

by the Bridger Batholith, which crops out to the west and is intrusive into the Mt. Helen Structural Belt. To the northeast of the MHSB there is an extensive body of Dry Creek Orthogneiss. Originally named by Barrus (1970), we have modified it to refer to the orthogneiss that is the dominant rock type along Dry Creek.

A moderate to strong northeast foliation in the Dry Creek Orthogneiss is cut by the mylonitic fabric of the Mt. Helen Structural Belt. Small ductile deformation zones with attitudes sub-parallel to the Mt. Helen zone are present within the Dry Creek Orthogneiss and become more numerous as the Mt. Helen zone is approached. Within the Mt. Helen zone enclaves of Dry Creek Orthogneiss are preserved as boudins ranging in width from approximately 2 m. to 25 m. and having lengths 2 to 3 times their width. The dominant mylonitic fabric of the structural zone anastomoses around these boudins of orthogneiss. The interface between orthogneiss boudins and mylonitic gneiss is typically gradational across a distance of·less than 30 cm. The abrupt variation in shear strain, and presumably fluid transport, at these interfaces make them ideal locations to study chemical and fluid changes associated with shearing.

3. PETROGRAPHY AND TEXTURE

3.1 Dry Creek Orthogneiss

The Dry Creek Orthogneiss has an estimated outcrop area of 400 km2 based on our reconnaissance mapping and rock descriptions of Perry (1965) and Barrus (1970). The dominant rock type within this area is equigranular, medium grained orthogneiss ranging compositionally from tonalite to granite. Although there is moderate compositional variation in the study area, the mineralogy of the Dry Creek Orthogneiss is relatively uniform consisting of plagioclase, k-feldspar, quartz, biotite, and hornblende. Accessory phases include sphene, apatite, alanite, zircon, Fe-Ti oxide, and scapolite. Tabular feldspars with well preserved subhedral to euhedral crystallinity are interpreted as relict igneous phases. A moderate to strong gneissic fabric is defined by alignment of biotite and, to a lesser degree, hornblende. Pervasive metamorphic segregation is not evident, and the gneiss·has a homogenous, massive appearance. Foliation in the Dry Creek Orthogneiss trends northeast and dips steeply to the southeast. Weak mineral lineations plunging down dip are locally observed (fig. 2).

3.2 Mylonitic Gneiss

Mylonitic gneiss constitutes the major rock type within the Mt. Helen zone and also crops out in narrow ductile deformation zones within the Dry Creek Orthogneiss. Mylonitic gneiss represents Dry Creek Orthogneiss that has undergone intense ductile deformation and dynamic recrystallization. Mineralogically, the mylonitic gneiss is nearly identical to its precursor and contains plagioclase, k-feldspar, quartz, biotite, and hornblende with accessory sphene, apatite, alanite, zircon, and Fe-Ti oxides. Hornblende is locally absent and scapolite is not observed in any samples of mylonitic gneiss.

Overall grain size reduction, particularly the breakdown of biotite folia, accounts for the darker color of the mylonitic gneiss.

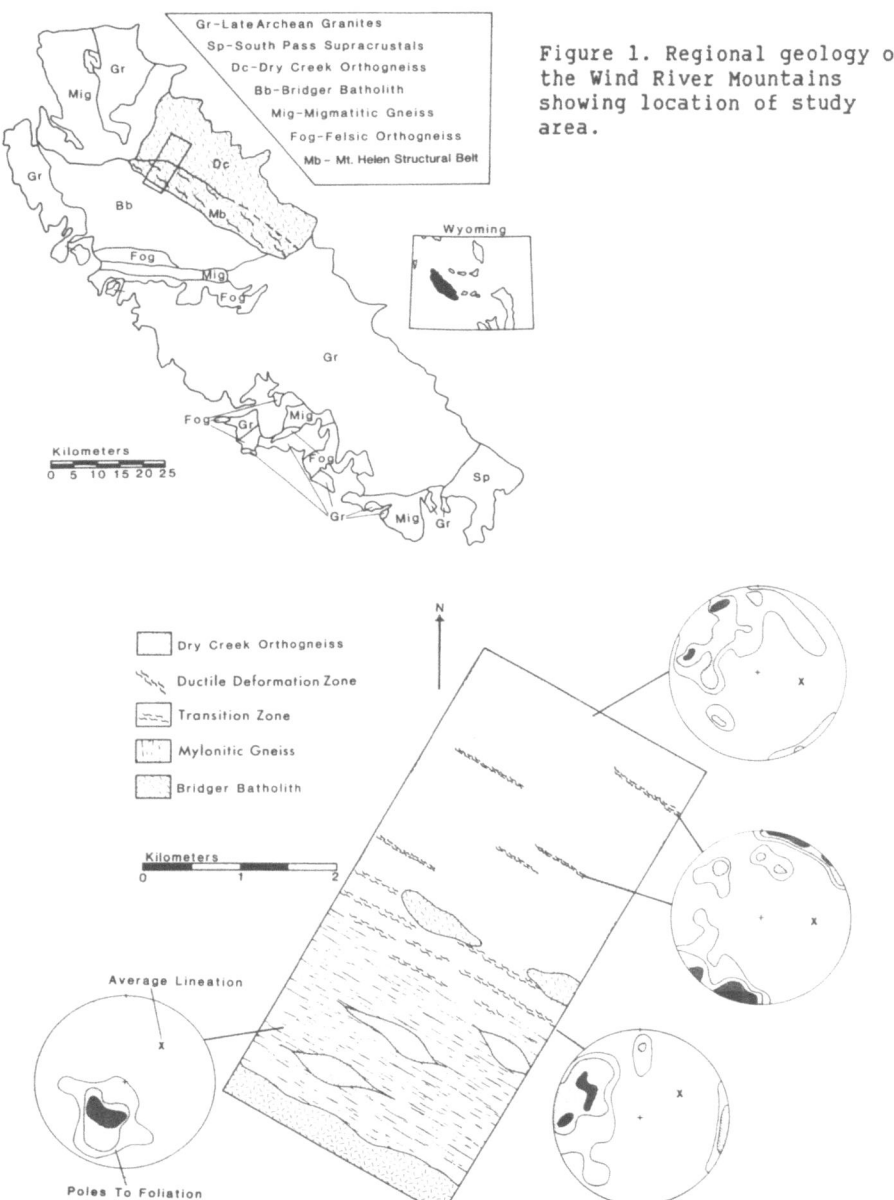

Figure 1. Regional geology of the Wind River Mountains showing location of study area.

Figure 2. Generalized geological map of the transition area from Dry Creek Orthogneiss to Mylonitic Gneiss of the MHSB.

Quartz and feldspar in the mylonitic gneiss form very elongate polycrystalline ribbons and give the gneiss a pinstriped appearance. A zone of transition between Dry Creek Orthogneiss and mylonitic gneiss (fig. 2) contains abundant mylonitic rock in ductile deformation zones that locally form up to 25% of the outcrop.

3.3 Granulite Enclaves

Enclaves of pelitic, mafic, ultramafic, and quartzose rocks occur in both the mylonitic gneiss and the Dry Creek Orthogneiss. Outside the MHSB these rocks form semi-coherent belts several kilometers long and up to 500 m wide. Within the MHSB, however, these belts have been dismembered so that the rocks now form attenuated and isoclinally folded boudins that range in size from 10 to 30 m². Both within the MHSB and in the less deformed rocks to the northeast these enclaves preserve granulite mineral assemblages and metamorphic fabrics that predate the regional metamorphic fabric of Dry Creek Orthogneiss. In the MHSB the cores of the larger boudins still retain granulite mineralogies. These high grade assemblages found in mafic, ultramafic and pelitic rocks consist of orthopyroxene-augite-plagioclase, often without hornblende, orthopyroxene-hornblende-olivine-spinel and k-feldspar-garnet-cordierite-orthopyroxene-biotite-quartz respectively. The sheared margins of the boudins are retrograded to upper amphibolite facies assemblages. These include hornblende-plagioclase in mafic rocks, olivine-anthophyllite-chlorite in the ultramafic rocks and k-feldspar-sillimanite-cordierite-biotite-quartz in the pelites. Quantitative geothermobarometry and phase equilibria for the assemblages in the retrograded boudin rims constrain conditions of shearing to approximately 650° C and 5 kb (Frost et al, 1984).

4. SAMPLING AND ANALYTICAL TECHNIQUE

Mylonitic gneiss - orthogneiss pairs were collected at three localities within the Mt. Helen Structural Zone where the interface between sheared and unsheared rock was sharp and well displayed. Differences in chemistry and fluid inclusions within each sample pair are attributed to changes brought about by shearing. Pre-shearing chemical heterogeneity was identified in weakly deformed orthogneiss samples collected throughout the MHSB and in the Dry Creek region to the northeast. The effect of pre-shearing heterogeneity is minimized by small sample spacing within each mylonitic gneiss - orthogneiss pair.

Samples weighing approximately 2.5 kg were crushed and powdered using tungsten - carbide apparatus. Separate splits of well mixed powder were analyzed at the University of Wyoming for major and trace elements using atomic absorption and instrumental neutron activation analyses (table 1).

5. FLUID INCLUSIONS

Fluid inclusions in doubly polished sections were observed on a Fluid Inc. freeze - thaw stage using synthetic inclusions and distilled water

Rock Type	OG	OG	TR	OG	MY	OG	MY	MY	OG
Sample #	1	4	7	8	9	10	11	12	13
SiO_2	62.88	62.70	69.86	72.75	73.57	66.97	66.17	70.26	67.79
TiO_2	.81	.86	.43	.30	.47	.80	.53	.47	.71
Al_2O_3	16.86	17.96	15.80	14.25	14.05	17.59	16.40	15.75	16.41
Fe_2O_3	5.57	4.35	2.71	2.29	2.06	3.29	4.07	2.46	3.06
MgO	2.03	2.01	1.07	.55	.95	1.63	2.00	1.11	1.76
CaO	4.44	4.19	2.86	2.08	2.78	4.12	4.02	2.93	3.97
Na_2O	4.01	4.48	3.76	3.54	3.89	4.15	3.70	3.87	3.89
K_2O	1.84	2.52	2.85	3.28	1.10	1.71	1.79	2.16	1.39
P_2O_5	.26	.21	.08	.07	.05	.15	.12	.09	.16
La	22.77	23.92	14.37	34.51	39.22	17.3	10.22	12.86	13.04
Ce	92.57	64.59	36.04	84.01	99.93	42.00	24.57	31.25	33.83
Sm	5.23	6.33	2.22	4.27	6.86	.516	.708	2.49	2.78
Eu	1.44	1.56	.504	.500	1.08	1.30	1.18	.761	.935
Tb	.737	.598	.981	1.08	1.96	.297	.297	.810	1.05
Yb	2.16	1.50	1.01	1.10	1.21	.506	.715	.918	1.62
Lu					.278	.086		.127	.146
Sc	11.79	8.85	5.81	5.34	8.18	7.10	10.74	6.13	6.21
Cr	40.93	28.56	21.19	17.08	9.54	48.95	47.87	19.24	28.48
Hf	6.46	4.87	2.54	3.94	5.73	4.59	1.85	4.27	4.00
Ta		.571	.481	.684	.628	.394	.211	.670	1.64
Th	8.18	6.21	3.48	15.21	19.22	.944	.841	2.37	1.45
Cs	1.05	1.32	.687	.781	.746		.131	1.03	.609

Table 1. Chemical analyses. Errors are +/- 2% for major elements and +/-
5%-10% for trace elements. OG=orthogneiss; MY=mylonitic gneiss;
TR=transition sample. Sample sets include: 7-8-9; Set 1, 10-11; Pair 2,
12-13; Pair 3

ice for calibration. A sharp contrast in fluid inclusion composition is
observed between orthogneiss and mylonitic gneiss throughout the study
area. Nearly pure CO_2 fluid inclusions dominate the orthogneiss, with
sparce aqueous inclusions occurring near the mylonite transition.
Crosscutting relationships between trails of aqueous inclusions and
clusters of CO_2 inclusions indicate that the CO_2 inclusions predate
trapping of aqueous inclusions. Melting temperatures for CO_2 inclusions
fall within 1o of -56.60 C indicating their nearly pure composition. The
average homogenization temperature for CO_2 inclusions is 220 C which
corresponds to a fluid density of approximately .80 gm/cc (Angus et al.
1976). CO_2 of this density could not have been trapped at the conditions
of shearing and must represent a fluid that was trapped in the
orthogneiss prior to shearing when P-T conditions were higher.
    All samples of mylonitic gneiss contain abundant, secondary aqueous
brine inclusions which appear to have been trapped during shearing. CO_2
inclusions are observed only in samples collected very near the actual
mylonitic gneiss - orthogneiss transition. Melting temperatures for
aqueous inclusions range from -16.90 C to -25.00 C, corresponding to
salinities of 20.4 to 25 weight % NaCl equivalent (Potter et al, 1978).
Hydrohalide is observed upon cooling in many inclusions indicating
salinities in excess of 23.3 weight % NaCl equivalent (Roedder, 1962).

Due to the lack of quantitative data an phase stability in saline systems
a detailed homogenization study was not performed on aqueous inclusions.

6. MINERAL CHANGES AND ELEMENT RE-DISTRIBUTION

Before chemical changes can be attributed to element mobility, the
chemical variability inherent within the unsheard orthogneiss must be
characterized. As can be seen from Table 2 , the gneiss is quite
inhomogeneous, with silica contents ranging from 63% to 73%. As would be
expected Al_2O_3, FeO, MgO, and CaO decrease with decreasing SiO_2.
Interestingly, Na_2O shows a minor decrease with increasing silica and K_2O
and total rare earths show variable changes. These unusual variations
indicate that the few samples we have from the unsheared orthogneiss
either were subjected to element mobility after crystallization or did
not originally represent a comagmatic suite. Consequently in these rocks
one can only make conclusions about element mobility by comparing
analyses of samples that are collected on a small scale.

To characterize accurately the bulk chemical change in a system, the
possible effects of volume change during shearing must be considered. To
do this we made modal density estimates using a standard point counting
procedure and published mineral densities of Deer et al. (1966). Density
changes within each sample pair are less than .05 gm/cc suggesting that
volume has remained nearly constant during shearing. A graphical
solution of Gresens (1967) equation (see Grant, 1986) can be used to
determine the relative mobilities of the various element, assuming that
the deformation was isovolumetric. These "isocon" diagrams (Fig. 3)
compare the concentration of elements in relatively unsheared orthogneiss
(C^0) to the concentration of the same element in corresponding mylonitic
samples (C^a) for each sample pair. Concentrations are in weight percent
oxide for major elements and ppm for trace elements. Scaling factors
have been applied where necessary to enhance graphical clarity. If the
assumption of constant volume is valid any element that has remained
immobile during shearing will plot on a line through the origin with a
slope of 1.0. Elements that plot above the line of constant volume have
been enriched in the mylonitic gneiss during shearing, while elements
that plot below the line have been depleted from the mylonitic gneiss
(Grant, 1986). Aluminum, which is generally considered immobile, plots
within analytical uncertainty on the line representing constant volume.
This observation supports the assumption of isovolumetric behavior during
shearing in the Mt. Helen zone.

Inspection of fig. 4 reveals that deformation has produced only small
changes in major elements while changes in trace elements are variable
and more pronounced. The most notable major element changes are a
depletion in K_2O in sample pair 1 and an enrichment in SiO_2 in sample
pair 3. Trace element redistribution is strongest for LREE, Sc, Cr, and
Th. These chemical changes can be related to minor mineral reactions
accompanied by local element mobility.

Textural evidence for a reaction involving sphene is observed where
sphene forms rims around grains of ilmenite with biotite and plagioclase
apparently being consumed. We believe that the hydration reaction
biotite + plagioclase + ilmenite + H_2O = k-feldspar + sphene + quartz is
responsible for the production of sphene. In sample pair 1 this reaction
is dominant in the mylonitic gneiss, while in sample pairs 2 and 3 sphene
is formed preferentially in orthogneiss relative to the mylonitic gneiss.

The production of sphene by this reaction is directly reflected in relative changes in REE, Hf, and Th, especially in sample pairs 1 and 2 where sphene production is much more evident that in sample pair 3.

Plots of chondrite normalized REE for the three sample pairs (fig. 5) illustrate local REE mobilization. The abundances and patterns of REE vary considerably between sample pairs reflecting original heterogeneity, but REE patterns within each sample pair are similar and differences in abundance are attributed to shearing. In the two sample pairs where sphene has grown in the orthogneiss relative to the mylonitic gneiss (pair 2 and 3) a general depletion in REE is noted. Where sphene has formed in the mylonitic gneiss (pair 1), a relative enrichment, particularly for the middle REE, is observed. Also plotted on fig. 5 are REE analyses for a sample which represents the transition from orthogneiss to mylonitic gneiss in sample set 1. The localized nature of element redistribution is well displayed in this sample set where REE enrichment in mylonitic gneiss relative to orthogneiss is accompanied by depletion in the transitional sample.

Changes in Cr, Hf, and Sc do not show any systematic relationship to degree of shearing or mineral reaction. These chemical changes must be related to either original heterogeneity or variable retention of enriched fluids. The depletion of K_2O in sample pair 1 and the enrichment of SiO_2 in sample pair 3 are artifacts of slight modal differences in k-feldspar and quartz respectively. These modal differences are interpreted as primary due to the lack of evidence for major reactions involving quartz or feldspar.

7. DISCUSSION

Shearing in the Mt. Helen Structural Belt was not accompanied by major modification of primary mineralogy which explains the isochemical behavior of major elements. However, trace element mobility during shearing did occur on a scale of 10 cm to 30 cm and this element mobilization must require the participation of a fluid phase. The localized nature and the extent of trace element mobility can be used to roughly constrain the relative volume of fluid that accompanied shearing, and the degree of fluid-rock chemical equilibrium.

The widespread occurrence of aqueous fluid inclusions and the pervasive hydration rims around the granulitic enclaves in the MHSB indicates that there was significant fluid flow during shearing. The fluid that accompanied shearing was capable of chemical transport only where minor mineral reactions were liberating the mobilized elements. This indicates that the fluid was not far out of chemical equilibrium with the country rock. Major fluid-rock disequilibrium would have resulted in considerable element mobility for both major and trace elements.

In the absence of major mineral reactions an unlimited volume of fluid in near equilibrium with country rock could pass through a shear zone without altering rock chemistry. In the Mt. Helen zone, hydration reactions in the rims of supracrustal boudins provide the only constraint on the volume of fluid that accompanied shearing. Since it requires 1 mole of H_2O to produce 1 mole of hornblende, hydration of a mafic granulite to an amphibolite with 50% hornblende would require a volume of H_2O that, at 650oC and 5 kilobars, is approximately 2 times the rock volume. This is a minimum estimate since fluid flux almost certainly was higher in the highly ductile quartzofeldspathic gneisses than it was in the less easily deformed mafic gneisses.

	Max. Original Variation	Max. Shear-Related Variation
SiO2	10.05	2.47
TiO2	.56	.27
Al2O3	3.34	1.19
Fe2O3	3.28	.78
MgO	1.48	.65
CaO	2.36	1.04
Na2O	.94	.65
K2O	1.89	2.18
P2O5	.19	.05
La	21.47	7.13
Ce	58.74	17.43
Sm	5.81	2.59
Eu	1.06	.77
Tb	.78	.88
Yb	1.65	.70
Sc	6.45	3.64
Cr	31.87	9.24
Hf	2.52	2.74
Ta	1.25	.97
Th	14.27	4.01
Cs	.71	.42

Table 2.
Maximum chemical variations observed in the original orthogneiss and related to shearing in wt% (major elements) and ppm (trace elements).

8. CONCLUSIONS

The conclusions of this study are: 1. Mineral reactions are necessary to mobilize elements during shearing, regardless of the presence of saline aqueous fluids. 2. Shearing in the Mt. Helen zone was essentially isovolumetric and isochemical with respect to the major elements. 3. Localized trace element redistribution during shearing is controlled by a minor mineral reaction involving sphene. 4. Fluid volumes on the order of 2 times rock volume infiltrated the MHSB and were nearly in chemical equilibrium with country rock.

These results indicate that considerable amounts of aqueous brine can penetrate the deep crust via zones of ductile deformation without producing major chemical modification of the crust.

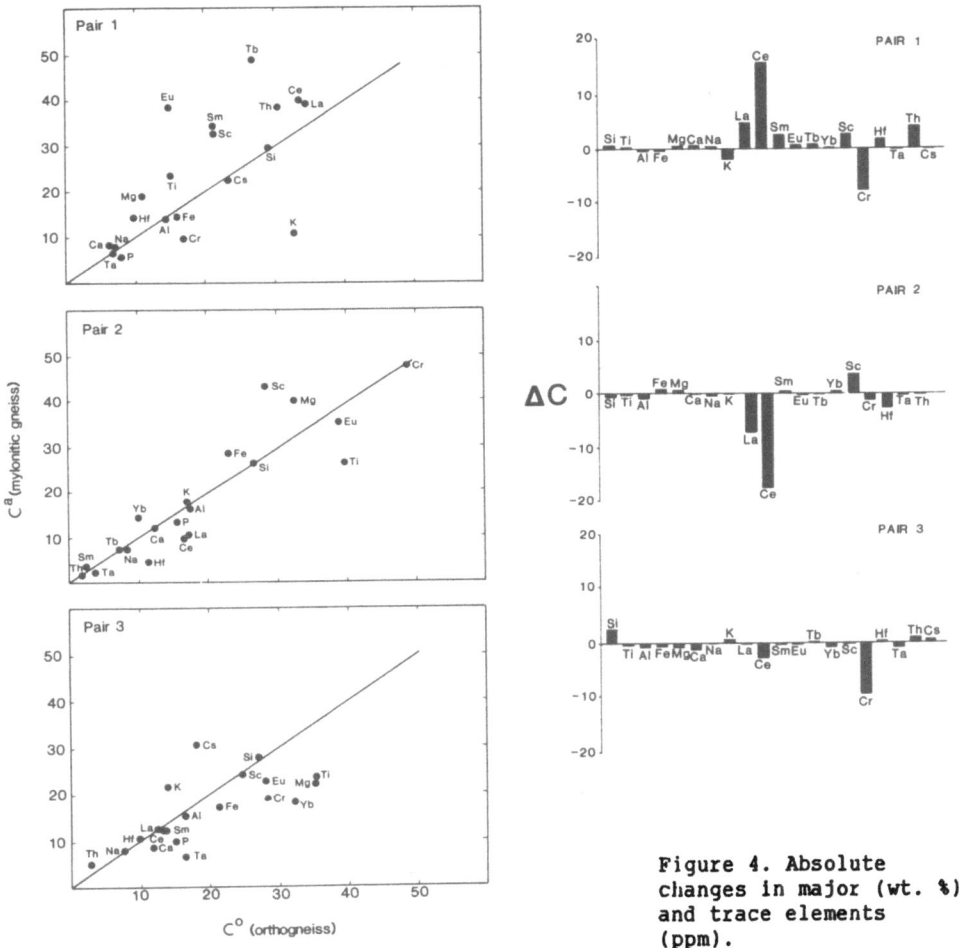

Figure 4. Absolute
changes in major (wt. %)
and trace elements
(ppm).

Figure 3. Isocon diagrams for orthogneiss - mylonitic gneiss pairs.
Concentrations are in wt.% for major elements and ppm for trace elements.
Scaling factors are: Si-.4, Ti-50, Al-1, Fe-7, Mg-20, Ca-3, Na-2, K-10,
La-1, Ce-.4, Sm-5, Eu-30, Tb-25, Yb-20, Sc-4, Cr-1, Hf-2.5, Ta-10, Th-2,
Cs-30.

148

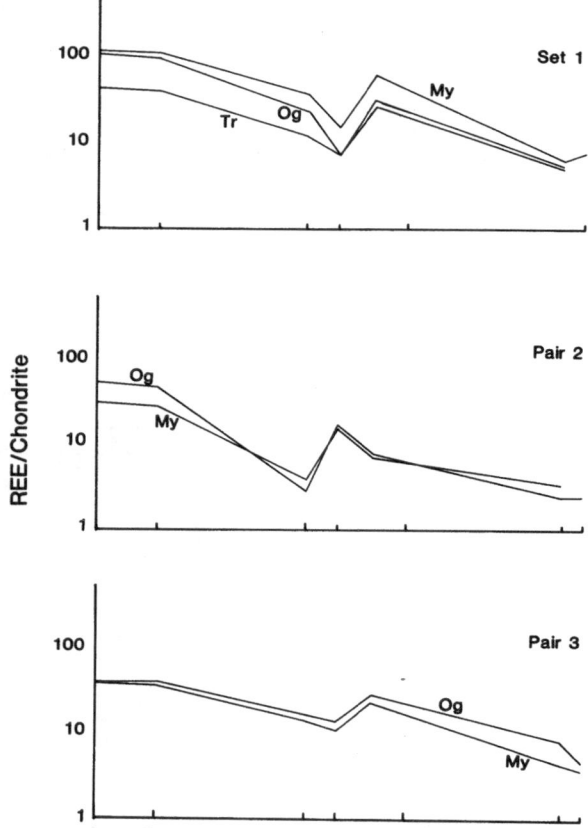

Figure 5. REE patterns
for 3 mylonite -
orthogneiss pairs.

9.ACKNOWLEDGEMENTS

This research was funded by NSF grants EAR-8408357 and EAR-8707296. We
would like to thank Dennis Geist and Carol Frost for their helpful
reviews of this paper.

10. REFERENCES

Angus, S., Armstrong, B., deReuck, K. M., Altunin, V. V., Gadetskii,
    O. G., Chapela, G. A., and Rowlinson, J. A., 1976, 'International
    thermodynamic tables of the fluid state. Vol. 3 Carbon Dioxide'
    Pergamon Press 385p.

Barrus, R. B., 1970, 'Petrology of the Precambrian rocks of the High
    Peaks area, Wind River Mountains, Fremont County, Wyoming' unpub.
    Ph.D. thesis Univ. of Washington.

Bridgwater, D., 1979, 'Chemical and isotopic redistribution in zones
    of ductile deformation in a deeply eroded mobile belt' U.S.G.S.
    Open File Report 79-1239, 505-526.

Deer, W. A., Howie, R. A., and Zussman, J., 1966, 'An introduction to
    the Rock-forming minerals' Longman Group Ltd. 528p.

Etheridge, M. A. and Cooper, J. A., 1981, 'Rb/Sr isotopic and
    geochemical evolution of a recrystallized shear (mylonite) zone
    at Broken Hill' Contrib. Min. Pet. 78, 74-84.

Frost, B. R., Hulsebosch, T. and Koesterer, M. E., 1984, 'Multiple
    granulite metamorphic events in the Archean rocks of the Wind
    River Mountains, Wyoming' Geol. Soc. Amer. Abstracts w. Programs
    16, 513.

Granger. H. C. McKay, E. J., Mattick, R. E., Patten, L. L. and
    MacIlroy, P., 1971, 'Mineral resources of the Glacier primitive
    area, Wyoming' U.S.G.S. Bull. 1319-F, 113p.

Grant, J. A., 1986, 'The isochon diagram - A simple solution to
    Gresens' equation for metasomatic alteration' Econ. Geol. 81,
    1976-1982.

Gresens, R. L., 1967, 'Composition-volume relationships of
    metasomatism' Chem. Geol. 2, 47-55.

Hickman. M. H. and Glassley, W. E., 1984, 'The role of metamorphic
    fluid transport in the Rb-Sr isotopic resetting of shear zones:
    evidence from Norde Stromfjord, West Greenland' Contrib. Min.
    Pet. 87, 265-281.

Kerrich, R., Allison, I., Barnett, R. L., Moss, S., and Starkey, J.,
    1980, 'Microstructural and chemical transformations accompanying
    deformation of granite in a shear zone at Mieville, Switzerland;
    with implications for stress corrosion cracking and superplastic
    flow' Contrib. Min. Pet., 73, 221-242.

Kerrich, R., Fyfe,W. S., Gorman, B. E., and Allison, I., 1977, 'Local
    modification of rock chemistry by deformation' Contrib. Min. Pet.
    65, 183-190.

Koesterer, M. E., Frost, C. D., Frost, B. R., Hulsebosch, T. P.,
    Bridgwater, D., and Worl, R. G., 1988, 'Development of the
    Archean crust in the Medina Mountain area, Wind River Mountains,
    Wyoming (U.S.A.)' Precambrian Res., in press.

Perry, K., 1965, 'High grade regional metamorphism of Precambrian
    gneisses and associated rocks, Paradise Basin Quadrangle, Wind
    River Mountains, Wyoming' unpub. Ph.D. thesis, Yale University.

150

Potter, R. W., Clynne, M. A., and Brown, D. L., 1978, 'Freezing point depression of aqueous sodium chloride solutions' Econ. Geol. 73, 284-285.

Roedder, E., 1962, 'Studies of fluid inclusions I: Low-temperature application of a dual-purpose freezing-heating stage' Econ. Geol. 57, 1045-1061.

Sinha, A. K., Hewitt, D. A., and Rimstidt, J. D., 1986, 'Fluid interaction and element mobility in the development of ultramylonites' Geology, 14, 883-886.

Vocke, R. D., Hanson, G. N., and Gruenenfelder, M., 1987, 'Rare earth mobility in the Roffna Gneiss, Switzerland' Contrib. Min. Pet. 95, 145-154.

THE ATANEQ FAULT AND MID-PROTEROZOIC RETROGRADE METAMORPHISM OF EARLY ARCHAEAN TONALITES OF THE ISUKASIA AREA, SOUTHERN WEST GREENLAND: REACTIONS, FLUID COMPOSITIONS AND IMPLICATIONS FOR REGIONAL STUDIES

A.P. Nutman[1], T. Rivers[1], F. Longstaffe[2] & J.F.W. Park[3]
1, Department of Earth Sciences, Memorial University of Newfoundland, Canada
2, Department of Geology, University of Alberta, Edmonton, Alberta, Canada
3, Derbyshire College of Higher Education, Derby, Derbyshire, England

ABSTRACT

The Isukasia area, southern West Greenland, is dominated by early Archaean rocks. These rocks are cut by major Proterozoic faults (mylonites), associated with which are greenschist facies retrograde metamorphism and metasomatism. Tonalitic gneisses are the dominant lithology of the area, and are used in this study to characterise the retrograde metamorphism and metasomatism. Retrogression took place between 350 and 450°c and involved a low δ-^{18}O, H_2O-rich, CO_2-bearing fluid of meteoric origin. Fluid access to the tonalitic gneisses occurred along the faults, and movement may have been by seismic pumping. Retrograde reactions adjacent to the fault zones resulted in leaching of SiO_2, mobility of alkalis on a scale of less than 100m, and some disturbance of trace element chemistry (e.g. Rb, Pb, Cl). The degree of SiO_2 leaching from retrogressed gneisses close to the faults is compatible with fluid:rock ratios of the order of 5:1. On the other hand, SiO_2 (as vein quartz) is commonly enriched in the faults, indicating fluid:rock ratios of at least 100:1, but probably much greater. These results, together with isotopic data (Rosing, 1983; Baadsgaard et al., 1986) clearly indicate that rocks affected by Proterozoic retrogression are too altered to provide reliable information on early Archaean crustal genesis. This rules out at least half of the ca. 3810 Ma Isua supracrustal belt.

151

D. Bridgwater (ed.), Fluid Movements – Element Transport and the Composition of the Deep Crust, 151–170.
© *1989 by Kluwer Academic Publishers.*

INTRODUCTION

The Isukasia area of southern West Greenland, which is dominated by very early Archaean rocks with amphibolite facies mineral assemblages (Table I, and see Nutman, 1986 for summary), is cut by numerous faults, some with major displacements (Fig. 1). Retrograde greenschist facies metamorphism and associated metasomatism spatially related with these faults is evident in the field (Nutman, 1982; Garde et al., 1983; Nutman et al., 1984). Most publications dealing with the geochemistry of these early Archaean rocks have either ignored the retrogression and metasomatism or only made passing reference to it (see Nutman, 1986).

Table I. Chronological sequence in the Isukasia area. Dates used are taken from Baadsgaard et al. (1986a,b).

10. Injection of crustally derived granite sheets; regional·thermal event. Faulting under greenschist facies conditions and hydrothermal activity. Injection of basic dykes 1,600–2,100 Ma.
9. Intrusion of pegmatites. ca. 2,500 Ma.
8. Local intrusion of granitic-granodioritic sheets. Polyphase deformation and amphibolite facies metamorphism. 2,600–3,100 Ma.
7. Intrusion of the Tarssartoq dykes, probably equivalent to the Ameralik dykes of Godthåbsfjord. >3,100 <3,400 Ma.
6. Deformation giving rise to upright folds.
5. d) Intrusion of pegmatitic gneisses. ca. 3,400 Ma. c) Intrusion of thin tonalitic sheets. b) Structural intercalation of supracrustal and gneiss units. a) Deformation, strongest in the south. It is possible 5a,b,c,d were contemporaneous.
4. Intrusion of the Amîtsoq white gneisses. ca. 3,600 Ma.
3. Intrusion of the Inaluk dykes.
2. Intrusion of the Amîtsoq grey gneisses, deformation. 3,700–3,750 Ma.
1. Formation of the Isua supracrustal rocks and intrusion of mafic and ultramafic rocks. ca. 3810 Ma.

In this paper we present field, petrographic, whole rock and mineral chemistry, and stable-isotope data in order to characterise the retrograde metamorphism and metasomatism associated with the faulting. This study concentrates on the mesocratic biotite tonalites of the 3750–3700 Ma Amîtsoq grey gneisses (unit 2, Table I), which are the dominant lithology east of the Ataneq fault (Fig. 1). The

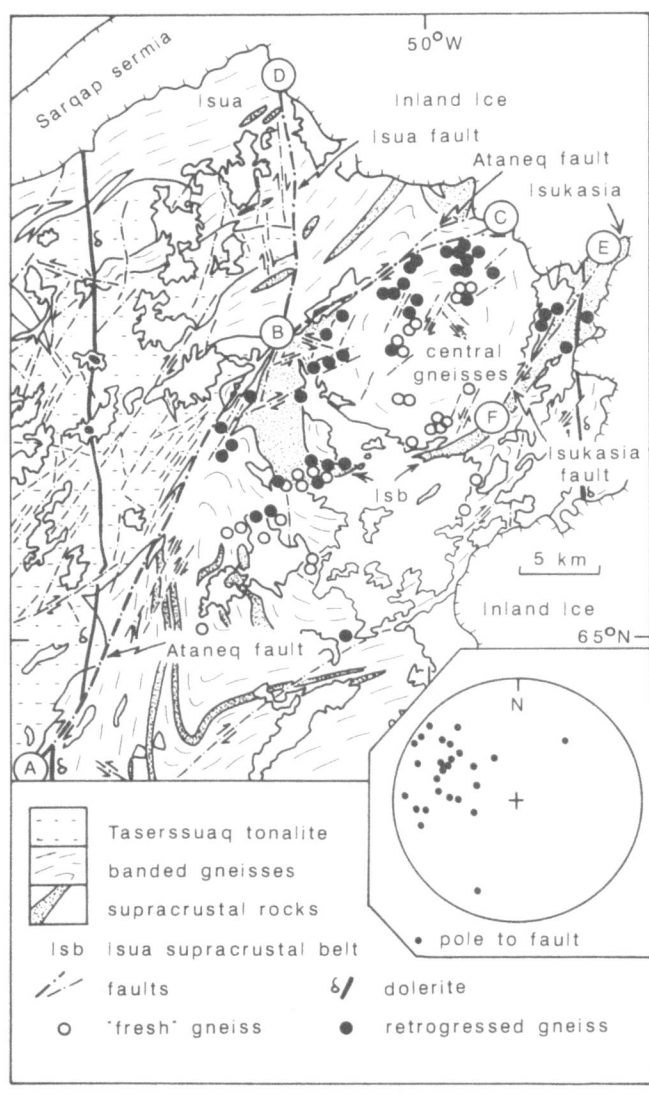

Figure 1. Geological sketch map of the Isukasia area

grey gneisses are associated with subordinate younger white
gneiss of granitic composition and older supracrustal
material (e.g. Baadsgaard et al., 1984, 1986a; Nutman &
Bridgwater, 1986). To the west of the Ataneq fault
tonalitic grey gneisses also predominate, but they are of
mid-Archaean age (Garde et al., 1983, 1986).

The aims of this study are twofold. Firstly to provide
information on the composition, minimum amount and origin
of the fluids present during the retrograde metamorphism,
together with inferences on fluid-rock reactions (see also
Rosing, 1983); and secondly, to evaluate the nature of the
metasomatism associated with the retrograde metamorphism
along the faults. Clearly this must be understood before
geochemical and isotopic data from the area are used to
constrain models of early Archaean crustal genesis (Gill et
al., 1981; Rosing, 1983; Nutman et al., 1984; Nutman &
Bridgwater, 1986; Baadsgaard et al., 1986b). In this study
we use the terms ´fresh´ to apply to rocks not influenced
by the faults and associated fluid movements; and
´retrograde´ to describe gneisses which have undergone
mineralogical and geochemical alteration adjacent to the
faults, but do not contain evidence of high strain; and
´mylonitic´ to describe the high strain rocks of the fault
zones which also contain evidence of geochemical
alteration.

FAULTS - AGE, STRUCTURE AND RETROGRADE METAMORPHISM

The western part of the Isukasia area is cut by numerous,
predominantly northeast trending faults (Fig. 1). Most have
small displacements, but the importance of the Ataneq fault
(McGregor, 1979) is shown by the truncation of the late
Archaean Taserssuaq tonalite (Garde et al., 1983, 1986). In
the eastern part of the area, there are fewer faults and
most have displacements of less than 50 m, apart from the
Isukasia fault (E-F, Fig. 1). A petrographic survey of more
than fifty samples of the gneisses (for locations, see Fig.
1) has revealed that retrograde metamorphism under
greenschist facies conditions (giving rise to the
assemblages albite + epidote + muscovite + biotite + quartz
± calcite ± chlorite ± K feldspar) is prevalent within 3 km
of the Ataneq fault and also around the Isukasia fault. As
much as half of the Isua supracrustal belt lies within the
´retrograde zones´ associated with these faults (Fig. 1).
The description of the faulting that follows is restricted
to the Ataneq fault, but other major faults have
essentially the same character.

The Ataneq fault is 50 to 100 m wide, dips to the
southeast and is flanked by minor subparallel faults (Fig.
1). Using basic dykes as markers, dextral strike slip

displacements of at least 4 km is inferred on the Ataneq
fault (Chadwick et al., 1983; Garde et al., 1983). Rare
stretching lineations suggest there is also a reverse
dip-slip component of movement on the Ataneq fault.
Although most faults in the Isukasia area have the same
sense of displacement as the Ataneq fault, there are also a
few small faults trending between 120° and 140° with
sinistral strike-slip displacement (Fig. 1). Garde et al.
(1983) suggested that the regional pattern of Proterozoic
faults may comprise a conjugate system resulting from
approximately E-W compression. However, the orientation of
the Ataneq fault does not fit precisely with this model. It
is possible that the section of the Ataneq fault marked A-B
on Fig. 1 exploited a zone of amphibolite facies sheared
rocks formed by thrust movements on the western limb of a
major late Archaean structure, the overturned Kangerssuaq
antiform (Nutman, 1986; Park, 1987). Its more northerly
orientation may be due to either a lowering of the critical
shear stress due to the strong anisotropy caused by late
Archaean shearing, or because the Isua supracrustal belt
and the central gneisses acted as an auge, deflecting
palaeostress trajectories around it (Park, 1987).

The Ataneq fault is dominated by ductile mylonites and
blastomylonites derived from grey gneisses. These rocks are
composed of quartz + albite + epidote + muscovite $\pm$ calcite
$\pm$ dolomite $\pm$ biotite, and are commonly enriched in quartz
compared to the adjacent gneiss complex. Carbonate is
present in both the mylonitic fabric and in later veins.
The excess quartz present in many of the mylonites was
probably derived from veins (see microtextural arguments
below). Locally, mylonitised supracrustal rocks including
banded iron formation (also quartz rich) can be recognised.
Evidence of brittle deformation in the form of cataclasite
bands and thin pseudotachylites is also present. However,
ductile deformation under greenschist facies conditions was
predominant.

The temperature range of retrograde metamorphism and
associated mylonitisation can be estimated by a number of
methods. K-feldspar - plagioclase thermometry (Whitney &
Stormer, 1977) on retrogressed gneisses using mutual
K-feldspar - plagioclase boundaries, albite developed along
cracks in microcline, and microscopic albite exsolution
from microcline has yielded temperatures between 300 and
400°C. Using the partitioning of carbon isotopes between
carbonate and graphite, Perry & Ahmad (1977) calculated
temperatures of 400 to 500°C from a part of the Isua
supracrustal belt where retrogression is widespread. Using
oxygen isotope quartz - magnetite thermometry on banded
iron formations in the Isua supracrustal belt Perry et al.
(1978) calculated temperatures of ca. 390°C. This latter
temperature comes from rocks which display amphibolite

facies assemblages with some evidence of superimposed retrogression under greenschist facies conditions. The temperature may reflect equilibration during faulting and retrogression and/or the ca. 1600 Ma thermal event.

The pressure at which retrogression took place is more difficult to estimate. However, Rosing (this vol.) records kyanite in Proterozoic shear zones in the Isua supracrustal belt. These data allow a reasonable estimate of the minimum P,T conditions during retrograde metamorphism of ca. 350 to 450°C with pressures of at least 3 kbars, assuming the kyanite crystallised during the faulting.

The timing of the faulting and retrograde metamorphism is constrained by the following field relationships and isotopic determinations. Much of the faulting post-dated intrusion of Proterozoic dolerite dykes, which have been dated at between 2100 and 1950 Ma (Kalsbeek et al., 1978; Wagner, 1982). Retrogressed margins of Archaean basic dykes in the area show evidence of disturbance of their Rb-Sr isotope systematics in the mid-Proterozoic, with one dyke showing evidence of disturbance in the Cambrian (Bridgwater, Nutman & Pedersen, unpublished data). Rosing (1983) reported mid-Proterozoic disturbance of Rb-Sr and Sm-Nd isotopic systems in rocks of the Isua supracrustal belt that show evidence of greenschist facies retrogression. Baadsgaard et al. (1986b) showed that the Pb-Pb isotopes of retrogressed gneisses in the area were disturbed during the mid-Proterozoic. The faulting thus either preceeded or was related to a 1600 Ma thermal event that reset the Rb-Sr isotopic system of biotite throughout the region (Baadsgaard et al., 1976, 1986a; Garde et al., 1986) and was associated with intrusion of sparsely distributed granite dykes (Kalsbeek et al., 1980). The faulting may possibly be related to middle-Proterozoic deformation in the Nagssugtoquidian mobile belt to the north. A similar history of faulting has been reported elsewhere in southern West Greenland (Berthelsen & Bridgwater, 1960; James, 1975; Chadwick et al., 1983; Garde et al., 1983; Smith & Dymek, 1983).

WHOLE ROCK GEOCHEMISTRY AND STABLE ISOTOPES

Some of the more significant bulk compositional differences between ´fresh´ grey gneisses and those affected by retrograde metamorphism (i.e. excluding mylonitic samples) are shown as a series of histograms (Fig. 2). The retrogressed samples show significant increases in volatiles (mostly H_2O, but also CO_2, Cl and N), Rb, K_2O and perhaps Pb (not all shown in Fig. 2) and loss of SiO_2 and Na_2O compared to the fresh rocks.

The increase of K_2O relative to Na_2O in the grey gneisses may be at least partly compensated for by the opposite change seen in interlayered granitic white gneisses (Fig. 3), suggesting that there may have been a re-distribution of these elements between grey and white gneiss units on a mesoscopic scale.

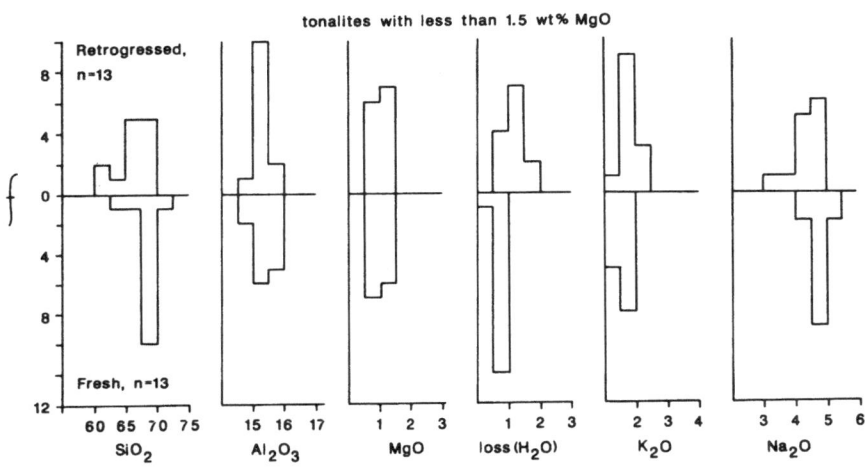

Figure 2. Major element compositional differences between fresh gneisses and retrogressed grey gneisses.

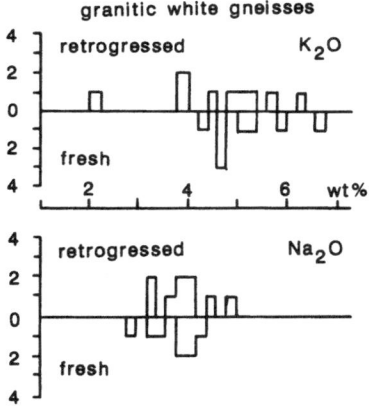

Figure 3. Differences in K_2O and Na_2O abundances between 'fresh' and retrogressed grey gneisses.

REE´s (rare earth elements) do not appear to be greatly fractionated in moderately retrogressed grey gneisses, but mylonite samples from the Ataneq fault showing significant enrichment in SiO_2 are depleted in REE´s in comparison to fresh samples (Fig. 4). This is attributed to dilution of the REE´s by silica addition. There has also been some depletion of light REE´s with respect to the heavy REE´s (Fig. 4).

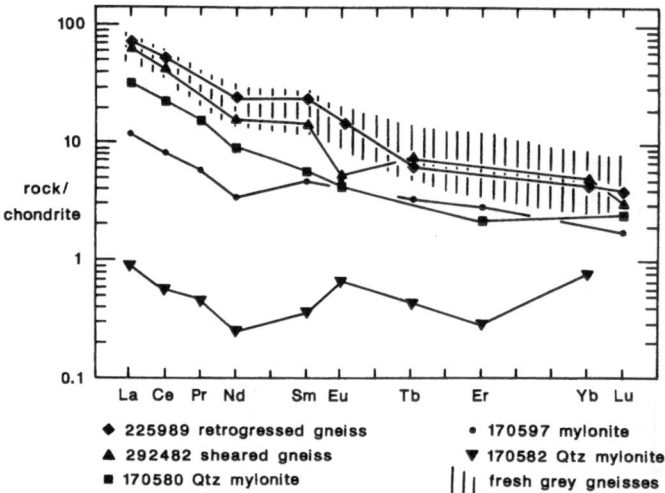

♦ 225989 retrogressed gneiss   • 170597 mylonite
▲ 292482 sheared gneiss   ▼ 170582 Qtz mylonite
■ 170580 Qtz mylonite ||| fresh grey gneisses

Figure 4. REE chemistry of fresh grey gneisses, retrogressed grey gneisses and mylonites.

Oxygen isotope results presented by Baadsgaard et al. (1986b) showed that grey gneisses with petrographic indications of retrogression have lower mean $\delta^{18}O$ values (+3.3 to +7.9‰; mean = +6.6‰SMOW) than fresh grey gneisses (mean = +7.6‰SMOW), with a single mylonite aqnalysed having a $\delta^{18}O$ value of -1.3‰SMOW. These results suggest that retrograde metamorphism was accompanied by incursion of a low $\delta^{18}O$ fluid phase. In this study we have analysed additional high strain gneisses, quartz-rich mylonites and carbonate minerals from mylonites (Table II, Fig. 5). The mylonites and high strain gneisses have whole-rock $\delta^{18}O$ values between -7.8 and +4.2‰SMOW. Calcite has $\delta^{18}O$ values of +3.4 to +7.0‰SMOW and $\delta^{13}C$ values of -4.6 to -4.4‰PDB. Fe-dolomite has a $\delta^{18}O$ of 12.1‰SMOW and $\delta^{13}C$ of -12.9‰PDB.

On the premise that the SiO_2-rich mylonites are dominated by hydrothermal quartz (see microtextural arguments below), it follows that the mylonites with the most negative $\delta^{18}O$ value (-7.8‰SMOW) most closely reflect the hydrothermal source. Applying the Friedman &

O´Neil (1977) oxygen isotope quartz-water thermometer at 350 and 450°C (likely temperature range of retrogression, see above) yields an estimate of the oxygen isotope composition of the hydrothermal fluid in the fault (Table III). The low $\delta^{18}O$ values obtained for the fluid (as low as -13.6, Table III) indicates that meteoric water must have been involved in the faulting and retrogression.

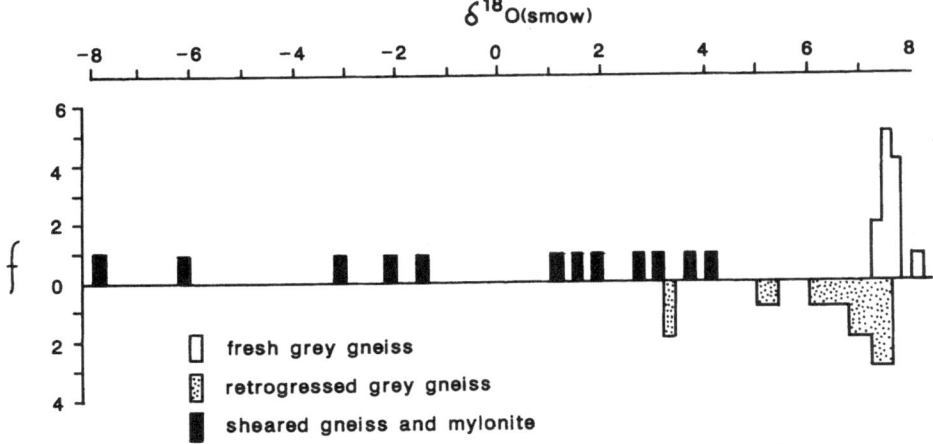

Figure 5. Whole rock oxygen isotope variation of fresh grey gneisses, retrogressed grey gneisses and mylonites expressed as $\delta^{18}O$ SMOW.

$\delta^{18}O$ values as low as those indicated in Table III for the fluid of meteoric origin are interpreted to indicate a cold climate at the time of faulting (see Dansgaard, 1964).Other studies (e.g. Kerrich et al., 1984) have shown that meteoric water can penetrate to substantial depths along faults.

From petrographic evidence it is clear that most of the carbonate which was employed in oxygen isotope determinations was located in veins in the mylonites, rather than disseminated through the matrix (Table II). Use of the oxygen isotope water-calcite and water-dolomite thermometers of Friedman & O´Neil (1977) has permitted the $\delta^{18}O$ values of fluids in equilibrium with the analysed carbonates at 350° and 450°C to be calcultated (Table III). The diverse $\delta^{18}O$ values calculated for water in isotopic equilibrium with calcite, dolomite and quartz at 350 and 450°C show that these minerals cannot have equilibrated with a fluid phase with a unique oxygen isotope composition.

The $\delta^{13}C$ values for the carbonates of -4.4 to -12.9% PDB is unusual (Table II) for mylonites. Such a range is

Table II. Oxygen ($\delta^{18}O$ SMOW) and carbon ($\delta^{13}C$ PDB) isotope data for grey gneisses and mylonites. Samples of the 2830xx series come from the extension of the Ataneq fault south of the area shown in Fig. 1.

fresh grey gneiss		undeformed, retrograde grey gneiss		sheared gneiss and mylonite		
sample	$\delta^{18}O$	sample	$\delta^{18}O$	sample	$\delta^{18}O$	$\delta^{13}C$
225887	+7.62	225841	+7.34	170580	+1.95	
225889	+7.65	225851	+6.12	170597	+3.09	
225892	+7.46	225858	+6.76	292478	+5.27	
225893	+7.46	225902	+5.00	170579	+2.75	
225909	+7.46	225966	+5.25	167633	+1.42	
225942	+7.46	225989	+3.30	225987	+3.57	
225943	+7.23	236907	+7.17	170587	+1.10	
229403	+7.25	236908	+7.49	283063	-7.83	
229444	+7.39	236933	+7.87	283068	-6.21	
229450	+7.52	236959	+7.24	283040	+4.16	
236946	+8.11	236961	+7.50	283062	-2.27	
236968	+7.75	237000	+7.08	283067	-0.18	
236991	+7.54	292115	+7.24	283002	+2.77	
236996	+7.38	292116	+6.88	283003	+3.09	
		292117	+6.34	calcite	+7.03	-4.38
		292483	+6.86	dolom.	+12.10	-12.90
				283010	-3.18	
				calcite	+3.44	-4.62

Table III. Calculated oxygen isotopic compositions ($\delta^{18}O$ SMOW) of fluid from measured isotopic compositions of the most negative ^{18}O quartz-rich mylonite (interpreted to be dominated by hydrothermal quartz) and carbonate from veins in mylonites at different temperatures.

	$\delta^{18}O$ measured	calculated $\delta^{18}O$ fluid		
		350°C	450°C	5°C
Quartz-rich mylonite				
283063	- 7.83	-13.6	-11.4	
283010 calcite	+ 3.44	- 0.9	+ 1.0	-30
283003 calcite	+ 7.03	+ 2.7	+ 4.6	-24
dolomite	+12.1	+ 4.6	+ 6.5	

common in carbonates precipitated from near-surface
groundwaters (ca. 5°C), as described by Keith & Weber
(1964), Salomons & Mook (1986). Assuming the carbonates
equilibrated with water at 5°C yields δ^{18}O values of
-24 to -30%SMOW (Table III), which are likely values for
recent meteoric water in the area (see Dansgaard, 1964).

It is concluded that either the analysed carbonate
veins are unrelated to, and later than the quartz that
dominates many of the mylonites, or they were initially in
isotopic equilibrium with the quartz, but subsequently
re-equilibrated with ground water at a low temperature.

REACTIONS DURING RETROGRADE METAMORPHISM

Where unaffected by retrograde metamorphism, the fresh grey
gneisses consist of plagioclase (oligoclase-andesine) +
quartz + biotite, with hornblende present in more mafic
varieties. Accessory minerals are K feldspar, apatite,
allanite mantled by epidote, sphene and zircon. The first
signs of retrogression are the breakdown of plagioclase
along grain boundaries and intra-grain fractures to albite
+ Fe-poor epidote + muscovite ± calcite ± biotite. This is
accompanied by corrosion of the margins of biotite grains,
resulting in finer grained biotite + albite. Biotite may
also be replaced by epidote, and these reactions together
result in widespread development of the product assemblage
albite + Fe-poor epidote + muscovite ± calcite ± biotite in
plagioclase grains (commonly concentrated along twin
boundaries) and further recrystallisation of biotite. These
stages of moderate recrystallisation are not accompanied by
deformation. The most extensive retrogression occurs close
to major faults such as the Ataneq fault, and is associated
with development of a new mica fabric and small gashes
infilled with quartz ± calcite. In these rocks plagioclase
grains are largely replaced by albite + epidote + muscovite
± calcite. Plagioclase grains may also show ductile
deformation and microfracturing, with the fractures
infilled by muscovite + calcite. Hornblende in more mafic
varieties is replaced by chlorite and biotite. The grey
gneisses are almost devoid of K feldspar, but the white
gneisses contain abundant K feldspar, which is generally
not replaced by muscovite. However, where K feldspar and
plagioclase grains are contiguous, the plagioclase has
encroached into the K feldspar with development of a thin
rim of albite along the contact, and additionally K
feldspars are traversed by small fissures infilled with
albite. This suggests Na and K exchange occurred via the
medium of a fluid phase, which is compatible with the bulk
changes in K_2O and Na_2O contents of these rocks (Figs.
2 & 3).

High strain grey gneisses adjacent to the Ataneq fault
have essentially the same mineralogy as their non-sheared,
but retrogressed equivalents. Many grey gneiss mylonites
are more quartz-rich than their assumed protoliths adjacent
to the faults, and have the mineralogy quartz + albite +
muscovite + epidote ± biotite ± calcite. Most of the quartz
forms ribbons within the mylonite fabric and a lesser
amount of quartz occurs as discordant, locally
ptygmatically folded veins cutting the mylonite fabric.
This suggests that SiO_2 enrichment of these rocks
occurred by repeated formation of quartz veins during
shearing.

Beach (1976) and several authors subsequently have
suggested that retrograde reactions in some shear zones can
be explained in terms of ionic equilibria, with reactant
mineral species being incongreuntly dissolved in a
volumetrically abundant fluid phase. However, it is not
necessary to interpret all reactions as ionic on a cm
scale. Carmichael (1969) pointed out that a balanced
chemical equilibrium may be the sum of two or more local
ionic equilibria on the scale of a thin section. Possible
ionic and balanced chemical equilibria in the retrogressed
grey gneiss are examined below.

In many of the moderately retrogressed grey gneisses
plagioclase is commonly replaced by muscovite + epidote +
albite, without calcite. Point counting the mineralogy of
some partially replaced plagioclase grains indicates that
muscovite and epidote occur in proportions between 2:1 and
4:1. Albite appears stable during this reaction, which can
therefore be qualitatively expressed as follows;

R1) anorthite$_{(plagioclase)}$ == 2-4muscovite + epidote.

On the basis of chemical analyses of fresh plagioclase,
epidote and muscovite, and keeping Al_2O_3 constant
(Carmichael, 1969), this reaction can be expressed in terms
of two mass balanced ionic equilibria:

R1A) plag + .271Mg + 3.43K + 0.022Ti + 1.046Fe + 5(OH) ==
== 2musc + epid + 8.191Na + 0.321Ca + 13.61SiO$_2$ +
1.878(O),

R1B) plag + 0.541Mg + 6.881K + .044Ti + 2.190Fe + 9(OH) ==
== 4musc + epid + 14.86Na + 2.281Ca + 24.93SiO$_2$ +
4.36(O).

The right hand sides of these reactions have a lower molar
volume than the left hand sides, even if all the SiO_2 on
the right hand side was deposited as quartz at the reaction
site. In fact, quartz was not formed in situ, and silica
depletion of the retrogressed gneisses away from the faults

(Fig. 2) suggests that SiO_2 was removed in solution, resulting in a larger volume loss for R1A and R1B when proceeding from left to right.

Recrystallisation of biotite occurred concomitantly with replacement of the anorthite component of plagioclase by muscovite and epidote. Two microstructural generations of biotite can readily be distinguished in most thin sections of the grey gneisses on the basis of grain size and habit, and it has been noted above that albite occurs locally in the embayments of biotite grains. However, individual grains from the two generations of biotite have similar compositions, indicating that re-equilibration occurred between the old and the new grains during or after retrogression. On account of this re-equilibration, the composition of the ´old´ biotite is not known and it is not possible to write a mass balanced ionic equilibrium to account for the recrystallisation of biotite. However, it appears reasonable to qualitatively consider that some of the components (Mg,K,Ti,Fe) on the reactant side of reactions R1A and B came from recrystallisation and modal reduction of biotite. Excess Ca on the product side of reactions R1A and B may have contributed to the growth of epidote in biotite domains. Therefore, the total geochemical flux added to or removed from the rock as a result of reactions such as R1A and B and the recrystallisation of biotite may be principally loss of SiO_2 and gain of H_2O (and other volatites), with subordinate losses of Na_2O and CaO and gain of K_2O (Fig. 2).

Evidence of another retrograde reaction is seen in some grey gneisses in which muscovite + calcite ($\pm$ biotite) formed in approximately equal proportions at the expense of plagioclase. Growth of muscovite + calcite at the expense of two feldspars may be expressed by the balanced chemical equilibrium:

R2) muscovite + calcite + 2quartz ==
    Kfeldspar + anorthite + CO_2 + H_2O,

proceeding from right to left. As most of the fresh grey gneisses are virtually devoid of K feldspar, significant production of muscovite + calcite (more than a couple of modal percent) could not have taken place by R2. However, in white gneisses, which intially contained modally abundant K feldspar as well as plagioclase, R2 potentially could have been important for the production of the assemblage calcite + muscovite. In the grey gneisses, an alternative reaction that involved input of K to the reaction site via a fluid phase which also contained both CO_2 and H_2O, could have permitted formation of the muscovite + calcite assemblage at the expense of anorthite.

In this case production of the calcite + muscovite
assemblage is not constrained by the small modal proportion
of K feldspar. Using microprobe analyses of plagioclase,
muscovite and carbonate from a grey gneiss close to the
Ataneq fault and keeping Al_2O_3 constant, the following
mass balanced ionic equilibrium is derived:

R3) 2.3 plag ($Ab_{92}An_8$) + 0.76Ca + 0.85K + 0.13Fe +
0.15Mg +0.01Mn + H_2O + CO_2 ==
== musc ($Mu_{92}Pa_8$) + calcite ($Cc_{97}Mn_1Sd_1Rd_1$)
+ 2.18Na + 3.407Si.

Throughout the retrogressed grey gneisses, muscovite is
considerably more abundant than calcite, thus muscovite
must have been generated by reactions other than R2 or R3,
namely together with epidote by reactions such as R1A and
B. It is thus concluded that on a regional scale, hydration
reactions such as R1A and B in the grey gneisses were more
significant during retrograde metamorphism than mixed
volatile equilibria such as R2 or R3. However, the
widespread occurrence of small amounts of calcite in the
the grey gneisses serves to show that the fluid phase
during retrograde metamorphism contained both H_2O and
CO_2.
If R2 or R3 took place in an environment with a limited
supply of the fluid phase, as is likely in the gneisses
away from the mylonite zones, and the fluid phase was
initially rich in H_2O, then as these reactions proceeded
the fluid phase would have been buffered by the bulk
composition of the rocks. The fluid would have evolved to a
more hydrous composition, until all the CO_2 was consumed.
At this point hydration reactions such as R1A and B could
have taken place, until CO_2 recharge (if any) of the
fluid by infiltration took place. In some rocks showing
only limited retrogression, plagioclase grains may be
associated with both calcite-bearing and calcite-free
parageneses formed during retrogression. This is evidence
of CO_2 recharge during repeated infiltration of the
fluid. On the other hand, rocks closer to the mylonites
zones were in contact with a larger volume of fluid, as is
demonstrated by the presence of quartz ± carbonate vein
arrays and the much greater degree of retrograde
recrystallisation and the abundance of calcite + muscovite.
Therefore it is likely that close to the faults, higher
fluid:rock ratios were achieved by repeated infiltration of
a hydrous, CO_2-bearing fluid that prevented depletion of
CO_2 from the fluid phase, thus allowing reactions such as
R2 or R3 to continue.

MECHANISM OF FAULTING AND RETROGRADE METAMORPHISM

The concept of seismic pumping as applied to wrench faults by Sibson et al. (1975) provides a mechanism which may account for the mineral-chemical changes associated with the faulting. These authors suggested that during build-up of shear stress on a section of a wrench fault, dilation occurs with the opening of extensional fractures normal to the least principal compressive stress (σ_3). This causes a drop in fluid pressure in the dilatant zone, which is compensated by an inwards migration (sucking-in) of fluid towards it. This is pertinent to the development of shear zones because it has been shown that ductile shear zones can nucleate on dilatant fractures (Segall & Simpson, 1986).

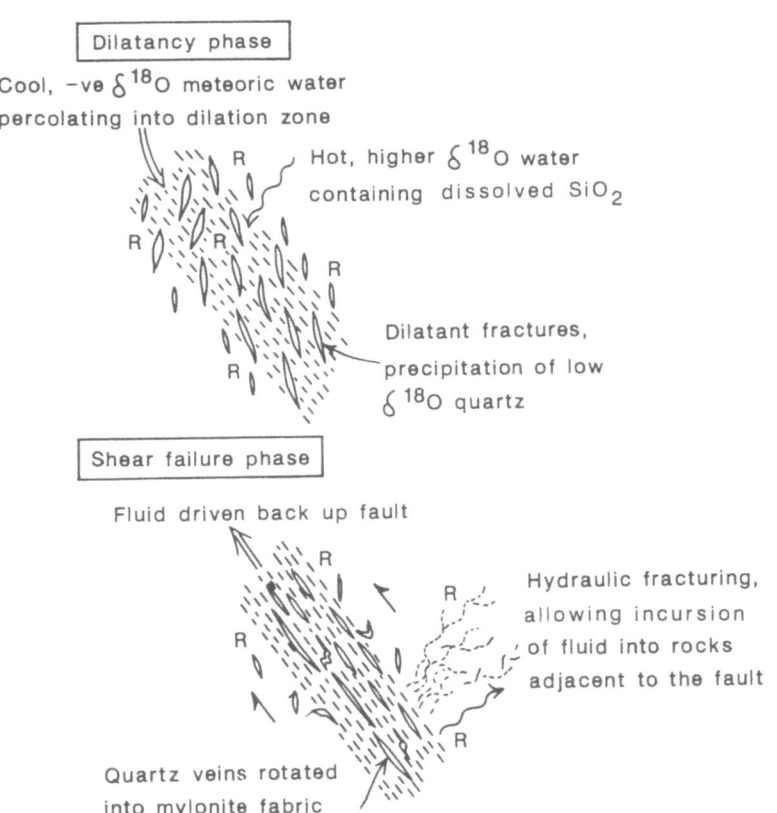

Figure 6. Schematic diagram illustrating faulting mechanism with seismic pumping and metasomatism.

In the model of Sibson et al., the infiltrating fluid infills the dilatant cracks, resulting in a rise in pore pressure, followed by shear failure, contraction and expulsion (pumping-out) of excess fluid. In the case of the stress system proposed for the Ataneq fault (Park, 1987), dilational cracks perpendicular to σ_3 were steeply inclined, thus permitting downward percolation of cool, negative $\delta^{18}O$ meteoric water (Fig. 6). At the present level of exposure in the Isukasia area, these waters interfaced with hotter, higher $\delta^{18}O$ fluid containing dissolved silica generated by reactions such as R1 to R3. As the solubity of SiO_2 in H_2O at the temperatures and pressures of interest is strongly temperature dependant (Fyfe et al., 1979, pp 73), the interfacing of the two fluids resulted in precipitation of low $\delta^{18}O$ quartz concentrated along the fault. Upon shear failure, the new vein quartz underwent ductile deformation and became incorporated into the mylonite fabric, with excess pore fluid being expelled up the fault into lower temperature regimes with further precipitation of low $\delta^{18}O$ quartz. Hydraulic fracturing may have allowed some of the low $\delta^{18}O$ fluid to penetrate laterally away from the shear zone through the adjacent rocks, furthering the progress of the retrograde reactions and providing a medium for transportation of ions in solution as documented by the changes in bulk composition.

Sibson et al. (1975) have shown that during repeated dilation and failure on wrench faults, very large amounts of fluid can pass through and react with the fault zone rocks at depth. At 350 to 450°C and 3 to 4 kbar the solubility of SiO_2 in H_2O ranges between 0.4 and 0.15 wt% (Fyfe et al., ibid). Thus for a mylonite containing 20 wt% precipitated ´hydrothermal´ quartz, the fluid:rock ratio must have been in excess of 100:1, and probably considerably higher. The marked changes of $\delta^{18}O$ values indicate that a large volume of ´fresh´ fluid passed through the faults, rather than a small volume of fluid being recycled numerous times. Areas distant from the faults that suffered only moderate retrogression appear to have had less than 1 wt% SiO_2 leached from them (Fig. 2). Based on the SiO_2 solubility data given above, this amount of SiO_2 depletion would require much lower fluid:rock ratios, perhaps in the order of 5:1. Thus we can demonstrate a transition from a strongly fluid-dominated system within and close to the faults to a less fluid -dominated system in areas of moderate retrogression.

Changes in bulk chemistry and in O, Pb, Sr and Nd isotopes can thus be correlated with fluid movement during retrogression along and associated with faults in this and other studies (Baadsgaard et al. 1986b; Rosing, oral comms, 1983, 1984; Nutman, 1986). The Isukasia area contains the best preserved remnants of some of the oldest terrestrial rocks and is therefore a target for geochemical studies to place constraints on terrestrial evolution. Evidence given here and in the aforementioned isotopic studies shows that the middle Proterozoic retrograde metamorphism has significantly altered the original geochemical signature of a large volume of these already restricted rocks. Clearly this alteration must be taken into account if future geochemical studies of crustal evolution are undertaken.

AKNOWLEDGEMENTS

A.P.N. was supported by the Royal Society of London, the Danish Natural Science Research Council and the Geological Survey of Greenland. Isotope research by F.J.L. at the University of Alberta and the research of T.R. are supported by grants from the National Sciences and Engineering Research Council of Canada. F.J.W.P. was supported by the British N.E.R.C. at the University of Exeter, and by the Geological Survey of Greenland (GGU). The Director of GGU is thanked for permission to publish this paper.

REFERENCES

Baadsgaard, H. Lambert, R.St.J. and Krupicka, J. 1976. Mineral isotopic age relationships in the polymetamorphic Amîtsoq gneisses Godthaab district, West Greenland. Geochim. Cosmochim. Acta, 40, 513-527.

Baadsgaard, H., Nutman, A.P., Bridgwater, D., McGregor, V.R., Rosing, M. and Allaart, J.H. 1984. The zircon chronology of the Akilia association and the Isua supracrustal belt, West Greenland. Earth Planet. Sci. Lett., 68, 221-228.

Baadsgaard, H., Nutman, A.P. and Bridgwater, D. 1986a. Chronology and isotope geochemistry of the early Archaean Amîtsoq gneisses of the Isukasia area, southern West Greenland. Geochim. Cosmochim. Acta, 50, 2173-2183.

Baadsgaard, H., Nutman, A.P., Rosing, M., Bridgwater, D. and Longstaffe, F. 1986b. Mid-Proterozoic metasomatic alteration of pre-3600 Ma Amîtsoq gneisses from the Isukasia area, southern West Greenland: Implications for early crustal studies. Geochim. Cosmochim. Acta, 50, 2165-2172.

Beach, A. 1976. The interelations of fluid transport, deformation, geochemistry and heat flow in early Proterozoic shear zones in the Lewisian complex. Philos. Trans. R. Soc. Lond., Ser. A280, 569-604.

Berthelsen, A. and Bridgwater, D. 1960. On the field occurrence and petrography of some basic dykes of supposed Pre-Cambrian age. Meddr. Grønland, 123 (3), 42 pp.

Carmichael, D.M. 1969. On the mechanism of prograde metamorphic reactions in quartz-bearing pelitic rocks. Contrib. Mineral. Petrol., 20, 244-267.

Chadwick, B., Crewe, M.A. and Park, J.F.W. 1983. Field work in the north of the Ivisartoq region, Godthåbsfjord, southern West Greenland. Rapp. Grønlands geol. Unders., 115, 49-56.

Dansgaard, W. 1964. Stable isotopes in precipitation. Tellus, 16, 438-468.

Friedman, I and O´Neil, J.R. 1977. Compilation of stable isotope fractionation factors of geochemical interest. 6th edition. United States Geological Survey, Professional Paper 440-KK, 12pp.

Fyfe, W.S., Price, N.J. and Thompson, A.B. 1978. Fluids in the Earth´s Crust. Elsevier, Amsterdam, 383pp.

Garde, A.A., Hall., R.P.H., Hughes, D.J., Jensen, S.B., Nutman, A.P. and Secher, O. 1983. Mapping of the Isukasia sheet, southern West Greenland. Rapp. Grønlands geol. Unders., 115, 20-29.

Garde, A.A., Larsen, O. and Nutman, A.P. 1986. Dating of late Archaean mobilisation north of Qûgssuk, Godthåbsfjord, southern West Greenland. Rapp. Grønlands geol. Unders., 128, 23-36.

Gill, R.C.O., Bridgwater, D. and Allaart, J.H. 1981. The geochemistry of the earliest-known metavolcanic rocks, at Isua, West Greenland: a preliminary investigation. Spec. Publ. Geol. Soc. Australia, 7, 313-325.

James, P.R. 1975. Field mapping of Bjørneoen and the adjacent coast of Nordlandet, Godthåbsfjord, southern West Greenland. Rapp. Grønlands geol. Unders., 65, 58-62.

Kalsbeek, F., Bridgwater, D. and Zeck, H.P. 1978. A 1950 ± 60 Ma Rb-Sr whole rock isochron age from two Kangamiut dykes and the timing of the Nagssugtoquidian (Hudsonian) orogeny in West Greenland. Can. J. Earth Sci., 15, 1122-1138.

Kalsbeek, F., Bridgwater, D. and Boak, J.L. 1980. Evidence of mid-Proterozoic granite formation in the Isua area. Rapp. Grønlands geol. Unders., 100, 73-75.

Kerrich, R, LaTour, T.E. and Willmore, L. 1984. Fluid participation in deep fault zones: Evidence from geological, geochemical and $^{18}O/^{16}O$ relations. J. Geophys. Res., 89, 4331-4343.

Keith, M.L. & Weber, J.N. 1964. Carbon and oxygen isotopic composition of selected limestones and fossils. Geochim. Cosmochim. Acta, xx, 1787-1816.

McGregor, V.R. 1979. Archaean grey gneisses and the origin of the continental crust: Evidence from the Godthåb region, West Greenland: In: F. Barker (editor), Trondhjemites, dacites and related rocks. Elsevier, Amsterdam.

Nutman, A.P. 1982. Further work on the early Archaean rocks of the Isukasia area, southern West Greenland. Rapp. Grønlands geol. Unders., 110, 49-54.

Nutman, A.P. 1986. The early Archaean to Proterozoic history of the Isukasia area, southern West Greenland. Bull. Grønlands geol. Unders., 154, 80pp.

Nutman, A.P., Allaart, J.H., Bridgwater, D., Dimroth, E. and Rosing, M. 1984. Stratigraphic and geochemical evidence for the depositional environment of the early Archaean supracrustal belt, West Greenland. Precambrian Res., 25, 365-396.

Nutman, A.P. and Bridgwater, D. 1986. Early Archaean Amîtsoq tonalites and granites of the Isukasia area, southern West Greenland: development of the oldest-known sial. Contrib. Mineral. Petrol., 94, 137-148.

Park, J.F.W., 1987. Fault systems in the inner Godthåbsfjord region of the Archaean block, southern West Greenland. Unpublished Thesis, University of Exeter.

Perry, E.C. and Ahmad, S.N. 1977. Carbon isotope composition of graphite and carbonate minerals from 3.8 Æ metamorphosed sediments, Isukasia, Greenland. Earth Planet. Sci. Lett., 36, 280-284.

Perry, E.C., Ahmad, S.N. and Swulius, T.M. 1978. The oxygen isotope composition of 3,800 m.y. old metamorphosed chert and iron formation from Isukasia, West Greenland. J. Geol., 86, 223-239.

Rosing, M. 1983. Metamorphic and isotopic study of the Isua supracrustals, southwest Greenland. Thesis (unpubl.) University of Copenhagen.

Rosing, M. this vol. Redistribution of ´immobile elements´ in the Isua supracrustals, West Greenland. In: D. Bridgwater (editor), Fluid movements, element transport and the composition of the deep crust. Reidel, Dordtrecht.

Salomons, W. and Mook, W.G., 1986. Isotope geochemistry of carbonates in the weathering zone. In D. Fritz and J. Ch. Fontes (editors). Handbook of environmental isotope geochemistry. Vol. 2. Elsevier, Amsterdam.

Segall, P. and Simpson, C. 1986. Nucleation of ductile shear zones on dilatant fractures. Geology, 14, 56-59.

Sibson, R.H., Moore, J.McM. and Rankin, A.H. 1975. Seismic pumping - a hydrothermal fluid transport mechanism. J. Geol. Soc. Lond., 131, 653-659.

Smith, G.M. and Dymek, R.F. 1983. A description and
    interpretation of the Proterozoic Kobbefjord Fault Zone,
    Godthåb district, West Greenland. Rapp. Grønlands geol.
    Unders., 112, 113-127.
Wagner, P.A. 1982. Geochronology of the Ameralik dykes at
    Isua, West Greenland. Thesis (unpublished) University of
    Alberta.
Whitney J.A. and Stormer J.C. 1977. Two feldspar
    geothermometry, geobarometry in mesozonal granitic
    intrusions: three examples from the piedmont of Georgia.
    Contrib. Mineral. Petrol., 63, 51-64.

DEFORMATION AND MASS TRANSPORT IN THE NORDRE STRØMFJORD SHEAR ZONE, CENTRAL WEST GREENLAND.

K. Sørensen[1] and J.D. Winter[2]

1. Statoil Efterforskning og Produktion Danmark
   Sankt Annæ Plads 13, 1298 København K, Danmark

2. Department of Geology
   Whitman College, Walla Walla, Wa 99362, U.S.A.

ABSTRACT.   The accumulated chemical data from high grade terranes indicate that granulites are commonly depleted in sialic and LIL components with respect to amphibolites. We report uniquely field controlled data from a major West Greenland shear zone in which the same chemical differences have developed between deformed and undeformed quartzo-feldspathic gneisses: it appears that K, Si, Rb and water have been added during deformation while Fe, Mg, Ca, Ti, and Sr were removed. We suggest a model 'of chemical mobility in which the mechanism is bidirectional, vertical diffusion through a fluid phase. The ultimate driving gradient appears to be the temperature gradient.

1. INTRODUCTION

Many studies have indicated that granulites are commonly depleted in K, Rb, Th, U, Pb, Cs, and Y, and enriched in Ca, Fe and Mg with regard to amphibolite facies rocks. However, the fundamental questions of transport mechanism and gradient remain unresolved (see Weaver and Tarney, 1983 and Rudnick *et al.*, 1985 for a review). It may also be argued that the chemical differences do not reflect transport phenomena at all, but rather original protolith differences.

Recent studies of shear zones and their surroundings (see below) have demonstrated chemical differences in settings for which protolith differences cannot reasonably be invoked. The chemical differences are highly variable, though, and have not led to any general model for mechanism and gradient. The model which we suggest for the Nordre Strømfjord shear zone is generally applicable to shear zones regardless of the particular chemical pattern. This account represents an outline of a more comprehensive work under preparation.

171

D. Bridgwater (ed.), Fluid Movements – Element Transport and the Composition of the Deep Crust, 171–185.
© *1989 by Kluwer Academic Publishers.*

172

2. FIELD SETTING AND METHODS

2.1. The Shear Zone

The Nordre Strømfjord shear zone ("NSZ", Bak *et al,* 1975; Sørensen, 1983) is situated within the Nagssugtoqidian

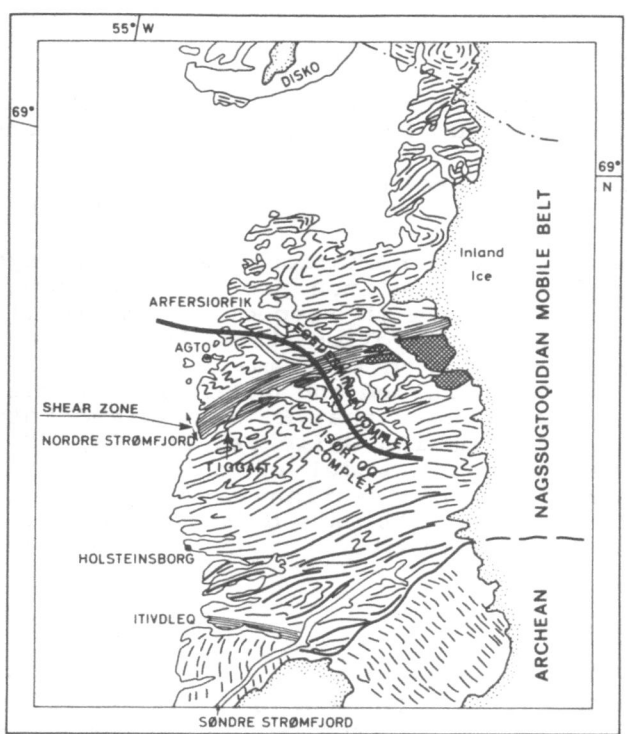

Figure 1. Regional setting of the Nordre Strømfjord shear zone

mobile belt of West Greenland (Escher *et al,* 1975; Geol. Surv. Greenl.,1985) and extends from the coast across the granulite facies Isortoq Complex and the amphibolite facies Egedesminde Complex to the inland ice (Fig.1) Formation of the shear zone was broadly contemporaneous with high grade metamorphism and occurred at 1.7 Ga. (Hickman, 1979; Hickman and Glassley, 1984).

At Tiggait, at the southern shear zone boundary (Fig.2), the rock units can be seen to rotate anti-clockwise as they enter the sinistral shear zone. We have investigated one particular unit of quartzo-feldspathic gneiss, the "Bugt Unit", which is representative of the dominant rock type of the region and spans the full range of shear strain at Tiggait (Sørensen, 1983). Due to the heterogeneity developed within

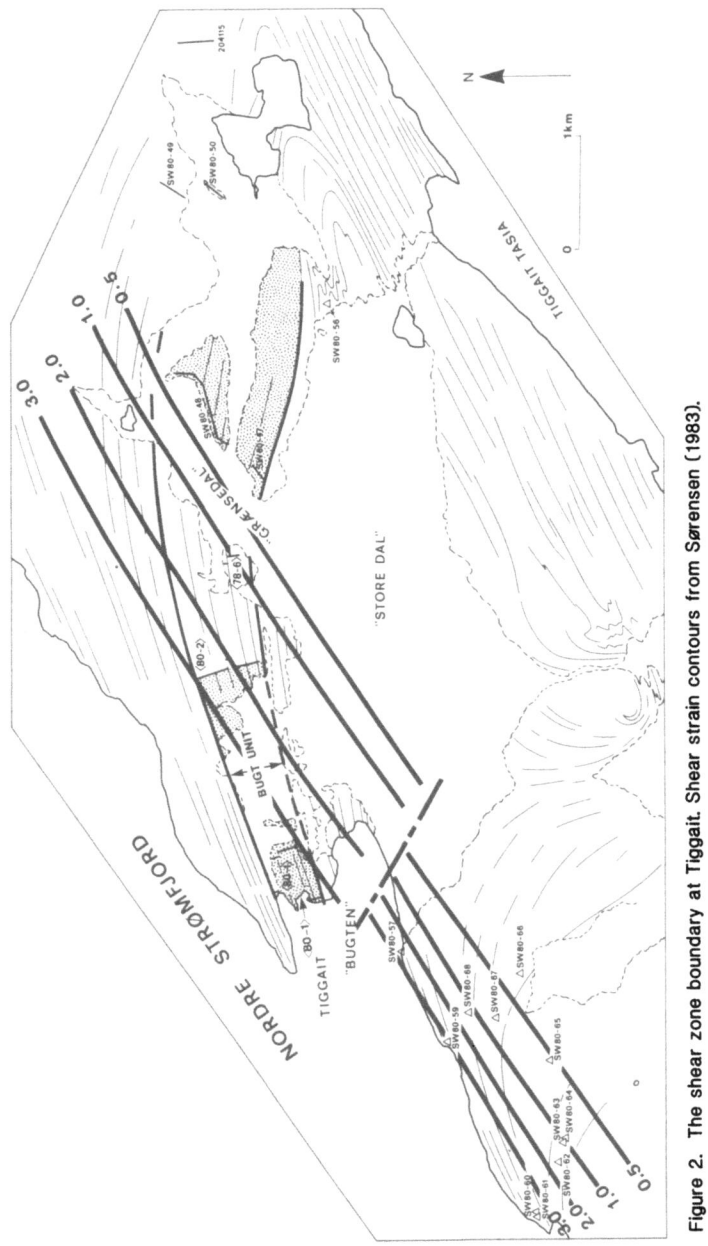

Figure 2. The shear zone boundary at Tiggait. Shear strain contours from Sørensen (1983).

the shear zone, it was deemed necessary to perform a systematic survey to document adequately the chemical and mineralogical changes of the Bugt Unit.

174

2.2. Survey and Sampling

Two profiles across the strike of the Bugt Unit were measured and sampled (Fig.2). Profile (80-1) represents the highest shear strain while profile (80-2) represents lesser strain. From profile (80-2) to profile (80-1) the Bugt Unit decreases in thickness from 700m to 400m as a result of shear. Shear strain contours are also indicated (Sørensen, 1983). Every band thicker than 5cm was measured and a sample collected for every 3m in (80-1), and for every 6m in (80-2). Equal proportions of 10 consecutive samples were combined to make "interval bulk samples" considered to represent the general composition of the respective intervals. 100g from each interval of a given profile were combined and this composite was analyzed as a representative of the composition of the profile as a whole, referred to as a "bulk profile analysis". The samples collected were quite representative of the respective intervals and of a sufficient density that we are confident that they represent the bulk composition of the profiles with reasonable accuracy. Outside the shear zone the Bugt Unit shows little variation and was sampled along five small profiles (SW80-47 to 50 and 204115).

3. RESULTS

The gneisses outside the shear zone are massive green-brown quartzo-feldspathic gneisses with 10-20% mafic constituents, mostly ortho- and clinopyroxene. Within the shear zone the overall character differs considerably, although relics of relatively unaltered host gneisses occur. The major difference is an increase in biotite and decrease in pyroxene content, accompanied by the development of a conspicuous banding and bleaching to a light grey color. The increase in biotite:pyroxene ratio occurs together with a marked decrease in total mafics.

Accompanying these changes is the conspicuous development of feldspar-rich, nearly mafic-free pegmatite gneisses which form concordant sheets traceable for up to hundreds of meters along strike in the highly deformed sections of the Bugt Unit. These pegmatitic gneisses make up 14% of profile (80-2) and 30% of profile (80-1). Thicknesses vary from less than 5cm to over 20m. They often contain gneiss schlieren or thin, nearly pure biotite selvages, thought to represent host gneiss relics depleted in felsic constituents. In the exposed walls SE of Grænsedal (Fig.2), at the macroscopic shear zone boundary, only a few pegmatitic gneisses occur and eventually none just beyond this area. The increase in average width of the pegmatitic gneiss sheets from (80-2) to (80-1) (despite an increase in strain) together with an overall correspondence between shear strain and abundance

very strongly implies synkinematic formation of the sheets.
Whole rock analyses for the bulk profile mixtures are
given in Table 1 together with a representative composition
of the gneisses of the Bugt Unit outside the shear zone.

OXIDE	1	2	3	D	4	D	5	D
Si	64.37	55.18	56.11	+0.92	59.43	+4.25	70.13	+5.76
Ti	0.59	0.52	0.39	-0.13	0.30	-0.22	0.32	-0.27
Al	16.26	18.20	17.86	-0.34	17.07	-1.13	15.83	-0.43
Fe(3)	1.93	1.30	1.40	+0.10	0.82	-0.47	0.83	-1.10
Fe(2)	4.40	3.43	2.78	-0.65	2.12	-1.31	1.74	-2.66
Mg	2.21	4.91	4.31	-0.60	2.94	-1.97	0.69	-1.52
Mn	0.09	0.09	0.08	-0.01	0.06	-0.03	0.03	-0.06
Ca	5.69	6.30	5.96	-0.34	4.20	-2.10	3.02	-2.67
Na	3.50	6.81	6.67	-0.14	6.56	-0.25	4.23	+0.73
K	0.63	1.31	2.36	+1.05	3.86	+2.55	2.36	+1.73
P	0.32	0.18	0.19	+0.01	0.18	0.00	0.13	-0.19
(H)	0.27	1.77	1.90	+0.12	2.47	+0.69	0.47	+0.20
TOTAL	100.26	99.99	100.01		100.01		99.78	
Rb (ppm)		38	53		100			
Sr (ppm)		628	592		583			

Table 1: Comparison of the composition of the bulk profile
analyses with the average composition of the gneisses
outside the shear zone and with Ramberg's (1951)
compositions.

    1 Ramberg's Isortoq Complex composition.
    2 Average reduced gneiss outside the shear zone
    3 Profile (80-2) bulk composition.
    3D Change of above composition from col. 2.
    4 Profile (80-1) bulk composition.
    4D Change of above composition from col. 2.
    5 Ramberg's Egedesminde Complex composition.
    5D Change of above composition from col. 1.

Comparison of the data indicates that deformation was accompanied by relative depletion in Al, Ti, Fe, Ca, and Mg, while
Si, K, water and Rb increase. Counting total Fe it is seen
that the changes in relative abundance, element for element,
are of the same sign for both (80-1) and (80-2). These data
suggest a relationship between chemical changes and deformation.

4. DISCUSSION

4.1. Chemical Changes

    The chemical changes that the Bugt Unit has undergone
are similar in sign and magnitude (except for Na) to those
found by Ramberg (1951) for the transition from the amphibolite facies Egedesminde Complex to the granulite facies
Isortoq Complex within the larger region which the NSZ tran-
sects (Fig.1). Following Ramberg (1951), more comprehensive
studies of granulite/amphibolite terranes generally agree
that granulite compositions are "more basic" than amphibolite
facies gneisses, being relatively depleted in Si, K, Ba,
and large ion lithophile (LIL) trace elements (Th, Rb, U,

Pb) while richer in Ca, Mg, Fe, Mn, Ti (and Sr)(see Raith *et al.*, Glassley *et al.*, and Friend, this volume, for some exceptions). These studies were based on comparison of many samples collected over large areas, so that initial compositional differences as an underlying cause cannot be ruled out. At Tiggait where the changes can be demonstrated within a 2-km profile of a single unit this alternative can be eliminated.

Of particular interest is the close spatial relationship between deformation and chemical mobility at Tiggait. Several studies have related chemical migrations to shearzones (Beach and Fyfe, 1972; Kerrich *et al.*, 1977; Brodie, 1980; Etheridge and Cooper, 1981; Floyd and Winchester, 1983). In these studies the chemical changes associated with deformation varied considerably, but hydration is a common factor. This indicates that fluids played a crucial role in the chemical transformations within shear zones. In the transformation of granulite to amphibolite facies chemistry, fluids, as well as melts, have been hypothesized as principal agents of change.

4.2. Relative vs. Absolute Changes

Direct comparison of the analyses representing the unaltered Bugt Unit and its altered equivalents by subtraction oxide-by-oxide (D's in Table 1) records the changes in <u>relative</u> abundances, although such comparisons are often misinterpreted to be absolute with the implicit assumption of constant mass or volume. The approach of Gresens (1967) has been applied to profile (80-1) to see if a plot of chemical changes vs. total mass suggests a solution to the question of absolute changes (Fig.3; Mn and Na are horizontal lines at approx. 0). We have left out the specific gravity terms in Gresens' Eq.14 which results in a plot with total mass, rather than volume as a variable in f. As the density terms comprise a constant, the only effect on the plot is a small shift in the abscissa values.

Assuming f=1 is equivalent to no net change in the total mass of the section at Tiggait during alteration. Intercepts with the abscissa at f=1 of lines representing oxide changes are the values of "D" for (80-1) in Table 1. Any other vertical slice gives the oxide flux corresponding to a different value of f (assuming a linear variation from unaltered gneiss to (80-1)).

In many studies involving chemical transport one component is assumed immobile, commonly Al or Ti, f then being determined from the intercept of that element with the ordinate and the corresponding fluxes then derived. We find that whatever element or element ratio (assuming mineralogical control) we pick as determinant, no singularly simple solution results. For example constant Al is very similar to constant

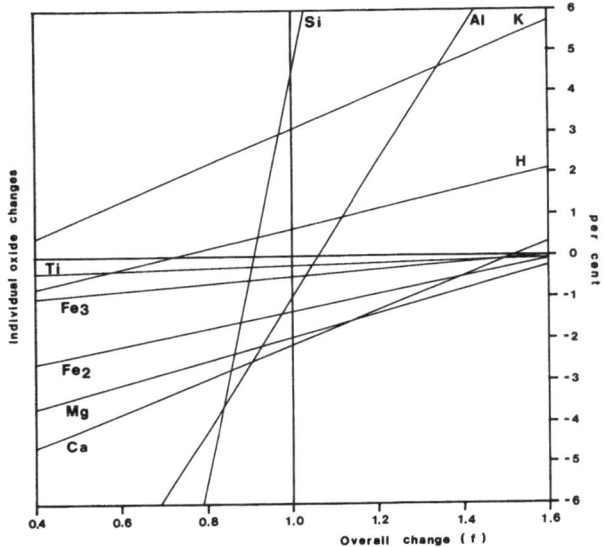

Figure 3. Gresens diagram for the chemical changes from the
shear zone surroundings to the bulk profile (80-1)
composition.

mass, but constant Ti requires huge increases in Si and Al
(Fig.3). Although several choices produce very unrealistic
results, it is notable that solutions requiring minimal
mass (and volume) changes have the least drastic fluxes.
Likewise, there is no mutual intersection of several oxide
lines to suggest a unique solution with several elements constrained.

The substantial amount of pegmatite gneisses of the
Bugt Unit raises the question of the bulk chemical changes
being accounted for by en masse injection of pegmatite material (melt) into the original gneisses. To test this possibility we have subtracted the amount of pegmatite in each
of the 2 profiles using the average pegmatite gneiss composition (an average of the compositions of 8 individual pegmatitic gneiss samples). They were subtracted by a weighted
amount equal to the measured abundance of pegmatite in the
(80-1) and (80-2) profiles.

If the compositional changes were due to the pegmatite
gneisses alone the calculated compositions of the gneisses
(columns 3 and 4 of Table 2) of profiles (80-1) and (80-2)
should compare directly to the unaltered gneiss (column 1).
A Gresens' variation diagram has been produced for (80-1)
for this case as well (Fig.4). As compared to the bulk changes
of Table 1 (inclusive of pegmatite gneisses) the overall
changes have greatly diminished. Fig. 4 does show a cluster

OXIDE	1	2	3	D	4	D
Si	55.18	68.84	54.04	-1.15	55.40	+0.21
Ti	0.52	0.06	0.45	-0.07	0.40	-0.12
Al	18.20	15.02	18.32	+0.12	17.95	-0.25
Fe(3)	1.30	0.37	1.57	+0.27	1.01	-0.28
Fe(2)	3.43	0.33	3.17	-0.26	2.88	-0.55
Mg	4.91	0.12	4.99	+0.08	4.14	-0.77
Mn	0.09	0.02	0.09	0.00	0.07	-0.02
Ca	6.30	1.18	6.73	+0.43	5.50	-0.80
Na	6.81	6.71	6.66	-0.15	6.50	-0.31
K	1.31	5.67	1.83	+0.52	3.08	+1.77
P	0.18	0.01	0.22	+0.05	0.25	+0.07
(H)	1.77	1.67	1.93	+0.15	2.81	+1.03
TOTAL	99.99	99.99	99.99		100.01	

Table 2: Calculated gneiss compositions. Obtained from the
bulk profile compositions of Table 1 by subtracting an
average pegmatitic gneiss composition.

1 Average reduced gneiss outside the shear zone.
2 Average pegmatitic gneiss.
3 Profile (80-2) calculated gneiss composition.
3D Change of above from col. 1.
4 Profile (80-1) calculated gneiss composition.
4D Change of above from col. 1.

of curves near f=1.1, indicating the possibility of a process
of pegmatite injection with another 10% (f=1.1) increase in
the country rock gneiss associated with very minor fluxes
for all elements but Si, Al, K, Na and water. However this

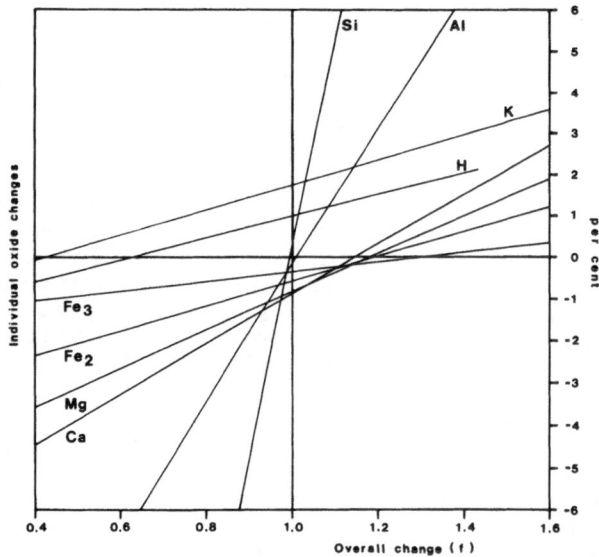

Figure 4. Gresens diagram for the chemical changes from the
shear zone surroundings to the calculated gneiss
composition of profile (80-1).

requires small <u>losses</u> of many elements, substantial Si addi-tions, and 140% increase in K in the gneisses.

Clearly, if pegmatites are assumed added to the gneisses in bulk (forcing a significant volume increase), the chemical effects are still manifest beyond the pegmatites themselves. In fact, only half of the K (and Rb) added to the bulk profile compositions resides in the pegmatitic gneiss sheets (compare col. 4 with col.1), the remaining half being added to the host gneisses of the most deformed (80-1) section.

4.3. Melt vs Fluid

The pegmatite gneiss compositions correlate well with igneous rocks of the granite rhyolite family suggesting a possible process of melt injection. For several reasons we find this mechanism unrealistic. In order to account for the pervasive host rock alterations we must assume a melt nearly saturated with volatiles. We are then forced to accept a mass increase in the gneisses of 20% (f=1.2 in Fig.4) in order to avoid the decreases in Ca, Fe and Mg; this in addi-tion to the 30% increase via pegmatite injection. Thus the pegmatites at Tiggait could not possibly have supplied the materials needed to change the host gneisses, an amount equal to 2/3 of their own bulk in solids.

From the data of Burnham and Jahns (1962) and Hamilton et al. (1964) we obtain a maximum content of water in granitic melts at 8 kb of 16%. Assuming the fluids to have contained a maximum of 10% of solutes implies that an amount of melt over 20 times the present pegmatite quantity should have defluidized completely in order to change the host gneisses. With considerably less than complete devolatilization and perfect cation exchange, we must deduce that enormous amounts of melt have passed through the crust.

Returning to the non-injection analogy of Table 1, the minimum amount of mass moved in order to transform the unal-tered gneiss to the (80-1) bulk composition is 16% (an ex-change involving the addition of 8% K, Si and water and an 8% loss of the others). By analogy with magmatic fluids and SiO_2 solubility data (Crerar and Anderson, 1971) a crude estimate of the solute content of deep crustal fluids is about 10%. This requires that a minimum quantity of fluid equal to the rock mass must have exchanged cations with the column. This estimate requires a 100% efficient exchange with a purely Si-K-H fluid entering and a purely Ca-Mg-Fe fluid leaving and must underestimate the actual fluid quantity by several orders of magnitude. A third process is also quite likely. Perhaps the water introduced caused local anatectic melts which produced the pegmatitic gneisses. This process is a hybrid of the previous two. However, as the melts are created <u>in situ</u>, this is really quite similar to the second process discussed above, for the ionic fluxes and arguments are the same.

This problem of "paradoxical volumes" arises for any process of chemical alteration assuming one-way transport, or for a purely fluid transport model, cf. the estimated fluid/rock ratio of 1000 to 10,000 of Etheridge and Cooper (1981). The same problem arises for a one-way fluid transport model for the Nordre Strømfjord shear zone. Already Ramberg (1951) noted the problem, when he calculated the amounts of fluid necessary to create a regional amphibolite to granulite facies transition, referring to the "ocean of water" required.

The only way to avoid this dilemma is for the ionic species to migrate through the fluid rather than with it. We find the large fluid/rock ratios rather imposing and prefer the more static fluid model. Some directed fluid flow is still possible (certainly water is introduced), however the rate must be limited or the paradoxical volume problem arises. The more static fluid model also eliminates the necessity for postulating an overall volume increase of the shear zone of approximately 30-40%, since it implies bidirectional transport and f approximately equal to 1 in Fig.3.

4.4. Gradient of Transport

It seems natural first to discuss whether the chemical composition gradient at Tiggait is a result of predominantly horizontal transport. If so, we estimate that K_2O must have been removed from the surroundings to a distance of approximately half the shear zone width, i.e. 10 km. Horizontal movement would be reasonable if a driving gradient could be demonstrated, especially if such a gradient implies decreasing viscosity towards the shear zone. As significant horizontal pressure gradients over such a short distance are prohibited, the obvious candidate for a gradient fullfilling this need is temperature.

Temperature may have changed appreciably over the shear zone boundary if the shear heating hypothesis of Turcotte and Oxburgh (1968) applies to this particular locality. It has been argued that the shear heating hypothesis cannot be reconciled with the evidence from the NSZ as such (Sørensen, 1983). We have attempted to assess possible temperature gradients at Tiggait using three independent geothermometers. The 2-pyroxene thermometer of Wells (1971) and Lindsley and Anderson (1983), and the garnet-biotite thermometer of Ferry and Spear (1978) and Ganguly and Saxena (1984) were of use in a number of samples. Garnet-biotite temperatures are considerably lower than those determined by pyroxene pairs. Finally, as the assemblage is so common, we used a geothermometer based on biotites coexisting with quartz and a Fe-Ti oxide. The formulation was developed by ISE Carmichael for silica-saturated volcanics and reported in Glassley (1983).

Each of the methods provides a distinctly different temperature estimate, but we are concerned with temperature differences rather than absolute values. Comparing various T estimates with sample localities indicate no apparent gradient in temperature across the shear zone boundary for any single thermometer. The lack of obvious horizontal gradients in T and P suggests that the driving gradient was more probably vertical rather than horizontal.

4.5. Mechanism

Relative mineral abundances of the granulitic and amphibolitic Bugt Unit compositions suggests that a reaction similar to the following occurred:

(1) $KMg_3Si_3AlO_{10}(OH)_2$ + $KAlSi_3O_8$ + Ca^{+2} =
           Bi   Ksp

  $3MgSiO_3$ + $CaAl_2Si_2O_8$ + SiO_2 + $2K^+$ + H_2O
     Opx   An   aq

The true reaction, not involving idealized end-members also consumed some Fe and Mg and produced clinopyroxene as well. A vertical temperature gradient will drive this reaction and will result in a vertical upward K-Si-water (KSW) flux and a downward Ca-Fe-Mg (CFM) flux. KSW is introduced to amphibolite facies gneisses (and possibly creates local anatexites while K, Si, and water are introduces more pervasively into the gneisses). CFM elements migrate downward, thus avoiding large volume increases, drastic KSW fluxes (Fig.3) and the fluid (or melt) volume dilemma.

In the activated state we thus expect a local concentration gradient in the fluid which decreases upward for KSW elements and decreases downward for CFM elements. These compositional gradients exist as a result of a multivariant reaction relationship spanning a temperature gradient. We thus see the temperature gradient as the ultimate gradient, mediated through activity gradients in a fluid phase set up by an ongoing mineral reaction.

The efficiency of this mechanism clearly depends on the diffusivity of the activated ions as well as on the magnitude of the compositional deviations. It seems reasonable to assume that both parameters increase with increasing deformation, for example due to localized stress gradients, in which case the compositional deviations and the mobility of the activated ions probably will be linked. The model applies specifically to reacting pyroxenes and biotites, but in instances where other multivariant reactions occur other fluxes will be set up. This model therefore can explain the wide variety of alterations documented in shear zones.

The PT conditions and the rock chemistry determine which mineral reaction will occur, but once started, deformation will enhance reaction kinetics and chemical mobility. We thus infer, that as soon as multivariant reactions are set up, chemical mobility will occur in the temperature gradient, and that the direction of transport of the elements is determined by reaction equilibrium, and not to any intrinsic tendency for upward or downward mobility. If this model is correct for Tiggait, it should also apply to the NSZ and its regional context, thus making a final tie with Ramberg's results.

4.6 Model

The mineral reactions at Tiggait are clearly "retrogressive" and we might inquire how they lead to a chemical alteration pattern similar to Ramberg's Isortoq-to-Egedesminde Complex pattern. At some point in time the regional metamorphic event leading to the metamorphism of the Egedesminde and Isortoq Complexes was "progressive", so that the amphibolite to granulite facies transition was established on amphibolite facies gneisses during a broad deformational event preceding the shear zone formation.

At the amphibolite to granulite facies transition there was no reaction bias, i.e. there existed an equilibrium between the build-up and the break-down of the phases participating in reaction 1. Downward from this level increasing biotite towards pyroxene bias must have existed over a zone due to variable biotite & pyroxene compositions. This would set up the gradient as discussed above. Below the "no bias level" an upward flux of KSW and a downward flux of CFM elements would be created in these gradients, leading to Ramberg's chemical stratification.

The gradient will not continue much above the "no bias level", where biotite (and perhaps amphibole) will be the only mafic minerals. Reaction (1) will gradually increase the KSW content in the vicinity of the "no bias level" and the reaction will locally reverse as the bias is a function of bulk composition as well as of temperature. We will thus, even with constant temperature, see a "no bias level" moving downward into the crust, from an original coincidence with the level of first pyroxene appearance (the granulite isograd). The band between the original isograd and the descending "no bias level" will be in reverse reaction bias, the mafics undergoing retrograde reations after first having formed progressively at the same temperature. These models explain the commonly reported retrograde nature of the amphibolite/granulite boundary.

Now suppose that deformation becomes confined, i.e.

the Nordre Strømfjord shear zone is established. Then the downward drift of the "no bias level" will be enhanced by deformation within the shear zone together with continued metasomatism. When the shear zone deformation ceases, we will over some crustal depth interval have a retrogressive level juxtaposed against a progressive level in the zone surroundings, as seen at Tiggait. The temperature may have been constant through the entire time interval. Of course gradually decreasing temperatures would only enhance the effect, but our thermometry suggests that this was not the case at Tiggait. This picture is also in accord with Hickman's (1980) good isochron for shear zone and non-shear zone rocks all from Tiggait. The rocks now within the shear zone at Tiggait will have had a history of progression and depletion, followed by retrogression and repletion in a time interval short enough to appear coeval with respect to the Rb/Sr isotopic system. The spreading of the shear zone inferred by Sørensen (1983) from geometrical reasoning may be explained as a result of depletion of the deeper lying reservoirs for water and K_2O.

5. CONCLUSIONS

We believe the alterations seen within the Nordre Strømfjord shear zone at Tiggait to have resulted from diffusion into the system from prograde reactions at deeper levels. Downward diffusion occurred as part of the same general mechanism. The process was similar to that on a regional scale which resulted in bulk chemical differences between granulite and amphibolite facies gneisses. We think diffusion was predominantly through an interstitial fluid, and that the rate was greatly enhanced by deformation.

ACKNOWLEDGEMENTS

Field work was initiated during the Agto 2 mapping project (1975-1978) sponsored by the Geological Survey of Greenland and run by the Dept. of Geology at the University of Aarhus. K.S. thanks his former colleagues N.Ø.Olesen and J.Korstgaard for their share in the planning and execution of the Agto 2 project. The subsequent detailed mapping and sampling at Tiggait in 1980 and reconnaissance in the shear zones between Nordre Strømfjord and Itivdleq in 1981 was made possible through a grant from the Carlsberg Foundation. The support from the Greenland Survey and the Carlsberg Foundation is gratefully acknowledged. We are also grateful to the Geological Museum in Copenhagen for the use of the JEOL 733 Superprobe (courtesy of the Danish National Research Council) and to Aarhus University for the whole-rock analyses. Thanks also to Bill Glassley for his review of the manuscript.

6. REFERENCES

Bak J, Korstgaard JA, Sørensen K (1975) A major shearzone within the Nagssugtoqidian of West Greenland. Tectonophysics 27:191-209.

Beach A, Fyfe WS (1972) Fluid transport and shear zones at Scourie, Sutherland: Evidence of overhtrusting? Contrib Mineral Petrol 36:175-180.

Brodie KH (1980) Variations in mineral chemistry across a shear zone in phlogopite peridotite. J Struct Geol 2:295-272.

Burnham CW, Jahns RH (1962) A method for determining the solubility of water in silicate melts. Am J Sci 260:721-745.

Crerar DA, Anderson GM (1971) Solubility and solvation reactions of quartz in dilute hydrothermal solutions. Chem Geol 8:107-122.

Escher A, Sørensen K, Zeck HP (1976) Nagssugtoqidian Mobile Belt in West Greenland. in Escher and Watt (eds) Geology of Greenland. Geol Surv Greenland, Copenhagen.

Etheridge MA, Cooper IA (1981) Rb/Sr isotopic and geochemical evolution of a recrystallized shear (mylonite) zone at Broken Hill. Contrib Mineral Petrol 78:74-84.

Ferry JM, Spear Fs (1978) Experimental calibration of the partitioning of Fe and Mg between biotite and garnet. Contrib Mineral Petrol 66:113-117.

Floyd DA, Winchester JA (1983) Element mobility associated with meta-shear zones within the Ben Hope amphibolite suite,Scotland. Chem Geol 39: 1-15.

Ganguly J, Saxena SK (1984) Mixing properties of aluminosilicate garnets: constraints from natural and experimental data, and applications to geothermobarometry. Am Mineral 69:88-97.

Geol Surv Greenland. Geological Map of Greenland: Agto Map Sheet. Copenhagen 1985.

Gresens RL (1967) Composition-volume relationships of metasomatism. Chem Geol 2: 47-65.

Hamilton DL, Burnham CW, Osborn EF (1964) The solubility of water and effects of oxygen fugacity and water content on crystallization in mafic magmas. J Petrol 5:21-39.

Hickman MH (1979) A Rb/Sr age and isotope study of the Ikertoq, Nordre Strømfjord, and Evighedsfjord shear belts, West Greenland- outline and preliminary results. Rapp Grønlands geol Unders 89: 125-128.

Hickman MH (1980) Rb-Sr age and isotope study of two Precambrian shear belts, West Greenland. (abstr) EOS 61:384.

Hickman MH, Glassley WE (1984) The role of metamorphic fluid transport in the Rb-Sr isotopic resetting of shear zones: evidence from Nordre Strømfjord, West Greenland. Contrib Mineral Petrol 87: 265-281.

Kerrich R, Fyfe WS, Gorman BE, Allison I (1977) Local modification of rock chemistry by deformation. Contrib Mineral Petrol 65:183-190.

Lindsley DH, Anderson DJ (1983) A two-pyroxene thermometer. Proc. Thirteenth Lunar Planetary Sci Conf, Part 2. Journ Geophys Res 88:Supplement A887-A906.

Ramberg H (1951) Remarks on the average chemical composition of granulite facies and amphibolite-to-epidote amphibolite facies gneisses in West Greenland. Bull Geol Soc Denmark 12:27-34.

Rudnick RL, McClennan SM, Taylor SR (1975) Large ion lithophile elements in rocks from high-pressure granulite facies terranes. Geochim Cosmochim Acta 49:1645-1655.

Sørensen K (1983) Growth and dynamics of the Nordre Strømfjord shear zone. J Geophys Res 88:3419-3437.

Turcotte DL, Oxburgh ER (1968) A fluid theory for the deep structure of dip-slip fault zones. Phys Earth Planet Inter 1:381-386.

Weaver B, Tarney J (1983) Elemental depletion in Archaean granulite facies rocks. In: Atherton and Gribble (eds), Migmatites, Melting and Metamorphism p250-263. Shiva, Nantwich.

Wells PRA (1977) Pyroxene thermometry in simple and complex systems. Contrib Mineral Petrol 62:129-139.

METASOMATIC ALTERATION OF ULTRAMAFIC ROCKS

Minik T. Rosing
Geological Museum, Øster Voldgade 5-7
DK-1350 København K
Denmark

ABSTRACT. Metasomatized ultramafic rocks from the 3800 Ma Isua supracrustal belt have been preferentially enriched in calcium, aluminium and silica, but show no or only little addition of potassium and rubidium. This unusual geochemical response to fluid infiltration can be explained in the light of a low buffered activity of aqueous silica during progressive silicification of the metasomatized units. Thermodynamic analysis shows that the selective element mobility was controlled by the stabilities of mineral assemblages in the ultramafic rocks during alteration.

Introduction

This paper presents an idealised thermodynamic analysis of a hydration – carbonation path of a suite of ultramafic rocks that have passed through a complex series of prograde and retrograde metamorphic events. This theoretical analysis is carried out in an efford to explain a large range in chemical compositions within individual ultramafic bodies. The variation, which is macroscopically observable as gross differences in mineral paragenesis, is spacially related to veins and to contacts with other lithologies. It is argued that most of the geochemical scatter is caused by secondary processes. Although some of the chemical differences between margins and interior of individual bodies may represent original variations in an igneous body, no concievable igneous fractionation process can account for the geochemical trends seen. Selective- or bulk-assimilation of the felsic country rock during emplacement is not supported by the geochemical data. As a first approximation the bodies are therefore taken as originally homogeneous. As can be seen in the following treatment large errors in estimates of the original major component abundances in individual samples would have little effect on the final conclusion.

The thermodynamic analysis is based on experimentally determined relations between pressure, temperature, mole fraction of CO_2 and stable mineral paragenesis in the system $MgO - SiO_2 - H_2O - CO_2$ (Greenwood, 1967), and is valid for ultramafic rocks in general. The analysis specifically deals with the redistribution of SiO_2, MgO, CaO, Al_2O_3 and

D. Bridgwater (ed.), Fluid Movements – Element Transport and the Composition of the Deep Crust, 187–202.
© 1989 by Kluwer Academic Publishers.

K_2O, and is relevant to amphibolite facies alteration of dunites, peridotites and serpentinites in most geological settings. To illustrate the specific metasomatic effects on the Isua ultramafics, a set of 3 samples have been chosen, which have unambiguous field relations. These are representative of a larger group of analysis (28).

The metasomatic path is modeled in terms of equilibrium states, although metasomatism in its nature is irreversible (Korzhinskii, 1970). A thermodynamic analysis must therefore build on the concept of local equilibrium (Thompson, 1959). This requires that an infiltrating fluid percolates through the rocks at a slow enough rate to allow each batch of fluid to equilibrate with its host minerals at all times. In nature, differences in kinetics for reactions between minerals and different fluid species, and for dissociation reactions in the fluid, will lead to differential deviations from the proposed equilibrium states. Although a steady state of constant affinity (De Donder and Van Rysselberghe, 1936) for individual reactions probably will be approached in a situation of constant rate of fluid flow, neither the thermodynamic nor the geological basis for analysis of such steady states exist for the system in question. For the lack of more accurate models, the assumption of equilibrium reactions gives the closest approximation to the natural processes.

Geology

The Isua supracrustal belt (ISB) is situated ca. 150 km northwest of the Greenland capital Nuuk. The belt comprises semipelitic and quartzo-feldspathic clastic sediments, carbonates, banded iron formation and basic and ultrabasic units (Nutman et al., 1984). Ultramafic rocks occur along the full length of the belt as layers and elongate pods, and are interpreted as representing dislocated dunitic to peridotitic sill-like intrusions. The ultramafic rocks show considerable variation in their present mineral assemblages ranging from almost pure dunites to talc schists.

The supracrustal rocks were isoclinaly folded and intruded by polyphase tonalitic to granodioritic gneisses in the period 3750 - 3600 Ma, and intruded by granitic pegmatites at 3400 Ma (Baadsgaard et al., 1986a).

All supracrustal units have been metamorphosed under amphibolite facies conditions (Boak and Dymek, 1982). Rosing (1983) found evidence for at least two amphibolite facies events followed by local greenschist facies retrogradation. Unpublished Sm - Nd mineral data (G. Gruau personal communication, 1987) indicate that penetrative amphibolite facies metamorphic recrystallization took place at 2800 Ma. Anatexis and regional pegmatite formation took place between 2500 and 2600 Ma (Baadsgaard et al., 1984) and is thought to reflect a widespread thermal event which affected the lead isotope system in titanites and apatites in the southern part of the ISB around 2500 Ma (Baadsgaard, 1983). Local recrystallization in shear zones and their immediate surroundings took place at 2000 - 1800 Ma (Wagner, 1982; Rosing, 1983; Nutman et al., this volume). At 1600 Ma the supracrustal rocks were locally intruded by

granitic sheets (Kalsbeek et al., 1980), and the Rb - Sr isotopic system
of biotites was reset regionally (Baadsgaard et al., 1986b).

The present variation in the mineralogy of the ultramafic rocks is
a result of this complex series of metamorphic events. Partial
replacement of early formed minerals by new ones suggest that the
present mineral assemblages do not represent equilibrium paragenesis.
The phase assemblages reflect both the present bulk compositions and the
latest metamorphic conditions under which a particular outcrop
recrystallized. In a polymetamorphic area such as the ISB, the phases
responsible for selective element retention during metasomatism are not
necessarily present in the observed mineral assemblages. A thermodynamic analysis of a hypothetical hydration / carbonation path at
amphibolite facies metamorphic conditions is carried out, rather than
attempting to use actual mineral compositions from the present phase
assemblages. These calculations are then applied to chemical results
from two samples from the margin and one sample from the interior of a
ultramafic body on the eastern shore of lake 678 m (cf. Nutman et al.,
1985) to illustrate the effects of metasomatism. This body is in contact
with metagreywacke lithologies of the "Felsic Formation" of Nutman et
al. (1983) in a 15 m wide and 50 m long exposure.

Mineralogy and geochemistry

The present mineralogy of the analysed samples are:

GGU 175564   : Tremolite , chlorite, opaque oxides.

GGU 175565   : Tremolite, chlorite, opaque oxides,

GGU 175566   : Olivine, talc, chlorite, tremolite, carbonate,
               fuchsite, opaque oxide.

Country   : Quartz, plagioclase, biotite, muscovite,
rock   carbonate, Fe-Ti-oxide.

Whole-rock geochemical analysis are presented in Table 1. The
geochemical signature of the tremolite-chlorite schists, which are the
most strongly altered metaperidotites, include enrichment in silica,
aluminium, titanium, calcium, sodium and zirconium. They also show
marked loss of magnesium and Ni. Total rare earth elements (REE) are
enriched in one marginal sample, and depleted in the other, relative to
the interior sample. These redistributions are not accompanied by
increases in potassium or rubidium, normaly noted as among the first
added to mafic lithologies during metasomatism in the ISB. Some of the
enriched components are elements which are believed to have low
solubilities in metamorphic fluids. Their increase could therefore be
caused by preferential leaching of all other components. However,
element ratios of these "immobile components" between least and most
strongly altered samples vary markedly (Fig. 1), and if an element such

Table 1

Element abundances determined by XRF at the Geological Survey
of Greenland, and provided by D. Bridgwater.

Wt. %	GGU 175564 margin	GGU 175565 margin	GGU 175566 center
SiO_2	48.48	49.38	42.63
TiO_2	0.58	0.51	0.27
Al_2O_3	5.42	4.77	2.81
Fe_2O_3	1.02	0.55	1.89
FeO	6.99	8.32	7.79
MnO	0.15	0.21	0.18
MgO	22.96	25.32	35.80
CaO	8.27	5.33	2.88
Na_2O	0.36	0.16	0.10
K_2O	0.02	0.00	0.36
LOI	4.78	5.13	4.83
P_2O_5	0.05	0.06	0.03
ppm			
Rb	1	1	31
Sr	15	11	55
Ba	0	0	19
Y	19	13	8
Pb	12	9	9
Zr	72	69	41
V	106	96	57
Co	90	99	148
Cr	1643	1471	1327
Ni	1354	1670	3328

as aluminium is used as a basis for comparison, other "immobile
elements" show marked changes. Further iron and chromium, which might be
expected to follow the "residual components" do not show significant
increases. Although some degree of Al, Ti, Cr, Zr and REE enrichment by
preferential leaching of other components is likely, mobility of at
least some of these components must be accepted to account for their
differential enrichment.

REE spectra (Fig. 2) from the two marginal samples have LREE
enriched patterns with negative Eu-anomalies of identical shape, but
total REE abundance of 175565 is 85 % higher than 175564. The pattern of
the central sample shows a marked HREE depletion relative to the two
marginal samples and lacks their negative Eu anomaly. This sample has
total REE intermediate between the two marginal samples. The negative La
anomalies of the marginal samples are probably analytical artefacts.

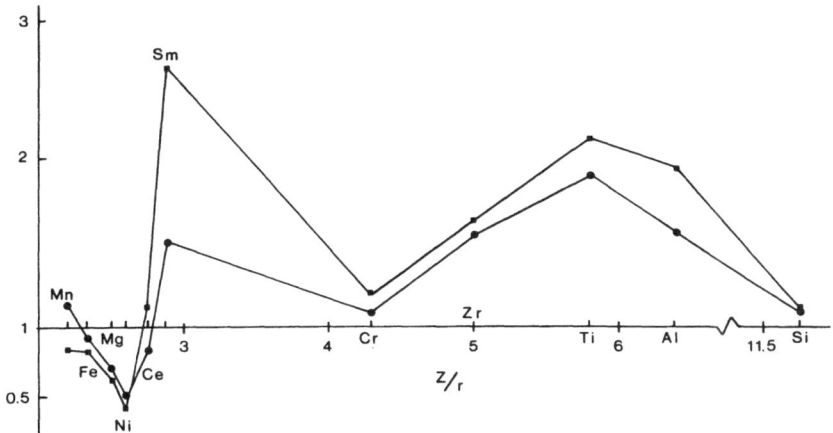

Figure 1 : Element weigh % ratios between marginal sample
175564 and interior sample 175566 (filled squares) and
between marginal sample 175565 and interior sample 175566
(filled circles) arranged in order of increasing ionic field
strength (formal charge divided by ionic radius from
Whittaker and Muntus (1970)).

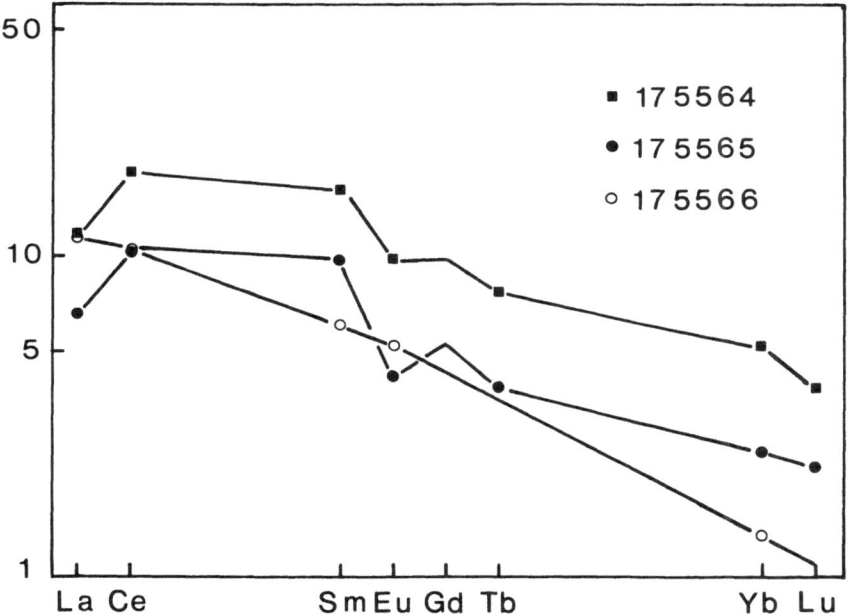

Figure 2 : Rare earth element abundances normalized to
chondritic values of Nakamura (1974) except for Tb which is
normalized to the value of Haskin et al. (1968).

Thermodynamic analysis

The stability conditions for minerals in equilibrium with a binary H_2O - CO_2 fluid at constant temperature and pressure, are expressed in terms of cation activities (a) corrected for differences in ionic speciation relative to that in pure H_2O (σ), and ratioed to the activity of hydrogen ions to the power of the formal charge of the cation. The speciation term (σ) is a formality of unknown numerical value (cf. Bowers et al., 1984).

The relationship between the aforementioned activity ratios and total dissolved matter in the fluids are unknown, but both petrologic evidence from the ISB (Rosing, 1983, 1987), and the general nature of deep hydrothermal solutions (Helgeson, 1967), suggest that the meta-somatizing fluid was a brine of high chlorinity. The dominant fluid species were probably highly associated chlorides. The total molality of individual elements are thus functions of the activity coefficients for their ions, the dissociation constant for their chlorides, pH, and the activity ratios depicted in the phase diagrams.

Thermodynamic data and conventions for minerals, ions, gasses and water are those of Helgeson et al., (1978). Activity coefficents for H_2O and CO_2 in non-ideal binary mixtures are from Bowers et al. (1984). Standard states for all minerals define unit activity of their pure phases at the pressure and temperature of interest. Standard states for the ionic species define unit activity in one molal solutions, referenced to infinite dilution. All minerals are represented by their pure end-member phases. Amphibolite facies metamorphic conditions of 500 °C and a total pressure of 5 kbar are used in the analysis of the meta-somatic event, in accord with Boak and Dymek (1982) and Rosing (1983).

The protolith for the hypothetical reaction path is an antigorite - olivine serpentinite. This assemblage reacts with CO_2 introduced in an aqueous fluid phase, to form the assemblage antigorite - forsterite - magnesite, which buffer the fluid at constant values of X_{CO_2}, aqueous silica activity and magnesium to hydrogen ion activity ratio. When all forsterite is consumed by the reaction the assemblage antigorite - magnesite controls the silica and magnesium activity values while X_{CO_2} increases, until antigorite reacts with CO_2 to form talc and magnesite. The assemblage antigorite - talc - magnesite again buffer the fluid at fixed values of X_{CO_2}, aqueous silica activity and magnesium to hydrogen ion activity ratio. When antigorite is exhausted univariant reactions between the assemblage talc - magnesite and the fluid phase define a functional relationship between X_{CO_2}, aqueous silica activity and magnesium to hydrogen ion activity ratio. Figure 3 shows saturation surfaces for the relevant mineral phases in the system SiO_2 - MgO - H_2O - CO_2. These surfaces are constructed from "law of mass action" expressions, using equilibrium constants for mineral hydration and carbonation reactions calculated using the "SUPCRT" computer program of Helgeson et al. (1978). From Figure 3 it can be seen that the activity of aqueous silica defined by the phase assemblages along the progressive hydration / carbonation path increase dramatically in the low X_{CO_2} portion of the diagram. This is a consequence of Mg fixation in carbonate. Increasing the SiO_2 / MgO ratio in the silicate system, increasingly stabilizes phases with high stoichiometric silica. These silicic phases buffer the fluid at progressively higher activity of aqueous silica as CO_2 is added to the system. The high MgO content of peridotites gives a high capacity for CO_2 fixation, such that the increase in the mole fraction of CO_2 in the fluid phase is damped relative to its increase in the system. This allows large fluid throughput within the X_{CO_2} stability range of individual mineral assemblages. The very high rate of change of silica activity relative to the rate of change of X_{CO_2}, together with the low molality of aqueous silica relative to that of H_2O and CO_2 in metamorphic fluids, allows X_{CO_2} in infiltrating fluids to effectively control the silica activity in ultramafic rocks. For low X_{CO_2} fluids this value is less than that of quartz saturation, which is the plane defined by the X_{CO_2} and a $Mg^{++}/\sigma Mg^{++}$ (a H^+)^2 axis in Figure 3. Due to the large stoichiometric coefficient of silica in most silicate minerals, the variable silica activity defined by the major phases effectively controls the stability of phases to accommodate minor components such as Al, K and Rb. The major phases shown in Fig. 3 therefore provide the framework for evaluation of the behavior of minor components in the system.

Figure 4 shows saturation curves for the indicated phases in the SiO_2 - MgO - CaO - Al_2O_3 - K_2O - H_2O - CO_2 system, with a $Mg^{++}/\sigma Mg^{++}$ (a H^+)^2 and a $SiO_{2(aq)}/\sigma$ $SiO_{2(aq)}$ defined by the assemblage talc - magnesite. This is along the intersect of the talc and magnesite saturation surfaces in Figure 3, from the point "C" toward higher X_{CO_2} values. From this figure the gradients in chemical potential which determine the direction of Al, Ca and K transport in the fluids can be derived, as well as the stability requirements for minerals to accomodate these components. The propensity for transport of a component in the fluid does not lead to enrichment in the rock, if this component is not accomodated by the stable phase assemblage.

In the case of the ultramafic rocks of the ISB, most external rock units are quartz bearing, and infiltrating fluids would be quartz -saturated. Externaly derived fluids with low X_{CO_2}, would deposit silica in the ultramafics through reactions such as Reaction 1.

Reaction 1:

$3 MgCO_3 + 4 SiO_2 + H_2O$ $- Mg_3Si_4O_{10}(OH)_2 + 3 CO_2$
magnesite   fluid   talc   fluid

In a similar manner the ubiquitous dolomite and calcite in the surrounding rock units controls the a $Mg^{++}/\sigma Mg^{++}$ (a H^+)^2 and a $Ca^{++}/\sigma Ca^{++}$ (a H^+)^2. The incoming fluid is thus undersaturated with magnesium relative to magnesite saturation, but supersaturated with calcium relative to the magnesite - dolomite assemblage. Thus leaching of magnesium and deposition of calcium will take place through exchange reactions such as Reaction 2, and the magnesite dissolution of Reaction 3.

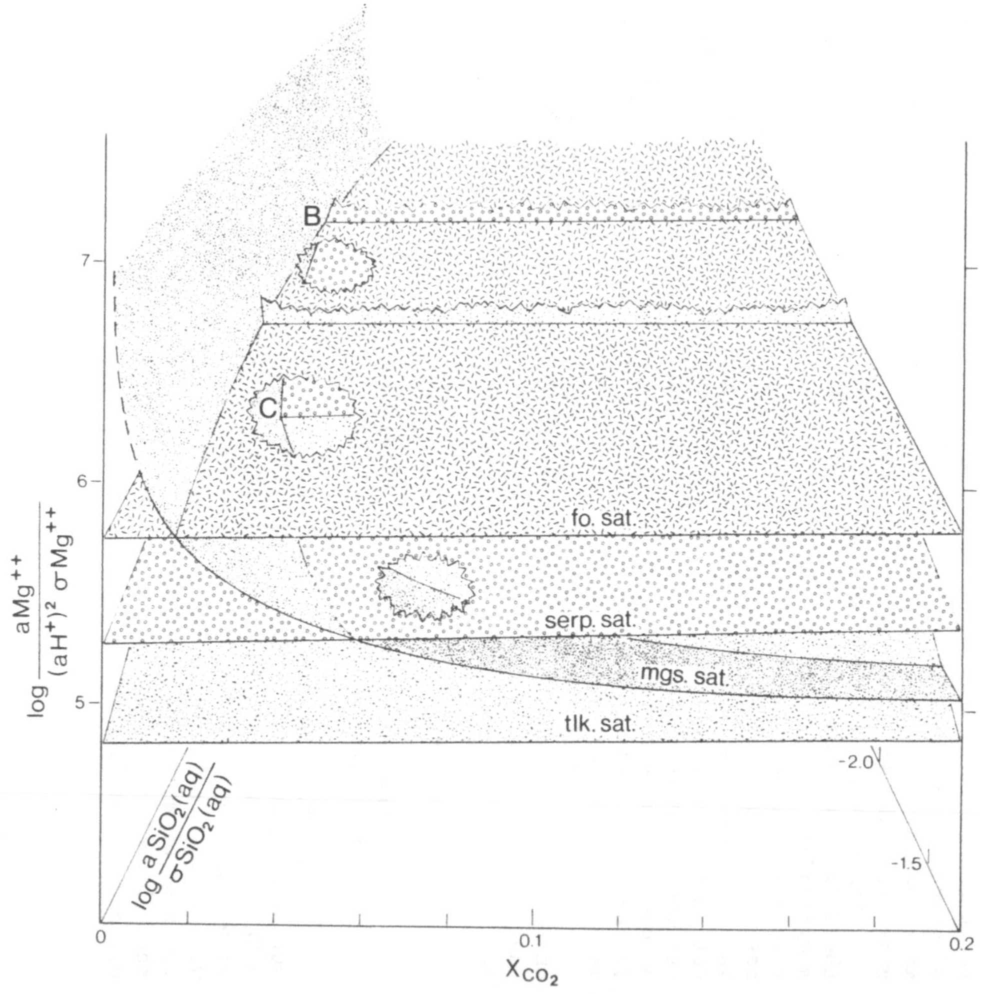

194

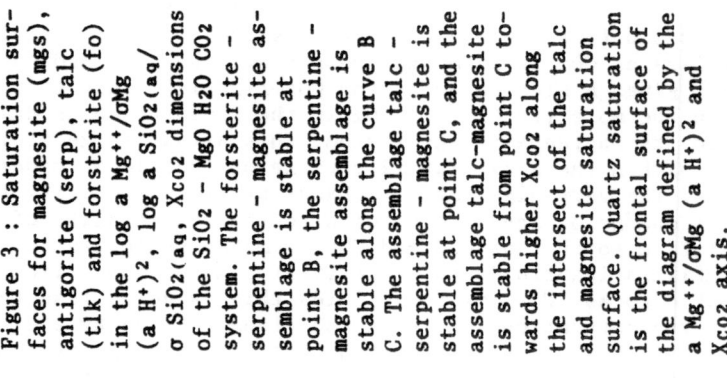

Figure 3 : Saturation surfaces for magnesite (mgs), antigorite (serp), talc (tlk) and forsterite (fo) in the log a Mg^{++}/σMg (a H^+)², log a SiO_2(aq/ σ SiO_2(aq, X_{CO_2} dimensions of the SiO_2 – MgO H_2O CO_2 system. The forsterite – serpentine – magnesite assemblage is stable at point B, the serpentine – magnesite assemblage is stable along the curve B C. The assemblage talc – serpentine – magnesite is stable at point C, and the assemblage talc-magnesite is stable from point C towards higher X_{CO_2} along the intersect of the talc and magnesite saturation surface. Quartz saturation is the frontal surface of the diagram defined by the a Mg^{++}/σMg (a H^+)² and X_{CO_2} axis.

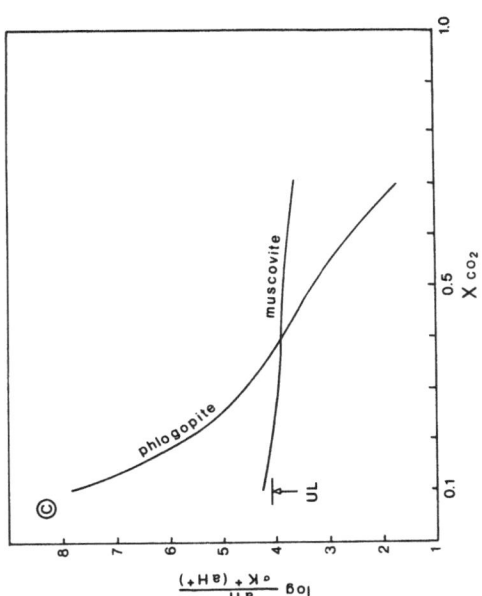

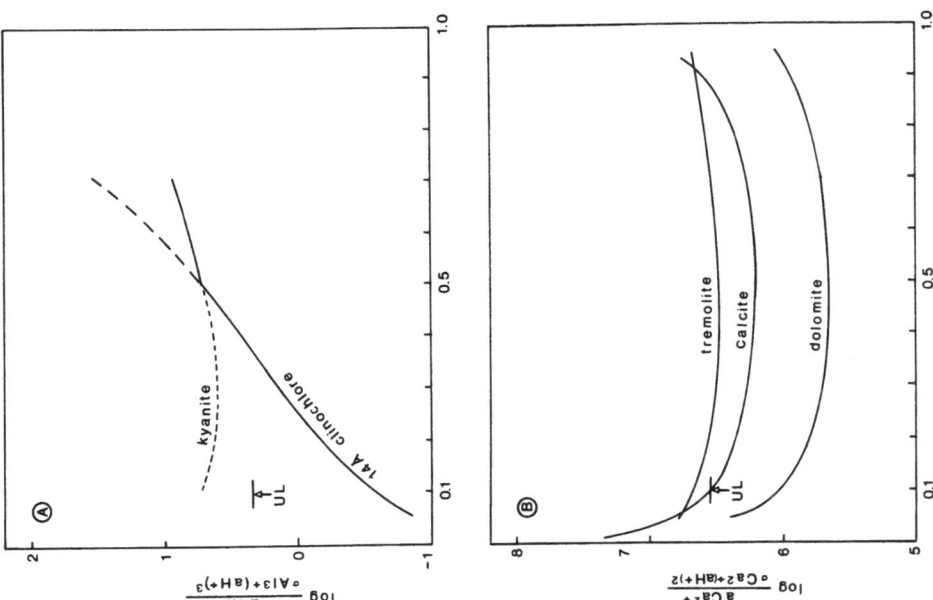

Figure 4 : Saturation curves for the indicated phases in the indicated systems with a Mg^{++}/σMg^{++} $(a\,H^+)^2$ and a $SiO_2(aq)$/ $\sigma\,SiO_2(aq)$ defined by the the major phase assemblages in Figure 3. In 4 C the a Al^{+++} / σAl^{+++} $(a\,H^+)^3$ is defined by the clinochlore curve in 4 A. UL denote the upper limis for the activity ratios in the country rock, namely A: kyanite quartz, B: calcite and C: K-feldspar - muscovite quartz.

Reaction 2 :

$$2 \; MgCO_3 \; + \; Ca^{++} \; - \; CaMg(CO_3)_2 \; + \; Mg^{++}.$$
$$\text{magnesite} \quad \text{fluid} \quad \text{dolomite} \quad \text{fluid}$$

Reaction 3 :

$$MgCO_3 \; + \; 2 \; H^+ \; - \; Mg^{++} + H_2O + CO_2$$
$$\text{magnesite} \quad \text{fluid} \quad \text{fluid}$$

The silicification due to Reaction 1 and the calcification due to Reaction 2 will proceed at constant a $SiO_{2(aq)}/\sigma SiO_{2(aq)}$ and a $Mg^{++}/\sigma Mg^{++}$ (a H^+)^2 until all magnesite is exhausted.

A consequence of the low buffered value of $SiO_{2(aq)}/\sigma SiO_{2(aq)}$ and the high a $Mg^{++}/\sigma Mg^{++}$ (a H^+)^2 in the fluid after equilibration with the ultramafics is that chlorite saturation define a very low a Al^{+++} / σAl^{+++} (a H^+)^3 activity ratio. This low silica and aluminium activity require a very high potassium ion activity to stabilize K_2O bearing phases such as muscovite, biotite and K-feldspar. At low X_{CO2} values in the metasomatizing fluid, the chlorite saturation level therefore effectively shields the ultramafics from K_2O additions, as shown in Figure 4. The low saturation levels of a Al^{+++} / σAl^{+++} (a H^+)^3 are an order of magnitude lower than levels defined by for instance, kyanite or the assemblage calcic plagioclase calcite in the quartz saturated exterior. This means that practically all aluminium will precipitate from an infiltrating fluid.

Mobility of Chemical Species

The thermodynamic analysis sets up the framework for discussion of additions to, and losses from the ultramafics, by defining gradients in activities of individual components, and defining the equilibrium requirements for stabilization of phases to accomodate these components.

Diffusion of components through a stagnant intergranular fluid body, and advection of component dissolved in a moving fluid phase are two possible endmember processes which can be responsible for the transfer of material between the ultramafic rocks and their environment. In the following treatment the effect of diffusion of components is considered insignificant relative to that of dissolution and precipitation of minerals from a percolating fluid.

As discussed earlier the transport characteristics of these fluids are controlled by a number of factors in addition to the cation to hydrogen ion activity ratios. These additional factors control the effectiveness of mass transfer, whereas the equilibrium requirements in terms of ionic activity ratios determine the direction of transfer. The largest uncertainties therefore lie in the estimate of the fluid/rock ratios necessary to cause the observed changes.

In order to get a first order impression of the range in ratio of fluid to rock necessary to produce the observed changes, the changes in the two major components SiO_2 and MgO are modelled. Assuming that activity coefficients and chloride dissociation constants for Mg and Ca are of comparable magnitude (Frantz and Marshall, 1982), the molality of dissolved calcium in the theoretical fluids are ca. two orders of magnitude larger than the molality of magnesium. This is comparable with the data for natural geothermal fluids of Helgeson (1967). The limiting factor for the Ca^{++} - Mg^{++} ion exchange reaction will therefore be the effectiveness of magnesium transport.

It is assumed that SiO_2 is introduced from the exterior, and causes the decarbonation reaction of Reaction 1. Magnesium is lost to the environment via the exchange Reactions 2 and 3. All losses and gains occur through reactions with a phase assemblage including stoichiometric talc and magnesite and a fluid of constant Xco_2. A low mole fraction of CO_2 in the fluid facilitate the observed changes and a value of 0.1 is arbitrarily assumed. The solubility of SiO_2 in pure H_2O, given by Anderson and Burnham (1965) is assumed to apply for the low Xco_2 fluid of this test.

The difference in MgO and SiO_2 between 175565 and 175566 can be produced by addition of 1.38 moles SiO_2 and removal of 2.23 moles MgO to each kg of 175566. The addition of SiO_2 can be regarded as independent of pH in a first order approach, because of the low acidity of silicic acid (Anderson and Burnham, 1965). For this reason the addition of SiO_2 is used to calculate the minimum fluid/rock ratio. The incoming, quartz saturated fluid has an activity of aqueous silica of $10^{-0.86}$ assuming that the speciation of silica in the low Xco_2 fluid does not deviate from that in pure H_2O. The talc - magnesite assemblage buffers the fluid at an activity of $10^{-1.625}$ at $Xco_2 = 0.1$. This is an 80 % decrease in the molality of SiO_2, which corresponds to a deposition of 0.11 moles of SiO_2 when one kg fluid reacts with the talc - magnesite assemblage. The observed addition of SiO_2 implies the reaction of 12 kg of fluid with each kg rock.

The only way that silica could be leached from the ultramafic rock is if there had existed a temperature gradient of more than 200 °C between the ultramafic body and its country rock during alteration. With a temperature change of this magnitude the change in SiO_2 solubility as a function of temperature would counterbalance the differences in buffered molality of aqueous silica between quartz saturation and that of the talc - magnesite assemblage.

The effectiveness of magnesium leaching from the ultramafics will be dependent on the relative importance of the exchange Reactions 2 and 3. If a total solubility of magnesium of $10^{-3.26}$ moles/ kg measured in hydrothermal fluid in equilibrium with dolomite and tremolite at 300 °C (Helgeson, 1967) is representative for the order of magnitude of magnesium concentration in the incoming fluid, and magnesite saturation defines a value that is 3 times higher, as predicted by activity ratios given by Bowers et al. (1984), a fluid to rock ratio of 4460 is required to account for the magnesium loss.

The gross discrepancy between the fluid/rock estimates from the MgO and SiO_2 variation, illustrate the strong model dependency of such calculations. The rate of MgO loss to SiO_2 gain in the ISB ultramafics is however comparable to losses and gains observed in studies of serpentinization of dunites (Condie and Madison, 1969), and the estimate of the total losses and gains in SiO_2 and MgO are probably realistic. It must therefore be assumed that actual magnesium solubility in the fluids interacting with the Isua ultramafics were considerably higher than the values found by Helgeson (1967). High chlorinity and low pH of the fluid would increase magnesium solubility. Changes in these factors would affect aluminium solubility an order of magnitude stronger than magnesium because of the higher valance of Al^{+++}.

The molality of total dissolved aluminium is linked to the a Al^{+++} / σAl^{+++} (a H^+)^3 activity ratio through unknown equilibrium constants for complexing reactions. A change in a Al^{+++} will however lead to changes in total dissolved aluminium of the same relative magnitude. As the saturation level for chlorite is an order of magnitude lower in a Al^{+++} / σAl^{+++} (a H^+)^3 than the level in the source of infiltrating fluid, and since magnesite dissolution (Reaction 3) will lead to an increase in pH, lowering the aluminium ion activity at constant a Al^{+++} / σAl^{+++} (a H^+)^3, aluminium will be effectively removed from the fluid upon reaction with the ultramafic rocks and the net-transport of aluminum will go into the ultramafic bodies.

The observed increase in Al of 0.4 moles per kg rock requires a molality of aluminium of 0.03 in the fluid, if the minimum fluid / rock ratio calculated from the gain in SiO_2 is considered. Regardless of the fluid / rock ratio, the molality of magnesium in the fluid must have been in the order of 15 times that of aluminium.

Discussion

As the metasomatic alteration of the ISB ultramafics is likely to have involved both gains and losses of material from the bodies as a whole, and since even isochemical metamorphic recrystallisation introduce volume changes, it is impossible to evaluate quantitatively the mobility of individual chemical components. The two major chemical components, SiO_2 and MgO, probably experienced the largest gains and losses, because of the high solubilities of their saturating phases in metamorphic fluids, and the large differences in chemical potential in these elements between the ultramafics and their silica-rich Mg-poor country rocks. Large variations in the concentration of the main components which correlate with the degree of metasomatism, impose a correlation between all other components and the degree of metasomatism. In studies of element mobility based on one set of empirical observations, the concept of immobile components, prevents an unbiased evaluation of the geochemical behavior of these components, and reduce the analysis to a trivial exercise. In the present case there is a fair correlation between the increase in Al_2O_3, TiO_2 and Zr. This could indicate that these components are passively enriched by an overall decrease in volume of the ultramafic body. Passive enrichment of Al by removal of other components would however require that SiO_2 is lost from the quartz-undersaturated ultramafics to the quartz-saturated environment. This would require unrealistic temperature gradients between the ultramafic units and their country rock, as shown above. Similarly REE,

Fe and Cr must be accepted as mobile components, to account for the changes in their abundances. The metamorphic assemblages present in the margins of the ultramafics show they have interacted with large masses of fluids, and that these fluids deposited and/or removed chemical components. If a "passive" model for the enrichment in Al_2O_3 is chosen, a mass of fluid which is several orders of magnitude larger than the mass of rock is required to remove the MgO from the rocks. As discussed above, the same factors that facilitate magnesium transport, increase the effectiveness of aluminium transport even more strongly. Even accepting that aluminium has a very low solubility relative to magnesium, the fluid volumes required to cause the observed changes by magnesium leaching, would be capable of transporting major amounts of Al, and the concept of immobility become meaningless. Very large fluid/rock ratios might be attained at the interface between rock units of contrasting competence, such as the metagreywacke - ultramafic contact. The loss of SiO_2 would remain paradoxical.

Although the presented model is a gross oversimplification, silicification and cation exchange reactions between the ultramafics and infiltrating fluids are suggested as the dominant causes of chemical alteration seen in the ultramafics of the ISB. Such processes can adequately explain the observed gains and losses in MgO, SiO_2, CaO and possibly Al_2O_3, without the necessity to call upon exotic transfer processes or fluids of extreme compositions or volumes. At least some of the increase in Al_2O_3 could be caused by introduction of aluminium from the environment via an aqueous fluid phase.

The mobility of aluminium has no influence on the modeling of potassium fixation in the ultramafics. Regardless of the source of Al in these rocks, the reduced silica activity combined with a high a $Mg^{++}/\sigma Mg^{++}$ (a H^+)^2, stabilizes chlorite at too low a value of a Al^{+++} / σAl^{+++} (a H^+)^3 to allow K-bearing phases to form at ambient a $K^+/\sigma K^+$(a H^+).

The abundances of K_2O and Rb in the interior sample 175566, are higher than expected magmatic values for peridotites. Potash and rubidium is probably hosted in fuchsite in this sample. The aluminium activity requirement for fuchsite stability, is lower than that of muscovite in Fig. 4 A, because of chrome substitution for Al(VI) in the white mica. Fuchsite growth in the interior sample can either be explained by a higher chrome activity relative to the margin during metasomatism, or by causes during the later metamorphic history.

A possible explanation for fuchsite stabilisation during the later metamorphic history could lie in the presence of carbonate in the interior in contrast to the margins. During prograde metamorphism decarbonation reactions would allow the interior to buffer X_{CO2} at values higher than the carbonate free margins. This would stabilize white mica in the presence of chlorite at ambient a $K^+/\sigma K^+$(a H^+) in accord with the diagram in Fig. 4 C. The geochemical behavior of the REE can not be modeled in the analysed thermodynamic system. Magmatic REE patterns of the ultramafics can be assumed to have been parallel to chondritic, and of comparable total abundance. If this is the case, the two marginal samples have been strongly enriched in all REE, whereas the central sample has been preferentially enriched in the light REE. This distribution may reflect a generally higher mobility of the light REE, relative to the heavy REE. If the REE distribution is accepted as secondary, the higher REE abbundance in the metagreywacke relative to the peridotite resulted in a higher chemical potential of REE in the fluid coexisting with this lithology than that defined by the ultramafic mineral assemblage.

Conclusions

Steep chemical potential gradients existed between the Isua metaperidotites and their metagreywacke county rock during amphibolite facies metamorphism of the ISB. The geochemical changes observed in the Isua metaperidotites were caused by interaction of these chemical gradients with mineral reactions taking place within the ultramafic bodies.

No chemical component can empirically be regarded as immobile in geologic systems.

The observed gross changes in REE patterns between least and most altered metaperidotite samples preclude detailed neodymium isotopic interpretations of their mantle source. Since the changes in REE geochemistry is caused by interaction with the country rock, alterations of the Sm-Nd isotope systematics in other units of the ISB must be expected, and Nd isotopic interpretations based on any unit in the ISB without demonstration that the system has been closed, must be seriously questioned.

Acknowledgements

Rock samples and geochemical data for this study were provided by D. Bridgwater. Discussions with D. Bridgwater, D.K. Bird and A.P. Nutman greatly improved this manuscript. Critique and suggestions from B.R. Frost have resulted in significant improvements. The director of the Geological Survey of Greenland is thanked for permission to publish this paper.

Financial support was provided by the Carlsberg Foundation.

References

Anderson, G.M. and Burnham, C.W., 1965. 'The solubility of quartz in supercritical water.' Amer. Journ. Sci. **263**, 494 511.

Baadsgaard, H., 1983. 'U-Pb isotope systematics on minerals from the gneiss complex at Isukasia, West Greenland.' Rapp. Grønlands Geol. Unders. **112**, 35-42.

Baadsgaard, H., Nutman, A.P, Rosing, M.T. and Bridgwater, D., 1984. 'A late Archaean pegmatite dyke swarm from the Isukasia area, southern West Greenland.' Rapp. Grønlands Geol. Unders. **125**, 48-51.

Baadsgaard, H, Nutman, A.P and Bridgwater, D., 1986a. 'Geochronological and isotopic variation of the early Archaean Amitsoq gneisses of the Isukasia area, southern West Greenland.' Geochim. Cosmochim. Acta, **50**, 2173-2183.

Baadsgard, H., Nutman, A.P, Rosing, M.T., Bridgwater, D. and Longstaffe, F.J., 1986b. 'Alteration and metamorphism of Amitsoq gneisses from the Isukasia area, West Greenland: Recommendations for isotope studies of the early crust.' Geochim. Cosmochim. Acta, **50**, 2165-2172.

Boak, J.L. and Dymek, R.F., 1982. 'Metamorphism of the ca. 3800 Ma supracrustal rocks at Isua, West Greenland: Implications for Early Archaean crustal evolution'. Eart Planet. Sci. Lett. **59**, 155-176.

Bowers, T.S., Jackson, K.J. and Helgeson, H.C., 1984. Equilibrium activity diagrams. 397 p. Spinger-Verlag, Berlin.

Condie, K.C. and Madison, J.A., 1969. 'Compositional and volume changes accompanying serpentinization of dunites from the Webster-Addie ultramafic body, North Carolina.' Amer. Mineral. **54**, 1173 - 1179.

De Donder, T. and Van Rysselberghe, P., 1936. Thermodynamic theory of affinity. Stanford University Press, Stanford, California.

Frantz, J.D. and Marshall, W.L., 1982. 'Electrical conductances and ionization constants of calcium chloride and magnesium chloride in aqueous solutions at temperatures to 600 °C and pressures to 4000 bars.' American Journal of Science, **282**, 1666-1693.

Greenwood, H.J., 1967. 'Mineral equilibria in the system $CO_2-MgO-SiO_2-H_2O$ In: P.H. Abelson ed. : Researches in geochemistry, volume 2, 542-567. Wiley, New York.

Haskin, L.A., Haskin, M.A., Frey, F.A. and Wilderman, T.R., 1968. Relative and absolute abbundance of the rare-earths. In: L.H. Ahrens ed. : Origin and distribution of the elements. 889-912. Pergamon Press. New York.

Helgeson, H.C., 1967. Solution chemistry and metamorphism.
In: P.H. Abelson ed. : Researches in geochemistry, volume 2,
362-404. Wiley, New York

Helgeson, H.C., Delaney, J.M., Nesbitt, H.W. and Bird, D.K., 1978.
Summary and critique of the thermodynamic properties of
rockforming minerals. American Journal of Science, 278A, 229 p.

Kalsbeek, F., Bridgwater, D and Boak, J., 1980. Evidence for
mid-Proterozoic granite formation in the Isua area. Rapp.
Grønlands Geol. Unders. 100, 73-75.

Korzhinskii, D.S., 1970. Theory of metasomatic zoning. 162 pp. Clarendon
Press, Oxford.

Nakamura, N., 1974. Determination of REE, Ba, Mg, Na, and K in
carbonaceous and ordinary chondrites. Geochim. Cosmochim. Acta, 38,
757-775.

Nutman, A.P, Allaart, J.H., Bridgwater, D., Dimroth, E. and Rosing,
M.R., 1984. 'Stratigraphic and geochemical evidence for the
depositional environment of the early Archaean Isua supracrustal
belt, southern West Greenland.' Precambrian Research, 25, 365-396.

Nutman, A.P., Rivers, T., Longstaffe, F. and Park, J.F.W., 1988.
'The Ataneq fault and mid-Proterozoic retrograde metamorphism of
early Archaean tonalites of the Isukasia area, southern West
Greenland: Reactions, fluid compositions and implications for
regional studies.' This volume.

Rosing, M.T., 1983. A metamorphic and isotopic study of the Isua
supracrustals, West Greenland. Unpublished Cand. Scient thesis,
Copenhagen University, Copenhagen.

Rosing, M.T., 1987. 'Redistribution of "immobile elements" in the
Isua supracrustals, West Greenland. In: D. Bridgwater ed.: Fluid
movements, element transport, and the composition of the deep
crust. Abstracts of the NATO advanced research workshop, Lindås
Norway.

Thompson, J.B. Jr., 1959. Local equilibrium in metasomatic processes.
In: P.H. Abelson ed. : Researches in geochemistry Wiley, New York.

Wagner, R., 1982. Geochronology of the Ameralik dykes at Isua, West
Greenalnd. M.Sc. thesis, University of Alberta, Edmonton, Canada.

Whittaker, E.J.W. and Muntus, R., 1970. 'Ionic radii for use in
geochemistry.' Geochim. Cosmochim. Acta, 34, 945-956.

MASS BALANCE OF A GABBROIC ROCK-AMPHIBOLITE TRANSITION

H.P. Zeck[1] and J. Toft[2]
[1] Geological Institute, Copenhagen University, 1350 Copenhagen
K, Denmark.
[2] Mærsk Olie og Gas A/S, Esplanaden 50, 1263 Copenhagen K,
Denmark.

ABSTRACT. Bulk chemical compositions of a series of gabbroic rocks and
their recrystallized equivalents have been compared with their specific
gravities, using the equation of Gresens (1967). By careful control of
parent-daughter rock matching the volume increase of the gabbroic rock-
amphibolite transition has been established at 2 ± 0.5 %. On that basis
the metasomatic effect accompanying the transition has been calculated.
   For SiO_2, TiO_2, Al_2O_3, FeO^*, MgO, P_2O_5, Sr, Y and Zr no mass
transfer could be shown by statistical testing (α = 0.05). The
following components were added to the basic rocks during amphibolite
formation: Cl (+ c. 230%), L.O.I, mainly H_2O (+ c. 50%), Th (+ c. 25%),
K_2O (+ c. 55%), Rb (+ c. 80%), Pb (+ c. 100%), Zn (+ c. 15%), MnO (+ c.
6%) and Na_2O (+ c. 5%). The following components were removed: S (- c.
60%) and CaO (c. 5%). Iron transfer could not be shown, but the increase
in $Fe_2O_3/(FeO + Fe_2O_3)$ ratio from c. 0.18 to c. 0.22 is statistically
highly significant.
   The element migration pattern is suggested to be characteristic for
the present type of setting: tholeiitic gabbroic sills/dykes intruded
into amphibolite facies gneisses and subsequently metamorphosed in
(lower) amphibolite facies.

INTRODUCTION

Knowledge of the metasomatic effect of a metamorphic transition provides
information on the hydrothermal regime during the metamorphism, and it
offers the possibility to reconstruct from the metamorphic rocks the
chemistry of the parent rocks. This information is important, for
example, when considering regional features such as the general ore
potential of large metamorphic rock complexes. And it is a critical
piece of information when the detailed magmatic chemistry of
meta-igneous rocks is used to reconstruct the large scale, plate
tectonic setting and development of a region.
   Metasomatism - transport of material in the context of petrological
processes - requires a fluid phase carrier. Low molality of natural
hydrothermal solutions requires large amounts of fluids, and/or an
effective circulation system, to register a noticeable result. The

D. Bridgwater (ed.), Fluid Movements – Element Transport and the Composition of the Deep Crust, 203–212.
© 1989 by Kluwer Academic Publishers.

feasibility of large scale, convective fluid regimes involving large
amounts of fluids has been proved by investigations in intra-oceanic
ridge systems (e.g., Corliss, 1971; Lister, 1972, 1981; Fyfe & Lonsdale,
1981) and continental areas with high level intrusions (e.g. Taylor,
1974).

The nature and extent of the bulk chemical effect of a metamorphic
event on a rock body will be dependent on the conditions under which the
recrystallization takes place. These conditions are given not only by T,
P and other intensive parameters, but also by extensive and kinetic
factors, e.g., the amount of the pertaining fluid phase, the structure,
permeability of the rock body and the country rock, their relative
amounts and the way they are intercalated. This would imply that each
metamorphic-metasomatic complex may show its own element migration
pattern.

An elegant and potentially very accurate way of evaluating mass balance
relations in metamorphic transitions has been suggested by Gresens
(1967). This author deduced a simple relation between the chemical
composition and the specific gravity of parent rock and metamorphic
daughter rock, and the volume change of the transition. For rock A
going to rock B:

$$f_v(g_B/g_A) \, c^B_n - c^A_n = x_n, \text{ where}$$

g_A = specific gravity of rock A
g_B = specific gravity of rock B
c^A_n = wt fraction of component n in rock A
c^B_n = wt fraction of component n in rock B
x_n = wt fraction of component n lost or gained
f_v = volume factor: volume of product divided by volume of parent
rock.

Each parent-daughter rock pair for which the specific gravities and
chemical compositions have been determined will thus provide an
independent estimate for the relation between the volume effect (f_v) and
the metasomatic effect (x_n) of the transition for each of the chemical
components considered. By considering a larger series of
parent-daughter pairs a very large body of data becomes available, which
by statistical reductions may give an evaluation of both the volume
effect and the metasomatic effect.

The purpose of the present study is to reach a precise evaluation
of the metasomatic effect of a common magmatic parent-metamorphic
daughter rock relation, the gabbroic rock-amphibolite transition, via
the Gresens equation. The sample material on which this study is based
was collected with a matched pair survey in mind. Consequently special
attention was paid to assuring correct parent-daughter rock matching.
Subsequent screening on the basis of composition-volume data was applied
to further improve the matching.

GEOLOGICAL SETTING AND PETROGRAPHY

The rock formation selected for the study is the Hyperite Suite in
Värmland, S Sweden, which consists of a partly recrystallized series of

tholeiitic sills and dykes, several hundred m thick, intruded at c.
1,500 Ma ago (Priem et al., 1968; Welin et al., 1980) into the
amphibolite facies gneisses N of lake Vänern (Morthorst et al., 1983),
Fig. 1. These basic bodies consist of gabbroic parent rocks, massive
hornblende metagabbros and amphibolites, plus rocks transitional between
these three rock types. The deformation and recrystallization producing
the metamorphic rocks were concentrated at the margins of the Hyperite
bodies and may at least in part be of Grenvillian age (Zeck and Wallin,
1980; Zeck and Hansen, 1988). Detailed information on the geological
setting and field appearance of the Hyperite bodies is given by
Morthorst et al. (1983).

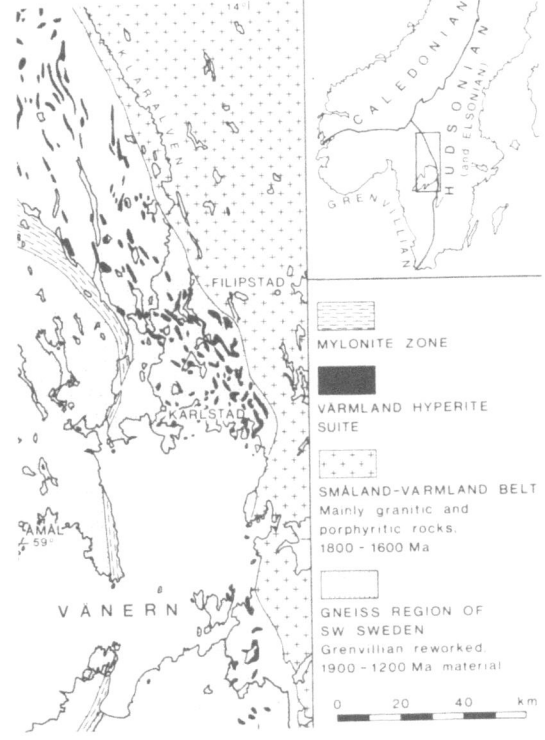

*Fig. 1. Location of the
Värmland Hyperite Suite, the
basic rock formation
comprising the gabbroic
rock - amphibolite transition
studied. From Zeck et al.
(1988); after Magnusson et al.
(1958).*

    Precise petrographical study reveals that 7 rock types can be
distinguished within the Hyperite Formation (Zeck et al., 1988). Sample
groups 1, 2 and 3 represent the gabbroic rocks; these show only a very
weak, inherent metamorphic imprint (Zeck et al., 1982). Group 5
represents the massive hornblende metagabbros and group 7 the
amphibolites. Group 4 contains rocks transitional between (1+2+3) and
5, and group 6 represents rocks transitional between 5 and 7. A
description of the 7 rock groups and information on sample locations,
chemical compositions and specific gravities are given in Morthorst et
al. (1983) and Zeck et al. (1988).

The present study will compare the gabbroic rocks (1+2+3) with the amphibolites (7). The transition (1+2+3)/(5) gives similar results. Comparisons involving the intermediate rock types (4) and (6) versus the gabbroic rocks indicate, not surprisingly, mass transfers which are less than those involving groups (5) and (7).

The gabbroic rocks (1+2+3) represent the magmatic parent rock within the Hyperite Suite, the amphibolites (7) their completely recrystallized and foliated derivatives, reflecting an anisotropic stress regime. Thirteen pairs of the (1+2+3)/(7) transition were available for study. The Hyperite Suite shows a reasonable degree of exposure. A series of large, fresh samples, weighing c. 5 Kg each, was collected. The magmatic protolith and the metamorphic product pairs were sampled as close together as the local conditions allowed. The sample distance for the 13 pairs varies from 5 to 80 m with an average of 25 m.

MASS TRANSFER CALCULATIONS

Mass balance evaluations on the basis of the Gresens equation are subject to two basic uncertainties. The first concerns the sample matching problem. The input data of the Gresens equation, g_A, g_B, c^A and c^B can be measured precisely, but their intrinsic precision is determined by how well the samples of parent and daughter rocks were matched in the field (Zeck et al., 1988). In a case of ideal matching the true parent rock of a metamorphic rock is identical to the rock actually sampled as such. In practice the variability within the protolithic rock suite, the less than perfect exposure, and demands on the quality and size of the samples will put a limit on how well this matching can be done.

The second uncertainty in mass balance evaluations concerns the volume effect of the transitions. After substituting measured g_A, g_B, c^A_n and c^B_n values in the Gresens equation, two variables remain, x_n, a measure for the mass transfer, and f_v, a measure for the volume effect of the transition. In practical mass balance evaluations (e.g., Babcock, 1973; Appleyard, 1980; Appleyard et al., 1979, 1981; Kerrich et al., 1980) it is far from obvious which of these two factors explains what proportion of the apparent parent-daughter rock differences. In most cases a certain f_v value is assumed or inferred on the basis of the assumption $x_{Al2O3} = 0$ or $x_{TiO2} = 0$. Substituting this f_v value in the equations for the various chemical components then yields the mass transfer pattern of the transition. For the specific metamorphic transition discussed in this paper a value $f_v = 1.02 \pm 0.005$ has been suggested, corresponding to a volume increase of $2 \pm 0.5\%$ (Zeck et al., 1988). This evaluation is based on the same set of samples and chemical analyses which were used in the present paper. The volume effect study showed that the f_v scores for the 13 available sample pairs were rather variable. It was argued that this was essentially caused by wrong matching of parent and daughter rock. Further screening on the basis of f_v/x_{SiO2}, x_{Al2O3} relations suggested that 6 pairs out of the 13 were less well matched, and consequently the volume increase of $2 \pm 0.5\%$ was calculated on the basis of the remaining 7 samples.

Par #	1	2	3	4	5	6	7	8	9	10	11	12	13
A	31715	31715	31728	31784	35112	35131	35205	31755	31755	31784	32030	32042	35215
B	31702	31714	31724	31780	35111	35132	35204	31749	31751	31778	32027	32040	35213
SiO2	-0.123	0.009	0.153	-0.940	-0.503	1.034	0.227	0.264	0.230	0.465	-0.914	-2.259	-2.194
TiO2	-0.036	0.003	-0.009	0.002	0.159	-0.046	0.036	-0.515	-0.263	0.297	0.383	-0.196	-0.239
Al2O3	-0.083	0.132	0.046	0.068	0.058	0.192	-0.009	0.887	0.573	-0.580	-1.277	1.727	2.679
Fe2O3	0.838	0.725	0.573	0.007	1.270	0.063	0.463	-0.097	1.093	0.854	1.609	-0.297	0.217
FeO	-0.910	-0.990	-0.439	-1.179	-0.544	0.217	-0.970	-1.938	-2.105	-1.053		0.926	-0.494
MnO	0.031	0.010	0.002	0.038	0.012	0.036	-0.007	-0.031	-0.029	0.030	0.018	0.024	0.015
FeO*	-0.155	-0.338	0.077	-1.173	0.599	0.274	-0.553	-2.026	-1.121	-0.285	1.761	0.659	-0.299
MgO	-0.339	-0.506	0.057	-0.935	-0.170	0.474	0.044	0.504	0.550	-1.487	2.039	1.039	0.593
CaO	-0.725	-0.252	-0.068	-0.603	-1.010	0.042	-0.065	-1.111	-0.903	-0.300	-0.398	-0.312	0.519
Na2O	-0.112	0.156	0.274	0.756	0.137	0.100	0.244	0.172	-0.291	0.176	-0.264	0.047	0.507
K2O	1.206	0.463	0.107	0.998	0.684	0.213	-0.062	0.156	0.516	0.478	0.461	0.703	-0.222
P2O5	0.061	0.030	-0.079	0.035	0.013	-0.054	-0.037	-0.061	-0.009	-0.031	0.068	0.044	-0.065
H2O	0.413	0.342	-0.042	0.559	0.000	0.000	0.555	0.527	-0.470	-0.520	1.108	0.625	0.000
LOI	0.958	0.685	0.313	0.351	1.262	0.372	1.040	0.744	0.366	0.278	1.286	1.001	0.603
FeO/MgO	0.067	0.083	0.011	0.134	0.146	-0.049	-0.055	-0.621	-0.466	0.588	-0.374	-0.172	-0.185
Rb	47.387	12.962	20.466	63.150	23.980	7.557	-4.157	9.358	20.489	19.944	21.093	48.471	-8.698
Cu	-43.168	-7.738	10.111	-23.693	3.745	5.589	19.060	-54.639	-69.654	-0.559	-8.087	-43.502	-18.444
Sr	-52.695	7.725	1.669	-9.232	-16.891	1.154	24.756	24.708	18.871	1.563	-57.949	-4.910	56.185
Ba	505.143	150.466	6.220	2.499	104.915	-19.416	-47.037	104.024	285.010	140.505	54.945	29.270	-31.484
Zn	25.027	11.263	5.522	86.176	14.041	15.322	4.552	-5.856	12.321	9.098	33.299	11.427	-2.958
Y	0.179	-2.195	2.355	15.081	4.615	1.558	0.212	-1.327	-1.076	3.687	6.618	-17.330	-4.985
Pb	0.309	-0.175	0.950	9.354	3.171	2.819	2.401	1.392	2.406		4.233	4.007	2.014
Zr	-23.646	-7.907	0.567	2.913	3.948	4.120	1.816	-8.617	-4.514	27.475	7.456	-35.877	-18.558
Th	0.058	-0.035	1.906	1.190	1.105	0.478	0.747	1.118	-0.320	1.874	0.321	-0.598	-1.018
S	-392.730	-115.646	-220.417	-925.206	-1130.458	-390.012	-1409.446	-1029.863	-1318.486	-493.476	-448.139	-1191.701	-1188.377
Ni	-7.172	-11.966	0.698	-15.600	-10.963	2.228	0.884	20.231	14.818	-31.277	53.041	45.428	6.157
Cl	567.090	465.653	361.966	472.209	450.770	468.655	474.929	770.348	695.482	854.706	376.057	418.488	432.572

Table 1. Gresens factor x_n scores for $f_v = 1.02$, for 13 matched gabbroic rock - amphibolite pairs, $(1 + 2 + 3)/(7)$. The first 7 pairs pass the screening $|f_v SiO2 - f_v Al2O3| \le 0.02$.

7 "best" pairs				all 13 pairs	
SiO_2	–	113.0			Sr	–	444.7
P_2O_5	–	17.13			P_2O_5	–	4.451
TiO_2	–	2.592			Y	–	4.369
Cu	–	2.299			TiO_2	–	3.842
Sr	–	2.020			MgO	–	2.843
Zr	–	2.014			FeO*	–	1.702
FeO*	–	1.404			Ni	–	1.602
MgO	–	0.795			Zr	–	1.240
Ba	–	0.590			Al_2O_3	–	0.683
Y	–	0.550			SiO_2	–	0.646
Al_2O_3	–	0.435					
		0.370	≡ α_{1-t} = 0.05	≡ 0.368			
Zn	–	0.330			MnO	–	0.355
Ni	–	0.322	X		Na_2O	–	0.344
Na_2O	–	0.319			FeO	–	0.285
Pb					Th		
		0.286	≡ α_{1-t} = 0.025	≡ 0.271			
CaO	–	0.276			Ba	–	0.247
MnO	–	0.267	XX		Cu	–	0.246
Rb	–	0.254			CaO	–	0.241
K_2O	–	0.245			Zn	–	0.230
		0.230	≡ α_{1-t} = 0.01	≡ 0.205			
Fe_2O_3	–	0.218					
S	–	0.214	XXX				
		0.205	≡ α_{1-t} = 0.005	≡ 0.176			
Th					Fe_2O_3	–	0.159
FeO		0.202			Oxid.r.*		0.157
Oxid.r.*		0.190	XXXX		Rb	–	0.141
LOI	–	0.178			K_2O	–	0.137
Cl	–	0.145			Pb		
					S	–	0.100
* Oxidation ratio:					LOI	–	0.096
$Fe_2O_3/(Fe_2O_3 + FeO)$, wt % data					Cl	–	0.083

Table 2. Sandler's A-statistic scores for all components, for the 7 "best" pairs and all 13 pairs, respectively. Significance levels for α = 0.05, 0.025, 0.01 and 0.005 (one-tailed test) are indicated. Scores for Pb and Th are not given, because analytical data for two pairs are lacking; their position in the two columns reflects their scores on a 4 to 10 degrees for freedom scale, respectively.

SiO_2, TiO_2, Al_2O_3

$FeO*$, MgO, P_2O_5 - no statistically significant mass transfer

Sr, Y, Zr

Ni, Cu, Ba   - uncertain status

Na_2O	x	+	5 %
MnO	x	+	6 %
Zn	x	+	15 %
CaO	xx	−	5 %
Th	xx	+	25 %
Pb	xxx	+	100 %
Rb	xxx	+	80 %
K_2O	xxx	+	55 %
FeO	xxx	−	7 %
Fe_2O_3	xxx	+	20 %
S	xxx	−	60 %
L.O.I.	xxxx	+	50 %
Cl	xxxx	+	230 %
Oxid.r.	xxxx	+	20 %

Table 3. Summary of the suggested mass transfer pattern of the gabbroic rock - amphibolite transition, (1 + 2 + 3)/(7). Crosses indicate level of significance (cf. Table 2). The percentage changes indicated are thought to be accurate within 10 - 20 per cent. Oxidation ratio (Oxid. r.) as in Table 2.

It would seem to follow that the mass transfer pattern should also be calculated on the basis of the 7 "best" sample pairs. However, it should be noted that the screening criterion in the volume effect study was only based on SiO_2, Al_2O_3 relations. These two components are well suited to provide a f_v screening criterion because their high concentrations yield good statistics. However, other components may be hosted in other minerals than the SiO_2, Al_2O_3 fit is based on. Consequently the 7 matched rock pairs selected as best fits on the basis of their SiO_2, Al_2O_3 relations are not necessarily the best choices for the other components. Therefore the mass transfer for the (1+2+3)/(7) transition has been calculated for both the 7 "best" pairs passing the screening f_{vSiO_2} $f_{vAl_2O_3} \leq 0.02$, and for the 13 pairs as a whole. It appears that in some cases the results of these two approaches are different, and a best choice has to be made in each individual case.

Another source for these discrepancies may be that both statistical samples were not truly representative of the natural rock body. This in reality is a common flaw in geological sampling.

The Gresens factor x_n scores for all pairs, for all 24 chemical components analysed constitute a large body of data which is given in Table 1. To evaluate these data Sandler's A-statistic (Haber and Runyon, 1977) has been applied to the scores for each of the components. The scores in this statistical test are presented in Table 2, both for the 7 "best" pairs and for all 13 pairs. The score of each component was related to standard confidence levels and the results of the two tests were compared. On this basis a mass transfer pattern is suggested (Table 3). For the components for which a statistically significant change could be shown ($\alpha = 0.05$) the suggested average change is indicated in Table 3; these figures are thought to be accurate within 10-20%.

SiO_2 and Al_2O_3 have a special status in the survey because the f_v matching quality control was based on these components, and consequently for these components the results of the 7 pairs test will be the more reliable. For both components the zero hypothesis holds at $\alpha = 0.05$ (onetailed test), but the low score for Al_2O_3 is surprising. Other components for which the zero hypothesis could not be negated at $\alpha = 0.05$, either in the 7 pairs test or in the 13 pairs test, are TiO_2, FeO^*, MgO, P_2O_5, Sr, Y and Zr. A significant increase could be shown in both tests for Na_2O at the $0.025 < \alpha < 0.05$ level. For Ni, Cu and Ba the results are ambiguous. Ni scores lower than the $\alpha = 0.05$ level in the 7 pairs test for a decrease of c. 7%, but clearly higher in the 13 pairs test, for an *increase* of c. 5%. These results are inconsistent and do not allow a conclusion. Unpublished results of a $X/MgO - TiO_2/MgO$ survey (see Zeck et al., 1983, for details of the method) do not indicate Ni mobility. For Cu the 7 pairs test scores clearly higher than the $\alpha = 0.05$ level for a decrease of c. 8% and for the 13 pairs the score is between the $\alpha = 0.025$ and 0.01 level for a decrease of c. 24%. This result is unsatisfactory and no firm conclusion can be drawn. Again, the $X/MgO - TiO_2/MgO$ survey does not indicate Cu mobility. For Ba the situation is similar, but here in both tests an average increase of c. 25% is indicated; a $X/MgO-TiO_2/MgO$ survey confirms Ba mobility. For CaO the zero hypothesis is negated at the $0.01 < \alpha \leq 0.025$ level in both tests. Pb and Th score rather differently in both tests, however there seems no doubt that both elements have been added during the metamorphic recrystallization. The same is true for the remaining components: MnO, Zn, Rb, K_2O, FeO, Fe_2O_3, LOI and Cl. S has been removed. The increase in oxidation ratio $Fe_2O_3/(Fe_2O_3+FeO)$, wt% data, from c. 0.18 to c. 0.22 is statistically highly significant.

CONCLUSIONS

The mass transfer pattern of a gabbroic rock-amphibolite transition has been determined by means of the Gresens equation. The survey shows that in spite of careful field control and subsequent composition-volume screening to improve parent-daughter rock matching, the results remain subject to a basic uncertainty which is mainly due to less than perfect matching and the fact that the statistical samples seem not truly representative of the rock complex. Other factors adding to the statistical dispersion around the suggested average mass transfers are local variations in the regional metasomatic pattern and true variations in the volume effect of the transition due to local mineralogical variation in the parent rock suite. The study shows that rather precise mass balance conclusions may be reached via the Gresens equation provided the sample material is specifically collected for the purpose and satisfactory parent-daughter rock matching is achieved.

In the suggested mass transfer pattern it is worth noting that Ti, P, Zr, Y and Sr, elements currently used for characterizing basic rock compositions in plate tectonic settings (Pearce and Cann, 1973), show an immobile behaviour. This suggests that amphibolites which developed from tholeiitic material in a similar setting - sills and dykes intruded in amphibolite facies gneisses and subsequently metamorphosed in (lower) amphibolite facies - have retained their Ti, P, Zr, Y and Sr compositional characteristics. The suggested immobility of Al is in agreement with suggestions for a number of different metamorphic transitions. Immobility of Si, Mg and FeO* is less commonly suggested. The depletion of Ca is unexpected, and may be characteristic for this type of setting.

ACKNOWLEDGEMENTS

The authors are greatly indepted to J.Morthorst who collected the samples on which this study is based in the course of his MSc thesis work. The authors wish to express their gratitude to Drs. J.C.Bailey and A.K. Higgins (Copenhagen) for commenting upon the manuscript, and to the Danish Research Council (SNF) for providing financial support to HPZ (J.No. 11-1970 and 81-3794). The paper was reviewed by Dr. F.Kalsbeek.

REFERENCES

Appleyard, E.C., 1980. Mass balance computations in metasomatism: meta-gabbro/nepheline syenite pegmatite interaction in northern Norway. *Contrib. Mineral. Petrol.*, 73: 131-144.
Appleyard, E.C. and Williams, S.E., 1981. Metasomatic effects in the Faraday metagabbro, Bancroft, Ontario, Canada. *Tscherm. Min. Petr. Mitt.*, 28: 81-97.
Appleyard, E.C. and Woolley, A.R., 1979. Fenitization: an example of the problems of characterizing mass transfer and volume changes. *Chem. Geol.*, 26: 1-16.
Babcock, R.S., 1973. Computational models of metasomatic processes. *Lithos*, 6: 279-290.
Corliss, J.B., 1971. The origin of metal-bearing submarine hydrothermal solutions. *J. Geophys. Res.*, 76: 8128-8138.
Fyfe, W.S. and Lonsdale, P., 1981. Ocean floor hydrothermal activity. In: C.Emiliani (ed.), *The Sea, Oceanic lithosphere*, vol. 7, Wiley-Interscience, 589-638.

Gresens, R.L., 1967. Composition-volume relationships of metasomatism. *Chem. Geol.*, 2: 47-65.

Haber, A. and Runyon, R.P., 1977. *General Statistics*. Addison-Wesley, Reading, 343 p.

Kerrich, R., Allison, I., Barnett, R.L., Moss, S. and Starkey, J., 1980. Microstructural and chemical transformations accompanying deformation of granite in a shear zone at Miéville, Switzerland; with implications for stress corrosion cracking and superplastic flow. *Contrib. Mineral. Petrol.*, 73: 221-242.

Lister, C.R.B., 1972. On the thermal balance of a mid-oceanic ridge. *Geophys. G: R. Astron. Soc.*, 26: 515-535.

Lister, C.R.B., 1981. Rock and water histories during sub-oceanic hydrothermal events. *Int. Geol. Congr., Paris, 1980, Coll. C4, Géol. Océans*, 41-46.

Magnusson, N.H., Asklund, B., Kulling, O., Kautsky, G., Eklund, J., Larsson, W., Lundegårdh, P.H., Hjelmqvist, S., Gavelin, S., and Ödman, O., 1958. Karta över Sveriges berggrund. *Sveriges Geol. Unders.* Ba 16.

Morthorst, J.R., Zeck, H.P. and Lundegårdh, P.H., 1983. The Proterozoic Hyperites in southern Värmland, western Sweden. *Sveriges Geol. Unders.* Ba 30: 104 p.

Pearce, J.A. and Cann, J.R., 1973. Tectonic setting of basic volcanic rocks determined using trace elemental analysis. *Earth Plan. Sci. Lett., 19: 290-300.*

Priem, H.N.A., Mulder, F.C., Boelrijk, N.A.I.M., Hebeda, E.H., Verschure, R.H. and Verdurmen, E.A.T., 1968. Geochronological and paleomagnetic reconnaissance survey in parts of central and southern Sweden. *Phys. Earth Planet. Int.*, 1: 373-380.

Taylor, H.P., 1974. The application of oxygen and hydrogen isotope studies to problems of hydrothermal alteration and ore deposition. *Econ. Geol.*, 69: 843-883.

Welin, E., Lundegårdh, P.H. and Kähr, A.M., 1980. The radiometric age of a Proterozoic hyperite diabase in Värmland county, western Sweden. *Geol. Fören. Förh.*, 102: 49-52.

Zeck, H.P. and Hansen, B.T., 1988. Rb-Sr mineral ages for the Grenvillian metamorphic development of spilites from the Dalsland Supracrustal Group, SW Sweden. *Geol. Rdschau*, 77: 683-692.

Zeck, H.P., Lou, S. and Ellgaard, L., 1983. Statistical evaluation of metasomatic effects in meta-igneous rock series. *Chem. Geol.*, 40: 51-63.

Zeck, H.P., Ottesen, O. and Toft, J. (1988). Volume effect of a gabbro-amphibolite transition. *Chem. Geol.*, 67, 141-153.

Zeck, H.P., Shenouda, H.H., Rønsbo, J.G. and Poorter, R.P.E., 1982. Hypersthene-ilmenite(/magnetite) symplectites in coronitic olivine gabbronites. *Lithos*, 15: 173-182.

Zeck, H.P. and Wallin, B., 1980. A 1220 ± 60 M.Y. Rb-Sr isochron age representing a Taylor-convection caused recrystallization event in a granitic rock suite. *Contrib. Mineral. Petrol.*, 74: 45-53.

MASS TRANSFER RELATED TO DUCTILE SHEAR ZONE DEVELOPMENT IN A METAGABBRO

J.L. POTDEVIN*, J.M. LARDEAUX**, D. COFFRANT**
*Laboratoire de Pétrologie, Université des Sciences et
Techniques de Lille, UA 719, 59655 VILLENEUVE D'ASCQ Cedex,
FRANCE. **Laboratoire de Pétrologie, Université LYON I, UA 726,
27-43 Bd du 11 Novembre, 69622 VILLEURBANNE Cedex, FRANCE

ABSTRACT. Metagabbros from the Rouergue area (French Massif Central),
show the synmetamorphic development of ductile shear zones under
amphibolite facies conditions. Shearing results in three kinds of rocks
(preserved, foliated, and mylonitic metagabbros) in which different
stages of deformation and reaction progress can be studied. Geochemical
methods and mass balance calculations have been applied to the analyses
of the three types in order to study the mass transfer controlled by
deformation and metamorphic processes. The mobility of some major and
trace elements have been shown in the shear zones only. The mobility
occured without significant changes in volume. The chemical changes are
related to progressive deformation and amphibolitization. Mechanisms and
limiting processes of mass transfer are discussed. The scale and the
character of this mass transfer requires the general percolation of
fluids through the metagabbros and the surrounding gneisses. The
limiting process of mass transfer seems to be the reaction kinetics
which are increased by deformation and grain size reduction in the shear
zones and not in the undeformed rocks.

Keywords: Mass transfer, Fluids, Ductile shear zone, Mass balance
calculations, Metagabbros, Chemical mobility.

1. INTRODUCTION

An important problem in orogenic terranes is to clarify the
relation between metamorphism and deformation. In particular,
ductile shear zones have been recognized as favourable structures in
which to study metamorphic processes, including mass transfer and
fluid-rock interactions (Beach, 1973; Kerrich et al, 1977; Hickman and
Glassley, 1984; Glassley and Bridgwater, 1985). This paper is concerned
with such problems and describes mass transfer occuring during the
development of ductile shear zones by progressive deformation in
metagabbros from the French Massif Central. In a previous study
(Lardeaux et al., 1988) the comparison of rock composition in the shear
zones and in the undeformed metagabbros showed a relation between strain

213

D. Bridgwater (ed.), Fluid Movements – Element Transport and the Composition of the Deep Crust, 213–230.
© *1989 by Kluwer Academic Publishers.*

intensity and the chemical variations of rock composition. In order to specify the mass transfer related to these chemical variations, some hypotheses must be made. Further some assumptions about elements mobility and volume changes are discussed in order to resolve mass balance calculations. The quantification and the description of this mass transfer, together with the relations with deformation and metamorphic reactions allow us to discuss the processes leading to the observed differences between the undeformed metagabbros and the sheared metagabbros.

2. GEOLOGICAL, PETROLOGICAL AND GEOCHEMICAL CONTEXT: A SUMMARY

2.1 Description of the samples studied:

The metagabbros studied have been sampled in the Rouergue metamorphic zone (southwestern part of the French Massif Central) within the "leptyno-amphibolitic group". The latter is a classical formation of the deep-seated zone of the European Variscan belt (Forestier, 1963; Vogel, 1967; Burg and Matte, 1978; Autran and Cogne, 1980; Matte, 1986). In the Rouergue area, the major tectonometamorphic imprint is related to polyphase ductile deformation during Hercynian amphibolite facies metamorphism (Collomb, 1963; Nicollet, 1978). Our samples are fine grained gabbros metamorphosed under conditions of about 550-650°C and 5-6 Kb (Lardeaux, 1985). In the field, and on a mesoscopic scale, the metagabbros range between mylonites, foliated amphibolites and metagabbros in which the original igneous textures are preserved. The different textural types are interpreted as representing different degrees of deformation. Transitions between these different stages of deformation occur.

2.2 Connection between deformation and chemical variation

In order to characterize the chemical variation within the metagabbros analysed, a multivariate statistical method, the principal component analysis (P.C.A.), has been applied in an earlier paper to the metagabbro samples (Table I), using both major elements and some trace elements which are generally considered immobile during metamorphism (Lardeaux et al., 1988). In this method, the weight percentages of the constituent oxides (major elements) and the trace elements (in ppm) are replaced by new parameters named factors. Only two or three of these factors are regarded as significant and describe the major part of the rock chemical variations. In the Rouergue metagabbros, the first factor, (F1) of the principal component analysis (a linear combination of major element percentages) explains more than seventy percent of the total variation in the system. F1 also discriminates between the three stages of deformation recognised in the gabbros thus showing a strong correlation between variations in rock chemistry and strain intensity (Fig 1a). When subdivided into separate components F1 allows the effect of change of each individual chemical constituent during deformation to be evaluated with respect to total change (Fig 1b).

TABLE 1

	1	2	3	4	5	6	7	8	9
SiO_2	52.85	51.54	52.82	53.72	51.78	52.86	55.30	54.82	55.78
Al_2O_3	16.25	15.47	17.07	16.42	16.72	17.33	16.27	17.05	16.07
FeO	8.51	9.46	8.32	8.17	8.47	8.10	7.69	8.08	7.77
MgO	5.33	6.39	5.42	4.66	5.68	5.07	4.70	4.36	4.13
CaO	8.43	9.22	8.15	8.58	9.24	8.63	7.82	7.52	7.56
Na_2O	3.08	2.37	2.90	2.86	2.55	2.92	2.95	3.10	2.75
K_2O	1.34	1.24	1.42	1.70	1.19	1.62	1.79	1.84	1.86
TiO_2	0.92	1.02	0.92	0.94	0.95	0.88	0.93	0.84	0.85
P_2O_5	0.18	0.11	0.20	0.40	0.21	0.27	0.21	0.18	0.21
MnO	0.16	0.19	0.16	0.15	0.15	0.16	0.15	0.15	0.13
LOI	0.74	0.97	0.59	0.64	0.88	0.81	0.70	0.79	0.96
Nb	7.00	5.00	6.00	7.00	6.00	5.00	6.00	5.00	7.00
Y	30.00	28.00	31.00	33.00	30.00	29.00	29.00	28.00	31.00
Rb	31.0	29.0	37.0	44.0	27.0	99.0	50.0	82.0	64.0
Sr	670	630	712	658	688	679	728	610	654
Zr	119	99	116	110	108	124	107	139	126

	10	11	12	13	14	15	16	17
SiO_2	54.00	55.21	54.87	54.92	53.78	52.54	54.69	54.66
Al_2O_3	16.50	17.97	17.12	16.48	17.39	16.39	16.68	17.09
FeO	7.30	7.72	7.80	8.02	7.18	8.59	7.91	7.60
MgO	4.70	3.98	4.13	3.72	4.05	5.50	4.56	4.12
CaO	6.10	6.22	5.93	6.07	6.51	8.72	7.88	6.17
Na_8O	3.22	3.36	3.07	3.59	3.43	2.75	2.93	3.33
K_2O	3.60	3.28	3.19	3.05	3.64	1.38	1.78	3.35
TiO_2	0.83	0.86	0.79	0.81	0.72	0.95	0.88	0.80
P_2O_5	0.23	0.27	0.25	0.32	0.27	0.22	0.22	0.27
MnO	0.14	0.13	0.13	0.12	0.13	0.16	0.15	0.13
LOI	1.28	0.96	1.14	1.23	0.98	0.76	0.82	1.12
Nb	6.00	5.00	6.00	6.00	7.00	6.20	5.75	6.00
Y	31.00	29.00	33.00	28.00	30.00	30.40	29.25	30.20
Rb	140.0	112.0	137.0	119.0	128.0	33.6	73.8	127.2
Sr	514	568	505	513	517	672	668	523
Zr	122	149	131	141	153	110	124	139

Table I - Major and trace elements contents for the metagabbros samples. Chemical analyses of the undeformed rocks (1-5), the little deformed rocks (6-9) and the highly deformed rocks (10-14). 15, 16, 17 are the mean chemical analyses of the undeformed, the little deformed and the highly deformed rocks.

Three groups are distinguished (Fig. 1b):
A) Chemical components which increase with strain (K_2O, Rb, Na_2O, Zr, SiO_2, LOI (loss on ignition), Al_2O_3, P_2O_5)
B) components which decrease with strain (CaO, TiO_2, MgO, MnO, FeO, Sr)
C) components which appear independant of strain (Y and Nb).
The changes in these components cannot be related to any initial magmatic trends in gabbroic rocks (Lardeaux et al., 1988). The cross-cutting nature of the shearing argues against coincidental

agreement between the suggested secondary changes and original
magmatic heterogeneities. As the metagabbros show neither pre-existing
magmatic variations (Marquer et al, 1985) to which the shearing can be
related nor later metasomatism controlled by the shearing (Floyd and
Winchester, 1983) they are well suited for mass balance calculations.

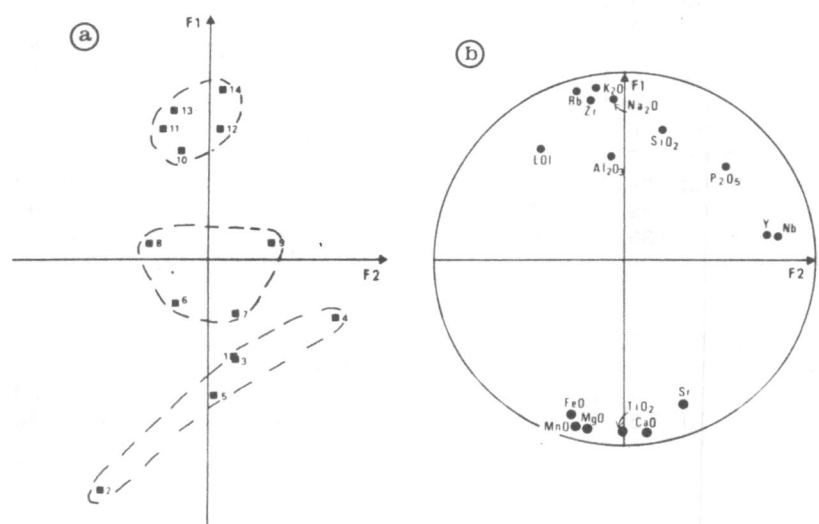

*Fig. 1 - Principal component analysis. F1 versus F2 (see Lardeaux et
al., 1988) a) Plot of chemical analyses from the metagabbros . Factor F1
(a linear combination of the chemical components) separates the three
deformation states. (Analyses Nos as Table 1, 1-5 undeformed, 6-9 little
deformed, 10-14 strongly deformed.) b) Contributions of chemical
elements to factors F1 and F2.*

3. MASS TRANSFER ESTIMATES

The mass balance method involves comparison between the rock
composition in an initial state with that in a final state, as a result
of processes such as weathering (Cramer and Nesbitt, 1983), metasomatism
(Gresens, 1967; Ferry, 1983) or deformation (Gratier, 1983).
Two major problems arise with mass balance calculations when trying to
specify mass transfer associated with deformation:

1) What was the initial rock composition?

2) Which reference frame can be used for mass balance calculations?

3.1 Initial rock composition

In units of initially homogeneous magmatic rocks, such as gabbros or granitoids, the chemical compositions of undeformed zones are generally taken as representative of the initial rock composition before deformation (Kerrich et al., 1977; Marquer et al., 1985). This assumption is however problematic:
- If mass exchange occurs between deformed and undeformed rocks.
- If the chemical variation is the result of differences in the degree of a mass transfer process which have affected both the undeformed and the deformed rocks.

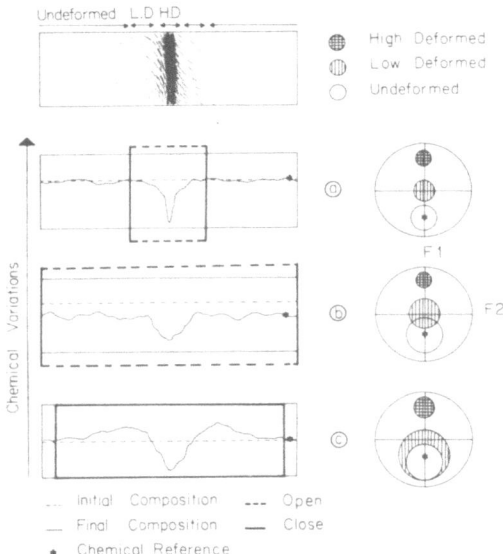

Fig. 2 - Chemical variation profiles through a shear zone and their effects on principal component analysis (P.C.A.). a) Chemical variations and related mass transfer occur only in the shear zone. The P.C.A. shows a straight correlation between the strain intensity and the chemical changes. Initial and undeformed rocks compositions are the same. b) Mass transfer acts differently in the shear zone and in the undeformed rock. The discrimination between little deformed and undeformed states is less evident. Undeformed rock composition is not the initial rock composition. c) Mass exchanges occur between the undeformed rock and the shear zone. In the P.C.A. no discrimination between undeformed and little deformed rocks can be made.

It is therefore necessary to discuss the distance over which mass transfer is likely to take place before choosing a chemical reference. In some cases (in particular pressure-solution phenomena) microstructures define the scale where the system can be regarded as

closed (Gratier, 1984; Potdevin, 1984). However in the Rouergue
metagabbros, in which metamorphic reactions occured throughout, such an
approach is impractical. A second approach to this problem is to look at
the chemical variations in sections across the shear zone and the
undeformed rock (Fig. 2), and at their implications for the principal
component analysis. For all the mobile elements two main chemical
profiles and subsequent chemical behaviours can be distinguished:

1) There is only chemical variation inside the shear zone. This
profile corresponds to two different cases:
a) Mass transfer has only affected the shear zone (Fig 2a). The
undeformed rock composition is the initial rock composition before
chemical mobility. The principal component analysis shows a straight
correlation between the strain intensity and the observed chemical
variation.
b) Mass transfer has affected rocks in the shear zone and
undeformed rocks differently (Fig 2b). The composition of the undeformed
rock does not represent the initial rock composition but can serve as
reference to estimate the difference in mass transfer between the shear
zone and the surrounding gabbros. In order for differences between
undeformed and deformed rocks to show up clearly on geochemical diagrams
it is necessary that any mass transfer which occurs in the undeformed
rocks should be small compared with those in the sheared rocks.

2) Mass exchange occured between the shear zone and the immediate
country rocks. Chemical differences can be related to the distance from
the shear zone (Fig. 2c). If the sampling scale is on a finer scale than
that of the mass exchange a careful examination of the chemical profile
can lead to a good estimate of the initial rock composition. No clear
distinction between little deformed and undeformed rocks can be made in
the component analysis.
In the Rouergue metagabbros, the principal component analysis
shows that the undeformed rock analyses display little variation while
there is a significant discrimination between the different deformation
states (Lardeaux et al., 1988; Fig. 1). Chemical variations are not
related to the distance from the shear zone suggesting there was no
material exchange between the shear zones and the adjacent country
rocks. The metagabbroic mass appears to have been an open system. As the
chemical changes related to the strain intensity are highly significant,
mass transfer in the undeformed rocks cannot have been very important,
and the composition of the undeformed rock can be taken to be close to
that of initial rock. The composition of the undeformed rocks can thus
be chosen as the chemical reference for mass transfer calculations,
regardless of the mechanism of transfer and their chronological or
geometrical relations with deformation.

3.2 Reference frame

In order to estimate mass transfer, two rock chemical analyses cannot
be compared directly (Fonteilles, 1978). Chemical analyses are given
as only weight percentages of the constituent oxides while gains or

losses depend on the volume variations associated with mass transfer
(Gresens, 1967). A reference frame is needed to convert variations in
chemical analyses into units of mass transfer (Carmichael, 1969;
Gresens, 1967; Cramer and Nesbitt, 1983; Gratier, 1984; Potdevin and
Marquer, 1987). This reference may be geometric: (assumption of constant
volume, Gardner, 1980; Sicard et al., 1986, Cotonian et al, 1988);
chemical: (assumption of immobile elements, Kerrich et al., 1977;
Chesworth et al., 1981; Ferry, 1983; Marquer, 1987; Trolliard et al ,
1987), or mineralogical: (assumption of inert minerals, Gratier, 1983;
Lelong and Souchier, 1979; Potdevin and Caron, 1986).

For the comparison of rock analyses, Gresens (1967) derived the general
relation:

$$\Delta Xn = XbFv(\rho b/\rho a) - Xa$$

where Xa and Xb represent the initial and the final rock respectively
and ρb and ρa represent their densities: ΔXn the gain or loss of a
constituent oxide: n (in mass percentage of initial rock): Xa and Xb
their weight percentages and finally Fv the volume factor (ratio of
final volume to initial volume). With this relation, the knowledge of
either the volume change or the geochemical behaviour of one component
fixes a unique solution for the different ΔXn. To discuss mass transfer
and volume change of the studied transformation, metagabbros analyses
can be used to construct composition-volume diagrams (Gresens, 1967)
where gains or losses of components are graphically shown as a function
of volume changes accompanying the deformation (Fig. 3a). In the present
paper a more recent composition-volume diagram, (Potdevin and Marquer,
1987) is used in which relative gains or losses of constituents oxides
(and consequently gains or losses of elements) are plotted with regard
to their initial contents. In this plot element mobility is defined as:

$$\Delta Xn/Xa = (Xb/Xa) \ Fv.(\rho b/\rho a) - 1$$

The slopes of the curves allow the comparison of the relative mobility
of both trace and major elements for the transformation studied. When
the slope value is greater than one, the element ratio increases with
the transformation intensity. If this value is lower than one, the
element ratio decreases with the transformation intensity. The behaviour
of two elements is similar if they are characterized by the same curve
in the composition-volume diagram (Fig. 3b). A further advantage
compared to the standard Gresens diagram is that the intersection
between curves and the Fv axis ($\Delta Xn=0$) are clearer.

3.3 Results

Figures 3, 4, 5 show composition-volume diagrams for the Rouergue
metagabbros. Mass transfers for two different chemical transformations
are shown: the general transformation of undeformed into highly deformed
rocks and the intermediate transformation of undeformed into little
deformed rock. Comparison is also made between little deformed and

highly deformed rocks. In the three cases, mean chemical analyses of the three deformation states (Table I) are used in order to reduce the effects of initial heterogeneities or uncertainties in individual analyses. As no significant rock density variations have been demonstrated for the metagabbros, the relative mass and volume changes should be equal (Potdevin and Marquer, 1987).

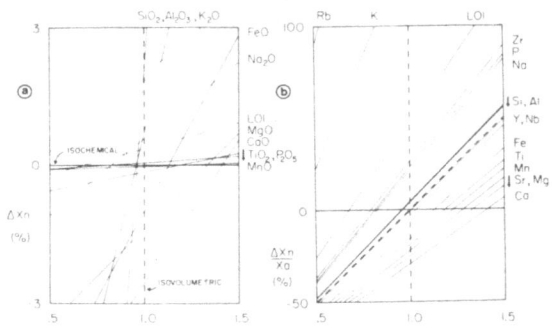

Fig. 3 – Composition volume-diagram for the change from undeformed to highly deformed rocks. a) Absolute gains or losses in oxides of major elements. b) Relative gains or losses of chemical elements. With ΔXn, gain or loss of a constituent n (in mass percentage of initial rock), Xa the weight percentage of this constituent in the initial rock, $\Delta Xn/Xa$ its mobility and Fv the volume factor.

Examination of the diagrams shows that, whatever the assumption for a reference frame, only few elements can be simultaneously regarded as immobile. The trace elements also exhibit differences in mobility and only two of them, Y and Nb, have the same behaviour. As these elements are among the least affected by metamorphic processes they can be used as a reference to estimate the mobilities of other components. In this case, Si and Al appear as the less mobile of the major elements. Al immobility is generally assumed during metamorphic process (Gresens, 1967; Carmichael, 1969; Fisher, 1970, 1973; Ferry, 1983) and is justified in many cases of mass transfer related to deformation (Kerrich et al., 1977; Gratier, 1983; Potdevin, 1984; Marquer et al., 1985; Sicard et al., 1986). If Al is regarded as strictly immobile (Table II), mass transfer estimations of Y and Nb lie in the uncertainty range of chemical analyses (this is not reciprocal because of a better precision for Al analyses). The assumption of Al immobility is consistent with the immobility of Y and Nb, and gives the reference frame we need for mass transfer calculations.
Using Al as a reference for the whole transformation, three kinds of element are distinguished (Fig. 3b and Table II).
- Immobile elements like Nb, Y and to a first approximation, Si.
- Mobile elements which are removed from the deformed rock: Ca, Mg, Sr, Mn, Ti and Fe.
- Mobile elements which are introduced into the deformed rock: Rb, K, Zr, P and Na.

The degree to which an element contributes to the total mass transfer depends not only of its mobility but also on its initial content in the rock. The greatest part of the observed mass transfer are seen in the major elements Ca, K, Mg, and Fe (Fig. 3a). The whole transformation is achieved by a small volume change and related mass variation ($\Delta M = \Delta V = -4\%$). As there are both loss and gain of material, mass losses are not balanced by mass gains in the shear zones.

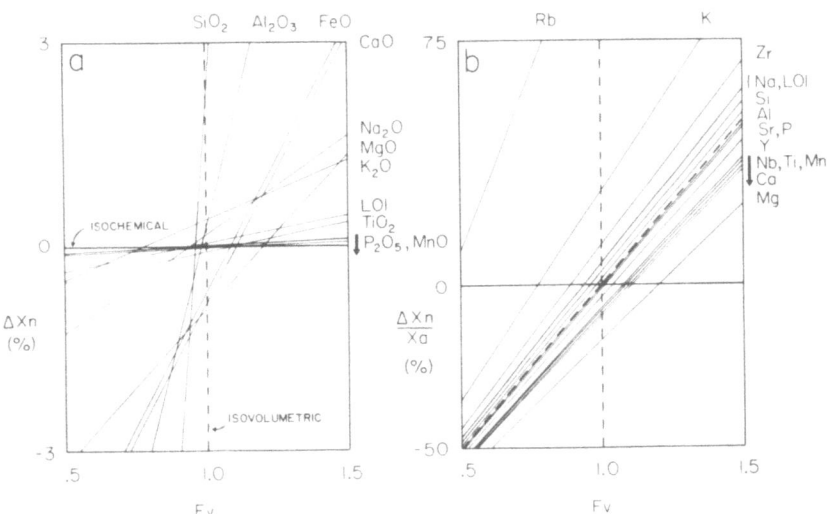

Fig. 4 - Composition volume diagram for the changes from undeformed to the little deformed rocks.

Element mobility may be related to strain intensity. For example, mass transfer and element mobility are less important for the change between the undeformed to the foliated rock than those estimated by the comparison between mylonitic and foliated rocks. In the composition-volume diagram, this is shown by a greater spread between the different curves in the second case (Fig. 5) than in the first (Fig. 4). This must be related to the difference in strain intensity between the final state (e.g. foliated or mylonitic rock) and the reference state (e.g. undeformed or foliated rock).

For all cases, the Al immobility assumption seems justified in spite of some apparent discrepancies in the behaviour of Y, Nb, Si and Al for the transition from the undeformed rock to the foliated rock (Fig. 4). This results in a higher relative uncertainty for the mass transfer calculations. Nevertheless, when the rock is little strained (foliated rock), significant mass transfer of Mg, Ca, Fe, K, Rb, Zr and Sr can be shown with $\Delta V = \Delta M = -2\%$ (Fig. 4). In general mass transfer is much more important although the total mass change (and volume change) do not show

222

the same evolution. The degree of element mobility is different for
different stages in the changes. For example, the mobility order of Mg
and Ca is not the same for the complete transformation as it is for the
change of undeformed into low deformed rock (Fig. 3, 4). The relation
between the transferred mass quantities of one element and the strain
rock intensity is not a simple one and is not the same for the different
elements. To account for this, we need to study the geochemical
processes which could have taken place in the rock during, or perhaps
after deformation.

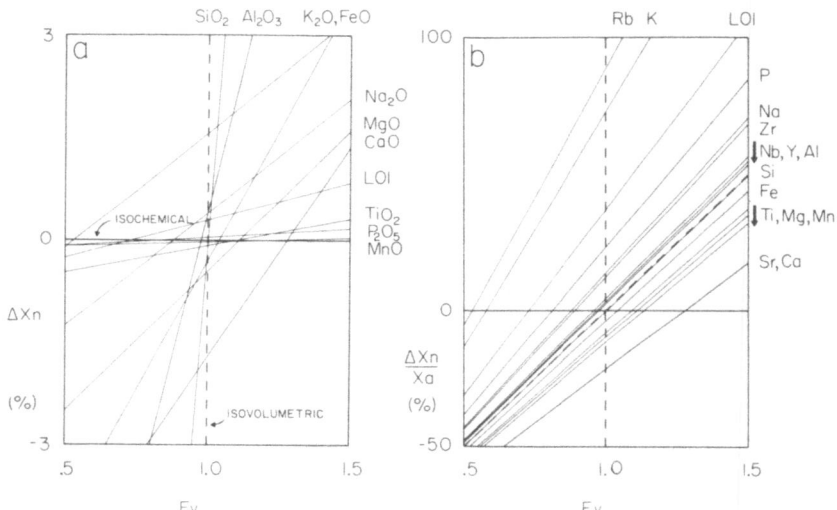

*Fig. 5 - Composition-volume diagram for comparison between the little
deformed and the highly deformed rocks.*

4. CONNECTION BETWEEN MASS TRANSFER, METAMORPHIC REACTIONS AND DEFORMATION

Up to now, we have estimated mass transfer which could account for the
observed rock chemical modifications between undeformed, little-deformed
and highly-deformed rocks. This mass transfer has been related to
strain intensity as a purely geometrical concept and has not taken
the processes occuring during deformation into account. From a
geometrical viewpoint a finite strain concentration within a shear
zone could allow the highly foliated rocks to be more susceptible to
mass transfer. In this case mass transfer should be independant of
deformation processes and could be later than the shear zone
development. In the Rouergue metagabbros, however the observed chemical

change is related to a mineralogical change which results from a progressive amphibolitization of the original magmatic paragenesis at 550°-650°C and 5-6 Kbar (Lardeaux, 1985; Lardeaux et al., 1988).

Cpx + Plag + FeTi oxides + H2O --> Amph + Zoï + Qtz + Biot

The K2O, Na2O, Rb and H2O enrichment within the shear zones is controlled by the breakdown of the primary magmatic Cpx to andamphibole biotite while it remains almost completely preserved in th   e undeformed rocks. Microstructural and mineralogical criteria show that the progressive deformation and amphibolitization of the Rouergue metagabbros are contemporaneous (Lardeaux, 1985; Lardeaux et al., 1988). This implies that mass transfer and the mineralogical and chemical changes it produced, are related to the metamorphic and the deformation processes which took place in the rock during the shear zone development. Furthermore since volume variations and deformation of the wall rocks are low, the shear zones development can be considered as if they were the result of a progressive heterogeneous simple shear (Ramsay, 1980). In this case chemical comparison of two rocks which have been affected by different degrees of shearing can give an estimate of the mass transfer and element mobility which accompanied metamorphic and deformation processes during the transition between a less strained to a more strained state of the same initial rock. Using the undeformed rocks for reference the estimates of mass transfer versus strain intensity show the geochemical evolution of the degree of transfer for individual elements with progressive rock deformation and amphibolitization (Fig. 6). It has been shown that there is an overall correlation between total mass tranfer and degree of deformation (Fig 1). As accurate strain measurements cannot be made, the first factor F1 of the principal component analysis can be used as a guide to the intensity of strain in individual samples (Fig. 6). In this diagram, the variation in the slope of the curves give the evolution of mobility for individual elements during progressive deformation. Two main types of chemical behaviour can be distinguished for the mobile elements. Fe, Ti, Mn, Rb, and K do not show any change in their mobility with increased strain. Their gain or loss in the shear zone increases constantly during deformation.   The mobility of others elements changes as deformation progresses. For example, Na and P are not seem to be affected by low deformation but increase markedly as deformation proceeds. Ca and Sr show the same increase in mobility as deformation increases although in this case they are lost from the shear zones.   Changes in water content during progessive deformation is more difficult to determine because of the large analytical uncertainties.   However the available results are consistent with an increase in both water content and mobility during deformation. Mg shows a contrasting behaviour; its mobility decreases during deformation. Mass transfer and the evolution of element mobility may be correlated with the metamorphic reactions accompanying rock deformation. For example, the pairs Rb and K and, Sr and Ca behave similarly as a result of their occupying similar sites in both magmatic and metamorphic minerals. In contrast the behaviour of Mg can be explained by a rapid destruction of pyroxene during the first stages of deformation. A continuous increase of mass transfer for the alkalis and the alkali earths could be the result of a continuous increase of recrystallization processes which affect plagioclases with a consequent modification of the chemical composition of the plagioclase. If the ionic equivalent of the mobile elements in the fluids is considered rather than their weight percent, problems arise with charge balance. This implies that oxidation reactions and consequent dissociation of water have taken place in the shear zone (Beach and Fyfe, 1972). Such an oxidation reaction of iron must accompany reactions which involve crystallization of biotite and amphibole (White and Yee, 1985) and introduction of water in the shear zone. A more complete study on the progressive changes in mineral and modal composition with strain is needed to explain mass transfer by the progress of amphibolitization reactions. Heterogenous deformation seems however to be a favourable setting to improve the understanding of metamorphic reactions.

5. DISCUSSION

The main results on the shear zone development in the Rouergue metagabbros are summarized bellow.

The geochemical study shows that chemical variations in shear zones are related to strain intensity and cannot be explained as the result of earlier magmatic or metasomatic processes (Lardeaux et al., 1986). Comparison of chemical variations between the deformed and the undeformed rocks implies that the shear zones and the rocks immediately surrounding them are an open system while composition of the undeformed rock must be very close to the initial composition.

Mass balance calculations demonstrate the significant mobility of some trace and major elements, with corresponding mass transfer the amount of which is strain dependant.

The mass transfer observed is consistent with the metamorphic reactions. They occur during the shear zone development by a progressive heterogeneous simple shear accompanied by progressive amphibolitization of the original gabbros. From these results, some interesting problems on shear zone development in the Rouergue metagabbros can be discussed.

Did metamorphic reactions enhance rock ductility or did deformation increase reactions kinetics?

What are the mechanisms which account for the observed mass transfer?

Finally, what are the limiting processes of the mass transfer mechanisms?

In the last few years, some attention has been paid to the softening processes that must accompany the concentration of deformation in shear zones. Among the different softening processes distinguished by White et al. (1980), geometric or fabric softening, continuous recrystallization, reaction softening (see also Rubie, 1983); and the effects of pore fluid, can have played a role during the development of shear zones in the Rouergue metagabbros. Lardeaux (1985) has shown that

deformation within the shear zone is accomodated by intense plastic
deformation and recrystallization of plagioclases, while the main
products of metamorphic reactions, amphibole and biotite, are less
deformed. In addition, amphibole and biotite exhibit brittle rather than
ductile behaviour. In this case metamorphic reactions do not seem to
have a significant effect as softening process during shear zone
development. Furthermore, although there are very few reaction products
in the undeformed rocks they have also been affected by metamorphic
reactions without any softening effect. Differences in reaction
progress must therefore be related to deformation processes which
enhance reaction kinetics.
The effects of pore fluid is more difficult to estimate because
the measured water content only reflects the amount of hydrated minerals
(and is thus an indicator of reaction progress) and not the fluid
distribution in the rock during deformation. As metamorphic reactions
involving hydration have taken place throughout the gabbros the
introduction of water must have been regional in the Rouergue
metagabbro, even if its rate of supply has been one of the limiting
factors on reaction progress. To understanding the fluid behaviour mass
transfer mechanisms need to be examined.
Two different mechanisms may account for mass transfer by fluids during
metamorphism (Beach, 1973; Fletcher and Hoffman, 1974; Fisher, 1978;
Ferry, 1983):
    Diffusion of elements in a fluid medium, through chemical potential
gradients.
    Percolation or infiltration of fluids through pressure or
temperature gradients.

    In the former, a continuous water supply is not needed, but the
distance of transfer is limited by the value of chemical potential
gradients. In the context of a shear zone the expected chemical
potential gradients may have been the result of differences in the
stress and the strain which have affected the minerals in the deformed
and undeformed rocks (Kerrich et al., 1977). However chemical potential
gradients are very small (Durney, 1976; Bosworth, 1981; Potdevin, 1984)
and are not effective over distances over one meter (Beach, 1973).
This implies that the rock must be regarded as a closed system at that
scale. As argued before, the metagabbro mass appears to have been
an open system at that scale and diffusion processes alone cannot
account for the observed mass transfer.
General fluid percolation through the rock is in agreement with
the following observations:
a)   Regional introduction of water:
b) mass transfer over distances greater than that of a single outcrop:
c) large increases in trace elements concentrations;
d) oxidation reactions.
    In this case the observed element transfer could be explained if we
consider the features of a fluid which flows and perhaps originates
from the acid country rocks which surround the metagabbro mass. An
influx of alkalis, and loss of alkali earth elements, Fe, Mg & Mn, is in
agreement with a process in which fluids coming from the gneisses have

percolated the basic rocks. In the area studied, the metabasites occur within large masses of leptynic rocks (e.g. alkali-rich formations), and abnormally K-rich amphibolites have been described (Collomb, 1970). The latter amphibolites are a common feature within deep-seated metamorphic terranes of the French Massif Central. They are interpreted as the result of a chemical exchange between basic rocks and surrounding gneisses (Suire, 1982; Mathonnat, 1983). According to these studies, the stronger the foliation in the basic rocks, the higher their alkali enrichment.

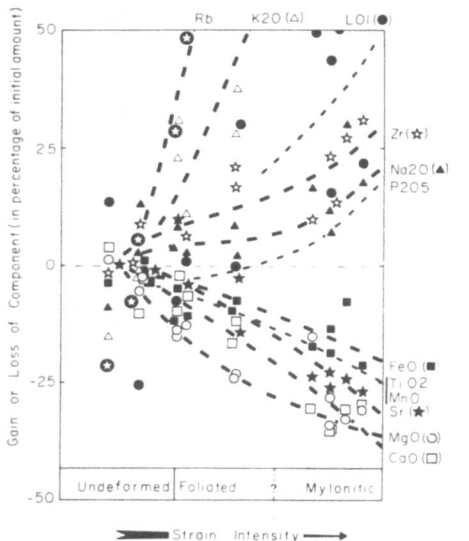

Fig. 6 - Element mobility versus strain intensity (F1 value). Relative gains or losses (mobilities) of all chemical elements (or oxides) have been calculated for each rock analysis.

TABLE 2

	1		2
SiO2	-0.21	Si	0.00
Al2O3	0.00	Al	0.00
FeO	-1.30	Fe	-0.15
MgO	-1.55	Mg	-0.28
CaO	-2.80	Ca	-0.32
Na2O	0.44	Na	0.16
K2O	1.83	K	1.33
TiO2	-1.18	Ti	-0.19
P2O5	0.04	P	0.18
MnO	-0.04	Mn	-0.22
LOI	0.31	LOI	0.41
Nb	-0.45	Nb	-0.07
Y	-1.44	Y	-0.05
Rb	88.39	Rb	2.63
Sr	-169.64	Sr	-0.25
Zr	23.10	Zr	0.21

Table II - Absolute (1) and relative (2) gains or losses of elements or oxides when Al immobility is assumed for the change of the undeformed rock into the high deformed rock.

The problem which remains is to understand how deformation can increases the reaction kinetics and associated gains or losses of material. One effect of deformation could be to increase the fluid flow within the shear zone. For example, permeability can be increased by hydrofracturing or grain size reduction. In this case the limiting process of mass transfer will appear to be the transport process itself. In order for the metamorphic reaction to proceed chemical components (e.g. water) must be supplied or removed from the rock. This hypothesis supposes that the different shear zones form a continuous network which allows the fluids to escape from the metagabbro mass with higher velocity in the deformed than in the undeformed rock. Transport kinetics in an isolated shear zone will be limited by the fluid velocity in the undeformed rock and no chemical transport will occur. Though the shear zones relations in three dimensions are difficult to measure, it is difficult to accept such a hypothesis in the Rouergue metagabbros. The effect of deformation must be to increase reaction kinetics themselves. Grain size reduction means that there are more grain surfaces to react with the fluids, while the plastic deformation of some minerals enhances the potential effects of metamorphic reactions. In this case the size of the shear zone is not a controlling factor and the same chemical changes with strain intensity will be observed.

In conclusion, our results demonstrate that under amphibolite facies conditions mass transfer in metamorphic rocks is strongly promoted by ductile deformation. Within deep-seated orogenic domains, ductile shear zones offer important pathways through which fluids can percolate giving rise to chemical exchange between contrasted lithologies.

ACKNOWLEDGEMENTS: This paper has benefited from review by David Bridgwater and referees. We would like to thank the organizers of the international meeting in Bergen for financial assistance. We are indebted to colleagues at the meeting and in Lyon university for useful discussions and to Didier Marquer for improvements to the manuscript.

REFERENCES

Autran A. & Cogné J.(1980). La zone interne de l'orogène varisque dans l'Ouest de la France et sa place dans le développement de la chaîne hercynienne. *26ème C.G.I, Paris, Colloque C6, Mém B.R.G.M.*, 108:90-111.

Beach, A. & Fyfe, W.S., 1972. Fluid transport and shear zones at Scourie, Sutherland: evidence of overthrusting ? *Contrib. Mineral. Petrol.*, 36: 175-180.

Beach, A., 1973. The mineralogy of high temperature shear zones at Scourie, NW Scotland. *J. Petrol.*, 14, 2: 231-248.

Bosworth, W., 1981. Strain induced preferential dissolution of halite. *Tectonophysics*, 78: 509-525.

Burg, J.P. & Matte P., 1978. A cross section through the French Massif Central and the scope of its Variscan geodynamic evolution. *Z. dt. Geol. Ges.*, 129, 429-460.

Carmichael, D.M., 1969. On the mechanism of prograde metamorphic reactions in quartz bearing pelitic rocks. *Contrib. Mineral. Petrol.*, 20: 244-267.

Chesworth, W., Desou, J. & Laroque, P., 1981. The weathering of basalt and relative mobilities of the major elements at Belbex, France. *Geochim. Cosmochim. Acta*, 45: 1235-1243.

Collomb, P., 1970. Etude géologique du Rouergue cristallin. *Mem. Serv. ex. carte France*, Paris, 419p.

Collomb, P., 1970. Orogenèse superposées et datations stratigraphiques dans les régions hercyniennes métamorphiques du Sud de la France. *Geol. Ass. Can., Spec. Paper*, 5, 89-104.

Cotonian, C., Potdevin, J.L., Bertrand H. & Lombardo, N., 1988. Pseudomorphoses coronitiques d'amphibole dans le trachyte de Monac (Velay oriental). Bilans de matière et réaction magmatique. *Bull. Minéral.*, 111, 89-95.

Cramer, J.J. & Nesbitt, H.W., 1983. Mass balance relations and trace element mobility during continental weathering of igneous rock. *Sci. Geol. Mem.*, 73: 63-73.

Durney, D.W., 1976. Pressure solution and crystallisation deformation. *Phil. Trans. R. Soc. Lond.*, A. 283: 229-240.

Ferry, J.M., 1983. Regional metamorphism of the Vassalboro Formation, South-Central Maine, USA: a case study of the role of fluid in metamorphic petrogenesis. *J. Geol. Soc. London,* 140: 551-576.

Fisher, G.W., 1970. The application of ionic equilibria to metamorphic differentiation: An example. *Contrib. Mineral. Petrol.*, 29: 91-103.

Fisher, G.W., 1973. Non equilibrium thermodynamics as a model for diffusion controlled metamorphic processes. *Am. J. Sci.*, 273: 897-924.

Fisher, G.W., 1978. Rate laws in metamorphism. *Geochim. Cosmochim. Acta*, 42: 1035-1050.

Fletcher, R.C. & Hofmann, A.W., 1974. Simple models of diffusion and combined diffusion infiltration metasomatism. In: Hofmann A.W. et al., (eds.) Geochemical Transport and Kinetics. *Carnegie Inst. Washington Publ.* 243-259.

Floyd, P.A. & Winchester, J.A., 1983. Element mobility associated with meta shear zones within the Ben Hope amphibolite suite, Scotland. *Chem. Geol.*, 39: 1-15.

Fonteilles, M., 1978. Les mécanismes de la métasomatose. *Bull. Mineral.*, 101: 166-194.

Forestier, F.H., 1963. Métamorphisme hercynien dans le bassin du Haut-Allier (Massif Central français). *Bull. Serv. Carte géol. France.*, 271, t. LIX.

Glassley W.E. & Bridgwater D., 1985. Fluids enhanced mass transport in deep crust and its influence on element abundancies and isotope systems. In: Tobi, A.C and Touret J.L.R. (eds.) *The deep Proterozoic crust in the North Atlantic province.* D. Reidel, Dordrecht, 105-117.

Gardner, L.R., 1980. Mobilization of Al and Ti during weathering: isovolumetric geochemical evidence. *Chem. Geol.* 30: 151-166.

Gratier, J.P., 1983. Estimation of volume changes by comparative chemical analyses in heterogeneously deformed rocks (folds with mass transfer). *J. Struct. Geol.*, 5, 3/4: 329-339.

Gratier, J.P., 1984. La déformation des roches par dissolution cristallisation. Aspects naturels et expérimentaux de ce fluage avec transfert de matière dans la croûte supérieure. *Thèse d'Etat, Université de Grenoble*, 315 p.

Gresens, R.L., 1967. Composition volume relations of metasomatism. *Chem. Geol.*, 2: 47-65.

Hickman, M.H.,and Glassley, W.E., 1984. The role of metamorphic fluid transport in the Rb/Sr isotopic resetting of shear zones: evidence for Nordre Stromfjord, West Greenland. *Contrib. Mineral. Petrol.*, 87, 265-281.

Kerrich, R., Fyfe, W.S., Gorman, B.E. & Allison, I., 1977. Local modification of rock chemistry by deformation. *Contrib. Mineral. Petrol.*, 65: 183-190.

Lardeaux, J.M., 1985. Ductilité du plagioclase et déformation des métagabbros dans le faciès amphibolite. *C.R. Acad. Sc. Paris*, 301, II, 11: 827-830.

Lardeaux, J.M., Potdevin, J.L. & Coffrant, D. , 1988. Geochemical modifications of metagabbros within an amphibolitized ductile shear zone, Chemical Geology, submitted.

Lelong, F. & Souchier, B., 1979. Les bilans d'altération dans les sols. *Sci. Sol.*, 267-279.

Marquer, D., 1987. Transfert de matière et déformation progressive des granitoïdes. Exemple des massifs de l'Aar et du Gothard (Alpes centrales suisses). *Mém. Docum. Centre Arm. Et. Struct. Socles*, Rennes, 10, 287 p.

Marquer, D., Gapais, D. & Capdevila, R., 1985. Comportement chimique et orthogneissification d'une granodiorite en faciès schistes verts (Massif de l'Aar, Alpes centrales). *Bull. Minéral.*, 108: 209-222.

Matte, P., 1986. La chaîne varisque parmi les chaînes paléozoiques peri-atlantique, modèle d'évolution et position des grands blocs continentaux au carbonifère. *Bull. Soc. géol. France*, 1, 9-24.

Nicollet, C., 1978. Pétrologie et tectonique des terrains cristallins anté-permiens du versant Sud du dôme du Lévezou (Rouergue, Massif Central). *Bull. B.R.G.M.*, I, 3, 225-263.

Mathonnat, M., 1983. La série métamorphique du Cézallier (M.C.F.). Lithologie et structure. Relations du groupe leyptino-amphibolique avec les autres formations de la région. *Thèse 3ème cycle, Université de Clermont*, 208 p.

Potdevin, J.L., 1984. Métamorphisme et tectonique dans les Schistes lustrés à l'Est de Corte (Corse). 3ème partie: Déformation par dissolution - cristallisation. *Thèse 3ème cycle, Université de Lyon*, 82 p.

Potdevin, J.L. & Caron, J.M., 1986 . Transfert de matière et déformation synmétamorphique dans un pli. I. Structures et bilans de matière. *Bull. Minéral.*, 109, 395-410.

Potdevin, J.L. & Marquer D., 1987. Méthodes de quantification des transferts de matière par les fluides dans les roches métamorphiques déformées. *Geodinamica acta*, 1, 3, 193-206.

Ramsay, J.G., 1980. Shear zone geometry: a review. *J. Struct. Geol.*, 2: 83-99.

Rubie, D.C., 1983. Reaction enhanced ductility: The role of solid-solid univariant reactions in deformation of the crust and mantle. *Tectonophysics*, 96: 331-352.

Sicard, E, Caron, J.M., Potdevin, J.L. & Déchomets, R., 1986 . Transfert de matière et déformation synmétamorphique d'un pli. Pseudomorphoses de lawsonite et caractérisation des fluides intersticiels. *Bull. Minéral.*, 411-422.

Suire, J., 1982. Signification du groupe leyptino-amphibolique de l'Artense (M.C.F.). Pétrographie, relations structurales, géochimie, géochronologie et origine. *Thèse 3ème Cycle, Clermont-Ferrand*, 168 p.

Trolliard G., Potdevin J.L., Lardeaux J.M. & Boudeulle M., 1987. Transfert de matière dans des roches métamorphiques non déformées. Exemple des métagabbros coronitiques du Rouergue. *Bull. Minéral.*, 110, 439-448.

Vogel, B.E., 1967. Petrology of an eclogite and pyrigarnit bearing polymetamorphic rock at Cabo Ortegal, N.W. Spain. *Leids geol. Meded.*, D40, 121-123.

White, S.H., Burrows, S.E., Carreras, J., Shaw, N.D. & Humphreys, F.J., 1980. On mylonites in ductile shear zones. *J. Struct. Geol.*, 2: 175-187.

White, A.F. & Yee, A., 1985. Aqueous oxidation reduction kinetics associated with sampled electron-cation transfer from ion containing silicates at 25°C. *Geochim. Cosmochim. Acta*, 49: 1263-1275.

INCORPORATION OF RB IN MUSCOVITE DURING PROGRESSIVE MEDIUM GRADE METAMORPHISM. AN INTERPRETATION OF EXPERIMENTAL RESULTS.

J.H.L. Voncken and J.B.H. Jansen
Department of Geochemistry and Experimental Petrology,
University of Utrecht
P.O. Box 80.021, 3508 TA Utrecht
The Netherlands.

EXTENDED POSTER ABSTRACT. Rb occurs as a minor element in a variety of igneous, metamorphic and sedimentary rocks. Muscovite is along with K-feldspar and biotite one of the most important host minerals for Rb. Voncken et al. (1987) describe a pure synthetic Rb-analogue of $2M_1$ muscovite. The Rb-analogue of muscovite can be formed stably at 600 $^\circ$C and 2, 3, 5 and 15 kbar (Voncken, unpubl. data). Several experimental studies on K-Rb exchange reactions between hydrothermal fluid and K-feldspar or phlogopite are reported (Beswick, 1973; Volfinger, 1974, 1976). There are relatively few exchange data for muscovite (Volfinger, 1969, 1976).

Ion exchange reactions of K and Rb between 2M alkali chloride solutions and synthetic muscovite and Rb-muscovite were carried out at 2 kbar total pressure between 400, 500 and 600 $^\circ$C. Golden charge capsules were placed in conventional cold seal pressure vessels. K and Rb contents in the cleaned and dried muscovite, which were dissolved in HF solutions and the coexisting hydrothermal fluid were analysed by AAS. The prepared muscovites were studied with XRD and SEM. Reaction rate experiments indicate that at 400 $^\circ$C, equilibrium is approached in 35 days (Fig.1).

The experimental results suggest a miscibility gap between the K- and Rb-muscovites with the mole fractions K/(K + Rb) of 0.06 and 0.83 at 400 $^\circ$C, 0.12 and 0.76 at 500 $^\circ$C and 0.32 and 0.69 at 600 $^\circ$C (Fig.2). The existence of an asymmmetric solvus is not surprising, as solvi appear to exist in pseudobinary systems between micas, for example phlogopite and Rb-phlogopite (Beswick, 1973). At a temperature of 675 $^\circ$C, almost exclusively alkali-feldspar and corundum is formed. Probably, the solvus is cut off by the feldspar stability field between 600 and 675 $^\circ$C before the consolute point is reached. The bracket for the boundary between muscovite and alkali-feldspar stabilities at 2 kbar, after Chatterjee & Johannes (1974), is plotted in fig. 2. Experimental work to determine the breakdown of the Rb-muscovite to feldspar, corundum and H_2O as well as distribution

D. Bridgwater (ed.), Fluid Movements – Element Transport and the Composition of the Deep Crust, 231–233.
© *1989 by Kluwer Academic Publishers.*

232

coefficients $(Rb/K_{fluid})/(Rb/K_{solid})$ between muscovite and vapor at 2 kbar at the temperatures involved is in progress.

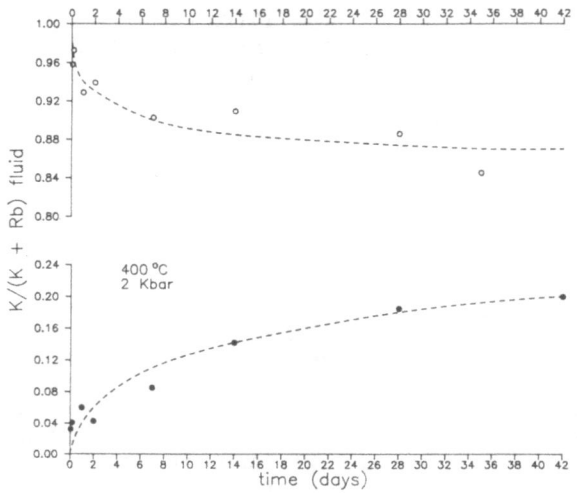

Figure 1. *Fluid composition in reaction rate experiments. Open circles represent compositions which were in contact with the Rb-analogue of muscovite, solid circles represent compositions which were in contact with muscovite.*

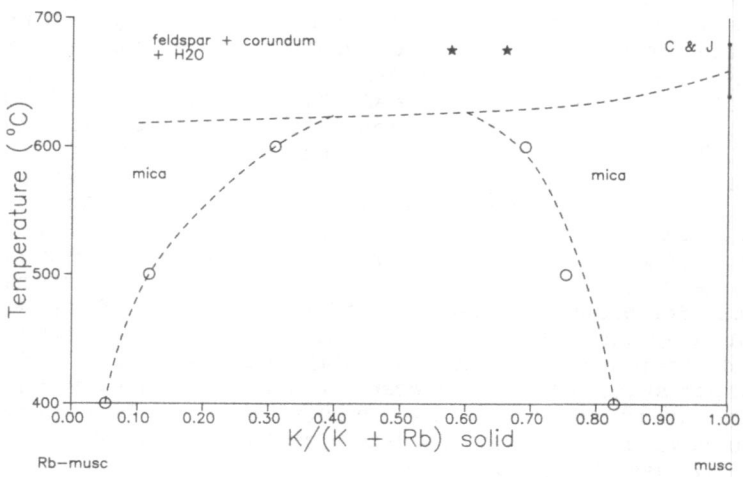

Figure 2. *Solid compositions after run times of 42 days indicate, when plotted against temperature, a solvus. The stars represent feldspar compositions.*

The experimental data indicate that at constant pressure with increasing temperature more Rb can be incorporated in muscovite in equilibrium with alkalichloride solutions, and that even at relatively high pressures Rb will be stably build in in the muscovite lattice. The enrichment of Rb in muscovite is limited to about 32 % at 620 °C and 2 kbar. Generally natural muscovites have Rb-contents of several thousands of ppm (e.g., Luecke, 1981), although Khanna (1977) described a (lithian) muscovite with 6.98 weight percent Rb. At the physical conditions of a medium grade metamorphism muscovite is an important phase for the storage of Rb, and it will keep Rb during progressive medium grade metamorphism untill its stability limit is reached in anatectic environments. It is an important mineral to achieve that Rb is collected in granitic magmas.

Beswick, A.E. (1973) 'An experimental study of alkali metal distributions in feldspars and micas'. *Geochimica et Cosmochimica Acta*, 37, 183-208.

Chatterjee, N.D. and Johannes, W. (1974) 'Thermal Stability and standard thermodynamic properties of $2M_1$ muscovite $(KAL_2Si_3AlO_{10}(OH)_2)$.' *Contributions to Mineralogy and Petrology*, 48, 89-114, 1974.

Khanna V.K. (1977) 'Note on the unusually high concentration of Rb in a lithium mica from Govindpal area of Bastar district.' *Journal of the Geological Society of India*, 18, 500-502.

Luecke W. (1981) 'Lithium pegmatites in the Leinster granite (southeast Ireland).' *Chemical Geology*, 34, 195-233.

Volfinger, M. (1969) 'Partage de Rb et Cs entre sanidine, muscovite et solution à 600°C-1000 bars.' *Comptes Rendus, Academie des Sciences, Paris*, 269, D1-D3.

Volfinger, M. (1974) 'Effet de la composition des micas trioctaedriques sur les distributions de Rb et Cs a l'etat de traces.' *Earth and Planetary Science Letters*, 24, 299-304.

Volfinger, M. (1976) 'Effet de la température sur les distributions de Na, Rb et Cs entre la sanidine, la muscovite, la phlogopite, et une solution hydrothermale sous une pression de 1 kbar.' *Geochimica et Cosmochimica Acta*, 40, 267-282.

Voncken, J.H.L., Van der Eerden, A.M.J., Jansen, J.B.H. (1987) 'Synthesis of a Rb-analogue of $2M_1$ muscovite.' *American Mineralogist*, 72, 551-554.

FLUID CONTROL ON EMPLACEMENT OF SIALIC MAGMAS DURING ARCHAEAN CRUSTAL ACCRETION

A. P. Nutman, Department of Earth Sciences, Memorial University of Newfoundland, St.John´s, Newfoundland, Canada.
A. A. Garde, Geological Survey of Greenland, Øster Voldgade 10, 1350 Copenhagen K, Denmark.

ABSTRACT

Archaean crustal accretion is marked at middle to deep crustal levels by predominantly grey gneisses (derived from diorites, tonalites and trondhjemites) intruded into supracrustal sequences dominated by mafic volcanic rocks and hypabyssal intrusions. The grey gneisses occur as inclined sheets and as diapirs. An example of middle Archaean crustal accretion is found in the Taserssuaq area, southern West Greenland. In this area, mafic volcanic rocks were intruded first by sheeted grey gneisses and then by the Taserssuaq tonalite complex, a more homogeneous body of grey gneisses interpreted to be a diapir. At early stages of accretion, the country rocks were cool, mafic volcanics containing perhaps ca. 5 wt% water. Heating of these rocks close to the magmatic protoliths of the grey gneisses would liberate water, which in combination with water released from the magmas themselves, increased the pore fluid pressure to the extent that hydraulic fractures formed in the supracrustal rocks. In an environment of horizontal extension or moderate compression, these fractures would be shallowly inclined and dilational, and would be exploited by the magma, giving rise to its sheeted form of intrusion. At a late stage of accretion, large batches of magma, such as forming the Taserssuaq tonalite complex, were intruded as diapirs into a thicker crust dominated by hot, rather ductile, drier (ca. 1 wt% water) sheeted grey gneisses. Interpretation of the geochemistry of grey gneisses intruded by these different mechanisms is discussed.

D. Bridgwater (ed.), Fluid Movements – Element Transport and the Composition of the Deep Crust, 235–243.
© *1989 by Kluwer Academic Publishers.*

INTRODUCTION

Archaean high grade gneiss terranes are commonly dominated by an association of grey gneisses derived from diorites, tonalites and trondhjemites, that intruded supracrustal sequences consisting mainly of mafic rocks, preserved as amphibolites. On a regional scale the grey gneisses are generally much more voluminous than the mafic rocks. The mafic rocks represent submarine volcanics and related intrusions formed in the Archaean equivalents of either extended rifts, back arc basins or an oceanic setting (Windley, 1984). The protoliths of some of the grey gneisses were intruded synkinematically into the mafic supracrustal sequences as moderately inclined to subhorizontal sheets, whilst protoliths of other grey gneisses were intruded as diapirs.

This type of terrane (grey gneisses and mafic supracrustal rocks) has been interpreted as the site of plate convergence, and may be the Archaean equivalent of modern island arcs (e.g. Myers, 1984; de Wit, 1986). Dehydration and melting at depth of the mafic supracrustal rocks and perhaps the upper mantle as well gave rise to magmas from which evolved the protoliths of the grey gneisses (e.g. Martin, 1986). Geochemical, isotopic and metamorphic studies show that the grey gneisses represent predominantly new sial rather than recycled sial; the emplacement of these rocks occurred over a short timespan and the repeated intrusion of the batches of magma which gave rise to the protoliths of the grey gneisses formed crust that was 60 to 80 km thick (e.g. Moorbath et al., 1986; Wells, 1980).

In order to examine the intrusion mechanisms of the protoliths of the grey gneisses, we have chosen the middle Archaean grey gneiss – mafic supracrustal rock terrane of the Taserssuaq area, north of Godthåbsfjord, southern West Greenland (Fig. 1).

GNEISS COMPLEX OF THE TASERSSUAQ AREA

Mapping of the Taserssuaq area (Fig. 1) was carried out by the Geological Survey of Greenland (Garde et al., 1983). The area is dominated by tonalitic, dioritic and granodioritic gneisses that we correlate with the 3070-2950 Ma type Nûk gneisses (McGregor, 1979; Baadsgaard & McGregor, 1981) of southwestern Godthåbsfjord. The magmatic protoliths of the grey gneisses in the Taserssuaq area intruded supracrustal units dominated by mafic volcanic rocks with subordinate amounts of ultramafic and sedimentary rocks, and were then in turn intruded by subordinate amounts of granite (s.s.), folded, sheared

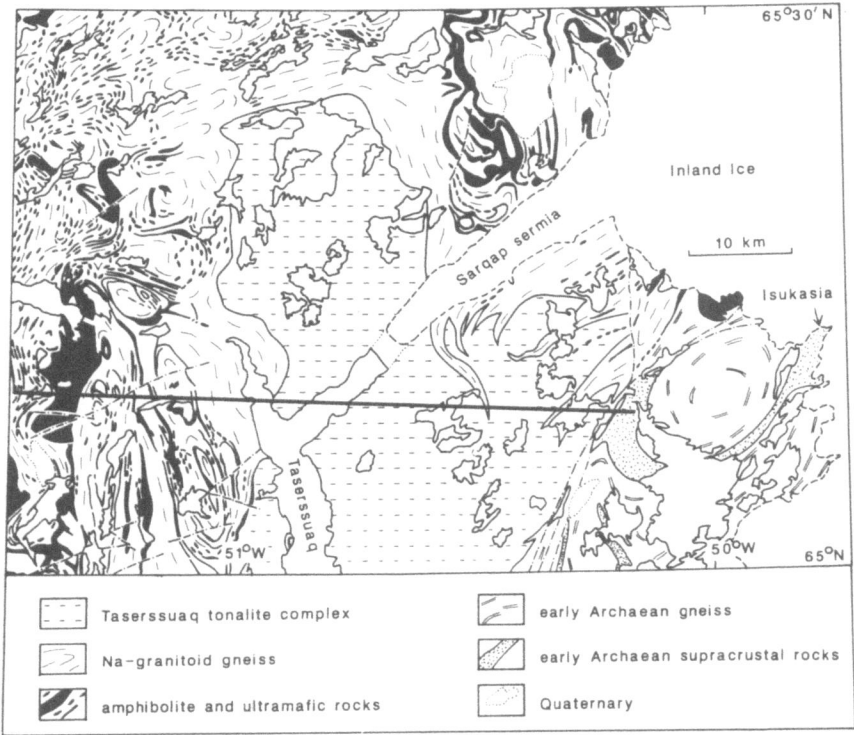

Figure 1. Geological sketch map of the Taserssuaq area.
Note - the Na-granitoid gneisses are the grey gneisses
discussed in the text.

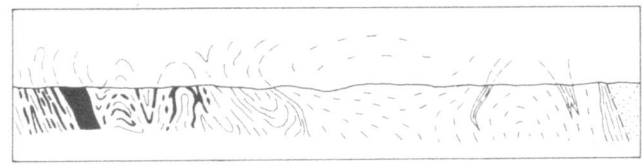

Figure 2. West-east cross-section, Taserssuaq area. See
Fig. 1 for location.

and metamorphosed under amphibolite to granulite facies
conditions (Table I). A near concordant U-Pb zircon age of
2982 Ma (R.T. Pidgeon in Garde et al., 1986) was obtained
from the Taserssuaq tonalite (a younger component of the
grey gneisses, Table I and Fig. 1). A Pb-Pb whole-rock age
of ca. 3000 Ma has been obtained for gneisses from the
west of the Taserssuaq area (P.N. Taylor in Garde, this
vol.). These dates are interpreted as the age of peak
metamorphism shortly after formation of the grey gneisses
(Taylor et al., 1980; Garde et al., 1986). During the ca.
3000 Ma event, granulite facies metamorphic grade was
attained in the west and upper amphibolite facies in the
east. In the late Archaean partial retrogression of the ca.
3000 Ma granulite facies assemblages occurred during
localised shearing and folding under amphibolite facies
conditions (Garde et al., 1983). Patchy retrogression of
the granulite facies assemblages also took place without
ductile deformation (Garde et al., 1983).

Table I. Archaean history of the Taserssuaq area.

4)	Injection of granitoid sheets and development of shear zones and folds under amphibolite facies conditions. 2500-2800 Ma.
3)	Peak of metamorphism; granulite facies in the west and amphibolite facies in the east. Injection of granites. 2950-3000 Ma.
2)	Injection of the magmatic precurssors of the grey gneisses, predominantly of tonalitic composition. ca. 3000 Ma.
1)	Formation of supracrustal sequence dominated by mafic volcanic rocks and associated hypabyssal intrusions. ca. 3100 Ma?

In the Taserssuaq area the original sequence of mafic
supracrustal rocks is divided into numerous screens,
ranging from several hundred to less than one metre thick,
separated by pegmatite-streaked grey gneiss sheets (Figs. 1
& 2). Large sheet-like units of grey gneiss can be traced
for considerable distances around early recumbent isoclinal
and later upright folds (Figs. 1 & 2). ´Removing´ these
folds shows that the grey gneiss sheets were originally
moderately to subhorizontally inclined. In the west and
northwest of the area the sheeted grey gneiss - amphibolite
terrane degenerates into an agmatite complex, consisting of
several generations of grey gneisses with jumbled,
irregularly distributed inclusions of the supracrustal
rocks (Fig. 1).

In the central part of the area (Fig. 1) a large homogeneous member of the grey gneisses named the Taserssuaq tonalite complex (Garde et al., 1983), consists of subordinate early gabbroic and dioritic components that are found as inclusions in large volumes of weakly foliated tonalite and late, subordinate granodiorite. The Taserssuaq tonalite complex contains inclusions of more foliated grey gneisses and also banded amphibolite. Locally, the margin of the Taserssuaq tonalite complex is discordant to foliated grey gneisses interlayered with strips of amphibolite (Garde et al., 1983). The western and central parts of the tonalite complex consist of polyphase tonalite, which locally shows well preserved igneous layering of shallow inclination (Garde et al., 1983). The northeastern part of the tonalite complex consists of several steeply inclined bodies of homogeneous tonalite. We interpret these homogeneous units as the stem of a mushroom-shaped intrusion, the upper part of which is represented by the western-central part of the Taserssuaq tonalite complex with its gently inclined igneous lamination (Fig. 2).

PROPOSED MECHANISMS OF INTRUSION OF THE GREY GNEISS PROTOLITHS

Early batches of grey gneiss magma were intruded into a host dominated by mafic supracrustal rocks. By analogy with well preserved sequences of submarine mafic rocks such as found in Phanerozoic ophiolite complexes, the Archaean mafic supracrustals were probably initially cold (sub-greenschist facies), and rather hydrous, with on average perhaps as much as 5 wt% H_2O (e.g. Laurent et al., 1979). Intrusion of hot (close to $1000^{\circ}C$) magmatic precursors of the grey gneisses would have heated up water and promoted dehydration reactions in the mafic supracrustals, and thus increased the pore fluid pressure. Fluid could have also exsolved from the cooling magmatic protoliths of the grey gneisses. The pore fluid pressure increased to such an extent that hydraulic fracturing occurred. These fractures would have been exploited by the magmas. Secor (1965) shows that with very small deviatoric stresses and σ_3 vertical, hydraulic fractures are tensional and subhorizontal, whilst with somewhat larger differences between σ_1 and σ_3, moderately inclined dilational shear fractures form. This is a possible mechanism for the intrusion of magmatic precurssors of grey gneisses to form shallowly-inclined sheet complexes, such as are interpreted to occur in the western part of the Taserssuaq area (Fig. 1). The high pore fluid during intrusion of the magmatic precurssors of the grey gneisses

in the form of sheets, could have promoted the development
of granitic pegmatite segregations, late in the
crystallisation and flowage of the magma. Such pegmatitic
banding is a very common feature of grey gneisses intruded
as sheets (e.g. McGregor, 1979).

Intrusion of the magmatic protoliths of the grey
gneisses by this mechanism would have thickened and heated
the crust. Wells (1980) suggests such a process would occur
by overaccretion. In overaccretion, each successive batch
of magma that will crystallise to form a body of grey
gneiss is emplaced at a high crustal level, because of the
low density of the magma relative to the crustal rocks.
This means that earlier sheeted bodies of grey gneiss with
their inclusions of supracrustal rocks are progressively
buried to deeper crustal levels, where they undergo
regional high grade metamorphism. Therefore, later batches
of magma, represented by grey gneiss bodies such as the
Taserssuaq tonalite complex, would have been intruded into
an already thickened crust, dominated by hot, perhaps
rather ductile and dry (probably ca. 1 wt% H_2O) grey
gneisses. At this later stage of crustal accretion the
magmatic protoliths of the grey gneisses may have been
emplaced by ductile deformation in a diapiric fashion,
giving rise to the more homogeneous bodies such as the
Taserssuaq tonalite complex.

Crust thickened by repeated addition of batches of magma
is out of isostatic equilibrium. Erosion continues until
depths where the rocks experienced regional upper
amphibolite and granulite facies metamorphism are exposed,
as is observed in the Taserssuaq area.

GEOCHEMICAL IMPLICATIONS OF MECHANISMS OF INTRUSION

Metasomatism associated with high grade metamorphism
(particularly the granulite – amphibolite facies
transition) in high grade gneiss complexes can
significantly modify the geochemistry of rocks on a
regional scale (e.g. Heier, 1978; Schiøtte et al., 1986).
Compositional changes of this type have also been detected
in the study area (Garde et al., 1986; Garde, this vol.).

However, these compositional changes will have been
superimposed onto any compositional changes to the magmatic
protoliths of the grey gneisses caused by interaction with
their host rocks during intrusion. Batches of magma that
gave rise to the first phases of the grey gneisses would
have been emplaced into predominantly supracrustal rocks,
in the presence of a hydrous fluid phase derived in part
from the supracrustal rocks, and probably in part from the
magma itself. The fluid would not have been pure water, but
a brine. Interfacing, and perhaps mixing of the brine with

the magmas could have significantly modified the ´igneous´ trace element geochemistry and possibly the isotopic systems of the grey gneisses. Such processes have recently been documented in the Superior craton, Canada (Fyfe, pers. comm., 1987). Therefore, even allowing for changes to the composition of the grey gneisses associated with high grade metamorphism (Garde et al., 1986; Garde, this vol.), their compositional variation is not purely magmatic, but may also be governed by fluid – host rock – magma interaction at the time of the emplacement of the magmatic protoliths of the grey gneisses. On the other hand, diapiric bodies of grey gneiss were intruded into a host dominated by hot grey gneisses. In this latter case, the host grey gneisses might have undergone local partial melting in the vicinity of large diapiric bodies of grey gneiss magma, to produce crustally derived granite s.s. magma. Evidence of crustal melting and magma mixing is seen in the Taserssuaq area, where the Taserssuaq tonalite complex is locally bounded by, and grades into granitic rocks, some of which are intrusive, whilst others were derived more or less in situ from grey gneisses (Garde, 1984; Garde et al., 1986). Thus late batches of grey gneiss magma could be modified by their interaction with crustally derived granite.

TECTONIC IMPLICATIONS

It has been proposed that Archaean tectonics involved some form of subduction (e.g. Myers. 1984; de Wit, 1986; Arndt, 1983; Nisbett, 1984; Pyle, 1987). If the proposed mechanism of intrusion of the early sheeted grey gneisses is correct, structural studies of well-preserved examples of these rocks may yield constraints on regional palaeostresses during subduction. In a tensional environment the magmatic protoliths of the grey gneisses would have been intruded as sheets into dilational fractures with no repetition of the mafic supracrustal sequence by thrusting or subsequent isoclinal folding. In a compressive regime, the magmatic protoliths of the grey gneisses would have been intruded as sheets into moderately inclined, dilational shear fractures. In this compressive regime, repetition of the mafic supracrustal sequence by thrusting and recumbent folding at depth as the crust was progressively thickened from above by emplacement of further bodies of grey gneisses would have been possible. The model with compressional palaeostresses, which allows repetition of the sequence by thrusting and recumbent folding during intrusion of tonalitic magma (e.g. Myers, 1984; de Wit & Wilson, 1986), fits descriptions of Archaean gneiss complexes that represent newly-accreted crust better than the model with tensional palaeostresses.

ACKNOWLEDGEMENTS

We thank Clark Friend for discussion of crustal accretion processes, and Tom Calon for reviewing the manuscript. The Director of the Geological Survey of Greenland is thanked for permission to publish this paper.

REFERENCES

Baadsgaard, H. and McGregor, V.R. 1981: The U-Th-Pb systematics of zircons from the type Nuk gneisses, Godthåbsfjord, West Greenland. Geochim. Cosmochim. Acta, 45, 1099-1109.
de Wit, M. 1986: What the oldest rocks say. In L. Ashwal, K. Burke, M. de Wit and G. Wells (editors). The Earth as a Planet, 14-17. Lunar and Planetary Inst. Tech. Rpt. 86-08.
de Wit, M. and Wilson, A.H. 1986: Felsic igneous rocks within the Barberton greenstone belt: high crustal equivalents of the surrounding tonalite-trondhjemite terrain, emplaced during thrusting. In M. de Wit and L. Ashwal (editors) Tectonic Evolution of Greenstone Belts, 89-92. Lunar and Planetary Inst. Tech. Rpt. 86-10
Garde, A.A. 1984: Field work between Fiskefjord and Godthåbsfjord, southern West Greenland. Rapp. Grønlands geol. Unders., 115, 20-29.
Garde, A.A. this vol.: Retrogression and fluid movement across a granulite-amphibolite facies boundary in middle Archaean Nûk gneisses, Fiskefjord, southern West Greenland. In D. Bridgwater (editor). Fluid Transport and the Composition of the Deep Crust. Riedel, Dordtrecht.
Garde, A.A., Hall, R.P., Hughes, D.J., Jensen, S.B., Nutman, A.P. and Stecher, O. 1983: Mapping of the Isukasia sheet, southern West Greenland. Rapp. Grønlands geol. Unders., 115, 20-29.
Garde, A.A., Nutman, A.P. and Larsen, O. 1986: Dating of late Archaean crustal mobilisation north of Qûgssuk, Godthåbsfjord, southern West Greenland. Rapp. Grønlands geol. Unders., 128, 23-36.
Heier, K.S. 1978: The distribution and redistribution of heat-producing elements in the continents. Phil. Trans. R. Soc. London., Ser. A288, 393-400.
Laurent, R., Hebert, R and Hebert, Y. 1979: Tectonic setting and petrological features of the Quebec Appalachian Ophiolites. In J. Malpas and R. W. Talkington (editors). Ophiolites of the Canadian Appalachians and the Soviet Urals. Memorial University

Department of Geology Report No. 8, contribution to IGCP project 39.

Martin, H. 1986: Effect of steeper Archaean geothermal gradient on geochemistry of subduction-zone magmas. Geology, 14, 753-756.

McGregor, V.R. 1979: Archean gray gneisses and the origin of the continental crust: Evidence from the Godthåb region, West Greenland. In F. Barker (editor). Trondhjemites, Dacites and Related Rocks, 169-205. Elsevier, Amsterdam.

Moorbath, S., Taylor, P.N. and Jones, N.W. 1986: Dating the oldest terrestrial rocks - fact and fiction. Chem Geol., 57, 63-86.

Myers, J.S. 1984: Archaean tectonics in the Fiskenæsset region of southwest Greenland. In A. Kroner and R. Greiling (editors). Precambrian Tectonics Illustrated, 95-112. E. Schweizerbart'sche Verlagsbuchhandlung, Stuttgart.

Nisbett, E.G. 1984: The continental and oceanic crust and lithosphere in the Archaean: isostatic, thermal and tectonic models. Can. J. Earth. Sci., 21, 1426-1441.

Pyle, B., Neal, C.R. and Taylor, L.A. 1987: Ancient oceanic crust subducted beneath the Kaapvaal craton: the genesis of eclogites in kimberlites. 18th Lunar and Planetary Science Conference Abstracts, 806-807. Lunar and Planetary Institute, Houston.

Schiøtte, L., Bridgwater, D., Collerson, K., Nutman, A.P. and Ryan, A.B. 1986: Effects of late Archaean high grade metamorphism and granite injection on 3800 Ma gneisses, Saglek - Hebron, northern Labrador. Element and isotopic redistribution across an amphibolite - granulite facies transition. Geol. Soc. Lond. Spec. Publ., 24, 261-273.

Secor, D.T. 1965: Role of fluid pressure in jointing. Am. J. Sci., 263, 633-646.

Taylor, P.N., Moorbath, S., Goodwin, R. and Petrykowski, A. 1980: Crustal contamination as an indicator of the extent of early Archaean continental crust: Pb isotopic evidence from the late Archaean gneisses of West Greenland. Geochim. Cosmochim. Acta, 44, 1437-1453.

Windley, B.F. 1984: The Evolving Continents (2nd edition), 48-65. Wiley, New York.

Wells, P.R.A. 1980: Thermal models for the magmatic accretion and subsequent metamorphism of continental crust. Earth Planet. Sci. Lett., 46, 253-263.

Wyllie, P.J. 1977: Crustal anatexis: An experimental reveiw Tectonophys., 43, 41-71.

THE ISOTOPIC CHARACTERIZATION OF AQUEOUS AND LEUCOGRANITIC CRUSTAL FLUIDS

Simon M.F. Sheppard
Centre de Recherches Pétrographiques et Géochimiques
B.P. 20, 54501 Vandoeuvre-lès-Nancy, FRANCE

ABSTRACT.The H- and O-isotope characteristics of the principal water types - sea, meteoric, formation, organic, metamorphic, magmatic and hydrothermal-are outlined. Sea and meteoric surface waters can be involved with deep crustal processes following their downward penetration, or by their upward infiltration after oceanic or continental subduction processes. Examples illustrate the identification of such water sources in deep crustal rocks.

The typically cold (< 30°C) Na-Ca-Cl brines ± CH_4 etc. encountered at deep levels (> 300 m) in Shields often plot to the left of the meteoric water line. Many of these brines are mixtures of recent meteoric waters with ancient brines. The most saline and therefore least contaminated brines may sometimes be modified Precambrian hydrothermal fluids. They underwent low temperature retrograde isotopic exchange with their host rocks and water/rock ratios were very small.

The importance of the release of hydrous fluids from sedimentary formations into overthrusted and hot crustal or mantle slabs is emphasized. Such fluids can become involved in fluid mixing processes and aid partial melting reactions. Coupled O-, Sr- and Nd- isotope variations in a High Himalaya leucogranite are shown to be related to comparable variations in the Tibetan Slab paragneisses which are considered to be equivalent to the underlying source rocks. These variations imply that (1) isotopic homogenization was ineffective on a submetric scale during diagenesis and metamorphism, (2) major convective circulation of fluids did not occur, and (3) the granite massif is composed of a large number of essentially indepenent batches of magma. They are a result of the differences in the mineralogical constitution of the sediments and the aging effects between sedimentation and anatexis.

1. INTRODUCTION

The complexity and variety of geological processes on Earth reflect in large part the role of fluids - aqueous, carbonaceous or silicate magmas. Fluids and their dynamics may contribute more or less dominantly to heat and mass transfer processes and profoundly influence chemical and physical reactions including fluid/fluid and fluid/rock interaction processes. The presence and nature of a fluid phase can also affect the mechanical properties of minerals and rocks. Knowledge of the origins, nature and amounts of infiltrating fluids within the crust are of crucial importance in order to understand crustal processes and their dynamics including metamorphic and melting processes and their relations to tectonic environment.

D. Bridgwater (ed.), Fluid Movements – Element Transport and the Composition of the Deep Crust, 245–263.
© 1989 by Kluwer Academic Publishers.

Water is the most abundant crustal liquid. Under given P-T conditions, its chemical composition is often essentially independent of its source whether it be of magmatic or meteoric origin for example. In a few situations, however, the presence of certain other constituents can be indicative of their source ; for example, the presence of nitrogen, either in micas or fluid inclusions, usually implies the involvement of organic matter bearing sediments (e.g. Dubessy and Ramboz, 1986). Nevertheless, it is parameters based on the water molecules themselves such as the D/H and $^{18}O/^{16}O$ ratios, which generally retain, in some more or less modified form, a memory of their source. Similarly for silicate magmas, stable isotope techniques have been applied to characterize the nature of the source region and the importance of fluid/fluid and fluid/rock interaction. During the past twenty years or so an enormous advance has been made to our understanding of the source, nature (chemical and physical) and amounts of infiltrating fluids in the upper continental and oceanic crust. In particular, coupled stable isotope studies using light (H, C, O, S) and/or heavy elements (Sr, Nd, Pb) and fluid inclusion studies have made a major contribution to this field (see for example Taylor, 1974; 1977; 1986; Zartman, 1974; Doe and Zartman, 1979; Allègre and Ben Othman, 1980; Hollister and Crawford, 1981; Roedder, 1984; Vidal et al., 1984; Sheppard, 1986a, b; Taylor and Sheppard, 1986; Deniel et al., 1987; Kerrich, 1987). The results of these studies have demonstrated the diverse origins of fluids and that surface waters of meteoric or seawater origin can penetrate and interact with crustal rocks to depths greater than 10km (see below). Our knowledge of the nature, origin and amounts of fluids at deeper crustal levels and how they evolved with time is, however, far less well developed.

This paper summarizes the H- and O-isotope geochemistry of aqueous fluids and how this has been applied to identify the source of some fluids within the crust. It also shows how combined light and heavy stable isotope studies have been used to characterize the source region of magma generation and the role of fluids during melting and magma emplacement. Many other aspects of the role, budget and isotopic nature of fluids during metamorphic processes cannot be discussed here (see for example : Rumble et al., 1982; Ferry, 1983; Graham et al., 1983; Tracy et al., 1983; Valley, 1986). Similarly , carbonaceous fluids (CO_2 and/or CH_4, etc), which are so widely observed in metamorphic rocks, are not considered here (see for example Hoefs and Touret, 1975; Kreulen, 1980; Pineau et al., 1981; Kreulen and Schuiling, 1982; Touret, 1981; Crawford and Hollister, 1986).

2. GENERAL PRINCIPLES

2.1. Isotopic Characteristics of Natural Waters

Pragmatically, it is convenient to define a number of water types which can be characterized isotopically (see review by Sheppard 1986a and references therein). Most, if not all, of these waters can be considered to be interrelated to each other through the dynamic exchange processes among the different reservoirs of the Earth.

Seawater. Waters of the oceans and open non-intracontinental seas.
Meteoric water. Waters that originated as precipitation.
Formation water. Brines or waters found in the interstices or pores of sediments.
Magmatic water. The separate aqueous phase that has equilibrated with magma regardless of its ultimate origin.
Metamorphic water. Waters that have equilibrated with or were liberated from metamorphic rocks undergoing dehydration.

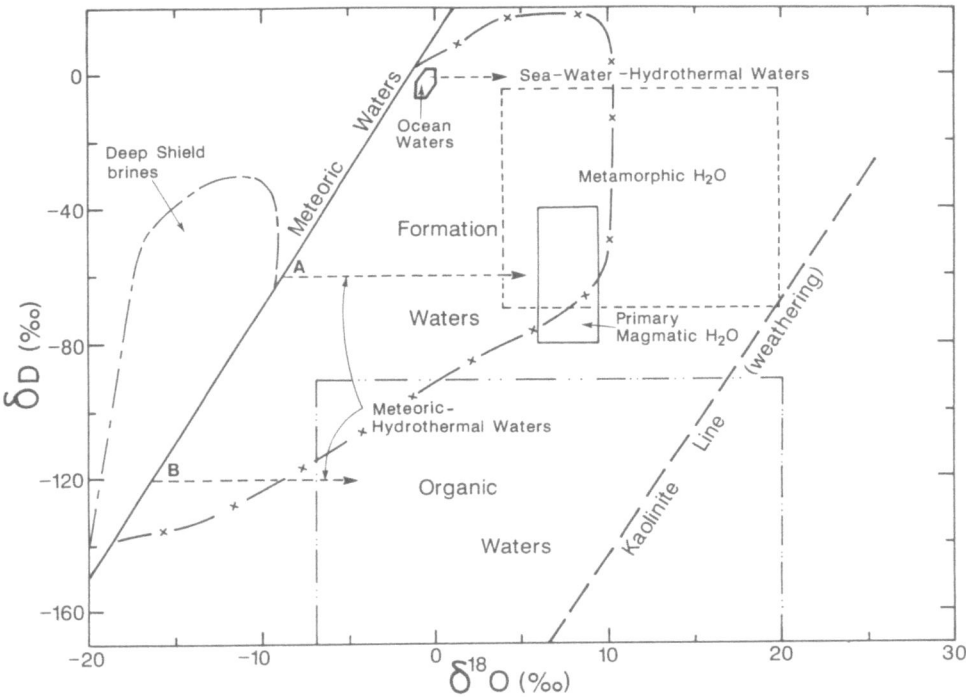

Figure 1. Hydrogen and oxygen isotope compositions or fields of the principal terrestrial water types (see text for definitions; modified after Sheppard 1986). The ^{18}O-shift trends of hydrothermal waters are shown for sea and meteoric waters of compositions A and B. The formation water field is generalized from data on basins from low to high latitudes; brines from a given basin plot in a relatively restricted area of this field. The kaolinite weathering line is given for reference (Savin and Epstein, 1970).

Hydrothermal water. Hot aqueous fluids of any origin.
Organic water. Water derived from the direct or indirect transformation of organic matter, kerogen, petroelum, natural gas, etc. by processes of dehydration, dehydrogenation, oxidation.

Although the concept of juvenile water or that water, released from the Earth's core or mantle, which has never been involved with the hydrosphere, can be defended, it has never been knowingly recognized; it is not considered further here. The non-genetic term formation water is used here in preference to connate water because the latter is strictly the water trapped in the pores of a sediment **at the time of its formation.** Many studies have shown that formation waters are very rarely the initial pore waters or connate waters that have undergone chemical and isotopic modifications (e.g. Clayton et al., 1966; Sheppard, 1984; Kharaka and Carothers, 1986). In many geological situations, it can be shown that a water from another environment has entered the system of interest. Where the origin of the waters cannot be defined, it is convenient to refer to them as **exotic waters**.

The H- and O-isotope characteristics of these principal water types are summarized in Figure 1. The data are given as δ values in per mil relative to Standard Mean Ocean Water or SMOW where

$$\delta_x = \left[\frac{R_x - R_{std}}{R_{std}} \right] 1000 \quad \text{in per mil}$$

and $R = D/H$ or $^{18}O/^{16}O$, x is the sample and std is the standard. The isotopic composition of a water may be measured directly on the water or fluid inclusions or determined indirectly by analysing hydrous minerals and applying the appropriate mineral-H_2O fractionation factors at the temperature of equilibration. The data base behind Figure 1 and review of the isotopic fractionation factors have been discussed most recently by Muehlenbachs (1986), Sheppard (1986a), Taylor (1986), Taylor and Sheppard (1986) and Kyser (1987). Attention is drawn to the following concerning Figure 1 :

- the range of isotopic compositions for seawaters and meteoric waters are for actual waters;
- although not yet well defined, ancient seawaters since about 2500 Ma may have had δD values between about -25 and O ‰ and $\delta^{18}O$ values between about -4 and O ‰;
- there are currently no compelling arguments that the systematics of ancient meteoric waters were radically different from the actual meteoric water line ($\delta D = 8 \ \delta^{18}O + 10$);
- the average isotopic composition of meteoric waters at a locality is usually simply related to the mean annual air temperature or its latitude, altitude and distance from the coast in the direction of movement of the air mass. Thus the isotopic composition of meteoric waters at a given locality can change with time;
- magmatic, metamorphic and organic waters are represented by fields whose boundaries are not absolute;
- the isotopic compositions of deep-seated or mantle-derived waters are taken to be within the magmatic water field;
- meteoric- and seawater-hydrothermal waters commonly show little variation in δD and the characteristic $\delta^{18}O$ shift to more positive values. This reflects mineral-H_2O exchange processes for both element and the distribution of mass of hydrogen and oxygen among the different exchanging reservoirs (overall water/rock ratio);
- the overlapping of fields implies that additional geological and geochemical arguments are necessary to arrive at a sound interpretation of the source of some waters ;
- the δD value is usually more definitive of the source of the water than the $\delta^{18}O$ value. The identification of the source of water that is based on the $^{18}O/^{16}O$ ratio alone must be treated with great caution except in special situations such as when $\delta^{18}O < 0$ ‰;
- certain waters such as pore fluids from the sedimentary oceanic crust and deep saline waters from shield areas can plot to the left or above the meteoric water line (see below).

It is essential to appreciate that the H- and O- isotope compositions of some waters cannot be simply interpreted by themselves to identify the source of the water. For example, a water with δD = -55 ‰ and $\delta^{18}O$ = + 7 ‰ could be from Figure 1, (1) magmatic, (2) metamorphic, (3) meteoric-hydrothermal, (4) evolved formation water, or (5) of mixed origin. In favourable circumstances other geochemical and geological information may add constraints to limit the choice (see below). Hot deep circulating meteoric or formation waters can readily attain such isotopic values through exchange reactions with minerals and because the most abundant meteoric water have $-70 < \delta D < + 10$ ‰.

2.2. Surface Water-Crust Interaction

Waters of surface origin such as meteoric, formation or seawater can interact with deep crustal rocks in two principal ways. They must either penetrate **down** through fractures and fissures in the crust or be carried down to subcrustal depths and then be released into **overlying** lower crustal rocks (Fig. 2).

2.2.1.Downward Penetrating Waters. A large number of stable isotope studies on high-level plutons and fossil and actual hydrothermal systems have documented the penetration of surface waters down to 10 km or more (e.g. Criss and Taylor, 1986; Muehlenbachs, 1986 and references therein). In some of these systems overall water/rock ratios may attain the energy-balance limit of about 2, or higher values in those systems associated with through-going magma conduits. Although the minerals are under a "lithostatic" pressure regime, the downward penetrating fluids are under "hydrostatic" pressure. The pores are interconnected and enable the fluids to more downwards from above. The penetration depth of surface waters into the crust is thus principally controlled by the permeability. This parameter will vary in space and in time during the evolution of the system. The basic physics of water penetration into rocks at high values of P and T is not yet well understood (Lister, 1983).

2.2.2.Upward Penetrating Waters. Water (hydroxyl and possibly interlayer and pore water) can be carried down to subcrustal depths during either the subduction of oceanic lithosphere (Fig. 2b) or the thrusting of upper continental crust under lower crust as a consequence of, for example, continent-continent collision processes (Fig. 2c). Part of the water in the underplate will be released along the thrust zone and into the overlying plate as the underplate undergoes compaction and heats up (e.g. Hilde, 1983; Fyfe and Kerrich, 1985). The released fluids are initially of seawater, meteoric, formation, metamorphic, etc. origin. They can become involved in prograde and retrograde metamorphic reactions and aid partial melting processes in the overlying formations. For the application of this principle, the age of thrusting must be shown to be contemporaneous with the movement of fluids and the hydration reactions, etc. (Turpin et al., 1988). It is probably no coincidence that most magmatic waters typically have $-85 < \delta D < - 40$ ‰, values which are within the range of many hydrous silicates in sediments, seawater-hydrothermally altered oceanic crust, and metamorphic rocks of $-100 < \delta D < -30$ ‰. A quantitative mass balance equation for the hydrogen cycle, however, cannot yet be made because of several partially or even completely unknown factors such as what fraction of the water is released into the overlying slab and with what fractionation factor, what is the contribution of mantle hydrogen, etc.. In addition, the δD values of many magmas have undergone modification during degassing and wall-rock interaction processes (e.g. Taylor, 1986); associated isotopic effects are difficult to quantify at present.

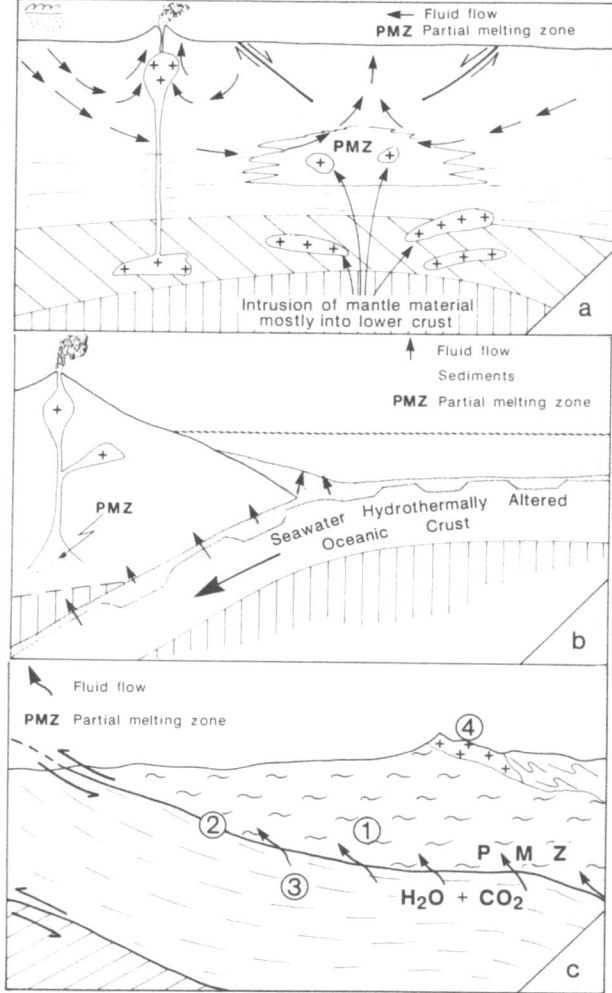

Figure 2. Schematic figures showing the interaction of waters, initially of surface origin, with the crust. **(a)** Downward penetration of waters which under certain circumstances may become involved with crustal anatectic processes; figure in part modified after Wickham and Taylor (1987) as proposed for the Trois Seigneurs area, Pyrenees where the partial melting zone was estimated to be about 12 km below the surface. **(b)** Upward infiltration of waters released from the dehydration of subducted seawater hydrothermally altered oceanic crust and sediments (modified after Hilde, 1983). **(c)** Upward infiltration of waters from dehydration of sedimentary formations overthrusted by a hot crustal slab which may undergo partial melting; the section could represent a N-S section of the Himalaya with **(1)** the Tibetan Slab paragneisses, **(2)** the Main Central Thrust, **(3)** the Lesser Himalaya sediments, and **(4)** the Manaslu granite.

Table I. Chemical and isotopic analyses of selected deep brines[a].

	Outokumpu	Centennial	Yellowknife	Sudbury	Copper Rand
	Drill hole	Minewater	Minewater	Minewater	Minewater
	Finland	Michigan	Northwest T.	Ontario	Quebec
Ref.[b]	1	2	3	3	4
Depth(m)	1,070	-	1,500	1,600	631
T ($^{\circ}$C)	16	-	23	22	~20
Ca (mgl^{-1})	5,700	62,900	57,300	63,800	67,600
Mg	1,100	179	920	78	2,267
Na	3,200	11,900	32,600	18,900	49,700
K	32	38	495	430	219
Cl	16,800	128,000	142,000	162,700	206,550
TDS	27,100	204,463	237,107	249,121	327,574
δD (‰)	-73	-25	-71	-39	-54
δ^{18}O (‰)	-13.0	-7.3	-14.4	-10.9	-16.0

a. See reference for more complete chemical analysis.
b. Reference : 1 = Nurmi et al., 1988; 2 = White, 1968; Kelly et al., 1986;
3 = Frape et al., 1984; 4 = Sheppard et al., in prep.

3. APPLICATIONS

Three case studies are reviewed to illustrate the application of stable isotope studies to crustal problems. They emphasize the advantages of integrating light stable isotope results with other data from, for example, fluid inclusions, radiogenic isotopes and/or the rock chemistry.

3.1. Crystalline Rock Hosted Brines

Brines with salinities between 10,000 to 400,000 mgl^{-1} total dissolved solids (TDS) are very widely distributed at depths greater than 300m or so in crystalline basements (Table I). They have been well documented from the Baltic (Finland, Sweden, USSR), and Canadian (Manitoba, Michigan, Northwest Territories, Ontario, Quebec), Shields (e.g. White, 1968; Fritz and Frape, 1982, 1987; Sheppard et al., 1981; Frape et al., 1984; Kozlovsky, 1987; Nurmi et al., 1988; Vetshteyn et al., 1981). The fluids are usually Ca-Na-Cl brines which are commonly associated with gases such as CH_4 and N_2 and, in some cases H_2. The brines are encountered in fault and shear zones in deep (< 1700 m) mine workings as, for example, in Canada or during moderate to deep (~ 12 km) drilling (Kola superdeep well; Kozlovsky, 1987). The volume of fluid in an individual reservoir is apparently relatively limited as the discharge, initially of up to many litres per minute, generally decreases with time and usually only lasts for a few weeks. These brines have attracted considerable attention recently because (1) their H- and O-isotope compositions

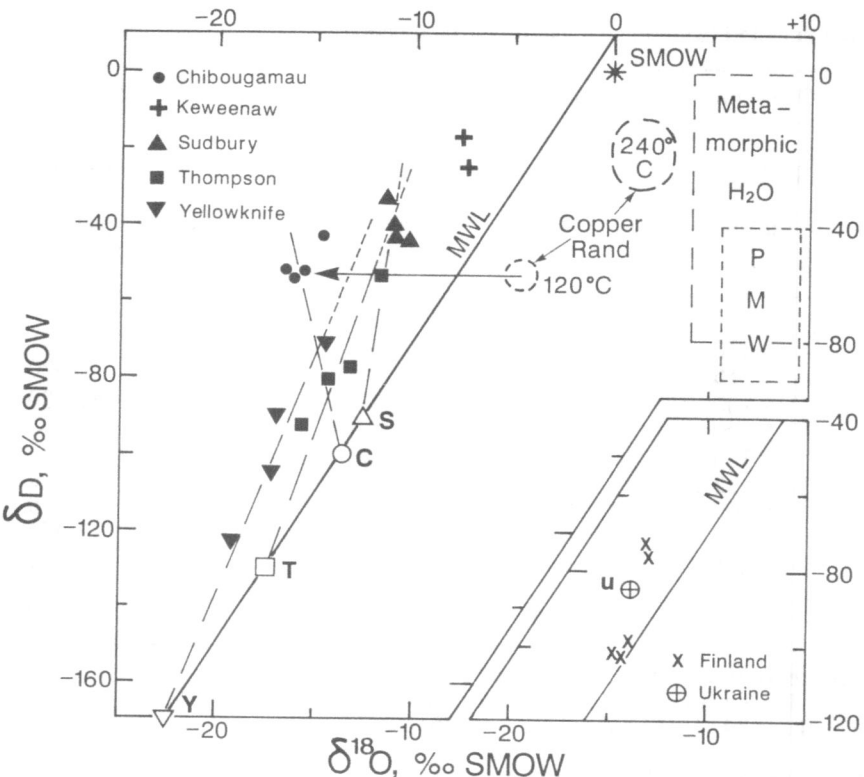

Figure 3 . Hydrogen and oxygen isotope compositions of brines in crystalline rocks from Precambrian Shields shown relative to the meteoric water line (MWL) and primary magmatic (PMW) and metamorphic water fields. Open symbols on the MWL are present-day meteoric waters at Sudbury (S), Chibougamau (C), Thompson (T) and Yellowknife (Y). Fields for the Copper Rand mine, Chibougamau are shown for Main-Stage mineralization at about 240°C and later pre-exchange vug filling fluids at about 120°C. See text for data sources.

typically plot to the left or above the meteoric water line (Fig. 3), (2) they are Ca-Na-Cl solutions with low Mg and K concentrations (Table I), and (3) methane and other gases may be significant constituents of the total fluid.

Taking the Canadian brines as an example, in a given mine δD values increase with increase in salinity from the present-day meteoric water value (Fig. 3). The most saline brines are depleted in $\delta^{18}O$ by up to 8 ‰ relative to meteoric water with the same δD value as the brine. Brines with salinities > 200,000 mgl⁻¹ TDS typically have δD > -55 ‰ whether they come from Quebec, Ontario, Manitoba or Michigan. (At such TDS values, minewaters from Northwest Territories are substantially more D-depleted (Table I). These results for deep minewaters imply that mixing has occurred between an extremely saline water with $\delta D > -55$ and recent meteoric waters. At least in part, this mixing phenomenon is probably related to the mine exploitation processes. In the drill hole samples from Finland, Nurmi et al. (1988) observed that beneath the fresh groundwater zone several sharply differentiated saline layers may occur. The Canadian brines may have been similarly stratified before being disturbed by the mine workings.

For Yellowknife (NWT), Thompson (Manitoba) and Sudbury (Ontario), Frape et al. (1984) calculated regression lines which intercepted the meteoric water line at the present-day composition and suggested that the brine pole had δD values of -50 to -20 ‰ and $\delta^{18}O$ values of -12 to -10 ‰, the locus of the convergence of the regression lines. Adding the data from Keweenaw (Kelly et al., 1986) and Chibougamau (Sheppard et al. in prep.) supports the proposed range of δD values but indicates that the $\delta^{18}O$ values must be much more variable (-17 to -7 ‰).

Secondary isotope exchange reactions at low temperatures between the fluids and their host rocks under low overall water/rock ratios have been proposed to account for the observed isotopic characteristics of the brines (e.g. Fritz and Frape, 1982; Frape et al., 1984; Kelly et al., 1986; Sheppard, 1986a). It was also assumed that the waters had isotopic compositions which plotted to the right of the meteoric water line before the secondary isotope exchange reactions took place. This has now been confirmed by Sheppard et al. (1981, in prep.) and Sheppard (1986a), based on fluid inclusion and isotopic studies of a zoned calcite crystal which grew in one of the brine-methane-filled vugs at the Copper Rand mine, Chibougamau. The vug calcite initially trapped brine at temperature of >120°C. Water in equilibrium with the core of the calcite has at this temperature $\delta^{18}O = -4.9‰$; it therefore plots to the right of the meteoric water line at $\delta D = -54‰$ (Fig. 3). During the subsequent cooling of the system to below 30°C (the calcite surface-brine O-isotope temperature is $< 28.5°C$), the O-isotope composition of the brine was dominated by the mineral oxygen reservoir and the temperature of exchange. The H-isotope composition of the brine probably remained essentially unmodified because of the extremely limited hydrous mineral reservoir in the vug and along the fractures. Thus the ^{18}O-depleted brines represent the result of retrograde alteration during cooling of the hydrothermal vein solutions under low water/rock ratios. The general variability of the magnitude of the ^{18}O-shift (3 to 8 ‰ to the left of the meteoric water line for the most saline waters) is probably reflecting a combination of small variations in the water/rock ratio, temperature decrease during retrograde alteration and wall rock mineralogy.

The age and origin of these brines is less evident. Sheppard et al. (1981, in prep) have shown that the gangue minerals of the main stage Cu-Au mineralization at Copper Rand were precipitated from fluids with $\delta D \sim -20$ ‰ and $\delta^{18}O \sim + 2$ ‰. The implied $\delta^{18}O$ value of the cold uncontaminated brine of about -17 ‰ could be in equilibrium with either the main-stage gangue minerals of quartz and ankerite at about 50°C or the later calcite at about 25°C. As the δD values decrease from main stage to post-main stage, the two hydrothermal fluids could conceivably have independent origins and be of different ages. Kerrich (1987) has also presented evidence that fluids infiltrated the Shield over a considerable time span during the Precambrian.

Kelly et al. (1986) have proposed a basinal water model for the Keweenaw Peninsula, Michigan brines where Phanerozoic formation waters penetrated down into the underlying basement rocks. They also applied this model to the Canadian Shield as a whole. However, there is no reason that all of the above mentioned Canadian waters must have the same origin or be of Phanerozoic age. Waters with similar chemical characteristics can be of very different origin or age. The main-stage mineralization at Copper Rand formed at about 2200 Ma (Thorpe et al., 1984) and could have been deposited from formation, meteoric or sea waters. The waters seem to be too ^{18}O-depleted to be metamorphic waters (Fig. 3). These Proterozoic fluids probably ascended along the shear zones in the meta-anorthosite host rocks. A formation water origin, which would also be consistent with the presence of CH_4, implies a thrusting environment.

3.2. Mixing Between Two Externally Derived Fluids

In the external section of the Western French Alps, the Hercynian La Lauzière plutono-metamorphic massif was affected by late Hercynian (~ 280 Ma) fractures (Negga et al., 1986). The initial dolomite-ankerite vein stage was followed by the calcite-pitchblende stage of uranium mineralization. By integrating, H-, C- and O-isotope studies with mineralogical and fluid inclusion data, the conditions of formation of the different stages were characterized. Because the U/Pb dating indicated a late Hercynian age for the mineralization followed by incomplete remobilization during the Alpine event, the fluid inclusion and isotopic data were carefully scrutinized for any evidence of reequilibration or modification during the Alpine event. No evidence was found for decrepitation of the fluid inclusions and a barren Alpine calcite vein which cuts the nearby Mesozoic formations has C- and O-isotope compositions distinctly different from the Hercynian vein carbonates. The microthermometric and isotopic characteristics of the fluids are thus interpreted to have preserved their late Hercynian signatures.

From the combined results Negga et al. (1986) showed that a hot (~ 350-400°C), highly saline (~ 20 wt % equiv. NaCl), reduced (fO_2 ~ 10^{-32}), deuterium - and ^{18}O-enriched fluid (δD ~ - 35 ‰; $\delta^{18}O$ ~ + 12 ‰), which precipitated the dolomite-ankerite stage, mixed with a less hot (~ 300-350°C), less saline (~ 8 wt % equiv. NaCl), less deuterium - and ^{18}O-rich (δD < - 50 ‰; $\delta^{18}O$ ~ + 8 ‰) but slightly more oxidizing fluid (fO_2 > 10^{-32}) during the calcite-pitchblende stage (Fig. 4). Both fluids appear to be equally ^{13}C-depleted, with $\delta^{13}C_{\Sigma C}$ ~ - 25 ‰ PDB; their composition therefore was controlled by the organic matter and/or graphite in the reservoir rocks. Fluid pressures decreased from ~ 2 ± 0.5 kb at the dolomite stage to ~ 1.5 ± 0.5 kb at the calcite stage. Both N_2 and CO_2 were detected in the fluids and CH_4 may also be present (Negga et al., 1986, p 180).

The isotopic compositions of these fluids are not distinctive of a single source so other geological arguments are needed to suggest whether the sources are formation, magmatic, metamorphic or meteoric waters. Because the associated metamorphic rocks were undergoing retromorphism, a local metamorphic source can be discounted. Similarly magmatic waters are excluded because no contemporaneous magmatism is known in the area. For these reasons, the more saline fluid was interpreted by Negga et al. (1986) to be formation or possibly metamorphic waters coming from a deeper level which later mixed with more oxidizing U-bearing meteoric waters (Fig. 5). This interpretation implies that the hot La Lauzière metamorphic complex was thrusted over sediments which subsequently underwent dewatering and/or dehydration. Although the relative importance of Hercynian

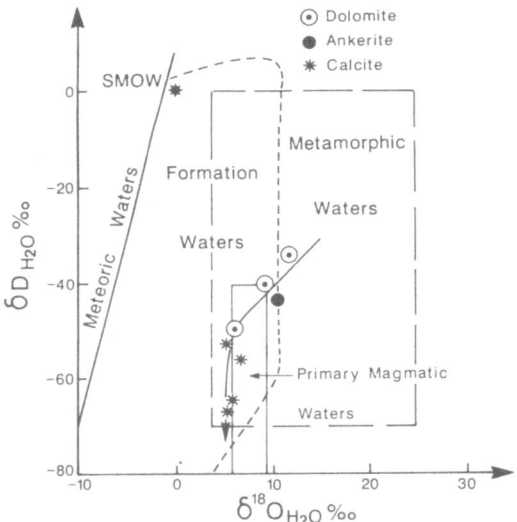

Figure 4. Plot of measured δD value of fluid inclusions versus calculated δ^{18}O value of water in equilibrium with the carbonate at 375°C for dolomite and ankerite and 325°C for calcite (modified after Negga et al., 1986).

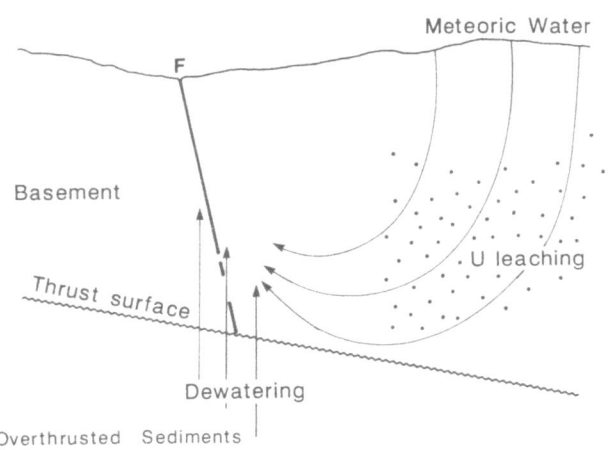

Figure 5. Schematic multi-stage model for the La Lauzière massif showing flow paths for hydrous fluids coming from the proposed dewatering of overthrusted sediments and their subsequent mixing with deep circulating uranium-bearing meteoric waters (modified after Negga et al., 1986).

and Alpine thrusting in this region is unknown, sediments of probable Carboniferous age are intercalated within the basement at less than 10 to 12 km depth in the nearby Belledonne massif.

The release of the sedimentary fluids into the inferred overlying slab by producing fractures and fracture permeability would also aid the subsequent influx of meteoric waters (Fig. 5). Since the δD values of the calcite stage meteoric fluids decreased from -50 to -65 ‰, Negga et al. (1986) proposed that the region was undergoing uplift during the evolution of the vein system. Although the meteoric-hydrothermal part of the system must have operated at near hydrostatic pressure conditions, the initial part of the dolomite stage could have been at near lithostatic pressure conditions.

3.3. Intra-Continental Collision and Granitic Magmatism

The genesis of the High Himalaya and North Himalaya peraluminous leucogranite belts has been related to the collision history between the Asian and Indian continents (Le Fort et al., 1987 and references therein). Based in particular on the High Himalaya Manaslu granite massif, combined geological and geochemical studies by Le Fort et al. (1987), Deniel et al. (1987), France-Lanord et al. (1988) and others have shown that it was derived by the partial melting of a thick sequence of quartzo-pelitic gneisses of the Tibetan Slab, which lie immediately above the Main Central Thrust (Fig. 2c). This shear zone separates the overthrusted gneisses from the underlying sediments of the Lesser Himalaya. Rocks equivalent to those which generated the granitic magmas at a greater depth can be sampled at outcrop, because of the nature of the N-S section of the Himalaya.

Based on the δD data on the Manaslu granite, Tibetan Slab gneisses and Lesser Himalaya sediments, France-Lanord et al. (1988) proposed that the low δD fluid (~ - 80 ‰), which infiltrated into the hot (~ 700°C) but dry gneisses and probably triggered or aided their partial melting, could have come from the dehydration of the Lesser Himalaya sediments at an inferred depth of about 40km or so (Fig. 6). The main Manaslu massif, which is about 30 km across and 8km thick, is characterized by a relatively uniform mineralogy and major element chemistry but highly variable $\delta^{18}O$, $(^{87}Sr/^{86}Sr)_{20Ma}$ and ϵNd values even down to a metric scale (Vidal et al., 1982; Deniel et al., 1987; Le Fort et al., 1987; France-Lanord et al., 1988). Trace element contents such as Ba, Sr and U are also extremely variable (Cuney et al., 1984). H- and O-isotope relations among the minerals indicate that high temperature equilibrium fractionations have been frozen in and that no extensive sub-solidus exchange with circulating fluids occurred in the main massif.

The correlations among the $\delta^{18}O$, $(^{87}Sr/^{86}Sr)_{20Ma}$ and ϵNd values of the granites can be matched to similar covariations in the source rock gneisses (Fig. 7), demonstrating that the isotopic variations in the granite were inherited from its source region. As Deniel et al. (1987) and France-Lanord et al. (1988) have emphasized, these relations imply that the granite massif is made up of a large number of different batches of melt which could have undergone no more than limited mixing between the source region and the emplacement level, some 10km or so above (Fig. 6). The isotopic age data indicate that the emplacement of the different batches of leucogranite probably lasted for more than 10 Ma (25 to 14 Ma).

By using the chemical-mineralogical parameter (A+B), where A = Al-(Na+K+2Ca) or the amount of aluminium in the rock that is not present as feldspar and B = Fe+Ti+Mg,

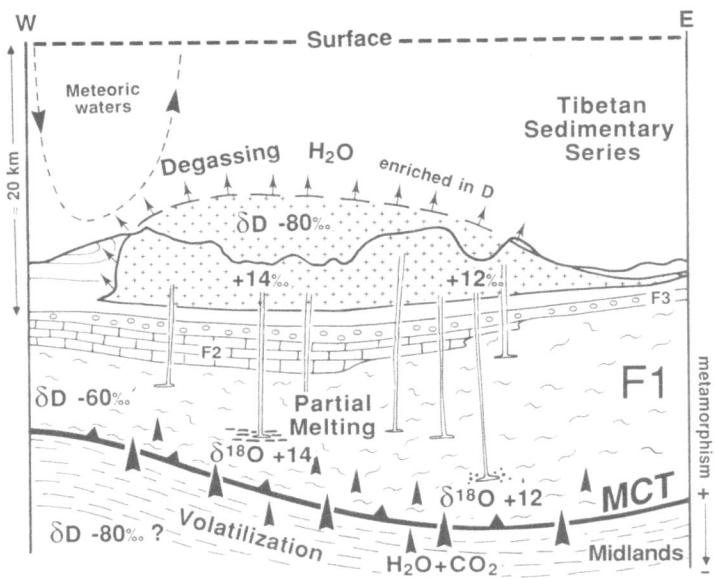

Figure 6. Schematic time-integrated model for the genesis of the Manaslu granite. Flow paths show (1) infiltrating H_2O+CO_2 fluids from the overthrusted Lesser Himalaya Midlands sediments into the hot high grade metamorphosed Tibetan Slab (F 1) where partial melting of the isotopically heterogeneous paragneisses occurs, (2) emplacement of magma batches with variable $\delta^{18}O$ values, (3) degassing of hydrous fluid from the granite, and (4) inferred meteoric water circulation system in the now eroded Tibetan Sedimentary Series. Note that the relation between the volume of the partial melting zone and that of the granite cannot be shown in such a section. Modified after France-Lanord et al. (1988).

France-Lanord et al. (1988) related the isotopic variations in the paragneisses (Fig. 8) to the proportion of phyllosilicates in their pre-metamorphosed state. These sediments could have been constituted from a mixture of 9-11 ‰ $\delta^{18}O$ quartz plus feldspar and 15-19 ‰ $\delta^{18}O$ phyllosilicates. At equilibrium, however, phyllosilicates are depleted in ^{18}O relative to quartz. The observed O-isotope variations therefore imply that the scale of equilibrium exchange among the different sedimentary units was not very extensive. Isotopic homogenization via a convecting fluid did not occur during the sedimentary, diagenetic or metamorphic evolution of the Tibetan Slab. Equally the infiltrating low δD fluid did not significantly modify the $^{18}O/^{16}O$ ratios of the Tibetan gneisses for two principal reasons : (1) the O-isotope compositions of the Lesser Himalaya sediments are broadly similar to those of the Tibetan gneisses (France-Lanord et al., 1988), and (2) the total quantity of oxygen introduced into the Tibetan gneisses as H_2O and CO_2 could not have been more than a few percent of the oxygen contained in the gneisses. These conditions are in contrast

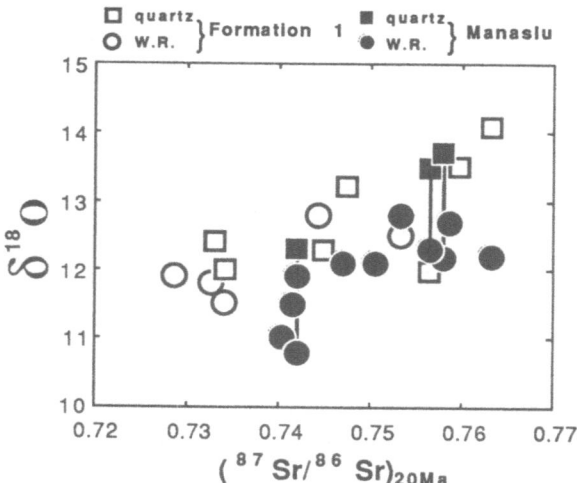

Figure 7. Plot of δ¹⁸O whole rock and/or quartz versus (^{87}Sr/^{86}Sr) ratio calculated for 20 Ma for the Manaslu granite and Formation 1 paragneisses of the Tibetan Slab. The field for the chemically variable Formation 1 is largely defined by quartz data and therefore plots slightly above the granite field which is principally based on whole rock data. Modified after France-Lanord et al. (1988).

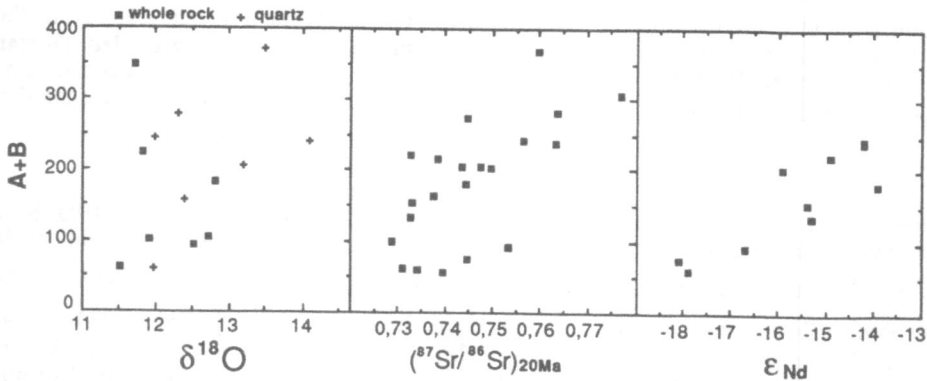

Figure 8. Plot of the chemical parameter (A+B) = [Al-(Na+K+2Ca)] + [Fe+Ti+Mg)]versus (a) δ¹⁸O quartz for Formation 1 paragneisses, (b) (^{87}Sr/^{86}Sr) calculated for 20 Ma, and (c) εNd (t = 500 Ma). Modified after France-Lanord and Le Fort (1988).

to those proposed for the generation of leucogranite in the Trois Seigneurs area of the French Pyrénées by Wickham and Taylor (1985 ;1987). They have argued that the pelitic series underwent pervasive exchange, depletion in ^{18}O and isotopic homogenization with a convecting sea water or meteoric water system and that partial melting occurred at 6-12 km in the presence of this surface water-rich fluid (Fig. 2a). The tectonic environments in the Pyrénées and Himalaya immediately before and during magmatism were, however, probably quite different.

The O-, Sr- and Nd-isotope variations in the High Himalaya leucogranites can therefore be understood in terms of inheritance from the source region. In turn, these variations reflect, in large part, the mineralogical constitution of the pre-metamorphosed sediments and, for Sr, the initial Rb content of the rock and long time interval (> 500 Ma) between sedimentation and partial melting (aging effect). The H-isotope differences between the granites and the paragneiss source rocks can be explained by introducing a D-depleted fluid ($\delta D \sim -80$) into the gneisses from the underlying Lesser Himalaya sedimentary formations followed by magmatic degassing processes (France-Lanord et al., 1988).

4. CONCLUDING REMARKS

Understanding the role of fluids and magmas in the crust requires knowledge of the origin or origins of their constituents in so far as all elements may not come from a common source. For aqueous fluids, combined H- and O-isotope studies have documented their diverse origins during crustal processes and the role of fluid mixing and fluid-rock interaction processes. The two principal surface water reservoirs, seawater and meteoric waters, can become involved with more or less deep crustal processes. Such waters must either percolate downwards and therefore be principally restricted to the outer brittle fracture zone of the crust or be carried down as hydroxyl, interlayer and pore water to deep levels via subduction or thrusting processes before being released into overlying formations. Although all of these situations have been documented, emphasis has been purposefully placed here on the overall upward percolation of exotic aqueous fluids (Fig. 2b and 2c). Such processes must be dominant in a wide variety of deep crustal environments where fluid pressures are similar to lithostatic pressures. In addition, prograde metamorphic reactions can be a major source of fluids which move towards the Earth's surface. For example, the H_2O and CO_2 liberated during the devolatilization of an average pelite at 500 °C and 5 kb occupies ~ 12 volume % of the rock and represents ~ 5 wt. % of the rock (Walther and Orville, 1982).

A number of isotopic studies have characterized the scale of equilibrium and related this to the nature of the fluid flow during regional metamorphism (see Valley, 1986, Appendix I for references). Examples are known which imply (1) essentially no fluid flow, (2) prevasive flow, and (3) highly focused or channelled flow. Invariably fluid flow was enhanced in those formations which were undergoing devolatilization reactions because of their higher permeability. More fluid often circulated through such zones than could have been locally produced. Quartz pods and segregations in metamorphic terrains represent a marker of highly channelled fluid flow. In most situations, however, the fluid-rock mass balance and its variation in space and with time during the P-T evolution of the system have not been quantified.

Combined light and heavy stable isotope studies have placed crucial constraints on relating igneous rocks to their source region as well as to some of the processes operating during the partial melting stage and emplacement of the magmas. Where infiltrating exotic fluids are involved during the anatectic processes, any thermal modelling of the system must take them into account. The High Himalaya case study has demonstrated that the coupled O-, Sr- and Nd-isotope variations in a plutonic massif can be related to comparable characteristics in the source rocks. Although chemical and isotopic variations in plutonic complexes can be produced by several other mechanisms including assimilation - fractional crystallization (see Taylor and Sheppard, 1986), several igneous systems deserve to be reexamined in the light of the principles of the High Himalaya case study.

The role of mantle derived fluids in deep crustal processes has not been considered here. In addition to magma, the mantle can be a source of H_2O, CO_2 and other volatiles in the lower crust. Nevertheless, the importance of this source of aqueous and carbonic fluids is still poorly constrained.

It has been shown above that in some systems fluids of magmatic origin cannot be distinguished from, for example, some metamorphic waters using isotopic methods alone. Studies of such systems need to integrate isotopic data with results from fluid inclusion, detailed field and other studies to arrive at more certain conclusions concerning the origin and role of fluids and their evolution in deep crustal rocks.

5. ACKNOWLEDGEMENTS

I am extremely grateful for stimulating discussions and collaborations with my colleagues M. Cuney, Ch. France-Lanord, J. Guha, P. Le Fort, J. Leroy, H.S. Negga and J. Rosenbaum. The manuscript has benefited from the review by M. Cuney. Contribution N° 767 Centre de Recherches Pétrographiques et Géochimiques.

REFERENCES

Allègre C.J. and Ben Othman D. (1980) Nd-Sr isotopic relationship in granitoid rocks and continental crust development: a chemical approach to orogenesis, Nature, **286**, 335-342.

Clayton R.N., Friedman I., Graf D.L., Mayeda T.K., Meents W.F. et Shimp N.F. (1966) The origin of saline formation waters, 1, Isotopic composition. J. Geophys. Res. **71**, 3869-3882.

Criss R.E. and Taylor H.P. Jr. (1986) Meteoric-hydrothermal systems. In Stable Isotopes in High Temperature Geological Processes J.W. Valley, H.P. Taylor Jr., and J.R. O'Neil (eds.). Rev.Mineral. **16**, 373-424.

Crawford M.L. and Hollister L.S. (1986) Metamorphic fluids: the evidence from fluid inclusions. In Fluid-Rock Interactions during Metamorphism. J.V. Walther and B.J. Wood (eds.) Springer-Verlag, New York, 1-35.

Cuney M., Le Fort P. and Wang Z. (1984) Uranium and thorium geochemistry and mineralogy in the Manaslu leucogranite (Nepal, Himalaya). In Geology of Granites and their Metallogenetic Relations. K. Xu and G. Tu (eds.) Science Press, Beijing, China, 853-873.

Deniel C., Vidal P., Fernandez A. et Le Fort P. (1987) Isotopic study of the Manaslu granite (Himalaya, Nepal) : Inferences on the age and source of Himalayan leucogranites. Contrib. Mineral. Petrol. **96**, 78-92.

Doe B.R. and Zartman R.E. (1979) Plumbotectonics, the Phanerozoic. In Geochemistry of Hydrothermal Ore Deposits 2nd Edition. H.L. Barnes (ed.), John Wiley, New York, 22-70.

Dubessy J. and Ramboz C. (1986) The history of organic nitrogen from early diagenesis to amphibolite facies : mineralogical, chemical, mechanical and isotopic implications. 5th Internat. Symp. Water-Rock Interaction, Reykjavik, Iceland, Extended abstracts 170-174.

Ferry J.M. (1983) Regional metamorphism of the Vassalboro Formation, south-central Maine, U.S.A.: A case study of the rock of fluid in metamorphic petrogenesis. J. Geol. Soc. London **140**, 551-576.

France-Lanord C and Le Fort P. (1988) Crustal melting and granite genesis during the Himalayan Collision orogenesis. Trans. R. Soc. Edinburgh : Earth Sci. ,in press.

France-Lanord C., Sheppard S.M.F. and Le Fort P. (1988) Hydrogen and oxygen isotope variations in the High Himalaya peraluminous Manaslu leucogranite: evidence for heterogeneous sedimentary source. Geochim. Cosmochim. Acta, **52**, 513-526.

Frape S.K., Fritz P. and McNutt R.H. (1984) The role of water-rock interaction in the chemical evolution of ground-waters from Canadian Shield. Geochim. Cosmochim. Acta **48**, 1617-1627.

Fritz P. and Frape S.K. (1982) Saline groundwaters in the Canadian Shield - a first overview. Chem. Geol. **36**, 179-190.

Fritz P. and Frape S.K. eds. (1987) Saline Waters and Gases in Crystalline Rocks. Geol. Assoc. Canada, sp. Pap. **33**, 259 pp.

Fyfe W.S. and Kerrich R. (1985) Fluids and thrusting. Chem. Geol. **49**, 353-362.

Graham C.M., Greig K.M., Sheppard S.M.F. and Turi B. (1983) Genesis and mobility of the H2O-CO2 fluid phase during regional greenschist and epidote amphibolite facies metamorphism: a petrological and stable isotope study in the Scottish Dalradian. J. Geol. Soc. London **140**, 577-599.

Hilde T.W.C. (1983) Sediment subduction versus accretion around the Pacific. Tectonophysics **99**, 381-397.

Hoefs J. and Touret J. (1975) Fluid inclusions and carbon isotope study from Bamble granulites (South Norway). Contrib. Mineral. Petrol. **52**, 165-174.

Hollister L. S. and Crawford M. L. (1981) Short Course in Fluid Inclusions : Applications to Petrology. Mineral. Assc. Canada **6**, 304 pp.

Kelly W.C., Rye R.O. and Livnat A. (1986) Saline minewaters of the Keweenaw Peninsula, Northern Michigan; their nature, origin, and relation to similar deepwaters in Precambrian crystalline rocks of the Canadian Shield. Am. J. Sci., **286**, 281-308.

Kerrich R. (1987) Stable isotope studies of fluids in the crust. In Stable Isotope Geochemistry of Low Temperature Processes T.K. Kyser (ed.) Mineralogical Assoc of Canada **13**, 258-286.

Kharaka Y.K. and Carothers W.W. (1986) Oxygen and hydrogen isotope geochemistry of deep basin brines. In Handbook of Environmental Isotope Geochemistry, vol. 2. Fritz P. and Fontes J. Ch. (eds) Elsevier, Amsterdam, 305-360.

Kozlovsky Y.A. (ed) (1987) The Superdeep Well of the Kola Peninsula. Springer-Verlag, Berlin 558 p.

Kreulen R. (1980) CO2-rich fluids during regional metamorphism on Naxos (Greece): carbon isotopes and fluid inclusions. Amer. J. Sci. **280**, 745-771.

Kreulen R. and Schuiling R.D. (1982) N2-CH4-CO2 fluids during formation of the Dome de l'Agout, France. Geochim. Cosmochim. Acta **46**, 193-203.

Kyser T.K. (1987) Equilibrium fractionation factors for stable isotopes. In Stable Isotope Geochemistry of Low Temperature Processes T.K. Kyser (ed.). Mineralogical Assoc. of Canada **13**, 1-84.

Le Fort, P., Cuney, M., Deniel, C., France-Lanord, C., Sheppard, S.M.F., Upreti, B.N., et Vidal, P. (1987) Crustal generation of the Himalayan leucogranites. Tectonophysics , **134**, 39-57.

Lister C.R.B. (1983) The basic physics of water penetration into hot rock. In Hydrothermal Processes at Seafloor Spreading Centers. P.A. Rona, K. Boström, L. Laubier and K.L. Smith Jr. (eds.) Plenum Press, New York, 141-168.

Muehlenbachs K. (1986) Alteration of the oceanic crust and the ^{18}O history of seawater. In Stable Isotopes in High Temperature Geological Processes. J.W. Valley, H.P. Taylor Jr., and J.R. O'Neil (eds) Rev. Mineral. 16, 425-444.

Negga H.S., Sheppard S.M.F., Rosenbaum J. and Cuney M. (1986) Late Hercynian U-vein mineralization in the Alps: fluid inclusion andC, O, H isotope evidence for mixing between two externally derived fluids. Contrib. Mineral. Petrol. 93, 179-186.

Nurmi P.A., Kukkonen I.T. and Lahermo P.W. (1988) Geochemistry and origin of saline groundwaters in the Baltic Shield. Applied Geochemistry, 3, 185-203.

Pineau F., Javoy M., Behar F. et Touret J. (1981) La géochimie isotopique du faciès granulite du Bamble (Norvège) et l'origine des fluides carbonés dans la croûte profonde. Bull. Minéral. 104, 630-641.

Roedder E (1984) Fluid inclusions. Rev. Mineral. 12, 644 pp.

Rumble D., Ferry J.M., Hoering T.C. and Boucot A.J. (1982) Fluid flow during metamorphism at the Beaver Brook fossil locality, New Hampshire. Amer. J. Sci. 282, 886-919.

Savin S.M. and Epstein S. (1970b) The oxygen and hydrogen isotope geochemistry of clay minerals. Geochim. Cosmochim. Acta 34, 25-42.

Sheppard S.M.F. (1984) Stable isotope studies of formation waters and associated Pb-Zn hydrothermal ore deposits. In Thermal Phenomena in Sedimentary Basin, B. Durand (ed.) Editions Technip, Paris 301-317.

Sheppard S.M.F. (1986a) Characterization and isotopic variations in natural waters. In Stable Isotopes in High Temperature Geological Processes J.W. Valley, H.P. Taylor Jr.and J.R. O'Neil (eds.). Rev. Mineral. 16, 165-184.

Sheppard S.M.F. (1986b) Igneous rocks: III. isotopic case studies of magmatism in Africa, Eurasia, and Oceanic Islands. In Stable Isotopes in High Temperature Geological Processes J.W. Valley, H.P. Taylor Jr.and J.R. O'Neil (eds.).Rev. Mineral. 16 319-372.

Sheppard S.M.F., Guha J. and Leroy J. (1981) La caractérisation géochimique des fluides du gisement Cu-Au de Chibougamau, Québec. Rapport Annuel 1980, Centre de Recherches Pétrographiques et Géochimiques, 68-69.

Taylor H.P. Jr. (1974) The application of oxygen and hydrogen isotope studies to problems of hydrothermal alteration and ore deposition. Econ. Geol. 69, 843-883.

Taylor H.P. Jr. (1977) Water/rock interaction and the origin of H_2O in granitic batholiths. J. Geol. Soc. Lond. 133, 509-558.

Taylor H.P. Jr. (1986) Igneous rocks II. Isotopic case studies of circum Pacific magmatism. In Stable Isotopes in High Temperature Geological Processes. J.W. Valley, H.P. Taylor Jr.and J.R. O'Neil (eds.). Rev. Mineral. 16, 273-317.

Taylor H.P. Jr. et Sheppard S.M.F. (1986) Igneous rocks: I. Processes of isotopic fractionation and isotope systematics. In : Stable Isotopes in High Temperature Geological Processes J.W Valley, H.P. Taylor Jr. and J.R. O'Neil (eds). Rev. Mineral. 16, 227-272.

Thorpe R.I., Guha J. and Cimon J. (1981) Evidence from lead isotopes regarding the genesis of ore deposits in the Chibougamau region, Québec. Can. J. Earth. Sci. 18, 708-723.

Touret J.L.R. (1981) Fluid inclusions in high grade metamorphic rocks. In Fluid Inclusions: Applications to Petrology L.S. Hollister and M.L. Crawford (eds) Mineral. Assoc. Canada 6, 182-208.

Tracey R.J., Rye D.M., Hewitt D.A. and Schiffries C.M. (1983) Petrologic and stable isotopic studies of fluid-rock interactions, south-central Connecticut. I The role of infiltration in producing reaction assemblages in impure marbles. Amer. J. Sci. **283A**, 589-616.

Turpin L., Maruejol P. and Cuney M. (1988). U-Pb, Rb-Sr and Sm-Nd chronology of granitic basement, hydrothermal albilites and uranium mineralization (Lagoa Real, South-Bahia, Brazil). Contr. Mineral. Petrol. **98**, 139-147.

Valley J.W.(1986) Stable isotope geochemistry of metamorphic rocks. In Stable Isotopic in High Temperature Geological Processes J.W. Valley, H.P. Taylor Jr.and J.R. O'Neil (eds.). Rev. Mineral.**16**, 445-490.

Vetshteyn V. Ye., Gavrish V.K. and Gutsalo L.K. (1981) Hydrogen and oxygen isotope composition of waters in deep-seated fault zones. Internat. Geol. Rev. **23**, 302-310.

Vidal Ph., Bernard-Griffiths J., Cocherie A., Le Fort P., Peucat J.J., et Sheppard S.M.F. (1984) Geochemical comparison between Himalayan and Hercynian leucogranites. Phys. Earth Planet. Inter. **35**, 179-190.

Vidal Ph., Cocherie A. et Le Fort P. (1982) Geochemical investigations of the origin of the Manaslu leucogranite (Himalaya, Nepal). Geochim. Cosmochim. Acta, **46**, 2279-2292.

Walther J. V. and Orville P. M. Volatile production and transport in regional metamorphism. Contr. Mineral. Petrol., **79**, 252-257.

White D.E. (1968) Environments of generation of some base-metal ore deposits. Econ. Geol. **63**, 301-335.

Wickham S.M. and Taylor H.P. Jr. (1985) Stable isotopic evidence for large-scale seawater infiltration in a regional metamorphic terrane ; the Trois Seigneurs massif, Pyrenees, France. Contrib. Mineral. Petrol.**91**, 122-137.

Wickham S.M. and Taylor H.P. Jr. (1987) Stable isotope constraints on the origin and depth of penetration of hydrothermal fluids associated with Hercynian regional metamorphism and crustal anatexis in the Pyrenees. Contrib. Mineral. Petrol. **95**, 255-268.

Zartman R.E. (1974) Lead isotopic provinces in the Cordillera of the Western United States and their geologic significance. Econ. Geol. **69**, 792-805.

STABLE ISOTOPE FRONTS AND CRUSTAL BUFFERING –
1-DIMENSIONAL MASS BALANCE AND KINETICS

P. Blattner,
NZ Geological Survey, (DSIR),
P.O. Box 30368, Lower Hutt, and
Institute of Nuclear Sciences (DSIR),
Private Bag, Lower Hutt, New Zealand

and

A. Absar,
Geology Department and Geothermal Institute,
University of Auckland,
Private Bag, Auckland, New Zealand

ABSTRACT. Zero dimensional and 1-dimensional models of isotope
exchange in fluid-rock interaction are discussed from a mass balance
and kinetic point of view. A rate-controlled 1-dimensional model
provides unique delta vs distance profiles for any given reaction rate
constant and interstitial fluid velocity. In principle, the model
allows to quantify even that fraction of fluid which, due to slow
exchange rates, fails to leave an effect in the rock of the aquifer it
has passed. All higher temperature aquifers or fluid trajectories
have an isotopic inversion point given by the porosity. This point
locates a more or less developed 'isotope front'. The fluid/rock
ratio over a whole aquifer is dictated by the porosity and fluid/rock
ratios $\gg 1$ are possible only when an aquifer discharges to the
surface. In the 1-dimensional view, most geothermal systems (e.g.
Wairakei) have reached a stage behind the inversion point, whereas few
geothermal systems (Ngawha) but most metamorphic terranes so far as is
known lie ahead of it and appear to be still rock-controlled in their
isotopic properties.

1.   INTRODUCTION

The 'oxygen isotope shift' of meteoric waters in geothermal systems,
ascribed by Craig (1963) to isotope exchange between water and the
reservoir rock, may be quantified in terms of an equivalent isotope
shift, opposite in sign, of the rock, using mass balance. Calculations
on a box (0-dimensional) model were offered by Clayton et al. (1968);
Taylor (1977) suggested a 'totally open' box model for applications in
mineral deposits and pointed out fossil hydrothermal systems.

265

D. Bridgwater (ed.), Fluid Movements – Element Transport and the Composition of the Deep Crust, 265–275.
© 1989 by Kluwer Academic Publishers.

Blattner (1985) and Blattner and Bunting (1980) gave a derivation for
a box model that allowed any desired degree of openness, and discussed
its merits in some detail. It was pointed out that a throughput of
water in idealised finite 'batches' is likely to simulate
inefficiencies of real systems, such as successive fracture filling
and seismic disruptions. Spooner et al. (1976) had already developed
a 1-dimensional model of oxygen isotope exchange between fluid and
rock. That model, consisting of successive boxes, may be extended to
take into account reaction rates (using eq. 5 below) and would then
serve much the same puropse as the rate controlled model presented
here. Temperature effects can be of importance when temperatures are
low, and near the fluid front in 1-dimensional models (Spooner et al.
1976, Cathles 1983). They will not be discussed here as the purpose
of this paper is to show the advantages of 1-dimensional as against
0-dimensional 'box' models and to highlight the fundamental difficulty
of achieving high surface-water to rock ratios in the lithosphere.
Colleagues at the Bergen meeting pointed out the similarity of the
1-dimensional isotope calculations to chemical calculations of
infiltration metasomatism. In the process of infiltration, oxygen and
hydrogen isotopes have the advantage, over chemical elements, of
providing isotopic labels of two essentially omnipresent constituting
elements of the lithosphere and its fluids.

2. BOX (0-DIMENSIONAL) MODEL WITH INSTANTANEOUS EXCHANGE

For equilibrium exchange in a closed box the maximum potential isotope
shift ('contrast' of some authors) is

$$P = \delta_{ri} - \delta_{wi} - \delta_{rf} + \delta_{wf} \qquad (1)$$

where r stands for rock, w for water, i for initial and f for final,
and $\Delta = \delta_{rf} - \delta_{wf}$ is a function of temperature[1]: isotope shifts may
be defined (e.g. for water) as $\sigma_w = \delta_{wf} - \delta_{wi}$. For exchange in a
closed box the mass balance is then

$$W\sigma_w = -R\sigma_r \qquad (2)$$

where W is the amount of oxygen in the water and R that in the rock.
The shift ratio with changed sign, $-\sigma_r/\sigma_w = W/R$, is thus a direct
measure of the water/rock ratio under these conditions.
  We may consider next a total amount of water W consisting of a
sequence of n batches of water w, that pass through the rock reservoir
'box', and equilibrate with the rock, quasi instantaneously. Setting

****

[1]As usual, δ is defined, e.g. in the case of oxygen isotopes, as

$$\delta^{18}O(°/_{oo}) = [\frac{(^{18}O/^{16}O) \text{ sample}}{(^{18}O/^{16}O) \text{ standard}} - 1]\ 10^3$$

W = nw the model is for finite n partially, and for n = ∞ totally open. Typical for the totally open box model is the maximum degree of exchange efficiency from the point of view of the water required, which is shown by the expression now replacing eq 2

$$(W/R)_\infty = \ln(1 - \sigma_r/\sigma_w) \tag{3}$$

The water W in the closed box is viewed as but one batch of the cumulative water throughput W in the open system, and the difference between the W/R of eq 2 and the $(W/R)_\infty$ of eq 3, which is always positive, gives the excess water required in order to achieve a specific shift ratio (σ_r/σ_w) in a closed as opposed to an open system. Use of eq 2 obviously leads to overestimates of the W/R ratio that can reach orders of magnitude for large $|\sigma_r/\sigma_w|$.

The 0-dimensional box model of water-rock interaction may be particularly suited to active geothermal systems because of the spatial limitations of their "high-temperature" zones (usually near 300°C) and the possible internal recycling of fluids. Such systems can be placed into an order of progressive hydrological evolution (i.e. increasing cumulative water/rock ratio) by use of eq 3 (Blattner, 1985). Total cumulative throughput is a factor of some possible significance in the assessment of the potential production of power from an individual active system, as well as in the interpretation of fossil systems that have now become mineral deposits.

3.   1-DIMENSIONAL FLOW AND GEOCHEMICAL FRONTS

In most general geological applications, and for many geothermal systems, it must obviously be preferable to model fluid throughput in at least one dimension. Still not taking into account limitations of reaction rates, a 1-dimensional oxygen isotope exchange model has first been proposed by Spooner et al. (1976). Such a model considers a batch of fluid to pass through a porous rock column (aquifer) which

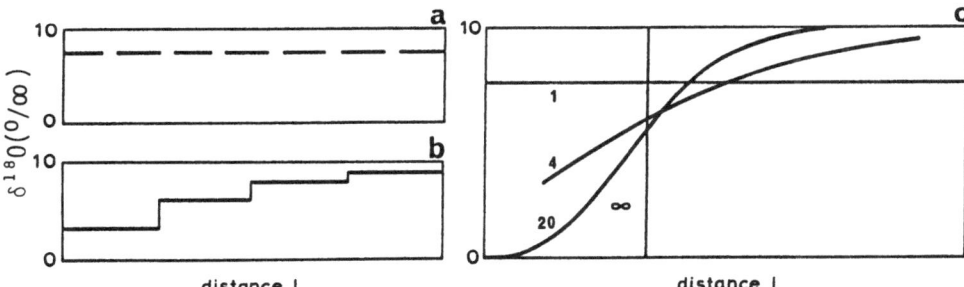

Fig. 1   Stable isotope profiles δ vs distance, e.g. 1 = 5000 m;
Φ = 0.25;
(a) no subdivision, complete longitudinal mixing $\sigma_r = 0.25$ P, $\sigma_w = 0.75$ P (closed box model)
(b) four compartments, four portions of water (n = 4)
(c) smoothed profiles for n = 1, 4, 20, ∞

268

consists of successive boxes or compartments, inside which total
mixing and equilibration occurs 'instantaneously' i.e. is achieved in
the time between the arrival and departure of each batch of water. In
spite of being similarly unrealistic in this respect as the previously
discussed 0-dimensional models, this 1-dimensional approach serves to
develop the concept of an isotope geochemical front. Simple examples
for a given length of aquifer (e.g. 5000 m) are shown in Fig. 1,
where it is assumed that $\Delta_{rw} = 0$ (this would be the case for oxygen or
hydrogen isotopes at high temperature, or for Sr isotopes at any
temperature). The porosity in terms of oxygen (Φ) is assumed to be
0.25. Fig. 1a shows the case of the closed system; in Fig. 1b the
same aquifer has been subdivided into four successive compartments
(n = 4). In Fig. 1c smoothed δ vs distance profiles are shown for 1,
4, 20 and infinitely many (n = ∞) compartments. The profiles can be
obtained analytically using

$$\delta_{c,p} = \delta_{ri} (1-\Phi)^p [1 + \sum_{k=1}^{c-1} \frac{(p + k - 1)!}{(p - 1)! \, k!} \Phi^k],$$

where c refers to the consecutively numbered rock compartments and p
to the equally numbered portions (batches) of fluid (M.R. Manning,
pers. comm.).

It is seen that with increasing subdivision of the aquifer, the
isotope shifting capacity of the water, $W\sigma_w$, is used more efficiently
by the rock, a geochemical front becoming increasingly well defined.
The perfect geochemical front for n = ∞, will lag behind the advancing
front of physical fluid in the proportion given by the effective
porosity, i.e. in the case of oxygen isotopes,

$$F_d = \frac{\Phi\rho_w\nu_w}{\Phi\rho_w\nu_w + (1-\Phi)\rho_r\nu_r} , \tag{4}$$

where ϕ is the volumetric porosity, ρ density, and ν the concentration
of oxygen by mass. F_d may be called a distance retardation factor. A
similar definition for trace elements has been given by Krauskopf
(1986), but from the point of view of time rather than distance. Fig.
2 illustrates the retardation factor for

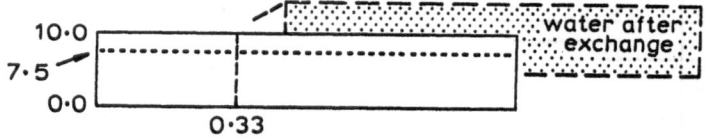

Fig. 2   Case of n = ∞, as in Fig. 1c, (i.e. with no longitudinal
mixing). Isotope ratios completely determined by fluid to as
far as 0.33 of trajectory length. This geochemical front is
preceded by column of completely exchanged (and therefore
geochemically no longer identifiable) fluid (fluid front at
distance 1.33).

oxygen isotopes and shows how an increasing column length of exchanged, i.e. geochemically completely adapted, fluid (of 'porosity' $1-\phi$) precedes the geochemical front.

4. KINETICALLY CONSTRAINED 1-DIMENSIONAL MODEL

The front shapes obtained by assuming fixed finite compartments in the previous 'instantaneous exchange' model seem intuitively more realistic than the vertical front for $n = \infty$, although the assumption is unrealistic. Similar flattened fronts may, however, be obtained by restricting the kinetic exchange rate between the minerals of a rock and fluid. McKibbin et al. (1986) developed a 1-dimensional isotope exchange model based on a finite difference method of solving transport equations. In this model a slightly simplified[2] isotope transport equation is formulated as

$$(\partial/\partial t)(\Phi'\delta_w + \delta_r) + (\Phi'/\phi)u\partial\delta_w/\partial l = 0$$

where t is time, l distance, and u the Darcy velocity, so that the interstitial velocity is $v_i = u/\phi$. The rate of exchange at a given time and distance is defined by

$$\frac{\partial \delta_r}{\partial t} = -\kappa(\delta_r - \delta_w - \Delta_{rw})_{t,l} \qquad (5)$$

A constant pressure gradient must exist to drive the flow. McKibbin et al. (1986) then specify a length L of aquifer, and express all lengths non-dimensionally as $Z = l/L$. Time is expressed non-dimensionally (in units of the fluid replacement time $t_v = L/v_i$ for the distance unit $Z = 1$), as $\tau = t/(\phi L/u)$. The unit time $\tau = 1$ thus corresponds to the time taken by a fluid particle to travel the length L of aquifer. Eq. 5 now reads

$$\frac{\partial \delta_r}{\partial \tau} = -C (\delta_r - \delta_w - \Delta_{rw})$$

and at fixed L the parameter

$$C \equiv C_L = \kappa(\phi L/u) = \kappa L/v_i \qquad (6)$$

is constant. However, we can also set the length L equal to the increasing distance reached by the physical front of fluid, so that $\tau = 1$, and L contains the whole altered δ-profile at any stage of the process; in this case let $C \equiv C_f$. In this way L is determined internally to a particular simulation, with no arbitrary or accidental input. For practical reasons, the computer program of McKibbin et al (1986) gives

****

[2]It is assumed that $^{18}O/(^{16}O + ^{18}O)$ be equivalent to $^{18}O/^{16}O$.

δ_r and δ_w profiles as a function of C_L and τ, so that for $\tau = 2$ the fluid front has advanced exactly to distance $l = 2L$ and the profiles displayed (over one L) are only the first halves of complete profiles for doubled L and $C_f = 2C_L$. It follows generally that $C_f = \tau C_L$ and the parameter C_f uniquely determines a shape of a δ-profile with distance and a degree of definition of a δ-front (Figs 3,4). The values δ_r must have been increasingly (δ_w decreasingly) affected by fluid/rock exchange, anywhere from the physical infiltration front upstream. It is seen in eq 6 that the important parameter C (hence-forth $C = C_f$) is the product of a ratio $c^* = \kappa/v_i$ on one hand, and the aquifer length L on the other, so that $C = c^*L$. It is the 'inter-active' ratio c^* which controls the inherent shape of the δ or compositional front for a given distance L of fluid progress, whereas the distance factor determines the degree of its linear 'longitudinal' magnification.

As in section 3, an arbitrary, high porosity has again been chosen for Figs 3 and 4. Fig. 3 shows both δ_r and δ_w profiles for $\tau = 1$, a plot length L and various C-values. Fig. 4 shows only the δ_r profiles. Both figures can be interpreted in two ways.
Interpretation I is for a fixed value of fluid progress L but different aquifers, so that C becomes a direct measure of the interactive para-meter $c^* = \kappa/v_i$. If v_i is also constant, C becomes a measure of the rate constant κ. Using Fig. 4 in this way it is seen how the δ_r vs distance profile becomes sharper with increasing C, i.e. now with increasing c^*. Values for κ are indicated for a given v_i and L. For ideal 1-dimensional flow with $c^* \to \infty$, the profile for $C = \infty$ in Fig. 4 shows a vertical δ-front at distance $L\Phi$, the physical front having reached exactly the end L of the plot. Progress of the vertical front for $c^* = \infty$ corresponds to the hypothetical progress of the pore fluid by a modified Darcy velocity (with $\Phi = 1$ rather than $\phi = 1$), the oxygen based porosity Φ giving the exact position of the limiting vertical line for the bunch of δ-profiles that tend to wrap around it in Fig. 4.

Obviously, large values c^* are possible not only when κ is large but also due to low interstitial velocities. Our mathematical approach (from McKibbin et al. 1986) neverthless favours low C values and due to numerical diffusion and finite exchange volumes, profiles with $C > 10$ in Fig. 4 would be less sharp than should be the case for ideal laminar flow without diffusion; the profiles for $C = \infty$ are of course sketched in directly.

Interpretation II. Moving to specific flow situations in an aquifer, with constant c^*, κ and v_i (Interpretation II of Fig. 4), the parameter C becomes a measure of the distance L traversed by the fluid

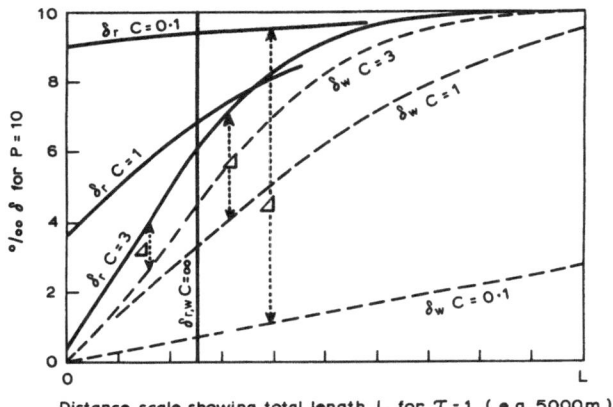

Fig. 3   Front shapes (δ-profiles) as a function of C; both δ_r and δ_w shown, $\Phi = 0.24$. This plot is obtained for an equilibrium ($\Delta^{eq} \cdot_{rw}$) value of zero. The non-equilibrium Δ values found have decreasing contents of the original contrast (here $\delta_{ri} - \delta_{wi}$) with increasing C.

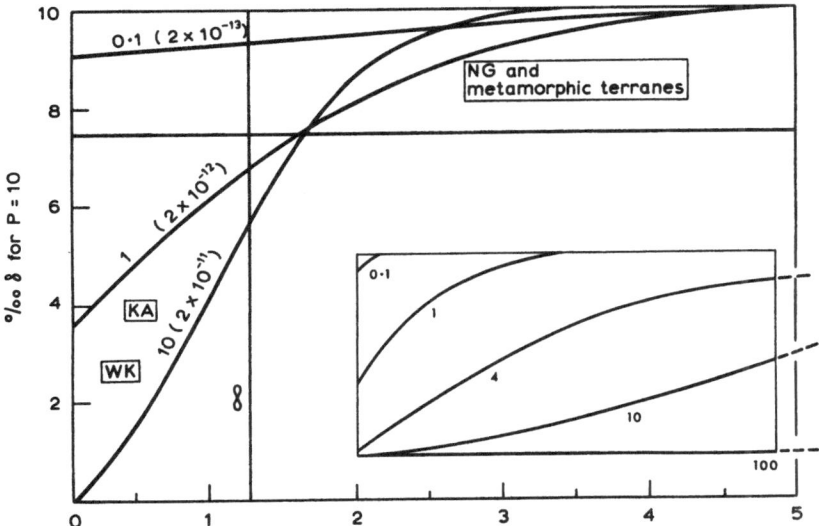

Fig. 4   Front shapes (δ_r profiles) as a function of C for arbitrary $\Phi = 0.24$. - Interpretation I : If plot length is fixed and $\tau = 1$, increasing C shows the effect of an increase in the interactive parameter $c^* = \kappa/v_i$ only; brackets show κ if $v_i = 10^{-8}$ m/s and L = 5000 m - Interpretation II as evolution in time and distance : c^* is constant and the plot (for $\tau = 1$) expands in proportion to C (Inset). WK = Wairakei KA = Kawerau, NG = Ngawha. Line at $7.5°/_{oo}$ as in Fig. 1.

in the time $\tau = 1$, the proportionality factor being c*. The successive shapes $C = f(t)$ now belong to an evolutionary sequence of which the interactive constant c* determines the pace in time and distance. In other words the δ-profiles that correspond to the successive C values become lengthened in proportion with the growth of C. Their succession in a fixed distance framework is shown in the inset of Fig. 4. A perfect vertical front is only possible with c* $\to \infty$ as any initially curved profile will become flattened with increasing L.

In summary, if a complete δ_r profile could be reconstructed in nature – and let us assume that the temperature was close to constant, and its effects allowed for, such as to comply with the condition $\Delta = 0$ for the purposes of presentation – then L, C and c* would be known. For a given porosity the profile would <u>have to be</u> one of a bundle such as shown in Fig. 4 and inset. Values for κ and v_i have been suggested for the Kawerau geothermal system at 1-2 km depth by McKibbin <u>et al.</u> (1986) as $\kappa \approx 10^{-12}s^{-1}$ and $v_i \approx 10^{-9}ms^{-1}$; in that case $C \approx 1$ for $L = 1000$ m, and $C \approx 10$ for $L = 10,000$ m, etc. For lower crustal levels c* would probably be larger, since κ must increase with temperature, and v_i could well be similar or smaller. Practical C values for $L = 10,000$ m might range upwards from 10^{-2} as a minimum at shallow depths, and not considering rocks with particularly low κ, such as quartzite.

5. TOTAL FLUID THROUGHPUT

The possibility to quantify exchange rates in this model allows, in principle, to estimate the total fluid throughput, however small the interactive parameter may be, as a function of C. For all C values at $\tau = 1$, all of the isotopic change is confined to the areas of Figs 3 or 4[3] and, provided the interaction ends sharply, e.g. by a drop in temperature, the exchange efficiency as a function of C can be expressed as $(100/P\Phi L) \int_0^L (P-\delta_r)dl$, the product $P\Phi L$ being the full shifting capacity of all the pore fluid, and the integral the sum of the actually achieved isotope shifts of all rock cross sections. Inspection of Fig. 4 shows that the efficiency drops from 100% at $C = \infty$ to less than a few percent for $C = 0.01$. A temperature corrected δ_r profile (i.e. a given C value) is therefore by itself an indicator of the excess fluid that has passed through an aquifer, namely the fluid that has <u>not</u> left a 'receipt' in the isotope shift of the rock. In this way it <u>is</u> possible to estimate real fluid through-puts where isotope mass balance can provide only minimum values. Fig. 5 shows the exchange efficiency plotted empirically against C. It has also been seen in Fig. 3, that low C implies large non-equilibrium, i.e., rate-controlled Δ_{rw} values, which in turn may offer an alternative approach to estimating C.

****

[3] In Fig. 2, by contrast, the total shifting capacity of all of the fluid originally in the pore space has been allowed to be used up, so that $\tau = 1.33$.

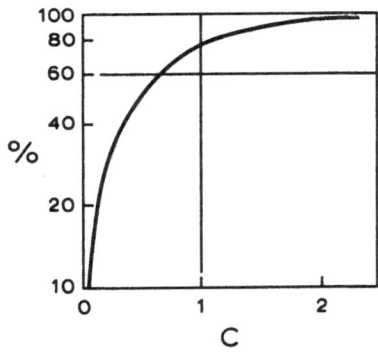

Fig. 5   Percentage of fluid
throughput, $(100/P\Phi L)\int_0^L(P-\delta_r)dl$,
accounted for in isotope record
δ_r as a function of C.

6.   CONCLUSION

The mass balance approach to crustal fluid throughput is fundamental, but often more realistic interpretations are possible with a one dimensional than with a box model. For 1-dimensional fluid paths in the lithosphere, and barring miracles, the fluid/rock ratio relative to the whole length of the aquifer must generally be that determined by the porosity. However, a successful aquifer must end in a sink for the fluid, and higher fluid/rock ratios are possible according to the capacity of the sink. Only if the sink is a surface discharge, are fluid/rock ratios $\gg 1$ possible and isotope fronts can 'break through' to the surface (geothermal systems).

In this paper we have attempted to demonstrate the power of a 1-dimensional model with kinetic exchange control. In order to do this as transparently as possible, a high porosity is selected and the model is followed rigidly, in that Φ, κ, and v_l are held constant for the length of the aquifer; temperature corrections are assumed to have been made. In terms of the present model each case of 1-dimensional fluid flow is characterised by only one value C, which determines the δ_r-profile (Fig. 4). Two typical geological situations can be distinguished: (1) only slightly exchanged rock with low and (2) strongly exchanged rock with large isotope shift $|\sigma_r/P|$. Case (1) is typical of most geological applications. The small shift does not allow to locate the position of the area concerned relative to the inversion distance ΦL, unless there is evidence for isotope equilibrium between fluid and rock; this would require $C > 1$ (Fig. 3), putting the area ahead of the 'front' at ΦL. Although more complete interaction with postulated initial water may have occurred further upstream in an aquifer, there is no evidence for this in the small isotope shift found, and the possibility of less negative input waters, of local or deeper crustal origin and isotope signature, must seem large. After all, fluids are the main agent of intracrustal equilibration. Metamorphic terranes typically belong to this rock dominated case and Ngawha is one of few geothermal systems which do. Its fluids may well be of magmatic or metamorphic origin (Blattner, 1985). Case (2) where a large isotope shift exists, is interpreted more readily. Here $C > 1$ must always hold (Fig. 4) and we are therefore located behind the inversion distance ΦL. The efficiency of the exchange and therefore the determination of fluid throughput by isotopes could be better than about 70% (Fig. 5). Most geothermal systems, if interpreted on a 1-dimensional basis, would approximate this case. Insofar as they are zones of high heat flow with a lateral low-temperature fluid input, they can be seen as beginnings of 1-dimensional aquifers that discharge the fluids soon afterwards onto the surface, the remainder of the distance L (for $\tau = 1$) having only theoretical significance. Since C for an aquifer is proportional L and therefore also measures the cumulative fluid/rock ratio up to a fixed distance such as ΦL, so perceived 1-dimensional systems could approximate the behaviour of a box model. The apparent absence of δ_r gradients with depth could be explained readily by a relatively low δ_r-gradient in real space. In geothermal systems the 1-dimensional approach could then dispose of the requirement to have mixing and recycling of fluids in a 'box-like' reservoir. It would not affect the order of evolution of systems suggested in Blattner (1985) and three typical cases from that study (Wairakei, Kawerau, and Ngawha) are shown at their estimated positions in relation to P in Fig. 4.

Deeper crustal flow paths that bring surface water into zones of high grade metamorphism, have occasionally been proposed (Wickham and Taylor 1985). Such 'aquifers' are likely to involve high temperatures, high κ, and high C values in terms of the present model, implying areas of almost totally exchanged rock ($\sigma_r/P = -1$) upstream from an inversion distance or 'modified Darcy front' and possibly justifying the use of the 'instantaneous' 1-dimensional model of eq 4 as an approximation. The area of total exchange may not be accessible in an aquifer active at the present time, but typical cases should be present in the geological record, although perhaps not in classic metamorphic terranes.

Acknowledgements

We thank Dr K. Lassey (INS) for critical reading, Dr R. McKibbin (Auckland University) and Dr M. Manning (INS) for significant advice, and Mrs A. Hepenstall (INS) for word processing.

7. REFERENCES

Blattner, P. (1985) Isotope shift data and the natural evolution of geothermal systems. Chem. Geol. 49, 187-203.

Blattner, P. and Bunting, D.G. (1980) Stable isotopes of minerals and rocks of the Ngawha Geothermal Field and Northland, and magmatic water revisited. NZ DSIR Geothermal Circ. PB/DGB-102, 24 pp.

Cathles, L.M. (1983) An analysis of the hydrothermal system responsible for massive sulfide deposition in the Hokuroku Basin of Japan. Econ. Geol. Monograph 5. 439-487.

Clayton, R.N., Muffler, L.J.P., and White, D.E. (1968) Oxygen isotope study of calcite and silicates of the River Ranch No. 1 well, Salton Sea geothermal field, California. Am. J. Sci. 266, 968-979.

Craig, H. (1963) The isotopic geochemistry of water and carbon in geothermal areas. In: E. Tongiorgi (Editor), Nuclear geology on geothermal areas; Spoleto; Cons. Naz. Ric., Lab. Geol. Nucl, Pisa, pp. 17-53.

Krauskopf, K.B. (1986) Aqueous geochemistry of radioactive waste disposal. Applied Geochemistry 1, 15-23.

McKibbin, R., Absar, A., and Blattner, P. (1986) The transport of oxygen isotopes in hydrothermal systems. Proc. 8th NZ Geothermal Workshop, pp. 29-36. Univ. of Auckland.

Spooner, E.T.C., Beckinsale, R.D., England, P.C., and Senior, A. (1976) Hydration, ^{18}O-enrichment and oxidation during ocean floor hydrothermal metamorphism of ophiolitic metabasic rocks from E. Liguria, Italy. Geochim. Cosmochim. Acta 41, 857-871.

Taylor, H.P., Jr. (1977) Water/rock interactions and the origins of H_2O in granitic batholiths. J. Geol. Soc. London 133, 509-558.

Wickham, S.M. and Taylor, H.P., Jr. (1985) Stable isotope evidence for large scale seawater infiltration in a regional metamorphic terrane; the Trois Seigneurs Massif, Pyrenees, France. Contrib. Mineral. Petrol. 91, 122-137.

Note added in press: The dimensionless number C used is this paper and in McKibbin et al. (1986) corresponds to a Damköhler-I number as used in chemical engineering. See also: Lassey and Blattner, "Kinetically controlled oxygen isotope exchange between fluid and rock in one-dimensional advective flow" (Geochimica et Cosmochimica Acta, August 1988).

THE EFFECT OF FLUID-CONTROLLED ELEMENT MOBILITY DURING METAMORPHISM ON WHOLE ROCK ISOTOPE SYSTEMS, SOME THEORETICAL ASPECTS AND POSSIBLE EXAMPLES.

D. Bridgwater, M. Rosing, L. Schiøtte,
Geological Museum (University of Copenhagen)
Øster Voldgade 5-7, DK-1350 København K., Denmark, and
H. Austrheim,
Mineralogical-Geological Museum
Sars'Gate 1, N-Oslo, Norway

ABSTRACT. The elements Rb, Sr, Sm, Nd, U and Pb are all redistributed during the movement of fluids associated with regional metamorphism. The effect of this redistribution on the isotopic systems commonly used for age determination depends on several factors. Rb, U and Pb are lost on a regional scale during prograde metamorphism to granulite facies assemblages and are regained during retrogression. Sr is redistributed on a cm scale during both prograde and retrograde metamorphism and is mobile on a scale of tens of meters during retrogression associated with shearing. As Rb is markedly more mobile than Sr, Rb/Sr ratios of individual samples can be modified without homogenisation of Sr on outcrop scale. Addition of Rb to a sample suite will displace an original Rb-Sr isochron to the right, giving rise to an anomalously low intercept $87Sr/86Sr$ ratio. Loss of Rb has the opposite effect. A high amount of Rb movement is required before isochron relationships are totally destroyed. Redistribution of Sm and Nd takes place on outcrop scale in response to the formation of partial melts and new mineral assemblages. This can lead to local resetting of the Sm-Nd whole rock isotopic system at the time of high grade metamorphism. The effect of regional metamorphism on the redistribution of the REE on a larger scale is not well constrained by bulk rock analyses. Addition of LREE is recorded in tonalitic gneisses up to tens of meters from adjacent granite sheets and pegmatites. This results in both changes in bulk REE patterns and in changes in the Nd isotopic signatures depending on the isotopic composition of the Nd introduced. Basic rocks and ironstones with low primary REE contents from the 3.8 Ga Isua supracrustal belt show variable amounts of LREE addition during late Archaean and Proterozoic metamorphism. Estimates of crustal residence times or early Archaean mantle compositions based on initial ϵ_{Nd} values from individual samples of these polymetamorphic rocks must be evaluated using field constraints. U loss during granulite facies metamorphism will arrest the radiogenic evolution of Pb. Straightforward interpretation of the resulting "palaeoisochrons" without allowing for secondary U loss leads to spuriously high ages and μ_1 values. U loss during granulite facies metamorphism is commonly accompanied by Pb loss, and in some examples by partial isotopic homogenisation of the remaining Pb. Addition of U to

277

D. Bridgwater (ed.), Fluid Movements – Element Transport and the Composition of the Deep Crust, 277–298.
© *1989 by Kluwer Academic Publishers.*

rocks with approximately isotopically homogenous Pb (for example during
retrogression from granulite facies) will yield ages close to that of U
addition. The effect of Pb addition is more complex as it depends on the
isotopic composition of the introduced Pb itself depending on the source
of the Pb. Under certain circumstances mixing of Pb from different
sources can give rise to Pb-Pb pseudoisochrons.

1. INTRODUCTION

In the last twenty years whole rock isotope studies have been highly
succesful in determining the age and origin of igneous rock suites.
Together with U-Pb age determinations on zircons they have provided
geologists with time framework of igneous and metamorphic events
extending at least as far back as 3.8 Ga. Changes in the initial ratios
of Sr and Nd from igneous suites through time have been used to
constrain theories on the timing and mechanism of crustal separation
from the mantle. The experience gained from studies of igneous rocks has
been widely applied to suites of high grade gneisses, and the results
used to offer the same constraints on their age and origin as those
obtained from igneous suites.

When whole rock isotopic methods and interpretations are applied to
gneisses, two major assumptions are made: Firstly that the samples used
are cogenetic and contemporaneous; secondly that secondary disturbance
of the isotopic systems has been on a scale which is smaller than
individual hand samples or, if this requirement is not satisfied, that
disturbance has not involved an overall loss or gain of either daughter
or parent elements from the suite as a whole. Apparently "reasonable"
results that agree with currently accepted models for the
differentiation of the Earth, and which show an internal consistency
between the results from different isotopic systems used for age
determination, are commonly used as a justification for making these
assumptions.

As field geologists who use isotopic results to control both the
age and petrogenesis of rocks which have complex post-magmatic histories
we have become critical when classical isotopic models are applied to
whole rock systems without a full appreciation of the possible effects
of secondary processes. In the following sections, the effects of
element mobility on whole rock isotope systems are discussed. Virtually
all our examples are taken from the North Atlantic Craton. Firstly we
have first hand knowledge of the field problems, and collected the
majority of the samples used in our examples (in some cases as part of
investigations of element mobility). Secondly there is excellent
geological and geochronological control (including ion-probe U/Pb
isotope data from zircons). Many of the examples cited have been
studied by a wide range of isotopic methods and show distinct episodes
of crustal formation and later reworking extending from the early
Archaean (3.8 - 3.5 Ga), the mid to late Archaean (3.3 - 2.5 Ga) and the
Proterozoic (ca. 1.8 Ga).

2. DEMONSTRATION OF ELEMENT MOBILITY DURING METAMORPHISM

There is a range of opinion about the degree to which elements become
mobile during regional metamorphism and possible mechanisms by which
transport could take place. A large part of the extensive literature on
the subject is based on whole rock chemical analyses of regional sample
sets of supposed common origin which have been affected by different
degrees of later metamorphism. While these studies are often useful
guides to what can take place, field controls are rarely perfect, and
the question of whether differences in chemistry on opposite sides of a
metamorphic facies boundary could be pre-metamorphic often remains open.
Conversely while differences in element concentrations across a
metamorphic facies boundary may point to bulk migration of a particular
element, the demonstration that the final average concentration is the
same on both sides of the boundary does not prove that the element in
question was immobile. As an extreme example: unless there are very
large changes in bulk rock composition, the total oxygen content of a
rock suite will commonly remain nearly constant when a major volume of
fluid has passed through the system. Extensive exchange of oxygen
between solid rock and fluid may, however, be demonstrated by
measuring the isotopic composition of the oxygen (see for example
Baadsgaard et al., 1986a).
Our contention is that all elements are potentially mobile during
the passage of a fluid through a rock system. The degree to which
individual elements become mobile is dependent on a large number of
factors including fluid/rock ratios, the composition of fluid involved,
the bulk composition of the rock from which they are derived and in
which they are eventually reprecipitated, the temperature and pressure
at the time of fluid movement and the mineral assemblages stable before
and during fluid movement. In the case of trace elements which do not
form independent minerals a major factor controlling their mobility
during metamorphism is the stability of suitable host minerals.

3. EFFECTS OF ELEMENT MOBILITY ON WHOLE ROCK ISOTOPIC SYSTEMS

3.1. The ageing effect on whole rock isochrons

Before discussing the effects of secondary element migration on
individual isotopic systems, it is important to outline the effect that
an increase in age has on the apparent quality of whole rock isochrons,
since when near perfect isochrons are obtained, this is commonly used as
evidence against either major primary variations or serious secondary
disturbances. As radiogenic daughter products develop in each sample
within a suite through time, any small deviations from a linear
relationship in the isochron diagram there may have been at an early
stage in the evolution of the system, become progressively less
important (Bridgwater & Collerson, 1976, 1977). As apparent from Fig.
1, when the sample suite ages, points that scatter around a hypothetical

isochron (Rb-Sr or Sm-Nd) retain their original distance from the isochron measured parallel to the vertical axis (that is to say variations in the initial 87Sr/86Sr or 143Nd/144Nd ratios remain constant). However, the distance of a particular data point measured at right angles to the reference isochron decreases as the slope of the isochron increases through time. Points that fall significantly away from an isochron at low ages will eventually move into the zone that is within analytical error from the isochron as the rock suite increases in age.

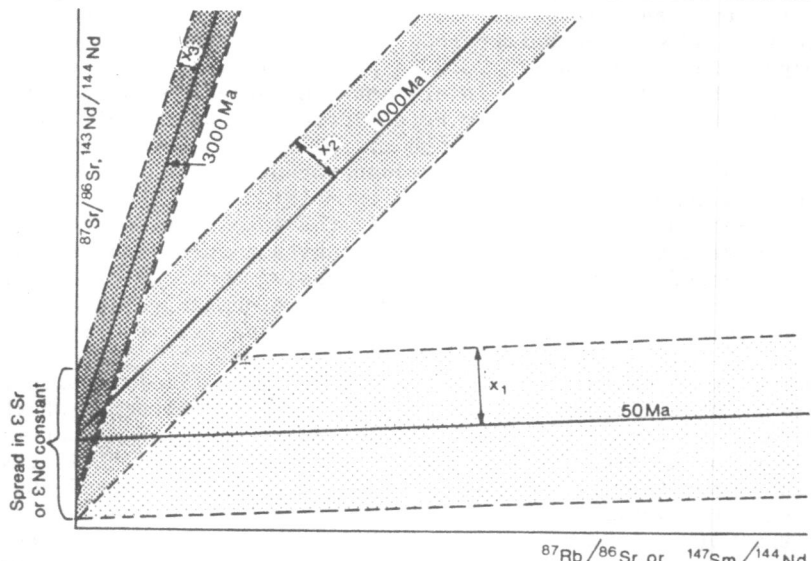

Fig 1. The change in scatter about an ideal isochron through time. As the age increases so does the slope of the isochron. Scatter measured parallel to the Y axis (ϵ_{Sr} or ϵ_{Nd}) remains constant for individual points whereas scatter measured at right angles to the isochron (X1, X2, X3) used to calculate MSWD values drops dramatically.

The quality of an isochron is measured by a standard statistical measure of the apparent scatter about the best fit line through the data points (MSWD value). The change in MSWD values through time was elegantly demonstrated in a Rb-Sr study by Vidal et al. (1984), who extrapolated the radiogenic evolution of a suite of late Tertiary granites from the Himalayas into the future: At present the sample suite shows so large a scatter that neither a reliable age nor initial ratio can be obtained. However, if the calculation is made assuming each sample has developed for a further 300 Ma, the suite yields an age of 383 Ma with an MSWD of 18. After 3000 Ma the suite yields an age of 3080 Ma with an MSWD of 0.27. Very few Archaean gneiss suites yield MSWD values close to or lower than 1. This implies either that the systems have been severely disturbed at a later stage or that the suites were initially highly inhomogeneous. In either case the average Sr and Nd initial ratios obtained must be interpreted with care.

3.2. The Rb-Sr system

Rb and Sr are geochemically very different elements and therefore liable
to fractionate during both primary igneous and metamorphic processes. Rb
concentrates in minerals such as biotite and K-feldspar that are common
in acid igneous suites and amphibolite facies gneisses. Rb is
particularly abundant in the granitic mid-and upper crust where it may
reach average concentrations of over 100 ppm; it is depleted in the
lower crust and mantle with average concentrations markedly below 10
ppm. There is now considerable evidence that this gross fractionation is
at least in part due to secondary redistribution during regional
metamorphism (see for example Bridgwater & Collerson, 1976, 1977;
Bridgwater, 1979; Wells, 1979; Griffin et al., 1980; Schiøtte et al.,
1986) although this idea did not have universal approval earlier among
geochronologists (for example Moorbath, 1977). There are still unsolved
problems related to the exact mechanism by which Rb is gained or lost
during metamorphism, but fluid transport is involved on scales of at
least tens of meters. Studies of migmatised amphibolite facies tonalitic
gneisses show Rb enrichment for many meters adjacent to granitic and
pegmatite sheets even when there is no evidence for the introduction of
a distinct younger phase as veins in the material sampled (Bridgwater &
Collerson, 1976, 1977; Nutman & Bridgwater, 1986).
   Although K is itself depleted during granulite facies metamorphism
regional changes in K/Rb have shown to be one of the easiest methods to
look for signs of secondary Rb movement. Order-of-magnitude changes in
this parameter across amphibolite/granulite facies boundaries occur.
Sr shows a much less marked variation within the crust. Regional studies
across amphibolite/granulite facies boundaries often give little
evidence for bulk changes in Sr content (but see Garde, this volume and
Fig. 2). This can in part be due to the difficulty in showing
statistically valid differences between two populations when the total
range in both groups is high and the change relatively small. It
probably also reflects the geochemical similarity of Sr to Ca and the
almost ubiquitous presence of Ca-bearing minerals in metamorphic rocks:
after break down of one assemblage of Ca-bearing minerals, Sr is
captured by Ca-bearing minerals in the new paragenesis. Both hornblende
and plagioclase feldspar which contain a major part of the Sr found in
metamorphic rocks are present over a wide range of crustal environments.
Sr is demonstrably mobile on a scale of a few centimeters. Springer et
al. (1983) and Collerson et al. (1984) showed that Sr contents and the
$87Sr/86Sr$ ratios of individual samples increased markedly in the
marginal 10 cm of a metamorphosed basic dyke cutting acid gneisses. This
was interpreted as the result of the release of radiogenic Sr during the
recrystallisation of biotite to chlorite during low grade metamorphism.
In this example at least part of the mobile Sr was incorporated in the
outer rims of hornblende. Pedersen & Bridgwater (1979) showed that
isotopic homogenisation of Sr took place in samples of Archaean gneiss
affected by Proterozoic fluid amphibolite facies metamorphism, as long
as the samples did not exceed 5 kg. Sr migration on a scale of meters or
tens of meters can be expected between lithologies with strongly

contrasting Sr contents and mineralogy such as marbles (originally Sr rich) and adjacent pelites (originally Sr poor) when the latter develop Ca-and Sr-bearing minerals during metamorphism. On a more regional scale an increase in Sr from an average of about 400 ppm to an average of 600 ppm has been noted as a result of fluid movements into gneisses adjacent to shear zones during the retrogression of Archaean granulite facies gneisses associated with Proterozoic thrust movements in the northern marginal zone of the Nagssugtoqidian mobile belt of SE Greenland (Fig.2).

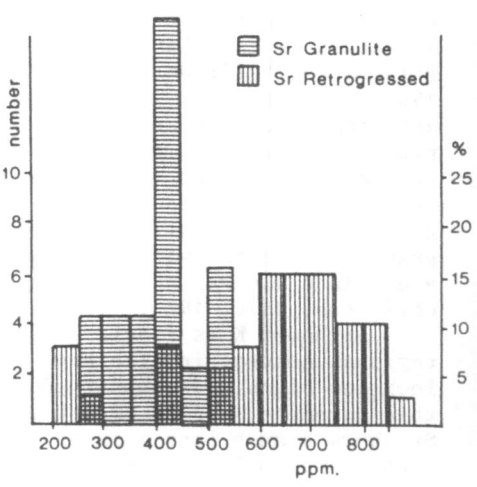

Fig. 2. Histogram showing both the general increase of total Sr and increase in spread of Sr contents during the early stages of retrogression of Archaean granulite facies gneisses from the northern margin of the Nagssugtoqidian mobile belt, SE Greenland. The least retrogressed gneisses, contain 20-30% less total Sr than their equivalents adjacent to shear zones which have higher proportions of amphibolite facies minerals and higher Cl and H_2O contents. The same transition is marked by an increase from an average of 9 ppm to 15 ppm Rb. In the center of shear zones Rb contents rise to 60 ppm, while Sr contents remain approximately constant.

3.2.1. Effects of element mobility on Rb-Sr isochrons. An average Rb loss during a secondary event from a suite of samples alone without Sr movement or re-equilibration will lower the 87Rb/86Sr ratios of individual points without changing their 87Sr/86Sr ratios at the time of Rb movement. If this takes place randomly at a measurable time after the whole rock system began to evolve, the first effect will be scatter on the left side of the original isochron. Provided the time from Rb loss to present is large compared with the time from initial formation to Rb loss, the ageing effect outlined above will tend to erase this scatter so that the overall effect will be to raise the initial ratio of the present day isochron without changing the measured age substantially. Conversely an average gain of Rb during a secondary event will raise the Rb/Sr ratio and yield a spuriously low initial ratio.

The type Uivak gneisses (amphibolite facies) and the Hebron gneisses (Uivak gneisses transitional to granulite facies) in northern Labrador have similar Rb-Sr isochron ages (ca. 3.6 Ga), but different apparent initial Sr isotopic ratios. Schiøtte et al. (1986) interpreted this as the result of late Archaean Rb gain and loss across facies boundaries (Fig. 3). Isochron relationships get erased when the element migrations are extreme (as exhibited by strongly LIL-element depleted granulites from northern Labrador in Fig. 3). The Rb/Sr results outlined in Fig. 3 from the Labrador gneisses are discussed in more detail by Schiøtte & Bridgwater (submitted). The Saglek-Hebron case must be taken as a strong warning against constraining isochrons by initial ratios estimated from models (as for example in the approach suggested by Cameron et al., 1981).

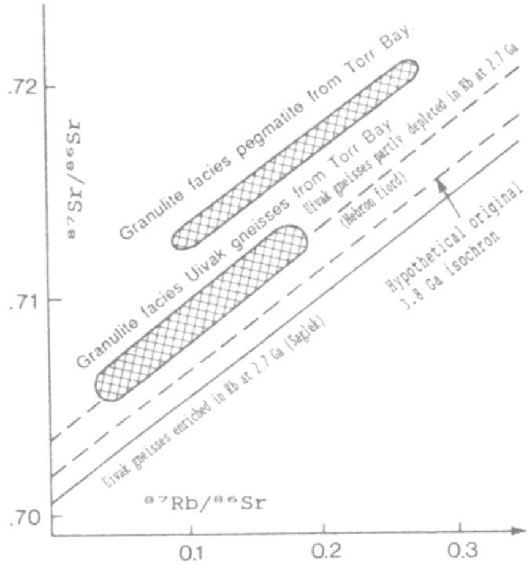

Fig. 3. Rb-Sr isotopic results from the Saglek-Hebron area to show the relation between apparent initial Sr ratios and metamorphic grade. Note that in the highly depleted Torr Bay granulite facies gneisses in which partial melting took place the hornblende-rich pegmatitic fraction has a markedly higher initial Sr ratio than the background gneiss, suggesting preferential migration of radiogenic Sr into the pegmatite.

3.3. The Sm-Nd system

Sm and Nd are closely related rare earth elements and exhibit a high degree of coherent behaviour during many geochemical processes, including igneous differentiation, metamorphism and sedimentation. They are concentrated in minerals such as sphene, allanite, apatite, zircon, garnet, hornblende and pyroxenes. Sphene, allanite and apatite show a strong preference for LREE to HREE. Garnet, and zircon, show a strong preference for HREE relative to LREE. The REE (and thus Sm and Nd) are commonly treated as immobile during metamorphism. This view has gained some support from the frequent preservation of Sm-Nd whole rock isochrons which are in agreement with U-Pb ages on zircons from the same suites of polymetamorphic rock but in which the Rb-Sr or Pb-Pb whole rock systems are disturbed. However, fluid movement during the formation of new mineral assemblages has a potentially disturbing effect on the Sm-Nd system with changes in Sm/Nd ratios and 143Nd/144Nd isotopic compositions in response to changes in fluid/mineral distribution coefficients. Any scatter developed about Sm-Nd isochrons, caused by secondary disturbances are subject to the same ageing effect as demonstrated for the Rb-Sr system, but the simplifying assumption that one of the elements is mobile whilst the other one remains immobile cannot be made for Sm-Nd. Below we discuss several cases where disturbance has taken place well above sample scale.

3.3.1. Disturbance apparent from changes in REE spectra.
a) Changes in bulk REE contents and the shapes of REE spectra can take place across metamorphic facies boundaries in response to the formation of new mineral assemblages particularly if the change is associated with the regional movements of fluids. It is, however, more difficult to generalise about the effect of metamorphism on REE spectra than for example Rb/Sr ratios across facies boundaries. Raith et al. (this

volume) have shown that in some localities the prograde formation of
charnockites with the formation of pyroxenes from hornblende can be
accompanied by enrichment in LREE. Raith et al. point out the importance
of accessory minerals in controlling the REE patterns of individual
samples; Burton & O'Nions (1988) suggest hornblende and pyroxenes as a
major factor controlling the fractionation of Sm/Nd during prograde
metamorphism. Data from SE Greenland (Glassley & Bridgwater, 1985)
suggests that the LREE content and La/Lu ratios of granulite facies
gneisses could increase during retrogression associated with fluid
introduction in agreement with petrological studies showing the
breakdown of a garnet-hornblende assemblage and the formation of biotite
gneisses. However, the changes in REE patterns observed across
metamorphic and litholigical boundaries are not always those which would
be predicted from the major element chemistry and main rock forming
minerals. We agree with Raith et al. that the presence of small amounts
of accessory minerals play at least as important a role in controlling
the REE contents of individual samples as the main mineral assemblage
both in igneous and metamorphic rocks.

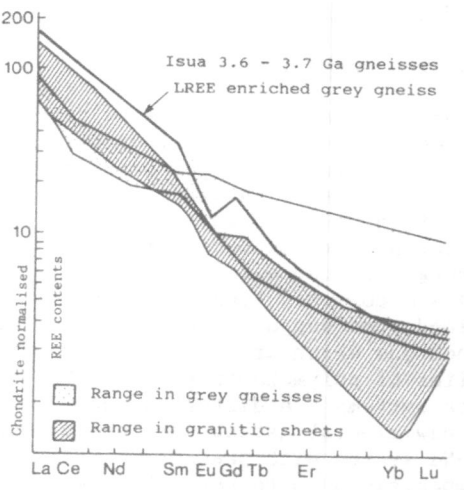

*Fig 4. REE patterns from the
3.6 - 3.7 Ga Isua gneisses,
southern West Greenland.
Metasomatised tonalitic grey
gneisses adjacent to granite
sheets show marked enrichment
in LREE.*

b) Changes in REE chemistry can be associated with the introduction of
migmatitic veins. The 3.7 Ga tonalitic and granodioritic grey gneisses
in the Isukasia area of southern West Greenland were invaded by major
swarms of granitic sheets at about 3.6 Ga. The geochemical and Sm-Nd
isotopic characteristics of the sheets are compatible with derivation
from a short-lived crustal source, for example partial melting of 3.7 Ga
tonalitic gneiss (Baadsgaard et al., 1986b). However, detailed
reexamination of Ce isotopic results by Shimizu et al. (1988) suggests
the involvement of Ce from a more LREE/HREE depleted source (or
alternatively a mineral controlled decoupling of the La-Ce and Sm-Nd
isotopic systems). The grey gneisses within several tens of meters of
the granitic sheets show a marked increase in LREE, with chondrite
normalised La/Lu ($(La/Lu)_N$) ratios increasing from an average of 10 in
the undisturbed grey gneisses to more than 50 in some examples of the

enriched grey gneisses (Fig. 5). The granitic sheets themselves commonly show lower REE contents and (La/Lu)ɴ ratios than the enriched grey gneisses adjacent to them (Nutman & Bridgwater, 1986; Shimizu et al., 1988), and the REE contents of individual samples of both the granitic sheets and their surrounding gneisses are interpreted as controlled by the preferential growth of an LREE-bearing phase in the country rocks. Єɴd(3.7 Ga) and Єcₑ(3.7 Ga) values of the LREE enriched gneisses are within the range of those from the unaltered rocks which we interpret as reflecting the short time interval (< 150 Ma) between formation and secondary disturbance.

c) The garbenschiefer from the Isua supracrustal sequence is a series of Mg and Al rich basic rocks (Gill et al., 1981) interpreted as intruded as a discordant sill into 3.81 Ga metasediments and volcanic rocks before the intrusion of the ca. 3.7 Ga Isua tonalitic gneisses (Nutman et al., 1984; Baadsgaard et al., 1984; Baadsgaard et al., 1986a; Compston et al., 1986). The garbenschiefer together with the surrounding metasediments and gneisses have been affected by several periods of later metamorphism including a regional late Archaean amphibolite facies event during which the main mineral assemblages in the supracrustal rocks were formed (Bridgwater et al., 1981; Rosing, 1983; Nutman et al., 1984). More localised upper greenschist to lower amphibolite facies assemblages developed during the Proterozoic when fluids pentetrated the supracrustal rocks as they were thrust over early Archaean gneisses. Renewed movement occurred along earlier fault planes such as that separating the two main sequences found in the supracrustal belt (Nutman et al., 1984). Considerable redistribution of trace elements in both the supracrustal rocks and those parts of the gneisses that are affected by Proterozoic recrystallisation has been demonstrated (Rosing, 1983; Baadsgaard et al., 1986a; Nutman et al., this volume). In the garbenschiefer there is a general coherence between whole rock REE spectra and major element chemistry (for example Mg number, Gill et al., 1981), with LREE/HREE depleted spectra from the most Mg rich members of the suite. We interpret this as an original igneous signature. Deviations from this pattern are seen in samples which recrystallised to chlorite schists with biotite, epidote and carbonate during Proterozoic shearing and which show high concentrations of K, Rb, Ba and Cl and high Єsᵣ(1.8 Ga) values. These samples show higher (La/Lu)ɴ and LREE contents which can be up to 10 times that seen in rocks with similar major element chemistry which were not recrystallised during the Proterozoic event (Fig.5).

3.3.2. Isotopic evidence for disturbance.
a) Sm-Nd data presented by Miller & O'Nions (1985) for layers of a sample of ironstone belonging to the Isua supracrustal suite shows evidence for considerable disturbance of the Sm-Nd isotopic system. The ironstone sample which weighed more than 20 kg was taken from a part of the supracrustal belt least affected by Proterozoic events. Data points scatter about a 2.8 Ga reference isochron. Єɴd(3.81 Ga) values range between +3.3 and -15.4, with an average of -6.4. This together with a Єɴd(3.81 Ga) value of -4.3 obtained on a second sample of the ironstone from the same group of outcrops is outside any reasonable estimate for

primary isotopic compositions of early Archaean sediments and contrasts with the ϵ_{Nd}(3.81 Ga) of +2.2 reported by Stecher et al. (1986) for the Isua ironstones (locality not published). The approximate 2.8 Ga isochron obtained by plotting the results from individual layers of one sample suggests internal redistribution of Sm and Nd on a scale of at least tens of centimeters during the formation of a late Archaean mineral assemblage dominated by Fe rich pyroxenes and amphiboles, magnetite and quartz. The markedly negative ϵ_{Nd}(3.81 Ga) values obtained by Miller & O'Nions can be explained as a late Archaean or Proterozoic interaction between the pyroxene-rich metamorphosed ironstone (with initially low total REE contents and low LREE/HREE ratios), and a LREE-enriched fluid derived from a source with a relatively low Sm/Nd ratio and long crustal history (for example the 3.7 Ga tonalitic gneisses or the adjacent more LREE rich metasediments in the sequence).

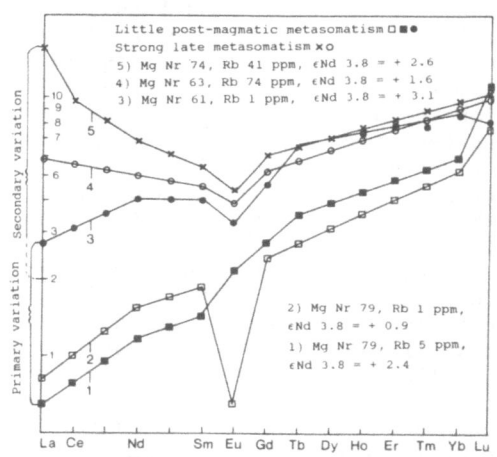

Fig. 5. REE patterns from the Isua garbenschiefer unit. Patterns 1-3 illustrate a range in REE patterns from rocks showing little effect of Proterozoic metamorphism. 1 (GGU 171756) & 2 (GGU 171758) are highly Mg rich rocks with little LIL-element enrichment and very low (La/Lu)$_N$ ratios. 3 (GGU 242744) is less Mg rich, but shows no sign of secondary LIL-element enrichment. LREE depletion is less marked than 1 & 2. 4 (GGU 171755)is similar in major element chemistry to 3 but shows high secondary enrichment in LIL-elements and a V shaped REE pattern suggesting moderate secondary LREE addition. 5 (GGU 171757) is similar in major element chemistry to 1 & 2 but shows marked secondary LIL-element enrichment. Sm and Nd concentration data by ID (Hamilton et al., 1983). Ce, Nd, Sm, Eu. Gd. Dy, Er, Yb for 3, 4 & 5 by ID (Gill et al., 1981). As there are slight differences between the abundances of Sm and Nd determined by Gill et al. (1981) and by Hamilton et al. (1983), the REE patterns by Gill et al. are adjusted so that element ratios remain the same but Sm values are those of Hamilton et al. La, Sm, Eu, Yb and Lu values in samples 1 & 2 by INA. INA Sm values are within error of those by Hamilton et al.

b) Sm-Nd determinations on the samples of the garbenschiefer studied
geochemically (collected before the complex metamorphic history of the
area was known) have yielded an age of 3.73 ± .15 Ga. Initial ϵ_{Nd} values
for individual samples vary between -0.3 and +3.0 (as the Sm-Nd age is
imprecise, the initial ϵ_{Nd} values have been recalculated at 3.81 Ga, the
U-Pb age obtained both from bulk zircon studies of acid volcanic units
in the sequence (Baadsgaard et al., 1984), and from ion-probe U-Pb
single grain studies (Compston et al., 1986)). The spread in ϵ_{Nd}(3.81
Ga) values was ascribed to secondary disturbance of the Sm-Nd system,
rather than mixing at the time of emplacement with different source
material, since parallel studies of the metasediments adjacent to the
garbenschiefer by Hamilton et al. (1983) showed no marked variation in
ϵ_{Nd}(3.81 Ga). The average ϵ_{Nd}(3.81 Ga) of ca. +2 obtained for the
garbenschiefer has been used subsequently as a fixed point on mantle
evolution curves. This is only justified if it can be safely assumed
that the secondary disturbance shown by the scatter did not
systematically displace the position of the data points to one side of
the isochron due to bulk introduction of REE from outside (in a manner
similar to the displacement of 87Sr/86Sr intercepts from Saglek-Hebron,
described above). In view of the abnormally low ϵ_{Nd}(3.81 Ga) results
obtained from the ironstones the assumption that the supracrustal rocks
are a closed system is debatable.

Petrographic and geochemical studies showed that LREE/HREE
enrichment in the garbenshiefer was correlated with the development of
new mineral assemblages and secondary LIL-element introduction during
Proterozoic shearing. 147Sm/144Nd ratios vary from 0.248 in samples with
little petrographic and geochemical evidence for secondary modification
at 1.8 Ga to 0.16 in samples with similar major element chemistry (Mg
nr) that are strongly affected by Proterozoic recrystallisation and LIL-
element addition (samples 1 & 5, Fig. 5). A change in the Sm/Nd ratio
of this magnitude at 1.8 Ga involving redistribution of REE within the
garbenshiefer unit would increase the apparent initial ϵ_{Nd} by more than
20 units. The much more modest variation in initial ϵ_{Nd} values observed
suggests interaction with less radiogenic Nd from outside. As in the
case of the ironstones, a suitable source is the ca. 3.7 Ga gneiss
over which the supracrustal rocks were thrust during the Proterozoic or
LREE enriched sediments within the succesion. We suggest that the main
reason why a Sm-Nd isochron is obtained from these highly-altered rocks
is that the LREE added was derived from a source with an age and initial
143Nd/144Nd isotopic composition close to that of the unaltered
garbenschiefer. In this case a secondary decrease in both Sm/Nd and
143Nd/144Nd ratios almost compensate for each other. The bulk effect is
a slight tilting of the isochron. Further work on the effects of
secondary movement of LREE on the isotopic composition of the
garbenschiefer currently being carried out by M. Rosing and G. Gruau
(University of Rennes) has yielded a range in ϵ_{Nd}(3.81 Ga) of 12
ϵ units.

c) In the ironstones and garbenshiefer examples cited above, one
important factor controlling the degree of apparent disturbance is the
low total REE content of the original rock. Problems of using ϵ_{Nd} values
to constrain crustal residence times of samples from polymetamorphic

terranes are reduced but not eliminated when rocks with high primary REE
contents are studied. Results reported by Jacobsen & Dymek (1988) from
the clastic metasediments within the supracrustal belt show these to
have LREE enriched REE spectra, with (La/Lu)$_N$ ratios between 5 and 20.
At the stratigraphic age controlled by the U-Pb zircon data (3.81 Ga)
they obtain ϵ_{Nd} values between 0 and +4, that is a markedly lower spread
than that obtained for the garbenschiefer and the ironstones. Jacobsen &
Dymek interpret the high positive $\epsilon_{Nd}(t)$ values obtained as evidence for
the presence of a highly depleted early Archaean mantle source and use
the spread in $\epsilon_{Nd}(t)$ values from different sedimentary units as evidence
for different sources including a component derived from a pre-3.81 Ga
sialic crust. Although they note that part of the scatter could be due
to local REE movement (controlled, for example, by the growth of garnet
in single samples) they do not consider the effects of regional REE
mobility during fluid movements. Considerable geochemical disturbance of
the samples used for the Sm-Nd work (many of which were collected close
to a major Proterozoic shear zone) is seen from the Sr model ages given
by Jacobsen & Dymek. Oxygen isotope studies from the same part of the
supracrustal belt (Perry et al, 1978; Read, 1976) show distinct local
variations within both the supracrustal rocks and the adjacent gneisses
which can be correlated with fault-controlled alteration. In view of the
evidence for REE mobility seen in samples from the supracrustal belt
specifically collected to minimise the effects of later regional
metamorphic and metasomatic events, we are sceptical about basing
calculations of either crustal residence times or degree of mantle
differentiation on the small variations in $\epsilon_{Nd}(t)$ values between
individual samples taken from highly disturbed sequences such as those
analysed by Jacobsen & Dymek.
d) The Kiyuktok gneisses, northern Labrador are migmatites consisting of
a tonalitic to granodioritic component derived from the regional 3.73 Ga
Uivak gneisses, invaded and mixed with granitiod material (see Fig.2,
Schiøtte et al., 1986), emplaced during the waning stages of late
Archaean granulite facies metamorphism between 2.7 and 2.8 Ga (Schiøtte
et al., in press). The whole rock chemistry of the granitiod sheets is
compatible with the partial melting of a crustal source. During the
emplacement of the late Archaean granitiod sheets there was considerable
geochemical interchange between the older matrix and the younger veins,
so that for example the matrix is enriched in LIL-elements compared to
less migmatised outcrops of the 3.73 Ga Uivak gneiss (compare the
effects of granitic sheeting in the Isua gneisses described above).
Collerson & McCulloch (1982) and Collerson et al. (1986) have shown that
the Kiyuktok gneisses are contaminated with varying amounts of Nd which
has an anomalously radiogenic composition (corresponding to a source
which was much less LREE enriched than the Uivak gneisses in the period
before 2.7 Ga). The granitoid sheets and the gneiss show similar spread
in ϵ_{Nd} at the time of granitoid emplacement. Growth of allanite in the
gneisses close to the granitoid sheets suggests REE exchange between the
two rock types and we interpret the introduction of the anomalously
radiogenic Nd as controlled by fluid exchange between granitoids and
gneiss.

3.4. The Pb-Pb whole rock method

The Pb-Pb method differs from other whole rock methods in that it uses a combination of two decay systems: 238U to 206Pb and 235U to 207Pb. This combination is necessary to obtain meaningful isochron ages since U is very mobile under oxidising conditions such as those which occur during weathering. Direct plots of either 207Pb/204Pb against 235U/204Pb or 206Pb/204Pb against 238U/204Pb ratios in an analogous manner to Rb-Sr isochrons are of little value because of recent U loss. Recent U-decay is dominated by the 238U-206Pb system. Since 235U has a short half-life compared to that of 238U, the percentage of 235U (and corresponding rate of production of 207Pb) has decreased steadily during the history of the Earth. Pb-Pb isotopic investigations are therefore of particular interest for Archaean rocks, since they have originated at a time when 235U decay played a significant role and gave rise to distinct 207Pb/204Pb signatures. For example in the North Atlantic Craton the early Archaean Amîtsoq gneisses from outer Godthåbsfjord are characterised by Pb with a pronounced low-207Pb signature. Owing to loss of U during a high grade metamorphic event between 3.5 and 3.6 Ga (Griffin et al., 1980), shortly after their separation from a presumed mantle-like source, the radiogenic evolution of Pb was retarded, leading to their characteristic isotope signature (Black et al., 1971; Moorbath, 1977; Taylor et al., 1980). Low-207Pb has shown to be an excellent tracer of early Archaean rocks in SW Greenland: Taylor et al. (1980) demonstrated that late Archaean gneisses in the Godthåbsfjord – Sermilik area contain varying amounts of low-207Pb derived from Amîtsoq gneisses present at depth. The mechanism by which this Pb has been introduced is discussed below. It should be noted that while the presence of Pb with a low-207 isotopic signature can be used to demonstrate the presence of a component derived from an older U depleted crust, absence of Pb with the characteristic low-207Pb signature does not preclude the involvement of an older crustal source which was less U depleted.

3.4.1. Effects of element mobility on Pb-Pb isochons. The mobility of U during metamorphism is well established (see for example Heier, 1973). Typical granulite facies acid gneisses from the North Atlantic craton contain less than 0.1 ppm U, a major part of which is held in stable accessory minerals such as zircon. Typical amphibolite facies gneisses contain between 0.5 and 2 ppm U, with occasional U rich mineralized pegmatites. Although the differences can be partly obliterated due to recent U loss (which occurs preferentially from amphibolite facies rocks), the differences are still apparent from methods such as airborn radiometric surveys (Secher, 1977). While recent U loss during weathering has no effect on the applicability of the Pb-Pb whole rock dating method, secondary loss of U (without major movement of Pb) at an intermediate stage in the evolution of a suite of rocks leads to so called palaeoisochrons (Jacobsen & Wasserburg, 1978): As the radiogenic evolution of Pb is arrested (or severely slowed down) at a time t>>0 Ma, data points in the 207Pb/204Pb vs. 206Pb/204Pb diagram remain close to the position they had reached at the time of U loss. The data points

will still fall along an apparent "isochron" but with a slope that is
higher than for the isochron for a corresponding undisturbed system.
Interpretation of a "palaeoisochron" without taking the U loss event
into consideration will lead to a spuriously high age and μ_1 value.
Elevated apparent μ_1 values, outside the range normally found in
undisturbed rock suites of the type studied, are a useful first guide to
the possibility that there has been secondary U loss. The most commonly
quoted example of a paleoisochron is from the Lofoten islands where ca.
2.7 Ga amphibolite facies gneisses were strongly depleted in U during
granulite facies metamorphism associated with the emplacement of deep
crustal intrusions during the Proterozoic (Jacobsen & Wasserburg, 1978;
Griffin et al., 1978). Pb isotope data for the early Archaean Uivak
gneisses at Hebron (Northern Labrador) which were affected by late
Archaean high grade metamorphism have also been cited as an example of a
paleoisochron (Moorbath et al., 1986) and are used to illustrate the
concept in Fig. 7. However, in this area interpretation of Pb-Pb data is
complicated by the movement of Pb and the results are discussed in more
detail at the end of this section. In general primary ages and ages of
secondary U depletion events cannot be accurately constrained from
palaeoisochrons, either because small amounts of U will remain in the
rock in mineral phases such as zircon, or because there has been some
later introduction of U during retrogression. In either case small
amounts of radiogenic Pb are formed after the depletion event leading to
transposition (and sometimes also slight rotation) of the paleoisochron
to the right.

Most regional collections of granulite facies gneisses used for
isotopic studies contain samples which have been partly retrogressed.
These show variable degrees of U enrichment compared to the least
retrogressed granulites. The degree to which U is introduced is likely
to be controlled by the oxidation state of fluids: U will remain in
solution in oxidising fluids, but will be deposited if reduced from the
U^{6+} state. This may explain why addition of U is sometimes poorly
correlated with addition of other LIL-elements during retrogression from
granulite facies and in other situations where fluid controlled movement
is important. Absence of evidence for U addition to rocks that have been
secondarily enriched in other LIL-elements does not necessarily mean
that the LIL-element bearing fluids were U poor.

The mobility of Pb during metamorphic processes has been considered
negligible by many of the isotope geologists working with the Pb-Pb
whole rock system (for example Moorbath & Taylor, 1985, 1986) and the
possible effects on Pb-Pb isochrons are therefore rarely considered. Pb
has, however, been shown to be scavenged by fluids circulating in
country rocks when these are intruded by plutonic bodies. The Pb from
the country rocks is mixed with the juvenile Pb within the plutons
during the last stages of igneous activity (Moorbath & Welke, 1969).
This model has been applied to explain the presence of variable amounts
of early Archaean isotopically `retarded' Pb in late Archaean gneisses
from the Godthåbsfjord area when these are found interleaved with older
crust (Taylor et al., 1980). There is no reason why fluids circulating
around igneous intrusions should be different from fluids associated
with metamorphic processes, and several cases of Pb removal in granulite

facies together with Pb introduction during retrogression from granulite
facies have been recorded (Schiøtte et al., 1986; Bridgwater,this
volume; Nutman & Friend, this volume; Garde, this volume; Schiøtte, this
volume). In the examples from SW Greenland it is suggested that Pb with
the characteristic low-207Pb signature used as a tracer for the presence
of an early Archaean crust by Taylor et al. (1980) in the late Archaean
gneisses (1980) was at least locally introduced by fluids during
post-magmatic fluid movements particularly during the retrogression from
granulite facies.

Contrasting Pb concentrations in rocks affected by different grades
of metamorphism are illustrated in Fig. 6. While it is difficult to
demonstrate conclusively that regional differences in total Pb contents
are due to Pb loss during prograde metamorphism (Fig. 6a) there is no
problem to show that total Pb increases during retrogression,both on an
outcrop scale across shear zones and regionally (Fig. 6b). Itis
impossible to generalise about the effect of this process on Pb-Pb
whole-rock isochrons as it depends on the source of the introduced Pb.
During depletion associated with granulite facies metamorphism some
degree of isotopic homogenisation of Pb can take place at least on
outcrop scale. This process may be of particular importance when partial
melting takes place (see for example the cluster of data points from an
outcrop of granulite facies gneisses, from Torr Bay, Labrador, Fig. 7).

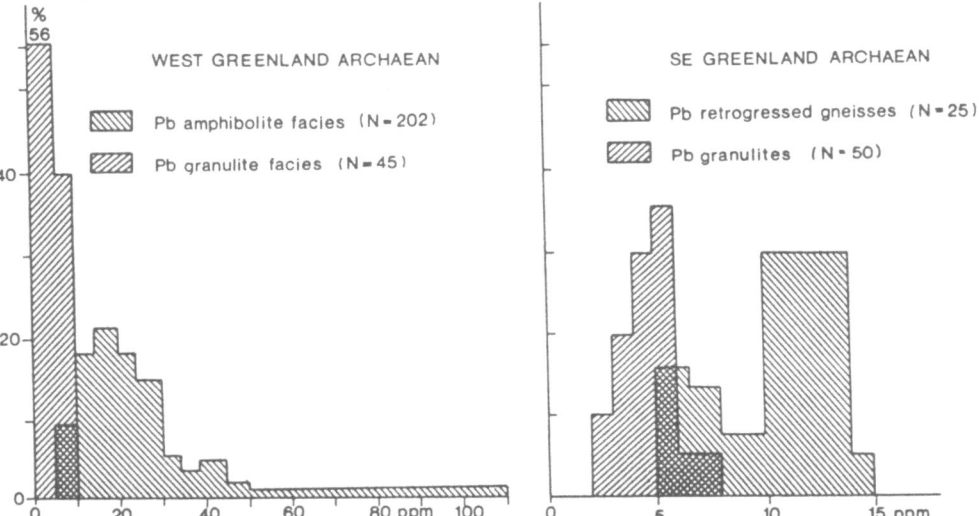

*Fig. 6 (a) Total Pb contents of amphibolite and granulite facies
gneisses from western Greenland. (b) Bulk Pb contents of granulites and
retrogressed granulites, northern margin of the Nagssugtoqidian mobile
mobile belt, SE Greenland.*

Addition of U (without Pb) to such outcrops during retrogression will
lead to Pb-Pb isochrons that date the retrogression event. Addition of
both U and Pb is more complex. In a discussion of currently available
isotope data from the Early Archean Uivak gneisses in northern

292

Labrador, Schiøtte (this volume) advocates cases where mixing of early
Archaean low-207Pb with foreign more radiogenic Pb and U during
retrogression from granulite facies has lead to chronologically
meaningless Pb-Pb pseudoisochrons.

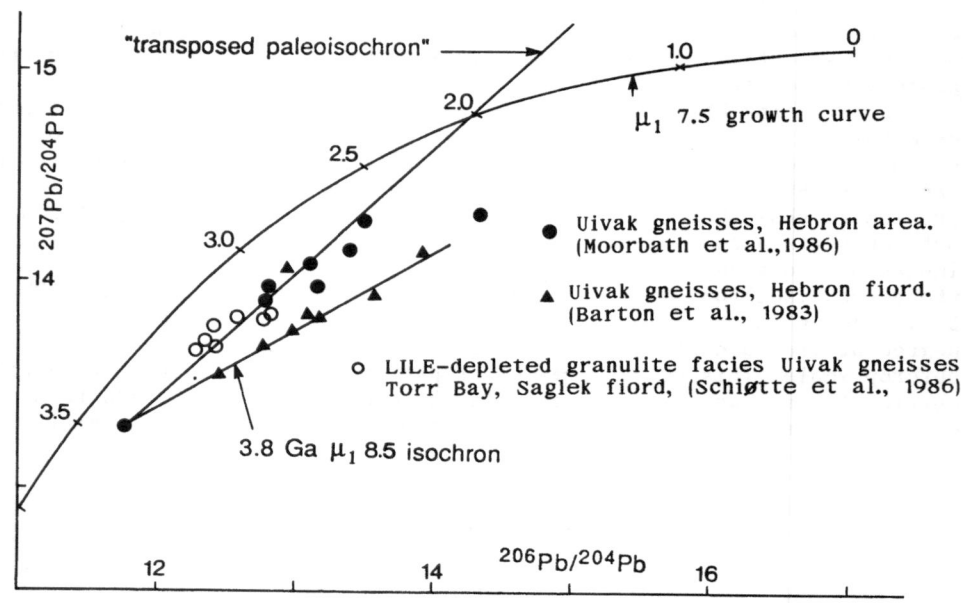

*Fig 7. Pb-Pb diagram for early Archaean Uivak gneisses from northern
Labrador to illustrate the palaeoisochron concept and the effect of
addition of Pb from a foreign source during late Archaean migmatisation.
Assumed ages for the main tonalitic component of the gneisses (3.73 Ga)
and the migmatisation event (2.74 Ga) are ion-probe and bulk zircon U-Pb
ages from Schiøtte et al. (in press) and Schiøtte (1988 unpublished).
Data for a monolith 7.5 km SW of Hebron village analysed by Barton et
al. (1983) constitute an imprecise 3.8 Ga isochron with a μ1 of 8.5
(this rather high μ1 value is reduced to values around 7.5 - 7.8 if the
3.8 Ga age is replaced by the zircon ages). One point from a highly
migmatised outcrop also analysed by Barton et al. (1983), at Hebron
village fall on the left side of the isochron. 7 out of 10 data points
for the collection analysed by Moorbath et al. (1986) fall on what was
originally interpreted as a transposed palaeoisochron with t1 at 3.6 Ga
and t2 (U depletion) at 2.5 Ga (straightforward interpretation of this
line without consideration of the secondary U loss leads to an age of
4.5 Ga with an absurdly high μ1). As outlined in the text we suggest
that the palaeoisochron interpretation is an oversimplification and that
the gneisses had a substantially more complex history, involving not
only U loss, but also Pb addition. The proposed palaeoisochron falls
on the right side of the majority of data points for a collection of*

highly LIL-element depleted granulite facies Uivak gneisses from the
Torr Bay area 40 km north of Hebron showing that the radiogenic
evolution of Pb was retarded rather than halted in the palaeoisochron
collection.

Addition of foreign more radiogenic Pb from fluids as suggested by
Schiøtte et al. (1986) and Schiøtte (this volume) may also be a more
geologically satisfactory interpretation of Pb data for some of the
samples belonging to the collection of Uivak gneisses from the Hebron
area analysed by Moorbath et al. (1986). Examination of the sample sites
(using the field notes made by A.B. Ryan who collected the original
samples) shows that they range from slightly retrogressed granulite
facies gneisses with very little evidence of post-granulite facies
migmatisation, to rocks which have been broken up and veined by syn-to
post granulite facies granitoid veins. The least radiogenic Pb recorded
by Moorbath et al. (1986) is from a locality north of Hebron village
which was affected by late Archaean granulite facies metamorphism but
not strongly migmatised. The corresponding data point falls on a Pb-Pb
isochron previously published by Barton et al. (1983) from a single
monolith of gneisses from an outcrop which shows well preserved early
Archaean structures and which is regarded by us as little affected by
element redistribution after ca. 3.6 Ga (apart from the slight Rb loss
during the late Archaean granulite facies metamorphism discussed by
Schiøtte et al. (1986) and earlier in the present paper). The other
samples from the Moorbath et al. collection fall on the left side of
Barton's isochron (Fig. 7). Those samples that yield the highest
$207Pb/204Pb$ ratios and which define the upper end of the "transposed
paleoisochron" are characteristically from retrogressed outcrops in
which the early Archaean gneisses are highly migmatised by late Archaean
granitoid sheets. For those outcrops addition of radiogenic Pb is
probably a more important factor than U loss. The situation of data
points from the "palaeoisochron collection" on the right side of data
points from more strongly LIL-element depleted (K/Rb > 450) dry
granulite facies Uivak gneisses from the Torr Bay area 40 km north of
Hebron shows that the radiogenic evolution of Pb in the "palaeoisochron
collection" was slowed down rather than halted. This also implies that
2.5 Ga (Fig. 7) is an underestimate for the age of the late Archaean
reworking event and that the 2.7 - 2.8 Ga ages suggested by zircons are
more likely.
    The Pb-Pb whole rock isochron dating method has been successfully
applied to late Archaean orthogneisses in SW Greenland (Taylor et al.,
1980) and has given ages that are broadly consistent with zircon ages.
However, Pb-Pb ages from high-grade terranes are typically rather
imprecise. Disturbances such as those outlined above would not
necessarily show up if the metamorphic events causing loss and gain of U
and Pb were not widely separated in age from the original igneous
emplacement of the gneiss precursors. In polymetamorphic gneiss terranes
in which the main components have short crustal residence times at the
time of metamorphism, Pb-Pb whole rock isochrons can be usefull as broad
guides to the age at which these separated from a mantle-like source,
but the ages obtained cannot be strictly interpreted in terms of well

defined igneous or metamorphic events.

4. CONCLUSIONS

We have suggested in this paper that isotopic whole rock systems can be
disturbed to varying degrees as a result of fluid transport during
metamorphism. It is not our aim to show that isotopic work in high grade
rocks is in any way suspect. On the contrary by integrating isotopic
investigations together with field and petrological studies a great deal
of information can be gained. However, caution is required when data are
used to justify theories of a more general nature (such as the rate of
crustal formation through time) without reference to the geological
limitations imposed by uncertainties in sampling and later alteration.
In many cases the net effect of a series of disturbances is a slight
rotation or displacement of isochrons which can modify but not
necessarilly destroy evidence of the original age and origin of a rock
suite. While a combination of as many different isotopic methods as
possible is recommended, the three systems in common use in whole rock
isotopic studies react in different ways to secondary disturbance and an
apparent lack of disturbance of one or more systems does not always
indicate that primary values are obtained. For the Rb-Sr and Sm-Nd
systems we warn strongly against simple interpretations of the average
initial isotope ratios in complex rock systems particularly when dealing
with Archaean rocks in which the ageing effect can mask real differences
in age, origin and metamorphic history. In the Pb-Pb system we agree
with earlier writers that Pb isotopes provide a powerfull tool to
demonstrate the involvement of older crustal material in younger rocks
but question whether mixing is always due to igneous processes. We note
several ways in which chronologically meaningless linear relationships
can develop during secondary events.

ACKNOWLEGEMENTS.
This paper summarises more than 15 years work with isotopic results from
rocks affected by high grade metamorphism, particularly in Greenland and
Labrador. We thank the Geological Surveys of Greenland and Canada and
the Department of Mines, Newfoundland for support in the field; the
National Research Councils of Denmark and Norway together with the
Carlsberg Foundation (Denmark) for financial support of both field and
laboratory work. NATO research grants are acknowledged for financing
field meetings in Greenland, Scotland and Norway and for extensive
funding for the research workshop at which this work was presented.
Individuals who have influenced our approach or helped us with
information include H. Baadsgaard, A.B. Ryan, P. Vidal., A.P. Nutman,
K.D. Collerson, C. Frost, R.C.O. Gill, G. Gruau, F. Kalsbeek and P.N.
Taylor. Acknowledgement of their help does not imply their agreement
with all the conclusions we make. J. Winter and T. Andersen read
earlier drafts and suggested considerable improvements.

REFERENCES

Baadsgaard, H., Nutman, A.P., Bridgwater, D., Rosing, M., McGregor, V.
R. & Allaart, J.H. (1984) The zircon geochronology of the Akilia
association and Isua supracrustal belt, West Greenland.
Earth Planet. Sci. Lett., 68, 221-228.
Baadsgaard, H., Nutman, A.P., Rosing, M., Bridgwater, D. & Longstaffe,
F.J. (1986a) Alteration and metamorphism of Amitsoq gneisses from
the Isukasia area, West Greenland: Recommendations for isotope
studies in the early crust.
Geochim. Cosmochim. Acta, 50, 2165-2172.
Baadsgaard, H., Nutman, A.P. & Bridgwater, D. (1986b) Geochronology
and isotopic variation of the early Archaean Amitsoq gneisses of
the Isukasia area, southern West Greenland.
Geochim. Cosmochim. Acta, 50, 2173-2183.
Barton, J.M.,Jr., Ryan, B. & Fripp, R.E.P. (1983) Rb-Sr and U-Th-Pb
isotopic studies of the Sand River gneisses, Central zone, Limpopo
mobile belt. *Spec. Publ. geol. Soc. South Africa*, 8, 9-18.
Black, L.P., Gale, N.H., Moorbath, S., Pankhurst, R.J. & McGregor,
V.R. (1971) Isotopic dating of very early Precambrian amphibolite
facies gneisses from the Godthaab district, West Greenland.
Earth Planet. Sci. Lett., 12, 245-259.
Bridgwater, D. (1979) Chemical and isotopic redistribution in zones of
ductile deformation in a deeply eroded mobile belt.
U.S. Geological Survey. Open file report. 79-1239: 505-512.
Bridgwater, D. (1989) Changes in the isotopic composition of
whole-rock Pb during different stages of retrogression of Late
Archaean granulite facies gneisses from Kangimut sammisoq, southern
West Greenland. In: Bridgwater, D. (ed.) *Fluid movements, element
transport, and the composition of the deep crust. Reidel,
Dordrecht,* (this volume).
Bridgwater, D. & Collerson, K.D. (1976) The major petrological and
geochemical characters of the 3600 m.y. Uivak gneiss from Labrador.
Contrib. Mineral. Petrol., 54, 43-59.
Bridgwater, D. & Collerson, K.D. (1977) On the origin of Early Archaean
Gneiss: A Reply. *Contrib. Mineral. Petrol.*, 62, 179-191.
Bridgwater, D., Allaart, J.H., Schopf, J.W., Klein, C., Walter, M.R.,
Barghoorn, E.S., Strother, P., Knoll, A.H. & Gorman, B.E. (1981)
Microfossil-like objects from the Archaean of Greenland: A
cautionary note. *Nature*, 289, 51-53.
Burton, K.W. & O'Nions, R.K. (1988) Isotope systematics and chronology
of granulite genesis in Sri Lanka. Abstract, *Chem. Geol.*, 70, EUG
Paris Meeting, 1988.
Cameron, M., Collerson, K.D., Compston, W. & Morton, R. (1981) The
statistical analysis and interpretation of imperfectly fitted Rb-Sr
isochrons from polymetamorphic terrains.
Geochim. Cosmochim. Acta, 45, 1087-1097.
Collerson, K.D. & McCulloch, M.T. (1982) The origin and evolution of
Archaean crust as inferred from Nd, Sr and Pb isotopic studies in
Labrador. *5th Intern. Conf. Geochronology, Cosmochronology and
Isotope Geology, Nikko, Japan*, extended Abstracts, 61-62.

Collerson, K.D., McCulloch, M.T. & Bridgwater, D. (1984) Nd and Sr
    isotopic crustal contamination patterns in an Archaean meta-basic
    dyke from northern Labrador. *Geochim. Cosmochim. Acta*, 48, 71-83.
Collerson, K.D., McCulloch, M.T., Bridgwater, D., McGregor, V.R. &
    Nutman, A.P. (1986) Strontium and Neodymium isotopic variations in
    early Archaean gneisses affected by middle to late Archaean
    high-grade metamorphic processes: West Greenland and Labrador. In:
    Ashwal, L.D. (ed.) *Workshop on Early Crustal Genesis: The World's
    Oldest Rocks*, 30-36. *LPI Tech. Report 86-04*. Lunar and Planetary
    Institute, Houston.
Compston, W., Kinny, P.D., Williams, I.S. & Foster, J.J. (1986) The
    age and Pb loss behaviour of zircons from the Isua supracrustal
    belt as determined by ion microprobe.
    Earth Planet. Sci. Lett., 80, 71-81.
Garde, A.A. (1989) Retrogression and fluid movement across a
    granulite-amphibolite facies boundary in middle Archaean Nuk
    gneisses, Fiskefjord, southern West Greenland. In: Bridgwater, D.
    (ed.) *Fluid movements, element transport, and the composition of
    the deep crust. Reidel, Dordrecht, (this volume).*
Gill, R.C.O., Bridgwater, D. & Allaart, J.H. (1981) The geochemistry
    of the earliest known basic metavolcanic rocks, at Isua, West
    Greenland: a preliminary investigation.
    Spec. Publ. geol. Soc. Australia, 7, 313-325.
Glassley, W.E. & Bridgwater, D. (1985) Fluid enhanced mass transport
    in deep crust and its influence on element abundances and isotope
    systems. In: Tobi, A.C. & Touret, J.L.R. (eds.) *The deep
    Proterozoic crust in the Northern Atlantic Provinces, Nato ASI
    series C.*, *158*, 105-117. *Reidel, Dordrecht.*
Griffin, W.L., Taylor, P.N., Hakkinen, J.W., Heier, K.S., Iden, I.K.,
    Krogh, E.J., Malm, O., Olsen, K.I., Ormaasen, D.E. & Tveten, E.
    (1978) Archaean and Proterozoic crustal evolution in
    Lofoten-Vesterålen, North Norway. *J. geol. Soc Lond.*, 135, 629-647.
Griffin, W.L., McGregor, V.R., Nutman, A.P., Taylor, P.N. &
    Bridgwater, D. (1980) Early Archaean granulite-facies metamorphism
    south of Ameralik, West Greenland.
    Earth Planet. Sci. Lett., 50, 59-74.
Hamilton, P.J., O'Nions, R.K., Bridgwater, D. & Nutman, A.P. (1983)
    Sm-Nd studies of Archaean metasediments and metavolcanics from West
    Greenland and their implications for the Earth's early history,
    Earth Planet. Sci. Lett., 62, 263-272.
Heier, K.S. (1973) Geochemistry of granulite facies rocks and problems
    of their origin. *Phil. Trans R. Soc. Lond.* A. 273, 429-442.
Jacobsen, S.B. & Wasserburg, G.J. (1978) Interpretation of Nd, Sr and
    Pb isotope data from Archaean migmatites in Lofoten-Vesterålen,
    Norway. *Earth Planet. Sci. Lett.*, 41, 245-253.
Jacobsen, S.B. & Dymek, R.F. (1988) Nd and Sr isotope systematics of
    clastic metasediments from Isua, West Greenland: Identification of
    Pre-3.8 Ga differentiated crustal components.
    J. Geophys. Res., 93, B1 338-354.
Miller, R.G. & O'Nions, R.K. (1985) Source of Precambrian chemical and
    clastic sediments. *Nature*, 314, 325-330.

Moorbath, S. (1977) Ages, isotopes and evolution of Precambrian
    continental crust. *Chem. Geol.*, 20, 151-187.
Moorbath, S. & Taylor, P.N. (1985) Precambrian geochronology and the
    geological record. In: Snelling, N. J. (ed.) The chronology of the
    geological record. *Mem. geol. Soc. Lond.*, 10, 10-28.
Moorbath, S. & Taylor, P.N. (1986) Geochronology and related isotope
    geochemistry of high-grade metamorphic rocks from the lower
    continental crust. In: Dawson, J.B., Carswell, D.A., Hall, J. &
    Wedepohl (eds.), The Nature of the lower continental crust.
    Spec. Publ. geol. Soc. Lond., 24, 21-220.
Moorbath, S., Taylor, P.N. & Jones, N.W. (1986) Dating the oldest
    terrestrial rocks - fact and fiction. *Chem. Geol.*, 57, 63-86.
Moorbath, S. & Welke, H. (1969) Lead isotope studies on igneous rocks
    from the island of Skye, northwest Scotland.
    Earth Planet. Sci. Lett., 5, 217-230.
Nutman, A.P. & Bridgwater, D. (1986) Early Archaean Amîtsoq tonalites
    and granites of the Isukasia area, southern West Greenland:
    development of the oldest-known sial.
    Contrib. Mineral. Petrol., 94, 137-148.
Nutman, A.P., Allaart, J.H., Bridgwater, D., Dimroth, E. & Rosing, M.
    (1984) Stratigraphic and geochemical evidence for the depositional
    environment of the early Archaean Isua supracrustal belt, southern
    West Greenland. *Precambrian Res.*, 25, 365-396.
Nutman, A.P. & Friend, C.R.L. (1989) Reappraisal of crustal
    evolution at Kangimut sanmmisoq, Ameralik fjord, southern West
    Greenland: Fluid movement and interpretation of Pb/Pb isotopic
    data. In: Bridgwater, D. (ed.) *Fluid movements, element transport,
    and the composition of the deep crust. Reidel, Dordrecht*, (this
    volume).
Nutman, A.P., Rivers, T. Longstaffe, F. & Park, J.F.W. (1989) The Ataneq
    fault and mid-Proterzoic retrograde metamorphism of Early Archaean
    tonalites of the Isukasia area, southern West Greenland: Reactions,
    fluid compositions and implications fro regional studies. In:
    Bridgwater, D. (ed.) *Fluid movements, element transport, and the
    composition of the deep crust. Reidel, Dordrecht*, (this volume).
Pedersen, S. & Bridgwater, D. (1979) Isotopic re-equilibration of Rb-Sr
    whole rock systems during reworking of Archaean gneisses in the
    Nagssugtoqidian mobile belt, East Greenland.
    Rapp. Grønlands geol. Unders., 89, 133-146.
Perry, E.C.,Jr., Ahmad, S.N. & Swulius, T.M. (1978) The oxygen
    isotope composition of 3,800 m.y. old metamorphosed chert and iron
    formation from Isukasia, West Greenland. *J. Geol.*, 86, 223-239.
Raith, M., Hoernes, S., Klatt, E. & Stähle, H.J. (1989) Contrasting
    mechanisms of charnockite formation in the amphibolite to granulite
    transition zones of southern India. In: Bridgwater, D., (ed.) *Fluid
    movements, element transport, and the composition of the deep
    crust. Reidel, Dordrecht*, (this volume).
Read, D.L. (1976) Oxygen isotope composition of the 3,800 m.y. old
    Isua gneiss of southwest Greenland.
    Unpublished M.S. thesis, Northern Illinois Univ.
Rosing, M. (1983) A metamorphic and isotopic study of the Isua

supracrustals, West Greenland.
Unpublished cand. scient. thesis, Copenhagen Univ.

Secher, K. (1977) Airborne radiometric survey between 63° and 66°N, southern West Greenland. *Rapp. Grønlands geol. Unders.*, 85, 49-50.

Schiøtte, L. (1988) En undersøgelse af metamorfe processers betydning for tidlig-Archaeiske bjergarters geokemi og isotop-geologi i nordlige Labrador, Canada.
Unpublished lic. scient. thesis, Copenhagen Univ.

Schiøtte, L. (1989) On the possible role of fluid transport in the distribution of U and Pb in an Archaean gneiss complex. In: Bridgwater, D. (ed.) *Fluid movements, element transport, and the composition of the deep crust. Reidel, Dordrecht,* (this volume).

Schiøtte, L. & Bridgwater, D. (submitted) Multi stage late Archaean granulite facies metamorphism in northern Labrador, Canada. To appear in NATO ASI Series volume covering work presented at the GRANULITES 88 congress in Clermont-Ferrand 1988.

Schiøtte, L., Bridgwater, D., Collerson, K.D., Nutman, A.P. & Ryan, A. B. (1986) Chemical and isotopic effects of late Archaean high-grade metamorphism and granite injection on early Archaean gneisses, Saglek-Hebron, northern Labrador. In: Dawson, J.B., Carswell, D.A., Hall, J. & Wedepohl, K.H. (eds.) The nature of the lower continental crust. *Spec. Publ. geol. Soc. Lond.*, 24, 261-273.

Schiøtte, L., Compston, W. & Bridgwater, D. (in press) Ion probe U-Th-Pb zircon dating of polymetamorphic orthogneisses from northern Labrador, Canada. *Can. J. Earth Sci.*

Shimizu, H., Nakai, S., Tasaki, S., Masuda, A., Bridgwater, D., Nutman, A.P. & Baadsgaard, H. (1988) Geochemistry of Ce and Nd isotopes and REE abundances in the Amîtsoq gneisses, West Greenland. *Earth Planet. Sci. Lett.*, 91, 159-169.

Springer, N., Pedersen, S., Bridgwater, D. & Glassley, W.E. (1983) One dimensional diffusion of volatile elements across the margin of a metamorphosed Archaean basic dyke from Saglek, Labrador. *Contr. Miner. Petrol.*, 82, 26-33.

Stecher, O, Carlson, R.W., Shirey, S.B., Bridgwater, D. & Nielsen, T. (1986) Nd-isotope evidence for the evolution of metavolcanic rocks from the Archaean block of Greenland and Labrador. *Terra Cognita*, 6, 236.

Taylor, P.N., Moorbath, S., Goodwin, R. & Petrykowski, A.C. (1980) Crustal contamination as an indicator of the extent of early Archaean continental crust: Pb isotopic evidence from the late Archaean gneisses of West Greenland. *Geochim. Cosmochim. Acta*, 44, 1437-1453.

Vidal, Ph., Bernard-Griffiths, J., Cocherie, A., Le Fort, P., Peucat, J.J. & Sheppard, S.M.F. (1984) Geochemical comparison between Himalayan and Hercynian leucogranites. *Physics of the Earth and Planetary Interiors*, 35, 179-190.

Wells, P.R.A. (1979) Chemical and thermal evolution of Archaean sialic crust, southern West Greenland. *J. Petrol.*, 20, 187-226.

ON THE POSSIBLE ROLE OF FLUID TRANSPORT IN THE DISTRIBUTION OF U AND Pb IN AN ARCHAEAN GNEISS COMPLEX

Lasse Schiøtte
Geological Museum (University of Copenhagen)
Østervoldgade 5-7, 1350 København K.
Denmark

ABSTRACT. The early Archaean gneiss complex of northern Labrador, Canada, was extensively reworked during the late Archaean, and correlations between whole rock isotopic signatures and late Archaean metamorphic grade suggest severe secondary disturbance of the isotopic dating systems. The frequent obtention of well fitted intermediate age whole rock Pb-Pb isochrons is enigmatic in this context. Thus a collection of the early Archaean Uivak gneiss taken on a 10 m^2 outcrop yields an apparently perfect 3443 Ma Pb-Pb isochron (μ_1=7.62) which, however, has no support from the established ion probe U/Pb zircon chronology of the same sample suite. Late Archaean disturbances of the Rb-Sr and Sm-Nd systems are apparent, in particular a high degree of Rb open system behaviour. It is therefore suggested that the Pb-Pb isochron is a mixing line with no direct age significance, between non-radiogenic Pb from an old (>3.7 Ga) crustal source and a more radiogenic juvenile Pb component added from a fluid phase. The mixing is thought to have taken place during retrogression from granulite facies at 2.7-2.8 Ga. The model requires addition of U and radiogenic Pb in a fixed proportion in order to explain the retention of a linear alignment until the present day. The example suggests that apparently plausible but meaningless Pb-Pb isochrons can show up in polymetamorphic terranes in spite of evidence for major disturbances of primary rock chemistry. In such areas Pb may be a more useful guide to the nature of metamorphic processes such as fluid movements, rather than an indicator of primary ages and origins.

1. INTRODUCTION

This paper discusses and interprets whole rock Pb isotopic data on Archaean gneisses in the Saglek-Hebron area of northern Labrador, Canada (Fig. 1) in the light of the recently established ion probe U-Pb zircon chronology of the same rocks. In addition comparisons with evidence from other isotopic systems are made wherever

D. Bridgwater (ed.), Fluid Movements – Element Transport and the Composition of the Deep Crust, 299–317.
© *1989 by Kluwer Academic Publishers.*

appropriate. The work is concentrated on the 3730 Ma old Uivak I
gneiss, the earliest orthogneiss unit recognised in the area, which
forms an estimated 60-70% of the total gneiss complex.

The presence of early Archaean (>3.5 Ga) gneisses in the
Saglek-Hebron area was originally noted by Hurst et al. (1975) and
Barton (1975). A preliminary U-Th-Pb chronology (including bulk
zircon, sphene, apatite and whole rock studies) was presented by
Baadsgaard et al. (1979), whilst later contributions to the
chronology have been made by various authors. A summary with
reinterpretations was given by Schiøtte et al. (1986). The
conventional isotopic dating systems (that is whole rock Rb-Sr,
Pb-Pb, Sm-Nd and even bulk zircon U-Pb) are generally highly
disturbed due to the polymetamorphic and migmatitic state of the
gneiss complex, and it is not always clear whether ages obtained by
these systems represent primary igneous events or metamorphic
overprinting. Therefore, in the present work, ion probe U-Pb age
determinations on separate generations of zircon growth are the
accepted geochronological framework of ages within which
conventional isotopic data are evaluated and comparisons are used
as a guide to secondary element migrations.

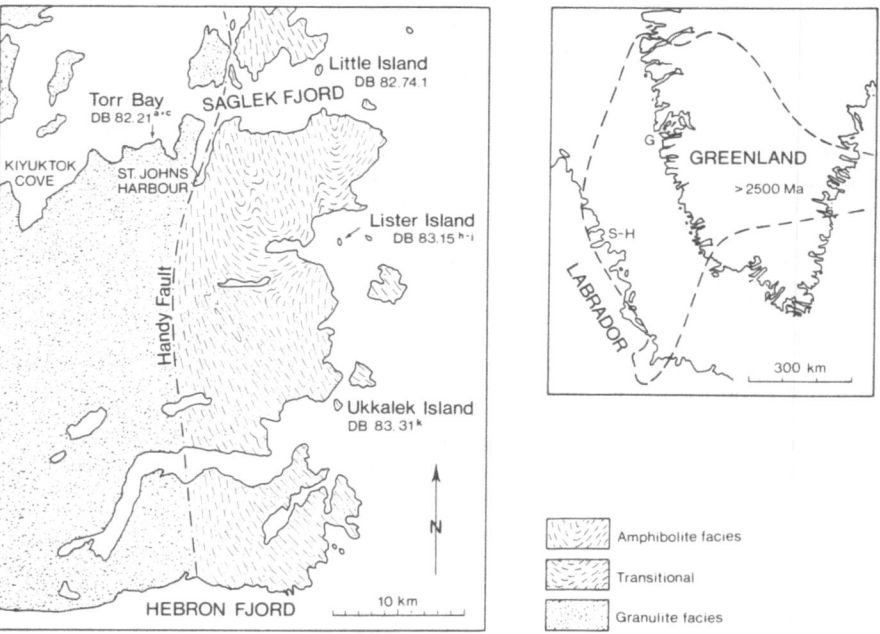

Fig. 1. Map showing the location of the Saglek-Hebron area (S-H)
within the Archaean craton of the North Atlantic, together with the
distribution of metamorphic facies within the area. Sample numbers
denote gneiss localities documented by ion probe single zircon work
(From Schiøtte et al. in press).

Due to post-Archaean faulting of the Saglek-Hebron area,
different levels of the Archaean crust are represented at the
present level of erosion (Bridgwater et al. 1975), and the area
offers unique opportunities for comparative isotopic/geochemical
studies across metamorphic facies boundaries. Thus in the northern
part of the area the N-S trending Handy Fault (Fig. 1) separates
granulite facies rocks (representing deeper levels) to the west
from amphibolite facies rocks (representing higher levels) to the
east. Further south the gneisses on the eastern block become
transitional to granulite facies. Retrogression zones are numerous
within the granulite facies terranes. This retrogression is partly
tectonically controlled, partly related to the emplacement of late
granitoids.

2. ANALYTICAL PROCEDURE

The U-Th-Pb dating of zircon was performed on the sensitive high
mass resolution ion microprobe SHRIMP at Australian National
University in Canberra, using the procedure described by Schiøtte
et al. (1988). The analysed spots were between 20 and 40 μm in
diameter. U and Th decay constants used are those of Jaffey et al.
(1971) and LeRoux & Glendenin (1963).
   Sample digestion for whole rock Pb studies was carried out in
Teflon bombs with a mixture of $HF-HNO_3-HCl$ concentrated acids. Pb
was separated in an ion exchange column loaded with an AG 1x8 resin
using HBr-HCl, with a second "clean up" using HCl. The Pb was
loaded on a single rhenium filament and run on the Cameca TSN 206A
mass spectrometer at Université de Clermont II in Clermont-Ferrand,
France. Isotopic analyses were corrected for mass fractionation
using an NBS 981 standard. The total Pb blanc was 6-8 ng. Errors
(1σ) on analyses are approximately 0.1%. Primeval Pb isotope ratios
used for calculation of single stage Pb growth curves are those of
Tatsumoto et al. (1973), assuming an age of 4.57 Ga for the Earth.
A modified version of the McIntyre et al. (1966) method for
calculation of isochron ages that takes correlated errors into
account has been developed at Australian National University in
Canberra; Pb-Pb isochrons obtained by the present author have been
calculated assuming a correlation coefficient of 0.9 for
$^{206}Pb/^{204}Pb$ and $^{207}Pb/^{204}Pb$.
   Usual techniques were used for separation of Rb and Sr. Sr
isotopic compositions were measured either on the Varian MAT TH5
mass spectrometer at Institute for Petrology, Geological Institute,
University of Copenhagen, using XRF for determination of Rb/Sr, or
on the Cameca TSN 206A mass spectrometer at Université de Clermont
II, using isotopic dilution techniques for measuring Rb/Sr.
$^{87}Sr/^{86}Sr$ was normalised to $^{86}Sr/^{88}Sr=0.1194$. 1σ errors are 1% on
$^{87}Rb/^{86}Sr$, on $^{87}Sr/^{86}Sr$ typically 0.01%. The ^{87}Rb decay constant
applied is 1.42×10^{-11} a^{-1}.
   Whole rock U analyses (for semi-quantitative estimates of

U/Pb) were made at Risø National Laboratory in Roskilde, Denmark using delayed-neutron counting following neutron irradiation. 1σ errors are typically less than 10%, however may reach 10-15% in very U poor depleted granulites. Major and trace element analyses cited in the text were made with XRF at the Geological Survey of Greenland and Institute for Petrology, Geological Institute, University of Copenhagen, respectively.

3. GEOLOGICAL BACKGROUND

3.1. Chronological framework

The ion probe U-Pb zircon chronology of the gneiss complex is presented in detail in a separate paper (Schiøtte et al. in press) Briefly, ca. 3730 Ma ages are obtained on euhedrally zoned igneous zircons from the Uivak I gneiss. A ca. 3620 Ma old generation of high-U low-Th zircons is thought to date early migmatisation by leucocratic veins. The Uivak II gneiss is an augen gneiss that intrudes the already migmatite banded Uivak I gneiss. It has not been dated by the present author, but field evidence in conjunction with ion probe zircon dates for older and younger rocks constrains the age to the time interval 3240-3620 Ma, which is consistent with provisional ion probe zircon data presented by Collerson (1983a). These early Archaean rocks are cut by the basic to ultrabasic Saglek dykes. A ca. 3240 Ma emplacement age has in turn been obtained on the volumetrically less important post-Saglek dyke Lister gneiss.

Whilst no late Archaean zircon growth has been recorded from Uivak I gneiss within the amphibolite facies terrane, new generations of U poor massive zircons of 2700-2800 Ma age have

Table I. Ion probe U-Pb zircon chronology (in Ma) for the Labrador gneisses according to Schiøtte et al. (in press).

3730-3860	Occasional pre-igneous rounded core zones in igneous zircons in the Uivak I gneiss. Indicate the (former?) existence of pre-Uivak gneisses in the area.
ca. 3730	Euhedrally zoned intermediate-to high-U igneous zircons in the Uivak I gneiss. Regarded as the emplacement age of the gneiss.
ca. 3620	High-U low-Th zoned zircons interpreted as having grown during migmatisation of the Uivak I gneiss.
ca. 3240	Euhedrally zoned intermediate-to high-U igneous zircons in the Lister gneiss. Regarded as the emplacement age of the gneiss.
2700-2800	Massive low-U zircons in gneisses subjected to granulite facies metamorphism. Interpreted as dating the major reworking event. Euhedrally zoned igneous zircons in pegmatitic samples date partial melting of the continental crust.

Zircon generations present in the Ukkalek Island sample (DB83.31k) are underlined.

grown in gneisses subjected to granulite facies metamorphism. Ages obtained on euhedrally zoned igneous zircons from pegmatitic samples show that partial melting of the terrane now in granulite facies took place in this time interval. Retrogression is in many cases intimately associated with the emplacement of leucocratic granitoid sheets of ca. 2740 Ma age (bulk zircon U-Pb (Schiøtte 1988 unpublished)), so that the whole sequence of events stretching from initial recrystallisation through partial melting and retrogression in these cases took place within <ca. 100 Ma. An abbreviated ion probe U-Pb zircon chronology is presented in Table I. It is emphasized that the major evolutionary divide between terranes now in granulite and amphibolite facies respectively is a late Archaean event affecting an approximately one billion years older continental crust.

3.2. Pb isotope characteristics in general

It has been noted that the early Archaean crustal evolution in northern Labrador shows many similarities to the semi-contemporaneous evolution in the Godthåbsfjord area of southern West Greenland (Bridgwater et al. 1976; Bridgwater et al. 1978; Collerson & Bridgwater 1979). However, whereas the early Archaean Amîtsoq gneisses in the type area of outer Godthåbsfjord are known for their extremely non-radiogenic Pb that has developed in a low-U environment since shortly after the separation of the gneiss precursor from the mantle (Black et al. 1971; Moorbath 1975, 1977), a large range of Pb isotope compositions is covered by the Uivak gneisses (Table II).

The straightforward interpretation of this difference is that the Uivak gneisses have had a larger range of U/Pb ratios from early Archaean time. However, noting that an important difference between the Godthåbsfjord area and the Saglek-Hebron area is a much higher degree of late Archaean reworking in the latter, Baadsgaard et al. (1979) suggested secondary introduction of juvenile Pb in the late Archaean as an alternative. This idea was reiterated by Schiøtte et al. (1986) who noted a correlation between apparent initial U/Pb ratios and late Archaean metamorphic grade:

It is apparent from Fig. 2 that the Pb typical of LIL element depleted dry granulite facies gneisses developed in a U-poor environment (similar to that indicated for the Greenlandic Amîtsoq gneisses) before the late Archaean high-grade metamorphism. Conversely the Pb of gneisses subjected to "wet" late Archaean high-grade metamorphism (that is proper amphibolite facies gneisses and retrogressed granulites) covers a large range of isotopic compositions. If the isotopic evolution of this Pb is extrapolated back to the time of the late Archaean high-grade event (along lines with ca. 2800 Ma slope in the diagram), it appears that the spread in isotopic compositions was considerable already at that time. Many of the "wet" gneisses actually contain Pb developed

Table II. List of currently available Pb isotope data on Uivak I
gneiss (extended from Schiøtte et al. 1986), together with data
from the late Archaean granitoids.

Sample	206/204	207/204	208/204	U	Pb		Sample	206/204	207/204	208/204	U	Pb	
					(ppm)							(ppm)	
Gneisses at Saglek (proper amph. fac.):							Gneisses on Ukkaiek Island (transitional to gran. fac.):						
74-161A	15.495	14.532	34.634			(1)	DB83.31C	18.22	15.52	37.39	0.286	17	(6a)
74-161C	13.711	14.205	35.622			(1)	DB83.31D	15.79	14.82	35.37		15	(6a)
74-161F *	13.962	14.117	34.533			(1)	DB83.31F	17.27	15.25	37.11		17	(6a)
74-40A	13.891	14.290	37.322			(1)	DB83.31G	13.57	14.15	35.77	0.275	21	(6a)
75-260C	16.005	14.469	39.525			(1)	DB83.31H	15.83	14.83	37.52		19	(6a)
74-161B	14.307	14.290	34.267			(1)	DB83.31J	16.07	14.89	35.98		19	(6a)
75-260B	13.986	14.188	38.688			(1)	DB83.31K *	16.18	14.92	35.80	0.102	16	(6a)
75-320	15.986	14.593	35.984			(1)							
75-296A	16.300	14.890	35.874			(1)	* Zircons were separated from this sample						
74-298	16.216	15.272	37.268			(1)							
75-291F	16.379	14.976	37.154			(1)	Depleted gran. fac. gneisses in Torr Bay:						
27SJ19	16.830	15.034				(2)	DB82.21A *	12.81	13.84	33.07	0.198	10	(6)
27SJ20	13.750	14.422				(2)	DB82.21B *	12.79	13.83	33.69		3	(6)
3SB3	13.862	14.153				(2)	DB82.21C	12.42	13.79	32.37	0.066	5	(6)
32SR42	14.060	14.363				(2)	DB82.21D	12.60	13.82	32.32		5	(6)
1SB127	15.481	14.886				(2)	DB82.21E	12.43	13.69	32.01		7	(6)
LSD136 **	14.416	14.416				(2)	DB82.21G	12.33	13.68	32.04		9	(6)
LSD138KM **	14.297	14.228				(2)	DB82.21M	12.37	13.71	32.10	0.048	6	(6)
LSD138P **	13.922	14.150				(2)							
3SB1 **	13.045	13.820				(2)	* Pegmatite samples (the others are restite)						
32SB43 **	16.031	14.638				(2)							
4B16 **	13.032	14.019				(2)	Kiyuktok gneisses (migmatised and retrogressed):						
DB82.74.1B	14.31	14.66	34.79	0.450	17	(6)	238A	12.44	13.56				(3)
							238K	13.37	13.90				(3)
* May be Uivak II							238L	13.80	14.03				(3)
** Classified as undiff. gneiss by Hurst et al. (1975)							238M	15.97	14.77				(3)
							238N	16.65	14.77				(3)
Gneisses at Hebron (transitional to gran. fac.):													
113A	12.78	13.72				(4)	Late Archaean granitoid sheets:						
113B	12.47	13.58				(4)	DB82.47P	12.75	14.00	32.39		21	(6)
114	12.99	13.79				(4)	DB82.47Q	12.66	13.87	32.27		14	(6)
115	13.16	13.85				(4)	DB82.47S	12.94	14.11	32.50		10	(6)
116	13.11	13.85				(4)	DB82.47T	12.66	13.91	32.71		19	(6)
117	13.93	14.16				(4)	DB82.47Y	12.63	13.97	32.41		18	(6)
118	13.58	13.96				(4)	LS83.2A	12.67	13.98	32.31		21	(6)
121	12.95	14.08				(4)	LS83.2B	12.78	13.99	32.29		21	(6)
DB83.20A	15.62	14.73	36.27		15	(6)	DB82.53D	13.02	14.05	32.52		9	(6)
DB83.20B	17.28	15.24	38.00		16	(6)	DB83.2A	15.10	14.43	35.63		57	(6)
DB83.41D	17.95	15.45	37.49		33	(6)	DB83.17A	15.38	14.74	34.84		35	(6)
KC33	13.115	14.097	35.844			(5)	DB82.15G	17.96	15.45	37.65		43	(6)
KC34	13.157	13.981	33.462			(5)	DB82.15H	18.35	15.57	37.99		49	(6)
KC35	13.002	14.045	35.569			(5)	DB83.6G	17.56	15.34	37.16		36	(6)
KC36	13.395	14.159	33.435			(5)	DB83.6H	15.45	14.72	35.18		31	(6)
KC37	12.655	13.827	33.545			(5)	DB82.25A	15.47	14.70	35.99		30	(6)
KC38	12.811	13.920	34.187			(5)	LS83.16	18.50	15.61	38.00		27	(6)
KC39A	13.496	14.309	33.709			(5)	DB83.41A *	17.94	15.40	37.36		24	(6)
KC40	14.351	14.333	33.346			(5)							
KC41	12.821	13.958	34.722			(5)	* Intimately mixed with the main gneiss, represented						
KC42	11.757	13.312	32.314			(5)	by DB83.41D from the Hebron area						

(1) Baadsgaard et al., 1979. (2) Unpublished data by P.N. Taylor. (3) Collerson et al., 1981. (4) Barton et al.,
1983. (5) Moorbath et al., 1986. (6) Data by the author, (6a) is the collection discussed in detail in the present
work.

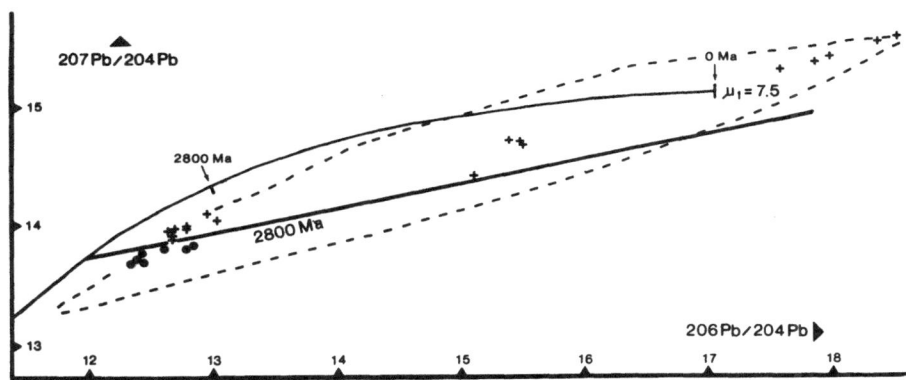

Fig. 2. Pb-Pb diagram showing the distribution of isotopic
compositions for the LIL element depleted dry granulite facies
gneisses (●) and the late Archaean granitoid sheets (+). The broken
curve encloses the field of Uivak I gneisses subjected to "wet"
late Archaean high-grade metamorphism. A line with 2800 Ma slope is
shown for reference. Rock samples falling above this line were more
radiogenic than those falling below already in the late Archaean.
Moreover, the single stage Pb growth curve for a hypothetical more
U rich reservoir (μ_1=7.5, currently regarded as mantle-like, e.g.
Taylor et al. 1980) is shown for reference.

in a moderately U-rich environment before the high-grade
event. Thus if the difference in Pb compositions were
attributed to a pre-metamorphic (primary?) variation in U/Pb
ratios, then a striking correlation between pre-late Archaean
characteristics and late Archaean metamorphic grade would
remain unexplained.

Schiøtte & Bridgwater (1986) suggested that the anomalously
radiogenic Pb was introduced to the "wet" gneisses by fluids during
emplacement of the late Archaean granitoid sheets that are often
intimately associated with retrogression from granulite facies. It
is apparent from Fig. 2 that the Pb typical of these sheets has a
crustal component. However, it is seen to plot above the 2800 Ma
reference line, showing that it was more radiogenic than typical
granulite facies Uivak Pb at the time of late Archaean high-grade
metamorphism. Similarly ϵ_{Nd} CHUR values at 2740 Ma are negative,
suggesting a crustal component, however numerically smaller than in
the main early Archaean gneiss. Thus the old continental crust
appears to be migmatised by intrusions of mixed origin.

It is a major challenge to the idea proposed by Baadsgaard et
al. and Schiøtte et al. that apparently plausible early to mid
Archaean Pb-Pb isochrons and errorchrons are frequently obtained on
sample collections incorporating the kind of Pb claimed to be

secondarily introduced. As an example Collerson et al. (1981, 1982) obtained an apparent five point 3505 Ma errorchron on migmatised and retrogressed granulites from Kiyuktok Cove (Fig. 1). In the 1982 paper the authors interpreted this as a mixing line without direct age significance. More recently Moorbath et al. (1986) have published what they interpret as a transposed palaeo-isochron with t_1 at ca. 3.6 Ga and t_2 (the time of U depletion) at ca. 2.5 Ga on a collection of gneisses from the transitional area in Fig. 1. The t_2 at 2.5 Ga actually coincides with the age of a post-tectonic granite in the area (zircon data by Baadsgaard et al. (1979)). A 3498±42 Ma (2σ) Pb-Pb isochron obtained by the present author on sample collections from two outcrops of the Lister gneiss which is zircon dated at 3240 Ma (Schiøtte et al. in press) gives the first hint that perfect isochrons do not necessarily have any direct age significance. Data for the Lister gneiss are presented in Table III. In the following section another such apparently plausible Pb-Pb isochron obtained on the 3730 Ma Uivak I gneiss is evaluated in the light of ion probe U-Pb zircon data for the same sample collection.

Table III. List of Pb isotope data obtained by the present author on the 3240 Ma Lister gneiss. The data points constitute a 3498±42 Ma (μ_1=7.62) isochron (not shown).

Sample	206/204	207/204	208/204	U (ppm)	Pb
DB83.29A	15.40	14.67	35.69	0.359	22
DB83.29B	17.88	15.45	37.70		14
DB83.29C	17.01	15.18	36.94		21
DB83.29D	18.23	15.53	37.77	0.376	15
DB83.15H	17.92	15.45	37.41		12
DB83.15I	18.51	15.62	38.05	0.450	13
DB83.15J	18.06	15.49	37.58		13

4. UKKALEK ISLAND: A CASE STUDY

An apparently perfect 3443±33 Ma (2σ) Pb-Pb isochron with a μ_1 of 7.62 was obtained on a collection of Uivak I gneiss samples taken within 10 m^2 on Ukkalek Island in the area transitional between granulite and amphibolite facies (Fig. 3). The gneiss is presently devoid of orthopyroxene, but this mineral (used as field indication of granulite facies metamorphism), is abundant in the surrounding supracrustal rocks. The Pb compositions (Table II) are all within the range of compositions claimed to be contaminated with juvenile late Archaean Pb, and the spread in isotopic composition (implying a large spread in U/Pb) is remarkable, so if the isochron has direct age significance, the idea of secondarilly introduced Pb seems unjustified.

However, the 3443 Ma age has no support from the zircon chronology (Table I). The maximum $^{207}Pb/^{206}Pb$ age obtained on

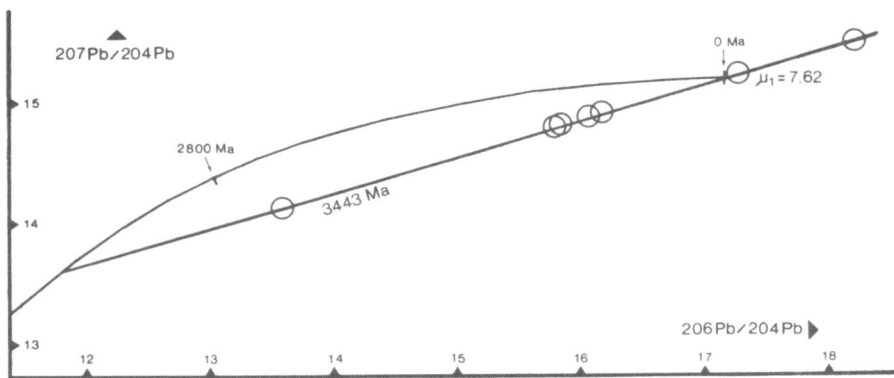

Fig. 3. 3443±33 Ma Pb-Pb isochron (μ_1=7.62) obtained on Uivak I gneiss from Ukkalek Island.

igneous zircons suggest an emplacement age of 3742±23 Ma (2σ), in accordance with the ca. 3730 Ma age obtained on Uivak I gneiss elsewhere. The early Archaean migmatisation is recorded by the 3620 Ma high-U low-Th zircon generation, whilst massive U poor zircons (separate grains as well as overgrowths on older zircons (Fig. 4)) have grown at about 2760 Ma. This late Archaean event also caused condiderable Pb loss from the older zircon generations. Rb-Sr data points cluster betweeen $^{87}Rb/^{86}Sr$ 0.17-0.32 and $^{87}Sr/^{86}Sr$

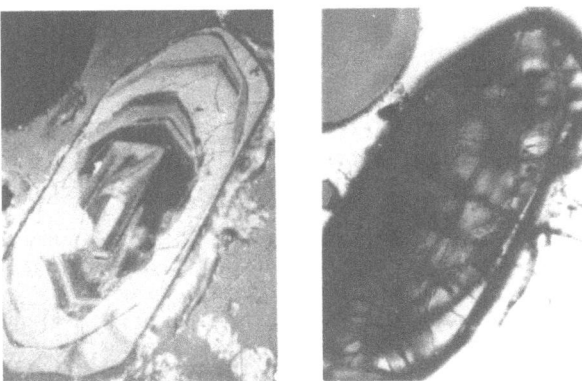

Fig. 4. Photomicrographs of zircon from sample DB83.31K; reflected light left, transmitted light right; field of view ca. 250 x 400 µm; internal crystal structures have been accentuated by HF vapour etching. The innermost euhedrally zoned (igneous) domain is partly recrystallised and is in turn surrounded by a massive U poor metamorphic rim zone. (From Schiøtte et al. in press).

308

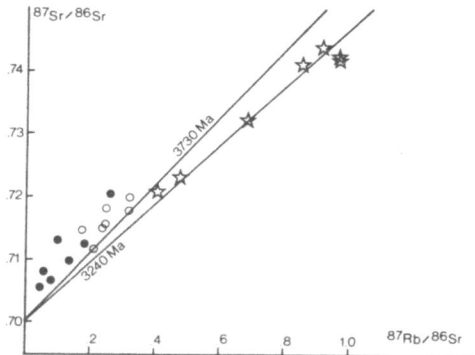

Fig. 5. Rb-Sr diagram for the Uivak I gneiss collection from
Ukkalek Island (O). Also shown are data for LIL element depleted
granulite facies Uivak I gneiss from Torr Bay (●) and data for the
Lister gneiss ✿. 3730 and 3240 Ma lines are shown for reference.

0.71-0.72; as apparent from Fig. 5 they do not constitute any
isochron. The following ϵ_{Nd} CHUR values at key points in the
evolution of the gneiss complex have been calculated from Sm-Nd
data for sample DB83.31K ($^{147}Sm/^{144}Nd=0.08222$, $^{143}Nd/^{144}Nd=$
0.510170 (Schiøtte 1988 unpublished)):

$\epsilon_{Nd}(3730) = + 7.0$; $\epsilon_{Nd}(3620) = + 5.4$; $\epsilon_{Nd}(3443) = + 2.7$
$\epsilon_{Nd}(2760) = - 7.6$; $\epsilon_{Nd}(0) = -48.1$

The positive ϵ_{Nd}-values obtained for the early and mid Archaean are
numerically too high to be geologically plausible, and a late
Archaean disturbance of the Sm-Nd system is suggested.
     Possible explanations for the discrepancies between whole rock
Pb dating and evidence from other isotopic systems (zircon U-Pb in
particular) are as follows:

1. 3443 Ma is an inaccurate determination of the age of early
   migmatisation, possibly due to slow cooling after 3620 Ma
   and hence prolonged Pb diffusion on outcrop scale.
2. 3443 Ma is the age of a separate plutonic/metamorphic event
   (such as the intrusion of the Uivak II gneiss) affecting the
   rock, with no other impact on the zircon U-Pb system than
   minor Pb loss from the already existing zircons.
3. The line is a palaeo-isochron with t_1 at 2.7-2.8 Ga and t_2
   (U depletion) at 1.5-1.7 Ga.
4. The line is a mixing line without direct age significance.

     Although the μ_1-value of 7.62 gives no hint of a prolonged
crustal history prior to 3443 Ma, the retention of 3.6-3.8 Ga

isotopic signatures in zircons shows that if the isochron has any
age significance at all, then it represents a secondary event
rather than the separation of the igneous precursor to the gneiss
from a mantle source. The existence of the very old zircon Pb also
shows that isotopic equilibration, strictly speaking, cannot have
taken place at 3443 Ma. However, if the zircon Pb only represents a
very minor fraction of the bulk rock Pb budget, then model (1) and
(2) may be considered.

Models (1) and (2) imply that the Uivak gneiss developed with
a remarkably juvenile U/Pb during the first ca. 300 Ma of
continental evolution, and they are both difficult to reconcile
with the strong impact the recrystallization event at 2760 Ma has
had on the zircons (and the Sm-Nd system). In particular the
presence of the late Archaean U poor massive zircon generation
typical of LIL element depleted granulites suggests that the rock
was depleted in U at this time, and that the moderately high U
contents, predicted from the present spread in the isotopic
composition of Pb are due to reintroduction of U by fluids during
retrogression from granulite facies.

Model (3) assumes that Pb was homogenised on outcrop scale
either during granulite facies metamorphism or during subsequent
retrogression. If U then was added to the gneiss from fluids, and
the U-Pb system developed undisturbed afterwards, a 2700-2800 Ma
Pb-Pb isochron would develop, most probably with a low μ_1-value.
If, however, U was lost again in the Proterozoic, the radiogenic
evolution of Pb would cease, and a palaeoisochron with a slope
corresponding to an age >2700-2800 Ma would result. In this model
the effect of early evolution in a low-U crustal environment and
the artificial enhancement of μ_1-values caused by Proterozoic U

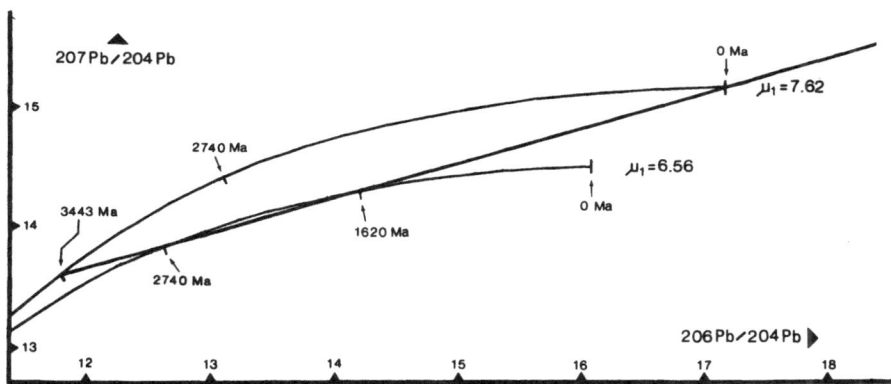

Fig. 6. Pb-Pb diagram illustrating how the radiogenic evolution of
Pb between a late Archaean event of isotopic homogenisation of old
crustal Pb (μ_1=6.56) and a mid Proterozoic U depletion event could
lead to an apparent 3443 Ma isochron with a juvenile μ_1-value of
7.62, a so called palaeo-isochron (model 3 in the text).

loss (the palaeo-isochron effect as outlined by Jacobsen &
Wasserburg 1978; Moorbath & Taylor 1985; Moorbath et al. 1986)
cancel each other, and an artificially juvenile μ1-value results.
The model which is illustrated in Fig. 6, is compatible with, but
does not require addition of anomalously radiogenic Pb (together
with U) during retrogression from granulite facies.

The Saglek-Hebron area is in fact weakly affected by the
Hudsonian (Ryan et al. 1983). The tectonical overprinting is
concentrated along distinct shear zones, and Collerson (1983b) has
shown examples of ca. 1.8 Ga secondary Rb-Sr isochrons. However,
this far east (Ukkalek Island) only greenschist facies conditions
are reached, which is not the metamorphic grade generally asociated
with major U loss. Moreover, the present day U contents of three of
the samples on the isochron have been measured (Table II) and shown
to be compatible with recent U loss only. The "residual" U/Pb
ratios are still so high that the post-Hudsonian radiogenic
evolution would have caused considerable scatter around a
palaeo-isochron.

Since none of the above considered models are entirely
satisfactory, it is suggested that the "isochron" could be a mixing
line (model 4). The endmembers in the mixing process would be a
relatively non-radiogenic crustal Pb (such as the Pb of LIL element
depleted granulites in Table II) and a more radiogenic Pb added
from a fluid during retrogression from granulite facies. The fluid
addition could be closely linked with the emplacement of the ca.
2740 Ma granitoids which carry a more radiogenic Pb than LIL
element depleted granulite facies gneisses (Table II). This model
requires homogenisation of the crustal Pb on outcrop scale during
granulite facies metamorphism, which is well in accordance with
findings on outcrops of LIL element depleted granulites in the
area. Furthermore, in order to avoid scatter caused by the
radiogenic evolution of Pb after mixing with the fluid, it is
required that U and Pb from the crustal and fluid sources are mixed
proportionately. The combined effect of mixing and subsequent aging
is then described by a set of relatively simple formulae. At a time
t after the mixing a given point on the mixing line will have the
following isotopic composition:

$$(6/4)=(1-X) \times ((6/4)_c+\mu_c(e^{\pounds 8 T}-e^{\pounds 8 t}))+ X \times ((6/4)_f+\mu_f(e^{\pounds 8 T}-e^{\pounds 8 t}))$$

$$(7/4)=(1-X) \times ((7/4)_c+\mu_c(e^{\pounds 5 T}-e^{\pounds 5 t}):137.88)+$$
$$+ X \times ((7/4)_f+\mu_f(e^{\pounds 5 T}-e^{\pounds 5 t}):137.88)$$

(where X and (1-X) represent the relative amounts of Pb from
crustal source (c) and contaminant fluid (f). £8 and £5 are the
decay constants of ^{238}U and ^{235}U. T is the time of mixing).

The ratio between ^{207}Pb and ^{206}Pb added to the gneiss as the
combined effect of mixing with the fluid at T and the subsequent U
decay between T and t will be

$$(7/6)_{add} = \frac{X \times ((7/4)_{dif} + (\mu_f(e^{\pounds 5 T} - e^{\pounds 5 t}) - \mu_c(e^{\pounds 5 T} - e^{\pounds 5 t})):137.88) +}{X \times ((6/4)_{dif} + \mu_f(e^{\pounds 8 T} - e^{\pounds 8 t}) - \mu_c(e^{\pounds 8 T} - e^{\pounds 8 t})) +}$$

$$\frac{+ \mu_c(e^{\pounds 5 T} - e^{\pounds 5 t}):137.88}{+ \mu_c(e^{\pounds 8 T} - e^{\pounds 8 t})}$$

(where $(7/4)_{dif} = (7/4)_f - (7/4)_c$ and $(6/4)_{dif} = (6/4)_f - (6/4)_c$)

For given values of T and t (in the present case 2740 and 0 Ma respectively) this simplifies to:

$$(7/6)_{add} = \frac{X \times Term_1 + Term_2}{X \times Term_3 + Term_4}$$

For LIL element depleted granulite facies rocks μ_c and hence also Term2 and Term4 are close to zero, in which case $(7/6)_{add}$ is constant (independent of X). This means that the points on the original mixing line will retain their linear alignment in the future (with different slopes for the line at different times). If a fluid with a relatively low U/Pb ratio is added, the radiogenic evolution of Pb will be slow, and the present day slope of the line will not be very different from the slope of the original mixing line. The line would then most probably be interpreted as a palaeo-isochron. The higher the U/Pb of the fluid (and the smaller the difference in isotopic composition of crustal and contaminant Pb), the closer the slope will be to a 2740 Ma isochron (or whatever the age of the mixing event is). Mixing with fluids of intermediate U/Pb will give lines with a range of slopes corresponding to apparently plausible but meaningless ages (Fig. 7).

At the present stage of research model (4) is considered the most plausible explanation for whole rock Pb-Pb "isochrons" such as the 3443 Ma line presently discussed that have no support from other isotopic systems. The fluids carrying the juvenile Pb are thought to derive from late Archaean granitoids. These granitoids could have picked up the juvenile Pb either from the mantle or from supracrustal rocks with higher U/Pb ratios than the orthogneisses. In the latter case the mixing process represents a somewhat less dramatic averaging out of isotopic characteristics between contrasting lithologies. For the sake of completeness it is noted that no linear alignment is achieved in $^{207}Pb/^{204}Pb$ versus $^{208}Pb/^{204}Pb$ plots (not shown). This is interpreted as due to more circumstantial Th (as compared to U) depletion during granulite facies metamorphism, making the subsequent mixing process more complicated.

312

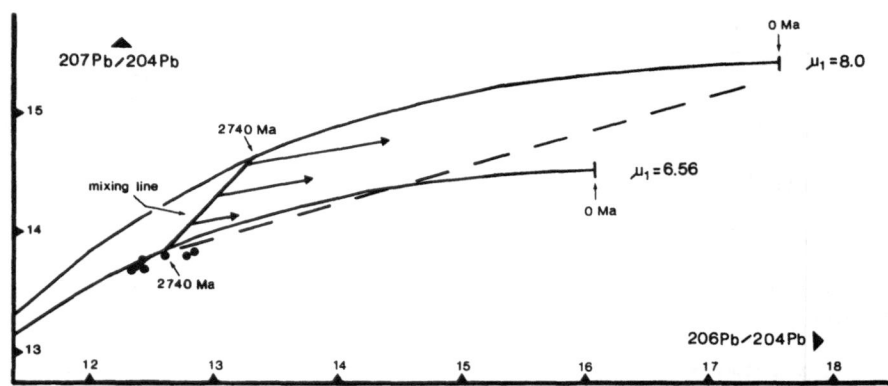

Fig. 7. Pb-Pb diagram illustrating model (4). At ca. 2740 Ma mixing
takes place between old crustal Pb (developed in a low-U
environment, μ_1=6.56) and a fluid derived from a moderately U rich
environment (arbitrarily chosen to be 8.0). Arrows illustrate the
post-2740 Ma radiogenic evolution of the Pb, when U and Pb are
added in a fixed proportion. The dry granulite facies gneisses ( )
are compositionally close to the U poor endmember in the mixing
process. The 3443 Ma line is shown broken for reference.

5. DISCUSSION

The primary aim of this paper has been to present an enigma, that
is the retention of well fitted Pb-Pb whole rock isochrons on
gneisses which otherwise show evidence for severe disturbance of
the primary chemistry. It has been shown that an apparently perfect
3443 Ma isochron obtained on a collection of Uivak I gneiss samples
with high apparent μ_2-values has no support from the single zircon
U-Pb chronology of the rock, and that the Sm-Nd and Rb-Sr systems
were disturbed during the late Archaean. It is argued that the
isochron is best explained as the combined effect of mixing between
non-radiogenic early Archaean Pb and late Archaean juvenile Pb and
subsequent aging. An important implication of this is that Uivak I
gneisses subjected to "wet" late Archaean high-grade metamorphism
may be contaminated with Pb from external sources. The fact that
apparently good early to mid Archaean isochrons and
palaeo-isochrons are obtained does not disprove that disturbances
of this kind took place.
    Although the proposed mixing process is simplified by the
geologically reasonable assumption of an isotopically homogenous
end member with a close-to-zero U content, the model can still be
critisised, as it requires U and Pb added in a fixed proportion.
The fact that the "dating" was made on a sample collection from a
single outcrop may play a key role. Regional collections are less

likely to be contaminated with fluids of constant U/Pb.

The case study illustrates some of the unique features of the Pb-Pb system when compared to other conventional isotopic dating tools: When early Archaean retarded Pb is around in an area, it will always have a strong impact when mixed with younger Pb. So whatever happens to such an area, it will always retain a certain "taste" of early Archaean. This is well in accordance with more classical Pb-Pb studies on igneous rocks from Skye, Scotland (Dickin 1981) and late Archaean gneisses in southern West Greenland (Taylor et al. 1980) where younger rocks are contaminated with old continental Pb, interpreted as assimilation during intrusion. A more controversial aspect of the present work is the concept of foreign Pb introduction the other way round, into a solid gneiss, in the metamorphic state. It is argued that various mixing lines having old retarded Pb as one endmember can easily be mistaken for isochrons when polymetamorphic rocks are dealt with, so on the whole Pb may be a more useful guide to the nature of metamorphic processes rather than indicating primary ages and origins.

It is very tentatively suggested that the situation described by model (4) could be relatively common. However, it could easily be overlooked in areas where the initial emplacement of a gneiss precursor and subsequent granulite facies metamorphism followed by retrogression were events in rapid succession, telescoped to within 100-200 Ma. Crustal and mantle sources in such areas would still not have developed distinct isotopic signatures at the time of Pb introduction and so contaminant and crustal Pb would be little different. For moderately U rich rocks Pb/Pb dating would in most such cases give good first order approximations to the age of the "mega-event".

6. SOME NOTES OF CAUTION

6.1. Comparisons with the main rock chemistry

K/Rb was used as an index of metamorphic geochemical modification of Uivak gneisses by Schiøtte et al. (1986). It was shown that granulites are alkali depleted with preferential loss of Rb relative to K, so that K/Rb is high, in the 500-2000 range. Conversely gneisses that did not reach granulite facies conditions during the late Archaean reworking, are alkali enriched with K/Rb ratios that are often below 200. The sample suite presently dealt with has intermediate K/Rb ratios, ranging from 268 to 338 which is consistent with the suggested sequence of events, that is depletion followed by reintroduction of Rb from fluids. A similar sequence of events could explain the Rb/Sr characteristics of the sample collection when compared to LIL element depleted granulites (Fig. 5). However, although K/Rb might have been expected to vary systematically from high values at the lower end of the 3443 Ma line to low values at the upper end, such a strict correlation with the amount of added fluid is not seen. If Rb contents did not fall

to virtually zero during the depletion event, the following mixing
process may have been more complicated than in the case of U where
$U_{pre-mixing} \ll U_{post-mixing}$. Similarly absolute contents of Pb are
intermediate between values typical of depleted granulites and
enriched amphibolite facies gneisses, without a strict correlation
along the 3443 Ma line.

Discussion has been concentrated about the Ukkalek Island
sample collection rather than the previously mentioned 3498 Ma
isochron that was obtained on the 3240 Ma Lister gneiss, because
the latter could also be explained in terms of contamination with
older U depleted crust during emplacement of the gneiss precursor.
K/Rb ratios of the Lister gneiss are very low, 110-160; but
theoretically this could reflect primary differences between the
Lister and the Uivak gneiss. The Rb/Sr system, however, shows
severe disturbance (Fig. 5). A secondary Rb enrichment is strongly
suggested: when extrapolated back to 3240 Ma, $^{87}Sr/^{86}Sr$ values for
the Lister gneiss are all in the 0.6960-0.7016 range, averaging
around 0.6995 which appears too low to be geologically plausible.
Thus the radiogenic evolution of Sr seems to have been enhanced at
a later stage, probably during one of the 2700-2800 Ma events.
Since Rb and Pb would probably show open system behaviour in the
same situations, it is considered likely that the Lister gneiss and
the Uivak I gneiss on Ukkalek Island have had similar histories: In
both cases U, Pb and Rb was introduced to the solid gneiss from
outside during the late Archaean reworking of the gneiss complex.

6.2. Regional significance

It has been argued that the Uivak I gneiss on Ukkalek Island may
have developed in a low-μ environment in the period between the
early Archaean and the late Archaean high grade metamorphism,
similar to what was the case for the semi-contemporaneous Amîtsoq
gneisses on West Greenland. The surprisingly radiogenic whole rock
Pb obtained on Uivak gneisses on this sample locality and elsewhere
in the Saglek-Hebron area is thought to be due to late Archaean Pb
introduction from external sources. However, it would probably be
premature to state that the pre-late Archaean Pb evolution always
took place in a low-μ environment.

The migmatisation event at 3620 Ma was in fact associated with
the growth of high-U low-Th zircons, indicating that U was
secondarilly added at a relatively early stage in the evolution of
the gneiss. In areas where this migmatisation was very pervasive,
an early Archaean enhancement of the radiogenic Pb evolution may
well have occurred. The only important difference between the 3620
Ma event and the late Archaean post-granulite facies migmatisation
event may after all be the 800-900 Ma that separate them from one
another. So whilst the conclusion reached in the present work, that
late Archaean redistribution of U and Pb took place, may be
essentially correct, the evidence could under certain circumstances
be blurred by the fact that a very similar redistribution process
occurred several hundred million years earlier.

Acknowledgements
Dr. D. Bridgwater, Geological Museum of Copenhagen and geologists
from Dept. Mines in Newfoundland are acknowledged for cooperation
in the field. The work is part of a major study set up by D.
Bridgwater and funded by Carlsbergfondet, to study the effects of
metamorphism and granite injection on early crust. Most of the
isotopic work was carried out under the excellent supervision of
Dr. Ph. Vidal, Université de Clermont II, CNRS LA 10,
Clermont-Ferrand where the author was funded by the French
government and the Danish Natural Science Research Council. The ion
probe zircon dating was supervised by Dr. W. Compston and others at
Research School of Earth Sciences, Australian National University,
Canberra, while the author was holder of a grant under the
Australian-European award program. The manuscript has been
critically reviewed by Dr. J. Kramers, University of Harare,
Zimbabwe and Dr. A. Baumann, Institut für Mineralogie in Münster,
BRD.

References

Baadsgaard H, Collerson KD & Bridgwater D (1979) The Archaean
    gneiss complex of northern Labrador: 1. Preliminary U-Th-Pb
    geochronology. Can. J. Earth Sci. 16, 951-961
Barton JM Jr. (1975) Rb-Sr isotopic characteristics and chemistry
    of the 3.6 B.Y. Hebron gneiss, Labrador. Earth Planet. Sci.
    Lett. 60, 337-338
Barton JM Jr., Ryan B & Fripp REP (1983) Rb-Sr and U-Th-Pb isotopic
    studies of the Sand River gneisses, Central zone, Limpopo molbile
    belt. Spec. Publ. Geol. Soc. South Africa 8, 9-18
Black LP, Gale NH, Moorbath S, Pankhurst RJ & McGregor VR (1971)
    Isotopic dating of very early Precambrian amphibolite facies
    gneisses from the Godthaab district, West Greenland. Earth
    Planet. Sci. Lett. 12, 245-259
Bridgwater D, Collerson KD, Hurst RW & Jesseau CW (1975) Field
    characters of the early Precambrian rocks from Saglek, coast
    of Labrador. Geol. Surv. Can., Paper 75-1, Part A, 287-296.
Bridgwater D, Collerson KD & Myers J (1978) The development of the
    Archaean gneiss complex of the North Atlantic region. In:
    Tarling DH (ed) Evolution of the Earth's crust. London Academic
    Press, pp 19-69
Bridgwater D, Keto L, McGregor VR & Myers JS (1976) Archaean
    gneiss complex of Greenland. In: Escher A & Watt WS (eds)
    Geology of Greenland. Grønlands Geol. Unders., Copenhagen,
    pp 18-75
Collerson KD (1983a) Ion microprobe zircon geochronology of the

Uivak gneisses: implications for the evolution of early
terrestrial crust in the North Atlantic craton. In: Ashwal LD
& Card KD (eds) Workshop on cross section of Archaean crust.
Lunar and Planetary Inst. Rep. 83-03, 28-33

Collerson KD (1983b) The Archaean gneiss complex of northern
Labrador 2. Mineral ages, secondary isochrons and diffusion of
strontium during polymetamorphism of the Uivak gneisses. Can.
J. Earth Sci. 20, 707-718

Collerson KD & Bridgwater D (1979) Metamorphic development of
early Archaean tonalitic and trondhjemitic gneisses: Saglek
area, Labrador. In: Barker F (ed) Trondhjemites, dacites and
related rocks. Elsevier, pp 205-273

Collerson KD, Kerr A & Compston W (1981) Geochronology and
evolution of late Archaean gneisses in Northern Labrador: An
example of reworked sialic crust. Spec. Publ. Geol. Soc.
Australia 7, 205-222

Collerson KD, Kerr A, Vocke RD & Hanson GN (1982) Reworking of
sialic crust as represented in late Archaean-age gneisses,
northern Labrador. Geology 10, 202-208

Dickin AP (1981) Isotope geochemistry of Tertiary igneous rocks
from the Isle of Skye, N.W. Scotland. J. Petrol. 22, 155-189

Hurst RW, Bridgwater D, Collerson KD & Wetherill GW (1975)
3600 m.y. Rb-Sr ages from very early Archaean gneisses from
Saglek Bay, Labrador. Earth Planet. Sci. Lett. 27, 393-403

Jacobsen SB & Wasserburg GJ (1978) Interpretation of Nd, Sr and Pb
isotope data from Archaean migmatites in Lofoten-Vesterålen,
Norway. Earth Planet. Sci. Lett. 41, 245-253

Jaffey AH, Flynn KF, Glendenin LE, Bentley WC & Essling AM (1971)
Precision measurement of the half-lives and specific activities
of U^{235} and U^{238}. Phys. Rev., C, 4, 1889-1906

LeRoux LJ & Glendenin LE (1963) Half-life of thorium-232. Nat.
Conf. Nucl. Energy Appl. Isotop. Radiat. Proc., 77-88

McIntyre GA, Brooks C, Compston W & Turek A (1966) The statistical
assessment of Rb-Sr isochrons. J. Geophys. Res. 71, 5459-5468

Moorbath S (1975) The geological significance of Early Precambrian
rocks. Proc. Geol. Assoc., 86, 259-279

Moorbath S (1977) Ages, isotopes and evolution of Precambrian
continental crust. Chem. Geol. 20, 151-187

Moorbath S & Taylor PN (1985) Precambrian geochronology and the
geological record. In: Snelling NJ (ed) The chronology of the
geological record. Geol. Soc. Memoir no 10, pp 10-28

Moorbath S, Taylor PN & Jones NW (1986) Dating the oldest
terrestrial rocks - fact and fiction. Chem. Geol. 57, 63-86

Ryan AB, Martineau Y, Bridgwater D, Schiøtte L & Lewry J (1983)
The Archaean/Proterozoic boundary in northern Labrador. Rep. 1
Geol. Surv. Can. Rep. 83-1A, 297-304

Schiøtte L (1988) Age, origin and significance of late Archaean
sheet intrusions in the early Archaean gneiss complex of northern
Labrador, Canada. Evidence from a multi-isotopic/geochemical
study. In: Unpublished Lic. Scient. thesis, Geological Museum,
Univ. Copenhagen.

Schiøtte L & Bridgwater D (1986) Intrusions of mixed origin
    migmatising early Archaean crust in northern Labrador, Canada.
    In: Ashwal LD (ed) Workshop on on early crustal genesis: the
    world's oldest rocks. Lunar and Planetary Inst. Rep. 86-04, 93-97
Schiøtte L, Bridgwater D, Collerson KD, Nutman AP & Ryan AB (1986)
    Chemical and isotopic effects of late Archaean high-grade
    metamorphism and granite injection on early Archaean gneisses,
    Saglek-Hebron, northern Labrador. In: Dawson JB, Carswell DA,
    Hall J & Wedepohl KH (eds) The nature of the lower continental
    crust. Geol. Soc. Spec. Publ. 24, 261-273
Schiøtte L, Compston W & Bridgwater D (1988) Late Archaean ages for
    the deposition of clastic sediments belonging to the Malene
    supracrustals, southern West Greenland: evidence from an ion
    probe U-Pb zircon study. Earth Planet. Sci. Lett. 87, 45-58.
Schiøtte L, Compston W & Bridgwater D (submitted) Ion probe U-Th-Pb
    zircon dating of polymetamorphic orthogneisses from northern
    Labrador, Canada. Can. J. Earth Sci.
Tatsumoto M, Knight RJ & Allegre CJ (1973) Time differences in the
    formation of meteorites as determined from the ratio of lead-207
    to lead-206. Science 180, 1279-1283
Taylor PN, Moorbath S, Goodwin R & Petrykowski AC (1980) Crustal
    contamination as an indicator of the extent of early Archaean
    continental crust: Pb isotopic evidence from the late Archaean
    gneisses of West Greenland. Geochim. Cosmochim. Acta 44,
    1437-1453

REAPPRAISAL OF CRUSTAL EVOLUTION AT KANGIMUT SAMMISOQ, AMERALIK FJORD, SOUTHERN WEST GREENLAND: FLUID MOVEMENT AND INTERPRETATION OF Pb/Pb ISOTOPIC DATA

Allen P. Nutman, Department of Earth Sciences,
Memorial University of Newfoundland, St.Johns,
Newfoundland, Canada
Clark R. L. Friend, Department of Geology,
Oxford Polytechnic, Oxford, England

ABSTRACT

Gneisses affected by ca. 2800 Ma granulite facies
metamorphism outcrop at Kangimut sammisoq, Ameralik fjord,
southern West Greenland. At ca. 2700 Ma these granulite
facies rocks were thrust over a terrane dominated by >3600
Ma Amîtsoq gneisses and a terrane dominated by 2800-2750 Ma
Ikkattoq gneisses, which had not undergone the granulite
facies metamorphism. Following the thrusting event, further
deformation gave rise to folds and subvertical shear zones
formed between ca. 2700 and 2550 Ma. Thrusting and
subsequent folding and shearing were accompanied by
retrogression of the granulite facies rocks under
amphibolite facies conditions. From Nd, Sr, Hf and U-Pb
isotopic systematics the rocks at Kangimut sammisoq have an
age of no more than 3000 Ma. However, they contain variable
amounts of unradiogenic Pb derived from early Archaean
rocks such as >3600 Ma Amîtsoq gneisses. Retrogression
involved an increase in the Pb content of the gneisses at
Kangimut sammisoq. Some or all of this Pb was introduced at
the time of retrogression via hydrous fluids that were
derived from, or passed through, the underlying tectonic
units of Amîtsoq and Ikkattoq gneisses.

INTRODUCTION

In Godthåbsfjord and Ameralik (G and A on Fig. 1) of the
North Atlantic Archaean Craton, early Archaean[1] Amîtsoq
gneisses are found in association with the middle Archaean
Nûk gneisses (McGregor, 1973). The Amîtsoq gneisses contain

---

[1] Division of the Archaean is into early (4000-3400
Ma), middle (3400-2900 Ma) and late (2900-2500 Ma).

D. Bridgwater (ed.), Fluid Movements – Element Transport and the Composition of the Deep Crust, 319–329.

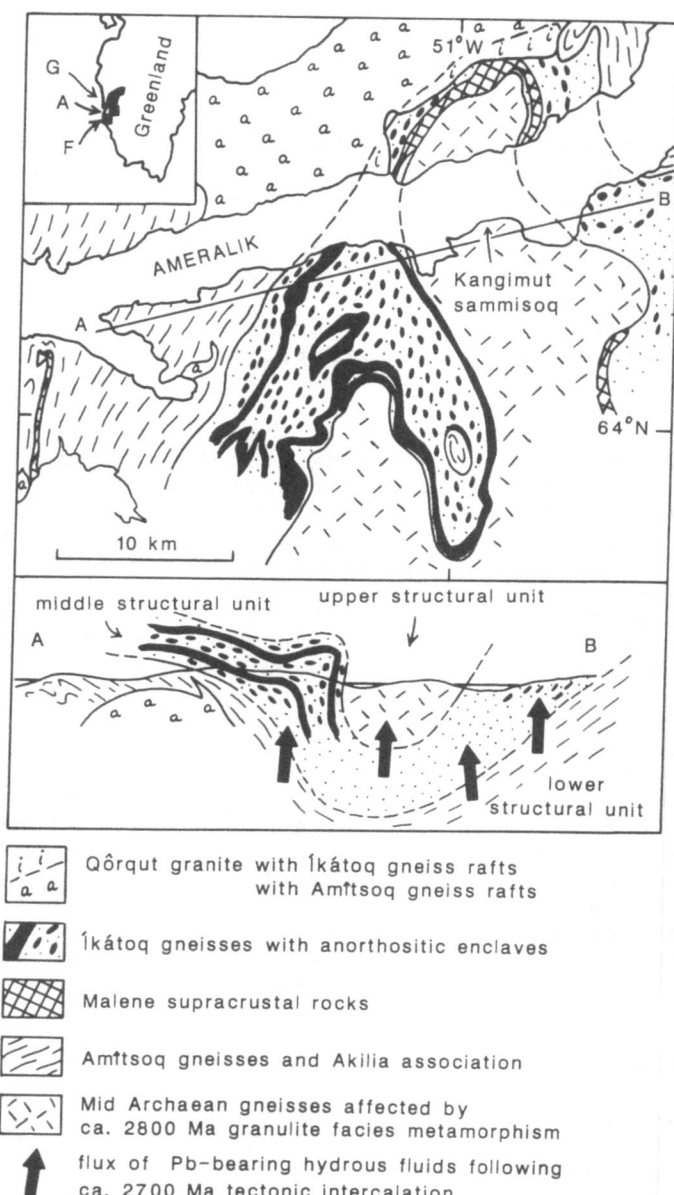

Figure 1. Geological map of the Kangimut sammisoq area, Ameralik fjord. In the inset, G shows Godthåbsfjord, A shows the Ameralik area and F shows the Færingehavn area.

Ameralik dykes, which are subconcordant strips of amphibolite, derived from basic dykes which cut the Amîtsoq gneisses prior to deformation and metamorphism later in the Archaean. All these rocks have amphibolite facies poly-metamorphic assemblages, and there is agreement between the field and isotopic identification of units as Amîtsoq or Nûk gneisses (e.g. Moorbath et al., 1972; Baadsgaard, 1973; Moorbath & Pankhurst, 1976; Baadsgaard & McGregor, 1981). To the south of Godthåbsfjord and Ameralik, there is an extensive region affected by granulite facies metamorphism at ca. 2800 Ma (Pidgeon & Kalsbeek, 1978; Pidgeon et al., 1976; McGregor et al., 1986). There has been lengthy debate as to whether the region affected by this 2800 Ma granulite facies metamorphism contains any early Archaean Amîtsoq gneisses (see McGregor et al., 1986 for summary). The debate has centred on units of heterogeneous gneisses, the largest of which reaches the coast of Ameralik at Kangimut sammisoq, formerly called Kangimut sangmissoq (Fig. 1). These heterogeneous gneisses contain tabular amphibolite bodies that resemble highly disrupted Ameralik dykes, which have been widely used as a field criterion to distinguish the Amîtsoq gneisses from younger lithologies such as Nûk gneisses of Godthåbsfjord, as established by McGregor (1973). The Amîtsoq-like field character of these heterogeneous gneisses (refered to in this paper as "old gneisses") is at variance with the results of Nd, Sr whole rock and U-Pb and Hf zircon isotopic studies which show no evidence that the "old gneisses" are significantly older than ca. 3000 Ma, and thus on the basis of the isotopic data cannot be interpeted at face value as Amîtsoq gneisses (e.g. Collerson et al., 1986; Jones et al., 1986; P. Kinny, pers. comm., 1986). However, the "old gneisses" contain variable amounts of unradiogenic Pb derived in part from an early Archaean source such as Amîtsoq gneisses (Taylor et al., 1980; Jones et al., 1986; D. Bridgwater, pers. comm., 1987).

LATE ARCHAEAN CRUSTAL EVOLUTION

Early models of crustal evolution in Godthåbsfjord, Ameralik and the region to the south suggested that the early Archaean Amîtsoq gneisses were reworked during a 3000 - 2800 Ma crustal accretion - differentiation event, marked by injection of the magmatic protoliths of the voluminous Nûk s.l. gneisses (e.g. Bridgwater et al., 1974; McGregor, 1979; Moorbath et al., 1986). This event was regarded to have culminated in a single regional metamorphic peak at ca. 2800 Ma which outlasted all

Table I. Geological evolution of the Ameralik - Færingehavn
area

**Events common to lower, middle and upper structural
units.**
3 Intrusion of the Qôrqut granite complex at ca. 2550 Ma
2 Formation of upright folds and sub-vertical shear
   zones.
   Amphibolite facies metamorphism, with retrogression of
   granulite facies assemblages in the upper structural
   unit (Tasiusarsuaq terrane).
   Intrusion of granitic sheets. 2550 to 2700 Ma
1 Tectonic intercalation of structural units/terranes,
   isoclinal folding of the lower and middle structural
   units/terranes.

**Development of upper structural unit (Tasiusarsuaq
terrane).**
2 Granulite facies metamorphism at ca. 2800 Ma
1 Intrusion of the protoliths of voluminous,
   predominantly tonalitic gneisses into a sequence of
   basic volcanic rocks and related anorthositic-gabbroic
   intrusions.
   Deformation and high grade metamorphism.
   2800 to 3000 Ma?

**Development of the middle structural unit (Tre Brødre
terrane).**
2 Intrusion of the protoliths of the predominantly
   granodioritic Ikkattoq gneisses, deformation and
   metamorphism between 2800 and 2750 Ma.
1 Formation of Malene supracrustal rocks and intrusion
   of anorthositic-gabbroic complexes into them.

**Development of the lower structural unit (Færingehavn
terrane).**
4 Intrusion of the Ameralik dykes.
3 Granulite facies metamorphism and intrusion of the
   protoliths of the Amîtsoq iron rich suite of augen
   gneisses and ferrodiorite. ca. 3600 Ma
2 Intrusion of the magmatic protoliths of the polyphase
   tonalitic gneisses with granitic layers forming the
   Amîtsoq gneisses, accompanied by high grade
   metamorphism and deformation. pre 3600 Ma
1 Formation of the Akilia (supracrustal) association.

---

significant ductile deformation, and was marked by
amphibolite facies assemblages throughout most of
Godthåbsfjord and Ameralik, and granulite facies in the
region to the south (e.g. Wells, 1979; Coe, 1980).

However, recent detailed mapping (Friend et al., 1987)
combined with U-Pb zircon dating (H. Baadsgaard, pers.
comm., 1986; P. Kinny, pers. comm., 1987) shows that the
Færingehavn and Ameralik areas (Fig. 1), consist of three
lithologically distinct terranes that were tectonically
juxtaposed between ca. 2750 and 2690 Ma (Table I). In this
paper, the juxtaposition of all the terranes is taken to
have occurred at 2700 Ma. Furthest to the south is the
Tasiusarsuaq terrane (structurally highest), affected by
2800 Ma granulite facies metamorphism and dominated by
tonalitic gneisses. The northernmost outcrops of this
terrane are in the core of a complex synform that crosses
central Ameralik and that includes Kangimut sammisoq (Fig.
1). This terrane overlies the Tre Brødre terrane, (middle
structural unit, Fig. 1), dominated by granodioritic
gneisses, which have never undergone granulite facies
metamorphism. These gneisses seem to be restricted to the
Tre Brødre terrane, and U-Pb zircon dating shows that they
are 2750 - 2800 Ma old (H. Baadsgaard, pers. comm., 1986).
These gneisses (previously regarded as Nûk s.l.
gneisses), are named here the Ikkattoq gneisses.
Structurally below the Tre Brødre terrane is the
Færingehavn terrane (lower structural unit, Fig. 1),
dominated by the >3600 Ma Amîtsoq gneisses. The rocks
outcropping in inner Godthåbsfjord and Ameralik belong to
the Tre Brødre and Færingehavn terranes. Following
juxtaposition of the terranes, folds and steeply-inclined
shear zones developed (Fig. 1; Table I). During its
tectonic juxtaposition with the other terranes and
subsequent folding, 2800 Ma granulite facies assemblages in
the Tasiusarsuaq terrane were extensively retrogressed
under amphibolite facies conditions, and the underlying
Færingehavn and Tre Brødre terranes underwent amphibolite
facies metamorphism (Friend et al., 1987).

ORIGIN OF THE GNEISSES AT KANGIMUT SAMMISOQ AND Pb ENRICHMENT BY HYDROUS FLUIDS

Taken at face value, the results from whole-rock Rb-Sr,
Sm-Nd (Fig. 2) and reconnaissance U-Pb and Hf zircon
isotopic data suggest that the "old gneisses" at Kangimut
sammisoq, and similar occurrences at Tinissaq and Akornga
to the south and east of Færingehavn, are unlikely to be
older than 3000 Ma (Jones et al., 1986; Collerson et
al., 1986; P. Kinny pers. comm., 1986). Pb/Pb isotope
whole-rock results (Taylor et al., 1980; Jones et al.,
1986) show that variably retrogressed and sheared granulite
facies gneisses near the northern and western tectonic
boundary of the Tasiusarsuaq terrane (such as at Kangimut
sammisoq, Fig. 1) produce a scatter on a Pb/Pb diagram

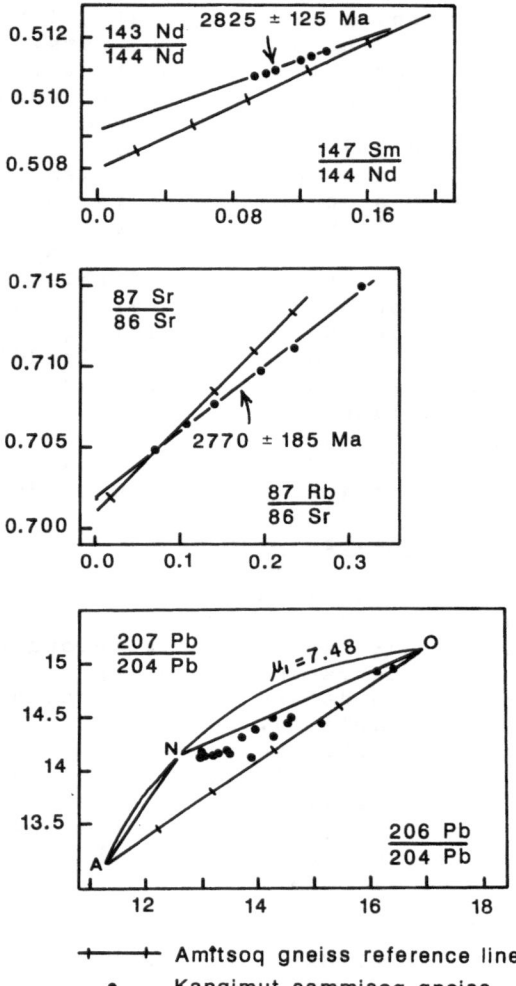

Figure 2. Rb-Sr, Sm-Nd and Pb/Pb isotopic data from Kangimut sammisoq (taken from Jones et al., 1986). The two samples marked by arrows were collected as middle Archaean (Nûk s.l.) gneisses (V. R. McGregor, pers. comm., 1987).

(Fig. 2). On the basis of this scatter Taylor et al. (1980) interpreted these middle Archaean rocks to contain Pb derived from an early Archaean source like Amîtsoq gneisses (A at 3600 Ma, on Fig. 2) mixed with Pb of the type expected in "juvenile" crust of middle Archaean age (such as N at 3000 Ma, on Fig. 2).

There have been two interpretations of these "old gneisses" with their Amîtsoq-type Pb component. Firstly, Taylor et al. (1980) suggested that <u>all</u> these gneisses are middle Archaean in age, and that they inherited their Amîtsoq-type Pb when their precursor magmas passed through, reacted with and hence were contaminated by, Amîtsoq gneisses at depth. However, Sr and Nd isotopic data for the Kangimut sammisoq gneisses show no evidence that their precursor magmas were contaminated by Amîtsoq gneisses (see Jones et al., 1986; Collerson et al., 1986; Moorbath et al., 1986).

Secondly, Bridgwater et al. (1981), Collerson et al. (1986) and Nutman et al. (1986) suggested that the "old gneisses" are Amitsoq gneisses, and that the Amîtsoq-type Pb component in them is more or less <u>in situ</u>, but had been mixed with more juvenile Pb during regional metasomatism associated with ca. 2800 Ma granulite facies metamorphism and/or the subsequent retrogression. This argument was supported by the structural interpretations of the region by Stainforth (1977) and Chadwick et al., (1982), which suggested that the Kangimut sammisoq gneisses were the lateral equivalents of the type Amîtsoq gneisses 20 km to the west, but at a higher metamorphic grade (see Fig. 26 in McGregor et al., 1986). However, interpreting these rocks as reworked, metasomatised Amîtsoq gneisses required that they had lost their Nd and Sr isotopic signature of a >800 Ma crustal residence prior to the granulite facies metamorphism at ca. 2800 Ma (Collerson et al., 1986). Ion-microprobe U-Pb and Hf isotopic studies of zircons from the "old gneisses" at Kangimut sammisoq have subsequently shown that the oldest domains in the zircons are only ca. 3000 Ma old (P. Kinny, pers. comm., 1986). This should be contrasted with cores of zircons in early Archaean rocks from Antarctica and Labrador affected by late Archaean granulite facies metamorphism, which <u>do</u> preserve the true antiquity of the protoliths (Black et al., 1986; Schiøtte, in press). Therefore, on the basis of the <u>current understanding</u> of isotopic systematics of rocks affected by granulite facies metamorphism and on the basis of currently available isotopic data, the authors accept the Nd, Sr and U-Pb zircon evidence that the "old gneisses" at Kangimut sammisoq do not contain a detectable Amîtsoq gneiss component. Instead, we suggest another explanation of the origin of the Amîtsoq-type Pb component in the Kangimut sammisoq "old gneisses", based on our own structural studies in the region.

Detailed structural studies (Friend et al., 1987 and unpublished data), show that the "old gneisses" at Kangimut sammisoq are tectonically underlain by terranes dominated by Ikkattoq and Amîtsoq gneisses (Fig. 1). Following juxtaposition of the terranes with burial of Amîtsoq and

Ikkattoq gneisses under the Tasiusarsuaq terrane, rocks at the margin of the Tasiusarsuaq terrane (such as at Kangimut sammisoq, Fig. 1), were deformed and almost totally retrogressed under amphibolite facies conditions (Table I). Gneisses in amphibolite facies shear zones at Kangimut sammisoq are enriched in Pb compared with adjacent unsheared, but retrogressed granulite facies gneisses, which in turn are enriched in Pb over granulite facies gneisses elsewhere in the region (Table II).

Table II. Pb abundances in middle Archaean tonalitic gneisses. Published XRF determinations on samples collected by Bridgwater, Nutman and Garde.

	Pb p.p.m.
Average of ten tonalitic Nûk gneisses from Nuuk town, unaffected by granulite facies metamorphism	14
Average of fourteen granulite facies gneisses, northwest of Godthåbsfjord	4
Average of sixteen gneisses, northwest of Godthåbsfjord, retrogressed from granulite facies	8
Average of four tonalitic gneisses from Tinissaq, southeast of Færingehavn, partially retrogressed from granulite facies	7
Average of eleven tonalitic gneisses from Kangimut sammisoq, totally retrogressed from granulite facies	8
Sheared gneiss, Kangimut sammisoq	17

This suggests that the retrogression involved introduction of Pb in a hydrous fluid, at least in part via ductile shear zones. Hydrous fluid (containing Pb) necessesary for the retrogression of the granulite facies assemblages may have been derived from, or passed through, the underlying Amîtsoq and Ikkattoq gneisses (Fig. 1). This would have disturbed the U-Pb systematics of the retrogressed granulites, giving rise to the scatter in the Pb/Pb plot (Fig. 2). This scatter could result from two processes; firstly, U loss or gain, and secondly, introduction of an exotic Pb component, that contained variable amounts of Amîtsoq-type Pb. The increase in the total Pb content of the gneisses upon retrogression and shearing shows that the second factor was significant. Support for this interpretation of the geology of Kangimut

sammisoq, which requires typical Amîtsoq gneisses at depth,
comes from the high I_{Sr} value (0.7060) for 2690 Ma
granite sheets (i.e. post-juxtaposition of the three
terranes) that cut the Tasiusarsuaq terrane (upper
structural unit) on the north side of Ameralik (Roberts,
1979).

DISCUSSION

The gneisses at Kangimut sammisoq are interpreted to show
introduction of exotic Pb associated with the influx of
hydrous fluids from underlying units following tectonic
juxtaposition. This may explain the origin of all the
unradiogenic Pb found in these rocks. Further Pb isotopic
studies are in progress to see if this could be the case,
or if some of the unradiogenic Pb was already present in
these rocks prior to tectonic juxtaposition of the terranes
(D. Bridgwater, pers. comm., 1987). This study is an
example of how Pb abundances and isotopic variations might
be used to show the movement, and perhaps source regions,
of fluids during medium to high grade metamorphism,
associated with major tectonic events (see also Schiøtte
this vol.).

ACKNOWLEDGEMENTS

We thank Vic McGregor, Brian Fryer, John Duke and David
Bridgwater for comments on an earlier draft of this paper
or a display on which this paper was based. George Jenner
is thanked for reviewing the paper. Field work was
supported by the Royal Society of London, the Carlsberg
Foundation, the Danish S.N.F., Memorial University of
Newfoundland, Oxford Polytechnic and the Geological Survey
of Greenland (GGU). The Director of GGU is thanked for
permission to publish this paper.

REFERENCES

Baadsgaard, H. 1973: U-Th-Pb dates on zircons from the
    early Precambrian Amîtsoq gneisses, Godthaab district,
    West Greenland. Earth Planet. Sci. Lett., 19, 22-28.
Baadsgaard, H. and McGregor, V.R. 1981: The U-Th-Pb
    systematics of zircons from the type Nûk gneisses,
    Godthaabsfjord, West Greenland. Geochim. Cosmochim. Acta
    45, 1099-1109.
Black, L.P., Willians, I.S. and Compston, W. 1986: Four
    zircon ages from one rock: the history of a 3930 Ma-old
    granulite from Mount Sones, Enderby Land, Antarctica.
    Contrib. Mineral. Petrol., 94, 427-437.

Bridgwater, D., McGregor, V.R. and Myers, J.S. 1974: A horizontal tectonic regime in the Archaean of Greenland and its implications for early crustal thickening. Precambrian Res., 1, 179-198.

Bridgwater, D., McGregor, V.R. and Nutman, A. 1981: Geological constraints on isotopic and thermal models for the consolidation of the late Archaean crust, West Greenland. Terra cognita, 1, 93 only.

Chadwick, B., Coe, K. and Stainforth, J.G. 1982: Magma generated structures and their subsequent development in the late Archaean evolution of northern Buksefjorden region, southern West Greenland. Geol. Rundsch., 71, 61-72.

Coe, K. 1980: Nûk gneisses of the Buksefjorden region, southern West Greenland and their enclaves. Precambrian Res., 11, 357-371.

Collerson, K.D., McCulloch, M.T., Bridgwater, D., McGregor, V.R. and Nutman, A.P. 1986: Strontium and Neodynium isotopic variations in early Archaean gneisses affected by middle to late Archaean high - grade metamorphic processes : West Greenland and Labrador. In Ashwal, L.D. (ed.) Early Crustal Genesis: The Worlds Oldest Rocks. Lunar and Planetary Inst. Tech. Rpt. 86-04

Friend, C.R.L., Nutman, A.P. and McGregor, V.R. 1987: Late-Archaean tectonics in the Færingehavn - Tre Brødre area, south of Buksefjorden, southern West Greenland. J. Geol. Soc. Lond., 144, 369-376

Jones, N.W., Moorbath, S and Taylor, P.N. 1986: Age and origin of gneisses south of Ameralik, between Kangimut sangmissoq and Qasigianguit. In Ashwal, L.D. (ed.) Early Crustal Genesis: The Worlds Oldest Rocks. Lunar and Planetary Inst. Tech. Rpt. 86-04

Moorbath, S. and Pankhurst, R.J. 1976: Further rubidium - strontium age and isotopic evidence for the nature of the late Archaean plutonic event in West Greenland. Nature, 262, 124-126.

Moorbath, S., O'Nions, R.K., Pankhurst, R.J., Gale, N.H. and McGregor, V.R. 1972: Further rubidium-strontium age determinations on the very early Precambrian rocks of the Godthåb district, West Greenland. Nature, 240, 78-82.

Moorbath, S., Taylor, P.N. and Jones, N.W. 1986: Dating the oldest terrestrial rocks - fact and fiction. Chem. Geol., 57, 63-86.

McGregor, V.R. 1973: The early Precambrian gneisses of the Godthåb district, West Greenland. Phil. Trans. R. Soc. Lond., A273, 343-358.

McGregor, V.R. 1979: Archaean gray gneisses and the origin of the continental crust: Evidence from the Godthåb Region, West Greenland. In Barker, F. (ed)

Trondhjemites, Dacites and Related Rocks. Elsevier, Amsterdam.

McGregor, V.R., Nutman, A.P. and Friend, C.R.L. 1986: The Archean Geology of the Godthåbsfjord Region, Southern West Greenland. In Ashwal, L.D. (ed.) Early Crustal Genesis: The Worlds Oldest Rocks. Lunar and Planetary Inst. Tech. Rpt. 86-04.

Nutman, A.P., Bridwater, D. and McGregor, V.R. 1986: Regional variation in the Amîtsoq gneisses related to crustal levels during late Archaean granulite facies metamorphism, southern West Greenland. In Ashwal, L.D. (ed.) Early Crustal Genesis: The Worlds Oldest Rocks. Lunar and Planetary Inst. Tech. Rpt. 86-04

Pidgeon R.T. and Kalsbeek F. 1978: Dating of igneous and metamorphic events in the Fiskenæsset region of southern West Greenland. Can. J. Earth Sci., 15, 2021-2025

Pidgeon, R.T., Aftalion, M. and Kalsbeek F. 1976: The age of the Ilivertalik granite in the Fiskenæsset area. Rapp. Grønlands geol. Unders., 73, 31-33

Roberts, I.W.N. 1979: Archaean evolution of inner Ameralik, south west Greenland, with special reference to mid-Archaean magmatism. Unpublished Ph.D. thesis, University of Wales, Aberystwth.

Schiøtte, L. in press: Redistribution of U and Pb by fluids suggested as a cause for a Pb-Pb pseudo-isochron obtained on old-Archaean orthogneisses. In Bridgwater, D. (ed.) Fluid movements, element transport, and the composition of the deep crust. Reidel, Dordrecht.

Stainforth, J.G. 1977: The structural geology of the area between Ameralik and Buksefjorden, southern West Greenland, PhD thesis, University of Exeter.

Taylor, P.N., Moorbath, S., Goodwin, R. and Petrykowski, A.C. 1980: Crustal contamination as an indicator of the extent of early Archaean continental crust: Pb isotopic evidence from the late Archaean gneisses of West Greenland. Geochim. Cosmochim. Acta., 44, 279-296

Wells, P.R.A. 1979: Chemical and thermal evolution of Archaean sialic crust, southern West Greenland. J. Petrol., 20, 187-226.

CHANGES IN THE ISOTOPIC COMPOSITION OF WHOLE-ROCK Pb DURING DIFFERENT
STAGES OF RETROGRESSION OF LATE ARCHAEAN GRANULITE FACIES GNEISSES FROM
KANGIMUT SAMMISOQ, SOUTHERN WEST GREENLAND

D. Bridgwater.
Geological Museum,
Østervoldgade 5-7 Copenhagen K,
DK 1350, Denmark.

ABSTRACT. Pb-Pb whole rock isotopic and total Pb and U concentration
data are presented from samples from a controversial outcrop of late
Archaean gneisses affected by circa 2.82 Ga granulite facies
metamorphism followed by different degrees of retrogression in the
period 2.82 to 2.4 Ga. The retrogression occured when the granulite
facies gneisses were thrust over an amphibolite facies terrane which
contains units of both early and late Archaean gneiss. Both U and Pb
were added during retrogression but in different proportions in
individual samples. The isotopic composition of the Pb from the least
retrogressed granulites is comparatively radiogenic. The small amount of
Pb present has a low proportion of Pb derived from an early Archaean
source. This is interpreted as due to post-granulite introduction of U
but little or no introduction of Pb in the first stage of retrogression.
Partly retrogressed gneisses contain higher total Pb. This is less
radiogenic than that seen in the least retrogressed rocks and a higher
proportion of the Pb is derived from older crust. The most highly
retrogressed rocks contain 5 - 10 times the total Pb present in the
least retrogressed rocks. The Pb in these rocks is comparatively
radiogenic and has a low proportion of early Archaean Pb. It is
suggested that the changes in Pb isotopic composition reflect different
proportions of Pb and U added to different samples during retrogression,
combined with different sources of the introduced Pb. This may be due to
more than one period of retrogression associated with different episodes
of thrusting or to changes in the source of the fluids causing
retrogression.

1. INTRODUCTION

1.1 History of research and development of ideas.
The age and origin of the Archaean gneisses from Kangimut sammisoq, a
peninsular on the south side of Ameralik fjord, southern West Greenland
(see Nutman & Friend, Fig. 1, this volume), have been a matter of
controversy between field geologists and geochronologists for over ten
years. Contrasting views are outlined bv different authors in

331

D. Bridgwater (ed.), Fluid Movements – Element Transport and the Composition of the Deep Crust, 331–344.
© *1989 by Kluwer Academic Publishers.*

Bridgwater et al., 1979, papers in Ashwal, 1986 (Jones et al., Collerson et al., Nutman et al.,) and in separate papers by Moorbath et al. (1986) and Nutman & Friend, (1989).

Field observations show that there are two main groups of gneiss at this locality: An older suite of migmatites cut by a series of metamorphosed porphyritic tholeiitic dykes; and a younger group of more mafic gneisses which range from diorites to mafic tonalites. The younger gneisses preserve original igneous layering and do not contain basic dyke remnants, and are regarded as intrusive into the older migmatitic gneisses. Both groups are affected by late Archaean granulite facies metamorphism and cut by post granulite facies granitic sheets regarded as equivalent to the circa 2.65 Ga. Qârusuk dykes (McGregor et al., 1983). The gneisses (particularly the older group) became partially mobile under granulite facies conditions and the basic dykes in the older group now outcrop as a series of separated inclusions which only locally show discordant contacts to an earlier migmatitic fabric. The outcrop is affected by late Archaean retrogression associated with local shearing concentrated along circa 20°/V structures (Bridgwater et al., 1981, 1985, Nutman et al., 1986, Nutman & Friend, this volume). Prior to the recent advances in the understanding of the tectonic framework of the area (Friend et al., 1987, see below) the interpretation prefered by many field geologists was that the older gneisses represented a direct continuation of the early Archaean Amîtsoq gneisses, the type locality of which is in the outer part of Ameralik fiord; while the younger gneisses were regarded as a rather mafic suite of orthogneisses emplaced at the same time as the circa 2.9-3.05 Ga Nûk gneisses of the Godthåbsfjord area (McGregor, 1973, Baadsgaard & McGregor, 1981). The transition between the amphibolite facies rocks in outer Ameralik fjord and the granulite facies rocks from Kangimut sammisoq was regarded as a prograde boundary (Wells, 1979), locally modified by late Archaean retrogression associated with shearing. This interpretation contrasted with those from isotopic studies summarised by several authors in Bridgwater et al. (1979). The isotopic work, which included U/Pb bulk zircon studies (Baadsgaard) and both Rb-Sr and Pb-Pb whole rock studies (Moorbath & Taylor) were reported as giving no evidence of a crustal history prior to 3.0 Ga for any of the rock units at Kangimut sammisoq. The Rb-Sr and Pb-Pb isotopic data was published in more detail together with Sm-Nd data by Jones et al. (1986) and Moorbath et al. (1986). Further Rb-Sr and Sm-Nd isotopic results were presented by Collerson et al. (1986). The isotopic results either implied that the field identification of early Archaean gneisses at this locality was incorrec¹ or that virtually complete erasure of the primary isotopic signature of the gneisses had taken place during late Archaean granulite facies metamorphism and subsequent retrogression.

The only isotopic indication that any early Archaean material is present is that seen in the Pb isotopes (Jones et al. 1986). These do not define a single late Archaean isochron but show considerable scatte¹ which cannot be attributed solely to variations in U/Pb ratios in magma from a single homogeneous source emplaced during the late Archaean. Varying amounts of a component with a low 207Pb/204Pb ratio characteristic of a depleted early Archaean lower crustal source such a the Amîtsoq gneisses of outer Ameralik fjord are present. Accepting the average isotopic values of Amîtsoq Pb given by Taylor et al. (1980) it can be calculated that up to 55% of the Pb in individual samples of the gneisses from Kangimut sammisoq was derived from an early Archaean U-depleted crust. There is no correlation between the Pb isotopic compositions obtained by Jones et al. and the geological divisions into "older" and "younger" gneisses. Samples from both show similar isotopic compositions, (V.R. McGregor, personal communication, 1987 and Fig 2). The presence of early Archaean Pb in the Kangimut sammisoq gneisses was interpreted by Jones et al. (1986) as the result of contamination of a late Archaean magma as it passed through an (unexposed) early Archaean unit during intrusion, (the model suggested by Taylor et al., 1980, to explain the presence of a component of early Archaean Pb in the late Archaean Nûk gneisses of the Godthåbsfjord area). It must be emphasised however that the proportion of older Pb present reported by Jones et al. in some samples from Kangimut sammisoq is far higher than that found in any of the 2.95-3.05 Ga Nûk gneisses from Godthåbsfjord. The only late Archaean orthogneisses with a component of early Archaean Pb as high as those found at Kangimut sammisoq are the circa 2.65 Ga Qârusuk dykes and equivalents interpreted as formed by partial melting of early crust (McGregor et al., 1983, Taylor et al., 1984).

Unfortunately the published isotopic determinations from Kangimut sammisoq were not accompanied by details of which units were studied or the metamorphic state and geochemistry of individual samples. In view of the potential interest for the interpretation of isotopic results from polymetamorphic high grade rocks, a programme of detailed collecting and both geochemical and isotopic work was started (see Bridgwater et al., 1981, 1985, Nutman et al., 1986 Collerson et al., 1986) using samples whose exact field setting was known and whose chemistry and mineralogy could be determined. Kinny (1987) confirmed the view held by geochronologists that there was no evidence for older crust in situ at Kangimut sammisoq when he obtained single zircon U-Pb ion-probe ages of 2.92 Ga for the protoliths of the earliest igneous rocks and circa 2.82 Ga for the age of granulite facies metamorphism. Kinny also reports Hf isotope results from zircons which precluded the presence of a major early Archaean crustal component in a sample of the early gneiss. Ion-probe U-Pb results reported from a comparable controversial outcrop 60 km to the south (Schiøtte et al., in press) are in agreement with the timing of events suggested by Kinny (1987) and also give no indication of early Archaean rocks outcropping in the granulite facies terrane south of Ameralik fjord. The lack of U-Pb and Lu-Hf zircon evidence for an early Archaean component in the gneisses, together with the re-interpretation of the field evidence by Friend et al. (1987), see below, means that the controversy about the age of the rocks is resolved (in favour of the geochonological interpretation). However in detail interpretation of the isotopic results from the outcrop still requires integration with the large amount of petrological and geochemical data which has become available because of the controversy. One of the main problems is why some samples of the late Archaean gneisses contain higher proportions of early Archaean Pb than those seen in the type Nûk gneisses which are still regarded as intruded through older crust. Is

this the result of contamination during intrusion as suggested by Jones et al. (1986), or could the mixed isotopic composition be the result of metamorphic redistribution of both Pb and U as suggested by Bridgwater et al. (1981, 1985) and in Labrador by Schiøtte et al.(1986) ?
The present paper reports results from samples specifically selected to test whether there is a clear link between degree of retrogression and Pb isotopic composition.

2. REGIONAL FRAMEWORK (adapted from Friend et al.,1987, Nutman & Friend, this volume).

The currently accepted geological model suggests that Kangimut sammisoq is part of a late Archaean granulite facies terrane formed from plutonic protoliths intruded during the late Archaean. The U-Pb ion probe data of Kinny (1987) and Schiøtte et al. (in press) constrain the igneous activity during which the protoliths to the gneisses formed to between 2.92 Ga, the age of igneous zircons from the earliest component in the complex and 2.82 Ga, the age of zircons grown during granulite facies metamorphism. The granulite facies gneisses (the Tasiusarsuaq terrane of Friend et al., 1987) were thrust over two major units of amphibolite facies gneiss, the 2.75-2.8 Ga Ikátoq gneisses which dominate the Tre Brødre terrane, themselves thrust over pre-3.6 Ga Amîtsoq gneisses, which dominate the Færingehavn terrane (Friend et al., 1987, Nutman & Friend, this volume). The age of thrusting is post the intrusion of the 2.75-2.8 Ga Ikátoq gneisses but earlier than late Archaean amphibolite facies metamorphism during which overgrowths formed on zircons at 2.65 Ga (Schiøtte et al., 1988). The gneiss complex was intruded by late Archaean granitic sheets at approximately 2.65 Ga and major late Archaean plutons (the Qôrqut granite) at 2.5 Ga (Baadsgaard, 1976). These late granitic rocks contain major components of melted crust (Taylor et al., 1984).

3.FIELD RELATIONS AND MAIN PETROLOGICAL FEATURES

The older gneiss unit at Kangimut sammisoq is a heterogenous medium to coarse grained quartzofeldspathic orthogneiss with inclusions of supracrustal material and broken-up fragments of basic dykes which locally cut across early migmatitic structures. The main heterogeneity in the older gneisses is due to partial melting and mobility during granulite facies metamorphism which led to separation of leucocratic and more mafic phases. The leucocratic phases range from mafic-poor diffuse segregations within the general body of the gneiss to distinct veins which cross cut earlier structures and break up the basic dyke remnants enclosed in the gneisses. The distribution of mafic minerals in the gneisses is patchy, with coarse-grained mafic rich segregations of orthopyroxene, garnet and Fe-Ti oxides partially replaced by orthoamphibole, hornblende and biotite. The younger gneisses are in general more homogeneous (except where they preserve original igneous layering). Both older and younger gneisses are affected by the circa

2.82 Ga granulite facies event. The younger gneisses develop coarse
pegmatites during granulite facies metamorphism. All the gneisses on
the outcrop have been affected by post-granulite facies retrogression.
The intensity of retrogression ranges from partial break-down of
orthopyroxene-garnet assemblages through to the total replacement of all
mafic minerals by biotite. Samples of the most retrogressed gneisses
contain no garnet, pyroxene or amphibole and develop interstitial
K-feldspar and muscovite. Retrograde P-T paths are given for the
Tasiusarsuaq terrane by Riciputi et al. (1988). Reconnaissance fluid
inclusion work (J. Konnerup Madsen, personal communication, 1983) showed
that quartz grains from the most retrogressed samples still contain some
CO_2 inclusions which agrees with the field interpretation that these
rocks have formed by the retrogression of earlier granulite facies
gneisses and are not younger intrusive veins. The degree of
retrogression varies markedly throughout the outcrop, in general it
appears to be due to the static introduction of fluids along a net-work
of fractures and grain boundaries with remnants of better preserved
granulite facies assemblages enclosed as islands in more retrogressed
material (see Bridgwater, 1979). Locally retrogression is concentrated
along small shear zones or along lithological boundaries such as the
contacts of basic dyke remnants.

Geochemical and petrological studies show that all the whole rock
isotopic systems are potentially disturbed, Rb, U and Pb are very
strongly depleted in both the felsic and mafic granulite facies rocks of
the outcrop compared to regional averages of either early or late
Archaean gneisses from the Godthaabsfjord - Ameralik fjord area (Nutman
et al., 1986). They are all reintroduced during retrogression. Sr
isotope studies (see for example Collerson et al., 1986, Jones et al.,
1986) give a scatter about a 2.77 Ga isochron with an ISr of 0.702. This
is consistent with post-granulite facies redistribution of elements with
marked Rb addition to rocks with a comparatively short crustal history
prior to the 2.82 Ga granulite facies event. Zircons show marked new
growth during granulite facies metamorphism. Homogeneous rims enclose
older zoned cores of zircon and earlier mineral phases such as quartz
and apatite. The REE patterns of individual gneisses are largely
controlled by the distribution of mineral phases developed when the
gneisses were partially melted during granulite facies metamorphism
The comparatively mafic-rich restites in the gneisses have flat REE
patterns with La/Lu chondrite normalised ratios of about 4, Felsic
partial melts are HREE depleted (La/Lu$_N$ 40) and show large positive Eu
anomalies. These patterns are similar to restite/partial melt patterns
seen in granulite facies gneisses from other parts of the North Atlantic
craton. The control of the REE patterns of individual samples by partial
melting and mineral assembalges developed during granulite facies
implies that Sm-Nd isotopic results reported from the outcrop by
Collerson et al. (1986), Jones et al. (1986) and Moorbath et al. (1986)
yield the age of granulite facies metamorphism rather than that of
original formation. (see Schiøtte et al., 1986 and Schiøtte & Bridgwater
in press). The positive ϵNd $_{2.8Ga}$ values obtained point to a short
crustal history prior to 2.8 Ga. The Pb-Pb whole rock data given by
Jones et al. (1986) and Moorbath et al. (1986) is thus the only evidence

336

for the involvement of early Archaean crust in any of the gneisses from these outcrops.

In view of the conclusive ion-probe U-Pb and Lu-Hf evidence from zircons that the igneous protoliths to the gneisses are younger than 3.0 Ga, the simplest way to explain the presence of Pb from an early Archaean crustal source in the late Archaean gneisses would be to adapt the geochemical model of Pb introduction during retrogression suggested by Bridgwater et al., (1981, 1985) so that it agreed with the more recent field framework. This model which is advocated by Nutman & Friend (this volume) suggests that the early Archaean Pb was introduced into depleted late Archaean granulite facies gneisses by fluids which had passed through older crust during retrogression associated with thrusting the Tasiusarsuaq terrane over terranes containing major units of early Archaean crust. In this case there should be a direct correlation between degree of retrogression, total Pb content and degree of contamination with Pb derived from older crust. A correlation of this type is seen in the Pb isotopic data from Buksefjord, a few kilometers to the south (Fig 1).

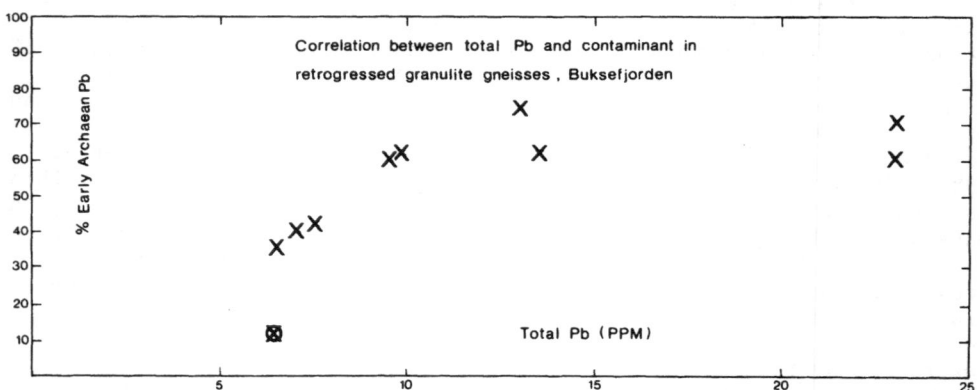

Fig. 1. Total Pb contents plotted against percentage of early Archaean Pb in partly retrogressed late Archaean granulite facies gneisses, Buksefjord, SW Greenland. Data from Taylor et al. (1980).

New results decribed below obtained as part of the programme of geochemical and isotopic studies following the debate presented in Bridgwater et al. (1979) suggest the simple model of contamination by Pb derived from a single source does not apply perfectly at Kangimut sammisoq.

4. SAMPLE SELECTION

The geochemical studies reported in abstract by Bridgwater et al. (1985) showed that both total Pb and U contents in the gneisses can be correlated approximately with degree of retrogression. The least retrogressed samples from the outcrop commonly contain less than 3 pmm Pb, the detection limit for accurate determinations using current XRF preceedures, while strongly retrogressed gneisses contain up to 15 ppm Pb. As there are no petrological or geochemical details published for the samples studied isotopically by Jones et al. (1986) their results are supplemented here by a small of new isotopic determinations from controlled gneiss samples selected to give the maximum range in geochemical and metamorphic characters.

As a subsiduary part of the element mobility study a series of samples was taken across the margin of a basic dyke fragment from the contact adjacent to completely retrogressed gneiss into the center of the enclave approximately 70 cm away from the margin. The dyke matrix, which is a medium to.coarse grained amphibolite is interpreted as recrystallised from an earlier granulite facies assemblage and shows no marked petrological changes from margin to center. A feldspar aggregate within the dyke, interpreted as recrystallised from an earlier igneous plagioclase phenocryst was also analysed. Total Pb contents in the dyke are well below 3 ppm throughout, and are given an estimated value of 1 ppm in Table 1. The Pb content in the feldspar aggregate lies between that in the dyke matrix and 3 ppm. There is marked drop in Rb and increase in K/Rb ratios from dyke margin to center suggesting introduction of LIL elements during retrogression.

All the samples examined in thin section show slight late static metamorphism with the formation of epidote superimposed on varying degrees of retrogression under amphibolite facies conditions.

Analyses. Pb separation for the first batch of samples was carried out at the University of Copenhagen, and for the second batch of samples at RHBNC by Dr M. Thirwell using standard techniques. The isotopic measurements were carried out under contract by RHBNC.

5. RESULTS

The new isotopic data is listed in Table 1 and plotted on Fig 2 together with the data reported by Jones et al. (1986).
The results fall within the range of results reported previously. The Pb isotopic composition of both the gneisses and the dyke varies considerably even between petrologically similar samples taken within less than a meter from each other. There is no marked correlation between total Pb content and the proportion of early Archaean Pb present (using a 2.65 Ga refererence isochron) such as might be expected if the Pb introduced during retrogression was extracted from a fluid with a isotopically homogenous composition. Samples 243029, respectively the most Pb depleted gneiss and 243003a, the most Pb enriched gneiss analysed yield Pb-Pb isotopic data points which are within analytical

338

error. The highest contents of early Archaean Pb are found in partly retrogressed samples which retain some mineralogical evidence of earlier granulite facies assemblages. The three dyke samples plot close to a 2.35 Ga reference isochron with a μ1 of 6.8 suggesting severe contamination by crustal Pb with a high early Archaean component. This is assumed to represent Pb introduced into the dyke during retrogression since there is no evidence that the dyke was emplaced through or into older crust. The major variation in the Pb isotopic composition of the dyke samples is seen in differences in the 206/204 ratios and is interpreted as controlled by relatively late movement of U, The feldspar aggregate contains the least radiogenic Pb, consistent with a low U/Pb ratio, the centre of the dyke shows the highest 206/204 ratio, consistent with preferential U movement compared to Pb into the central part of the dyke.

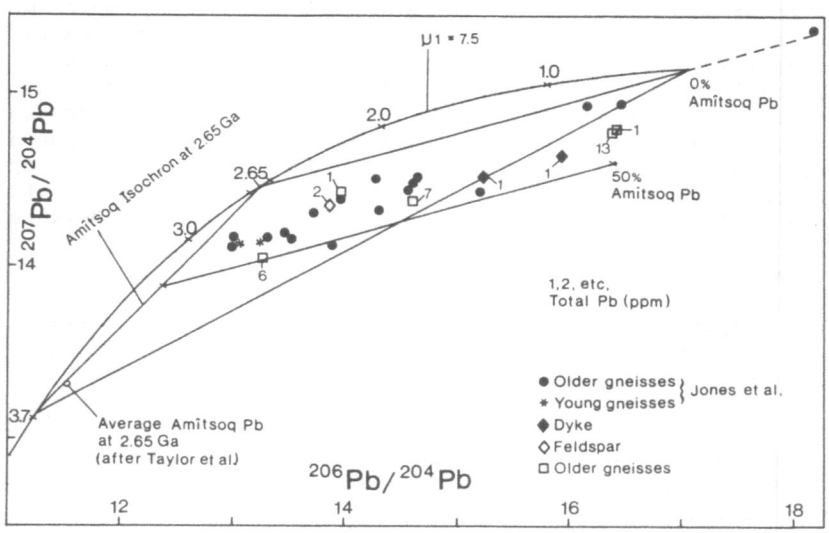

Fig 2. Pb-Pb whole rock isotopic data from Kangimut sammissoq.

6. DISCUSSION

The low total Pb contents of the granulite facies gneisses and the marked increase in total Pb in the retrogressed gneisses established earlier (Bridgwater et al., 1981, 1985, Nutman et al., 1986, Nutman & Friend, this volume), suggest that the contamination during intrusion model put forward by Jones et al. (1986) as the main reason why gneisses at Kangimut sammisoq have a component of early Archaean Pb needs revision. Whether or not there was a component of early Archaean in the gneisses at the time of granulite facies metamorphism is uncertain. There is now no direct field or isotopic evidence of early

Archaean rocks within the granulite facies Tasiusarsuaq terrane of Friend et al. (1987). It should perhaps be noted that the main occurences of rocks with a high proportion of early Archaean Pb within the Tasiusarsuaq terrane come from outcrops from which earlier geological interpretations had suggested the presence of highly migmatised early gneisses. The lack of correlation between either total Pb concentrations or metamorphic petrology and degree of contamination with early Archaean Pb shown in Table 1 also suggests that a single mixing model between a late Archaean component present in the granulite facies gneisses before retrogression and addition of Pb of constant isotopic composition from an underlying amphibolite facies slab during retrogression is too a simple an explanation. The present results show that the highest concentrations of early Archaean Pb are found in rocks in which retrogression has been moderately severe but has not totally obliterated granulite facies geochemical and petrological characters. Gneisses in which retrogression has gone to completion and in which total Pb contents have increased by up to an estimated 5 to 10 times that of the granulite gneisses contain a higher proportion of late Archaean Pb.

A possible explanation for the scatter seen on Fig. 2 and geochemical data in Table 1 is that during retrogression both the U/Pb ratio and the isotopic composition of the Pb in the fluid changed. During the first stages of retrogression, marked by the break down of orthopyroxene to orthoamphibole, the main component added to the gneiss was an aqueous fluid which introduced U but little or no Pb or Rb. This resulted in a present day relative enrichment in radiogenic Pb with high 206/204 ratios, such as seen in 243029. It is likely, but unproven that the small component of early Archaean Pb seen in samples 243029 and 243007a represent introduction of a small amount of early Archaean Pb during the first stages of retrogression rather than representing a small fraction of Pb mixed at the time of the formation of the protolith and retained in the rock at the peak of granulite facies. Increase in the degree of retrogression was accompanied by introduction of Pb at a greater rate than U, resulting in a general concentration of points with comparatively low 206/204 ratios. Assuming that the present concentration of 5-8 ppm Pb seen in these partly retrogressed gneisses (compared to the 1-3 ppm in the least retrogressed rocks) is largely introduced, a major part of the contaminant appears to be derived from source with a 50% component of older gneiss. This is in agreement with the model advocated by Nutman & Friend, (this volume) and with the results shown from Buksefjorden in Fig 1. During the final stage of retrogression in which potash feldspar and muscovite formed the total Pb contents increased to over 10 ppm and there was an increase in the U/Pb ratio. The proportion of early Archaean Pb introduced decreased. This resulted in Pb isotopic compositions at the present day with both high 206/204, and 207/204 ratios such as in sample 243003a which has an isotopic composition within error of the least retrogressed samples. The variation in the proportion of early Archaean Pb present in the Pb introduced suggests that the source of the fluid causing retrogression may have changed from one with Pb derived from a crustal source with about 50% of early Archaean gneiss, to a source dominated by Pb with a

TABLE 1: Pb isotopic data from geologically controlled samples Kangimut sammisoq.

GGU Nr	Pb ppm	$^{206}Pb/^{204}Pb$	$^{207}Pb/^{204}Pb$	$^{208}Pb/^{204}Pb$	K/Rb	Uppm	% Amitsoq Pb*
243001	7	14.587 ± 2	14.384 ± 2	34.074 ± 6	460	0.4	36.7
243002	6	13.266 ± 1	14.045 ± 1	32.632 ± 2	560	0.13	48.9
243003a	13	16.399 ± 1	14.807 ± 1	34.761 ± 6	285	0.37	25.0
243003e	1**	15.234 ± 2	14.520 ± 2	33.570 ± 5	651	n.d.	34.3
243003g	2**	13.861 ± 2	14.340 ± 2	33.570 ± 5	580	n.d.	26.3
243003i	1**	15.930 ± 2	14.639 ± 2	24.792 ± 5	610	n.d.	35.0
243007b	1**	13.973 ± 1	14.387 ± 1	33.251 ± 4	2600	0.07	23.1
243029	1**	16.420 ± 1	14.824 ± 1	34.572 ± 3	3000**	0.07	23.4

** Estimated contents of Pb and K/Rb ratios when below limits of accurate determination using XRF, (3 ppm Pb, 1 ppm Rb).

* Assuming an age of 2.65 Ga as time of mixing, a μl value for the reservoir of 7.48 and an average composition of Amitsoq Pb in outer Ameralik at 2.65 Ga to be: $206Pb/204$ Pb = 11.532 $207Pb/204Pb$ = 13.308, and that of the late Archaean reservoir to be $206Pb/204$ Pb = 13.222 $207Pb/204Pb$ = 14.444 (adapted from Taylor et al., 1980).

Sample description: Samples 243001 to 243003a are non-remobilised gneisses from a zone within 1 meter from the margin of a basic dyke remnant now with an amphibolite facies mineralogy. 001 and 002 are about 50cm from the dyke. 003a is at the dyke contact (0-10 cm into the gneiss). At the contact the gneiss develops a slight new foliation. There is no evidence of a new phase of intrusion along the dyke contact. None of this group retain granulite facies assemblages. 001 and 002 are quartz plagioclase gneisses with biotite and hornblende with local relics of orthoamphibole, and sphene replacing Fe-Ti oxides. The mafic minerals are grouped together in patches, interpreted as replacing earlier orthopyroxene - garnet Fe-Ti oxide assemblages. Zircons are strongly zoned with homogeneous rims surrounding zoned centers, which is interpreted as overgrowths of new zircon developed during granulite facies metamorphism (see Kinny, 1987, Schiøtte et al, in press). In 003a the patchy distribution of mafic minerals is destroyed by the formation of a more pronounced foliation. Hornblende, and oxides are replaced by biotite and sphene. There is some interstitial K-feldspar and a colourless mica plus epidote. Zoned zircons similar to those seen in 001 and 002 survive as do $CO2$ fluid inclusions.

243003e, basic amphibolite dyke 3-7 cm in from contact. Hornblende, plagioclase rock with minor interstitial sphene, biotite and late epidote. Local formation of chlorite.

243003g Aggregate of plagioclase after original phenocryst in amphibolite dyke circa 20 cm from margin.

243003i Amphibolite dyke 45-90 cm from contact.

243007b. A medium to coarse grained facies gneiss with a partially preserved relic granulite facies assemblage. Interpreted as a felsic partial melt under granulite facies with mafic segregations which may represent recrystallised mafic material which did not form part of the melt. One of the least retrogressed gneisses from the outcrop, with patches of orthoamphibole interpreted as replacing orthopyroxene. This is turn is replaced by hornblende and locally biotite. Individual garnet crystals brecciated and partly replaced by biotite and hornblende. Fe-Ti oxides locally rimmed by biotite. Very little geochemical evidence of the introduction of K, Rb or Pb.

243029. Partly retrogressed gneiss similar to 243007b. Shows the most LIL depleted chemistry observed from the outcrop. Rb less than 1 ppm.

composition comparable to that of rocks with a much shorter crustal
history and a Pb isotopic composition close to that of the late Archaean
mantle. Two geologically feasable explanations can be put forward: 1)
The change in Pb composition represents changes in the fluid during a
single retrogressive event: the first fluids introduced equilibrated
with gneisses from the Tre Brødre and Færingehavn terranes a relatively
short distance below the outcrops studied. In this case the first
introduced Pb represents the average composition of the underlying
gneisses as shown in Fig 1 of Nutman & Friend (this volume). As fluid
movement continued, it either tapped deeper crustal levels containing a
higher proportion of late Archaean gneisses or possibly the fluids were
derived from a juvenile magma at depth. 2) The change in Pb compositions
is in response to changes in the tectonic control. The two tectonic
slices underlying the Kangimut sammisoq outcrop are dominated by
different proportions of early and late Archaean gneiss. The Tre Brødre
terrane now in direct contact with the granulite facies rocks is
dominated by late Archaean rocks. The Færingehavn terrane with a high
proportion of early Archaean rocks is at present separated from the
gneisses at Kangimut sammisoq by the Tre Brødre terrane. There is at
present no geological evidence about the relative timing of the tectonic
interleaving in the late Archaean. It is possible that the Kangmimut
sammisoq gneisses may have overlain older gneisses directly at one stage
in their post-granulite facies history and were then later thrust over a
terrane dominated by younger rocks. Once established the shear zones and
lithological boundaries which acted as pathways for the fluids
controlling the first stage of retrogression would continue to act as
pathways during a second episode but the average isotopic composition of
the introduced Pb would change.

ACKNOWLEDGEMENTS.

The samples used for this study were collected together with A.P. Nutman
in 1981 as part of a joint programme sponsered by the Geological Survey
of Greenland to study the effects of polymetamorphism on geochemistry
and isotopic systems in complex gneisses which have been affected by
granulite facies metamorphism. Permission to publish from the Geological
Survey of Greenland is ackowledged. Field expertise and discussions with
A.P. Nutman and V.R. McGregor as part of ongoing investigations of this
complex area have been highly appreciated. Interpretations of the Pb
isotopic results have been under the guidance of L. Schiøtte, who also
reviewed this paper. The responsibility for interpretation however
remains mine. The samples used have been studied extensively both
geochemically and isotopically by several groups including K.D.
Collerson, K.H. Wedepohl and others, and further geochemical and
isotopic work is either in press or in advanced stages of preparation. I
would like to thank M. Thirwell for taking over the analytical work when
facilities for Pb separation at Copenhagen University broke down at a
critical stage. Total Pb analyses were provided by J. Bailey, University
of Copenhagen, U analyses by the Danish research reactor center, Risø.
Analytical costs were paid by grants from the Carlsberg foundation and
the Danish Scinetific research council.

REFERENCES

Ashwal, L.D. ed. (1986) Workshop on Early Crustal Genesis: The World's
  oldest rocks.
  LPI Tech. report, 86-04. *Lunar and Planetary Institute, Houston.*
Baadsgaard, H. (1976) Further U-Pb dates on zircons from the early
  Precambrian rocks of the Godthåbsfjord area, West Greenland.
  Earth Planet. Sci. Lett., 33, 261-267.
Baadsgaard, H. & McGregor, V. (1981) The U-Th-Pb systematics of zircons
  from the type Nûk gneisses, Godthåbsfjord, West Greenland.
  Geochim. Cosmochim. Acta, 45, 1099-1109.
Bridgwater. D. (1979) Chemical and isotopic redistribution in zones of
  ductile deformation in a deeply eroded mobile belt. Part 1, Chemical
  redistribution. *USGS Open-file report*, 79-1239, 505-526.
Bridgwater, D., Allaart, J.H., Baadsgaard, H., Collerson, K.D.,
  Ermanovics, I., Gorman, B.E., Griffin, W., Hanson, G., McGregor, V.,
  Moorbath, S., Nutman, A.P., Taylor, P., Tveten, E., & Watson, J.
  (1979) International field work on Archaean gneisses in the
  Godthåbsfjord-Isua area, southern West Greenland.
  Rapp. Grønlands geol. Unders., 95, 66-71.
Bridgwater, D., McGregor, V.R. & Nutman, A. (1981) Geological
  constraints on isotopic and thermal models for the consolidation of
  the late Archaean crust, West Greenland.
  Terra cognita, spec. issue., 93.
Bridgwater, D., Nutman, A.P., Rosing, M. & Schiøtte, L. (1985)
  Polymetamorphism, element mobility and modifications of isotopic
  systems in Archean gneisses. *Terra Cognita*, 5., 205.
Bridgwater, D., Rosing, M., Schiøtte, L. & Austrheim, H. (1989) The
  effect of fluid-controlled element mobility during metamorphism on
  whole-rock isotope systems, some theoretical aspects and possible
  examples. In: Bridgwater, D. (ed.) *Fluid movements, element
  transport and the composition of the deep crust. Nato ASI Series C:
  Mathematical and Physical sciences, series C., Reidel, Dordrecht*,
  (this volume).
Collerson, K.D., McCulloch, M.T., Bridgwater, D., McGregor, V.R. &
  Nutman, A.P. (1986) Strontium and Neodymium isotopic variations in
  early Archean gneisses affected by middle to late Archean high-grade
  metamorphic processes. West Greenland and Labrador. In: Ashwal, L.D.
  (ed.) *Workshop on Early crustal genesis: The World's oldest rocks.
  LPI Tech. report*, 86-04, 30-36. Lunar and Planetary Institute,
  Houston.
Friend, C.R.L., Nutman, A.P. & McGregor, V.R. (1987) Late-Archaean
  tectonics in the Færingehavn-Tre Brødre area, south of Buksefjorden,
  southern West Greenland. *J. geol. Soc. London*, 144, 369-376.
Jones, N.W., Moorbath, S. & Taylor, P.N. (1986) Age and origin of
  gneisses south of Ameralik, between Kangimut sangmissoq and
  Qasigianguit. In: Ashwal, L.D. (ed.) *Workshop on Early crustal
  genesis: The World's oldest rocks. LPI Tech. report*, 86-04, 59-62.
  Lunar and Planetary Institute, Houston.

Kinny, P. (1987) An ion-probe study of Uranium, Lead and Hafnium
  isotopes in natural zircons. *Unpublished PhD thesis, Australian
  National University, RSES, Canberra*, 1-160.
McGregor, V.R. (1973) The early Precambrian gneisses of the Godthåb
  district, West Greenland.
  Phil. Trans R. Soc. Lond., A, 273, 343-358.
McGregor, V.R., Bridgwater, D. & Nutman, A.P. (1983) The Qârusuk
  dykes: post-Nûk, pre-Qôrqut granitoid magmatism in the Godthåb
  region, southern West Greenland.
  Rapp. Grønlands geol. Unders., 112, 101-112.
Moorbath, S., Taylor, P.N. & Jones, N.W. (1986) Dating the oldest
  terrestrial rocks - fact and fiction. *Chem. Geol.*, 57, 63-86.
Nutman, A.P., Bridgwater, D. & McGregor, V.R. (1986) Regional variation
  in the Amîtsoq gneisses related to crustal levels during the late
  Archean granulite facies metamorphism, southern West Greenland.
  In: Ashwal, L.D. (ed.) *Workshop on Early crustal genesis: The
  World's oldest rocks. LPI Tech. report*, 86-04, 78-83. Lunar and
  Planetary Institute, Houston.
Nutman, A.P. & Friend, C.R.L. (1989) Reappraisal of crustal evolution
  at Kangimut sammisoq, Ameralik fjord, southern West Greenland:
  Fluid movement and interpretation of Pb/Pb isotopic data.
  In: Bridgwater. D. (ed.) *Fluid movements, element transport, and the
  composition of the deep crust. Riedel, Dordrecht*, (this volume).
Riciputi, L.R. Valley, J.W. & McGregor, V.R. (1989) Retrograde P-T-paths
  from two Archean granulite facies domians, S.W. Greenland.
  Geol. Soc. Am. Annual meeting Denver, A343.
Schiøtte, L., Bridgwater. D., Collerson, K.D., Nutman. A.P. & Ryan,
  A.B. (1986) Chemical and isotopic effects of late Archaean high
  grade metamorphism and granite injection on early Archaean gneisses,
  Saglek-Hebron, northern Labrador. In: Dawson, J.B., Carswell, D.A.,
  Hall, J. & Wedepohl, K.H. eds. The Nature of the Lower Continental
  Crust, *Geol. Soc. Lond. Spec. Publ.*, 24, 261-273.
Schiøtte, L., Compston, W. & Bridgwater D. (1988) Late Archaean ages
  for the deposition of clastic sediments belonging to the Malene
  supracrustals, southern West Greenland: evidence from an ion probe
  U-Pb zircon study. *Earth Planet. Sci. Lett.*, 87, 45-58.
Schiøtte, L., Compston, W. & Bridgwater, D. (1989) Single zircon U-Pb a-
  ge for the Tinissaq gneiss, southern West Greenland, A controversy
  resolved. *Isotope geoscience*, in press
Schiøtte, L., & Bridgwater, D. (submitted), Multi stage late Archaean
  granulite facies metamorphism in northern Labrador, Canada. To
  appear in NATO ASI Series volume covering papers presented ast the
  GRANULITES 88 congress, Clermont-Ferrand, 1988.
Taylor, P.N., Moorbath, S., Goodwin, R. & Petrykowski, A.C. (1980)
  Crustal contamination as an indicator of the extent of early
  Archaean continental crust: Pb isotopic evidence from the late
  Archaean gneisses of West Greenland.
  Geochim. Cosmochim. Acta, 44, 1437-1453

Taylor, P.N., Jones, N.W. & Moorbath, S. (1984) Isotopic assessment of relative contributions from crust and mantle sources to the magma genesis of Precambrian granitoid rocks. *Phil. Trans. R. Soc. Lond.*, A310, 605–625.

Wells, P.R.A. (1979) Chemical and thermal evolution of Archaean sialic crust, southern West Greenland. *J. Petrol.*, 20, 187–226.

DIFFUSION AND/OR PLASTIC DEFORMATION AROUND FLUID INCLUSIONS IN SYNTHETIC QUARTZ: NEW INVESTIGATIONS.

Anne-Marie BOULLIER (*), Gérard MICHOT (&), Arnaud PECHER (*/$)
and Odile BARRES (#).

* Centre de Recherches Pétrographiques et Géochimiques, B.P.20, 54501
VANDOEUVRE LES NANCY CEDEX, FRANCE.
& Laboratoire de Physique du Solide, Ecole Nationale Supérieure des Mines et
de la Métallurgie, Parc de Saurupt, 54000 NANCY, FRANCE.
$ Laboratoire de Géologie structurale, Ecole Nationale Supérieure des Mines et
de la Métallurgie, Parc de Saurupt, 54000 NANCY, FRANCE.
Laboratoire de Spectrométrie de Vibrations, Université de Nancy I, B.P.239,
54506 VANDOEUVRE LES NANCY CEDEX, FRANCE.

ABSTRACT. Synthetic quartz containing fluid inclusions (H_2O + NaOH, 0.5N) was
annealed at high temperature (T=448°C) and under confining pressure (P_c=200 or
350MPa). Changes in the shape of the inclusions were observed together with variations
of their filling densities which depend on the value of the internal pressure, P_i; the latter
tends to equilibrate with the confining pressure P_c either by decrepitation or by progressive
evolution. X-ray topography after treatment reveals contrast around the modified fluid
inclusions. T.E.M. investigations show some dislocations around the inclusions after
experiment. However, IR microspectroscopy does not show any visible change in the
water absorption band in samples before and after annealing.
    These results are discussed and some interpretations are proposed. As demonstrated
by previous authors (Gratier and Jenatton, 1984) the changes in the shape are due to
solution - deposition processes. Plastic deformation around fluid inclusions is probably
largely responsible for the changes in density, but diffusion processes (positive and
negative exchanges between quartz and fluid inclusion) cannot be entirely excluded. The
driving force may be the elastic strain energy due to the difference in pressure between the
fluid inclusion and the confining medium (Doukhan and Trépied, 1985).

1. INTRODUCTION.

Recent experimental work has shown that fluid inclusions in synthetic quartz underwent
strong shape and density modifications when submitted to P-T conditions which are
different from those on their trapping isochore line and where the internal pressure is larger
than the confining pressure (Sabouraud, 1981; Pêcher, 1981; Gratier and Jenatton, 1984;
Pêcher and Boullier, 1984). Complementary experiments have been performed on the
same material for the case where the internal pressure is lower than the confining pressure.

D. Bridgwater (ed.), Fluid Movements – Element Transport and the Composition of the Deep Crust, 345–360.
© *1989 by Kluwer Academic Publishers.*

This work presents the results of these experiments and of some investigations with X-ray topography, transmission electron microscopy and infrared microspectroscopy.

2. THE EXPERIMENTS.

2.1. Starting material.

Synthetic quartz crystal (S.I.C.N., Annecy, France), grown hydrothermally in a 0.5N NaOH solution following the Régrény's method (1973) was used. The seed contains numerous large cylindrical fluid inclusions which are orientated in a direction near the c axis of the crystal (figure 1). X-Ray topographs before experiments (figure 2) reveal that the fluid inclusions are localized on straight dislocations (Wilkins and McLaren, 1981; Michot et al., 1984); consequently, these fluid inclusions probably originated by corrosion of the cores of the dislocation before saturation conditions were attained in the autoclave. When saturation is reached (160MPa, 365°C), quartz growth blocks the tubes which are filled with a fluid with a density equal to that within the autoclave (0.8).

2.2. The initial homogenization temperatures.

The samples (0.8x2.5x5mm) are cut and two faces polished which contain a and c. Homogenization temperatures (Th) are measured before the experiments with a Chaix-Meca stage (Poty et al., 1976). They are very homogeneous in each sample (figure 3) but vary slightly from sample to sample (for instance 244.6°C for 2A10 and 247.6°C for 2A6). Controls by calibrations have shown that these differences are not due to measurements errors.

Only few P-V-T data are available on the system H_2O + NaOH + SiO_2. Régrény (1973) has shown that the isochores are subparallel to isochores in the pure H_2O system, but slightly shifted towards higher temperatures. Consequently, we will use the slopes of the isochores given by Burnham et al. (1969) for pure H_2O which are very close to the P-V-T data of Zang and Frantz (1987).

2.3. Principle and conditions of the experiments.

Fluid inclusions can be considered as micro-autoclaves in which the internal pressure (P_i) is a function of the temperature and molar volume of the fluid (figure 4). The confining pressure (P_c) is an external thermodynamic parameter of the system but it also prevents decrepitation of the fluid inclusions. Only one type of experiment has been performed so far; it corresponds to P_i higher than P_c (argon) (Pêcher and Boullier, 1984):

P_c(argon) = 200 MPa, T = 448°C (Figure 4)

These experiments have now been supplemented by experiments with a confining pressure P_c lower than P_i:

P_c(argon) = 350 MPa, T = 448°C.

To test the influence of a chemical gradient of water between the inclusion and the confining medium, experiments have been performed with water (rather than argon) as a fluid medium generating the confining pressure P_c (either higher or lower than P_i):

P_c(water) = 200 or 350 Mpa, T = 448°C.

All the experiments are reported in table 1.

3. EXPERIMENTAL RESULTS.

3.1. P_c(argon) lower than P_i.

A comparison of samples before and after the experiments shows important modifications of the shape and the density of the fluid inclusions (see Pêcher and Boullier, 1984). We distinguish two different behaviours: 1) the inclusions decrepitate as demonstrated by microfracturation or 2) the fluid inclusions do not decrepitate, but their shape and density are modified.

3.1.1. Changes of shape. They have already been described in detail for such experiments by Pêcher and Boullier (1984); in this paper, we will only refer to the most striking features.

a - decrepitated inclusions: microfractures contain numerous small fluid inclusions (secondary fluid inclusions or "decrepitation cluster", Swanenberg, 1980), the size of which decreases towards the external limit of the crack (20 to 2 μm); they have homogeneous filling ratios. They appear during decrepitation and then are healed by growth of quartz from the fracture walls (Lemmlein and Kliya, 1954; Smith and Brians, 1984).

b - undecrepitated fluid inclusions: the most striking feature of the undecrepitated fluid inclusions is a decrease of their length/width ratio towards an equilibrium value (2.7, figure 5), together with the appearance of rational walls (simple pyramidal and prismatic crystalline faces). Their evolution with time depends on the orientation of the initial fluid inclusion relative to the c axis of the quartz crystal. Generally, they tend towards the shape of a negative crystal shape (for details see Pêcher and Boullier, 1984).

3.1.2. Changes of homogenization temperatures. The temperatures of homogenization measured after experiments are reported in figure 6. The Th measurements are grouped into two sets for each sample: 1- the first one with slightly scattered but high values (up to 280°C) corresponds to decrepitated primary inclusions or secondary inclusions formed by decrepitation of the former ones. 2- the second set with values slightly higher than the initial Th, corresponds to the undecrepitated primary inclusions. The difference in the homogenization temperature increases with the duration of the experiment (figure 7): for the longest experiment (2A6), the difference is 13°C corresponding to a density decrease of 0.019.

3.2. P_c(argon) higher than P_i.

As for the above experiments, important changes of shape and density of the fluid inclusions are observed after annealing.

3.2.1. Changes of shape. In the case where the confining pressure P_c(argon) is greater than the internal pressure P_i, the changes of shape are somewhat similar to those observed in the previous case ($P_c<P_i$) except that the equilibrium shape ratio is lower (1.8, ;figure 5) and corresponds to a bipyramid without a prism.

3.2.2. Changes of homogenization temperatures. Figure 8 shows histograms of the homogenization temperature after experiment in the case where P_c is higher than P_i. Th is clearly shifted towards lower values (increase of the density), but the values are much more scattered than for the previous experiments and there is no well defined maximum. However, it should be noticed that the larger the inclusion volume, the bigger the difference between initial Th and final Th (figure 9).

3.3. Experiments with water as a confining medium.

Two experiments have been performed to test the influence of a possible chemical gradient of water between the inclusion and the confining medium. A sample was placed in a gold capsule containing water; the capsule was sealed and placed in the autoclave where the pressure medium is argon. Considering the solubility of quartz in water in the experimental conditions, the samples were polished again after the experiment for observations and for Th measurements. No damage is caused by that polishing.

These experiments do not show any difference to the previous ones (with argon as confining medium). The changes of shape and density are fully comparable to those observed for corresponding experiments (P_c lower or higher than P_i). Consequently, experimental results are given on the same figures (figures 5 to 8).

4. X-RAY TOPOGRAPHS.

4.1. Before experiment.

On X-ray topographs (Lang chamber), the samples show long straight dislocations as observed by Wilkins and McLaren (1981) in comparable synthetic quartz crystals. Fluid inclusions are not visible on X-ray topographs recorded at room temperature (figure 2). However, one series of topographs was obtained with the LURE synchrotron (Orsay, France), the sample being heated to 302°C (Michot et al., 1984). As soon as Th is reached, contrast appears around fluid inclusions which becomes visible in this way. If the temperature decreases below Th, the contrast around the fluid inclusions disappears except around some of those which have decrepitated. Consequently, in the experiments of Michot et al. (1984), the contrast around fluid inclusions is related to elastic deformation of the quartz crystals, due to an increase of pressure in the fluid inclusion.

4.2. After experiments.

X-ray topographs were performed on some samples after experiments. In every case, a permanent contrast is observed around fluid inclusions at room temperature, for experiments at P_c either lower or higher than P_i (figure 10). The contrast around fluid inclusions is very strong. If it is due to dislocations, then their density is higher than $10^5 cm^{-2}$. The shape of the contrast seems to be intermediate between the initial and the final shape of the fluid inclusions. This would indicate that the contrast is due to a mechanism which operates during the whole experiment and is not due to the final stage of the experiment (decrease of T and P_c).

5. PRELIMINARY TEM OBSERVATIONS.

After experiment, standard polished thin section (30μm thick) were made which were then thinned to less than 1μm using an ion thinner. As the probability of obtaining a thin area exactly around a fluid inclusion is very small, only one ion thinning was successful. The sample was studied with a JEOL 200CX transmission electron microscope at the Service Commun d'Analyses of the Nancy I University. Dislocations are observed at the tip of the fluid inclusion (sample 2A4, figure 11); they are roughly parallel to the c axis but their Burgers vector has not been characterized. Their density is approximately $10^7 cm^{-2}$. Some very small defects are also observed but their exact nature has not been identified; they could be very small bubbles, such as those observed by Kirby and McCormick (1979) in experimentally deformed synthetic quartz crystals and by McLaren et al. (1983) in heated synthetic quartz crystals, due to precipitation of water above the equilibrium concentration (Doukhan and Paterson, 1986; Cordier et al., in press).

6. INFRARED MICROSPECTROSCOPY DATA.

Infrared spectra were recorded with a Brucker IFS 88 Fourier transform spectrometer which is provided with a microscope. The samples investigated were unmodified and modified (before and after experiments), double polished and 800 μm thick. Two types of spectra were obtained for the specific seed under study for wavenumbers between 3600 and 3000cm^{-1}.This region corresponds to water or OH absorption bands in quartz (Kats, 1962; Aines and Rossman, 1984; Kronenberg et al., 1984). Flat spectra were observed for the major part of the seed, and spectra with high absorption for a region close to numerous and unusually large fluid inclusions. These discrepancies could be related to defects pre-existing in the seed before the hydrothermal growth of the synthetic quartz.

However, in the same area, spectra are identical before and after the experiment, even if performed very close to a fluid inclusion (figure 12). Consequently, it seems that any modification of the water distribution is too small to be detected with such a method.

7. DISCUSSION.

7.1. Changes of shape.

It is characterized by a decrease of the shape ratio (l/w) of the fluid inclusions (figure 5). In the case where P_c is lower than P_i (P_c = 200MPa), the decrease of l/w is faster for undecrepitated fluid inclusions where P_i is high (282MPa) than for decrepitated fluid inclusions where P_i is roughly equal to P_c (P_i= 200MPa). This phenomenon was described by Gratier and Jenatton (1984) for experiments realized at atmospheric pressure. These authors demonstrated that the change of shape is due to a solution - deposition process. The variation of the chemical potential at different points of the liquid - solid interface, caused by variation of the surface energy with the curvature, is the driving force and the rate of the l/w decrease is controlled by the rate of dissolution in the median part of elongated inclusions.

In this study (P_i = 282MPa, T = 448°C), the equilibrium shape is reached after only 350h; it corresponds to a configuration with minimum surface energy and depends on the confining pressure.

7.2. Changes of the homogenization temperatures.

7.2.1 Decrepitated inclusions. Decrepitation of fluid inclusions has been experimentally reproduced by Leroy (1979) without confining pressure and by Pêcher (1981) in synthetic quartz and by Sabouraud (1981) in fluorine with a confining pressure. In our experiments, the occurrence in the same sample of decrepitated and undecrepitated fluid inclusions with roughly identical volume and shape indicates that P_i-P_c (P_i>P_c) was near the decrepitation overpressure value; actually, considering the P_c-V-T conditions, the overpressure (82MPa) is similar to the value obtained at atmospheric pressure and 280°C by Leroy (1979) for large fluid inclusions (V > 2 x 10^4 μm^3) in synthetic quartz (85MPa).

We may consider that there is a competition during crack development between the confining pressure P_c and the internal pressure P_i and that the crack stops when both pressures are almost equal. In this case, the density of secondary inclusions in the healed crack would correspond to the P_c-T conditions. Actually, Th measurements on secondary inclusions (271°C for 2A7, 269°C for 2A1, 272°C for 2A6) indicate pressures of the order of 200 to 220MPa for T = 448°C.

7.2.2. Undecrepitated fluid inclusions. Even for the undecrepitated fluid inclusions, variations of Th are observed which are independent of the changes of shape (compare figures 5 and 7). These variations are either positive ($P_c<P_i$) or negative ($P_c>P_i$) and correspond to reverse variations of density of the fluid in the inclusions. The experiments performed in a water confining medium showed the same density modifications as for the argon experiments. Consequently, we believe that the chemical gradient of water between the inclusion and the confining medium is not the driving force for variation in the density. The driving force could be the volume dependent elastic strain energy due to the difference in pressure between the fluid inclusion and the confining medium (Doukhan and Trépied, 1985).Three possibilities may be envisaged: i) the included fluid diffuses along the long and straight dislocations on which the inclusions are located; ii) there is a variation of the volume of the inclusion due to plastic deformation of surrounding quartz, related to the difference between P_i and P_c; iii) the volume of the inclusion is constant but water diffuses away from or towards the fluid inclusions through the surrounding quartz.

Comparing the two types of experiments, the first hypothesis is rejected because it cannot explain the increase of density where $P_c>P_i$.The occurrence of dislocations around the fluid inclusions indicates that some plastic deformation has occurred, and that the volume of the inclusion varied during the experiments. The observed dislocation density ($\approx 10^7$ cm^{-2}) is compatible with the volume change required to justify such a density variation. If plastic deformation alone is operative, we cannot explain why the density variation is volume independent where $P_i>P_c$ and volume dependent where $P_i<P_c$.

The similar IR spectra obtained on samples before and after the experiments indicate that any diffusion of water is too small to be detected by this method. However, exchange of water between the quartz matrix and the fluid inclusion cannot definitely be excluded, because of the low sensitivity of the IR microspectrometric measurements.

Actually, such a process might explain better some discrepancies between both types of experiments:

-at $P_i>P_c$, Th variations are very homogeneous (volume independant); that would correspond to easy diffusion of H towards the quartz matrix because of its relatively low content in H/Si defects (less than 300ppm, D.Mainprice, comm. pers.).

-at $P_i<P_c$ Th variations are scattered and volume dependent. This would correspond to less easy diffusion of H from the quartz matrix towards the inclusion (long distances due to low content in H/Si defects). Large inclusions have greater elastic energy and are thus more able to attract H ions.

8. CONCLUSIONS.

The experiments described in this paper have shown that important changes of shape and density occur when fluid inclusions are submitted to P_c-T conditions different from their trapping ones. As demonstrated by previous authors (Gratier and Jenatton, 1984), the changes in shape are due to solution - deposition processes. Plastic deformation around fluid inclusions is probably responsible for the changes in density, but diffusion processes (positive and negative exchange between quartz and fluid inclusion) cannot be entirely excluded. More systematic and precise investigations are needed to understand fully the changes in density of the fluid inclusions. In any case the driving force for deformation or diffusion could be the elastic strain energy due to the difference in pressure between the fluid inclusion and the confining medium (Doukhan and Trépied, 1985).

We have to be cautious when extending our observations to natural fluid inclusions because of the peculiar characteristics of the material used for these experiments (relatively wet synthetic quartz crystal, very long and rectilinear dislocations, NaOH + H_2O filling of the inclusions...). However, these results can explain the often observed discrepancies in natural samples between fluid inclusion data and thermobarometric calculations using mineralogical assemblages, and the difficulty in finding fluid inclusions corresponding to the climax of metamorphism (Coolen, 1982; Pêcher, 1979, 1984; Sauniac and Touret, 1983). However, the observed modifications imply high temperature processes (plastic deformation and perhaps diffusion). Consequently, fluid inclusion studies are still meaningful and useful for the knowledge of the nature of the fluids and the P-T conditions operating during low temperature geological phenomena.

ACKNOWLEDGEMENTS. The authors wish to thank especially J.C.Doukhan for stimulation and constructive remarks and A.George, C.Ramboz, A.Burneau and Y.G.Zang for fruitful discussions. They are also grateful to M.Pichavant and A.Rouillier for help in experimental techniques and to W.L.Brown who kindly improved the english text. This project was supported by the C.N.R.S. (A.T.P. Physique des Hautes Pressions). Contribution C.R.P.G. n°798

REFERENCES.

Aines R.D. and Rossman G. (1984). 'Water in minerals? A peak in the infrared.' J. Geoph. Res., 89, 4059-4071.
Burnham C.W., Holloway F.R. and Davis N.F. (1969). 'Thermodynamic properties of water to 1000°C and 10000bars'. Geol. Soc. America Sp. paper, 132, 96p.
Coolen J.J.M.M. (1980). 'Chemical petrology of the Furua granulite complex, southern Tanzania.' G.U.A. papers of geology, Amsterdam, 1, 13, 258p.
Cordier P., Boulogne B. and Doukhan J.C. (1988). 'Water precipitation and diffusion in wet quartz and wet berlinite AlPO4.' Bulletin de Minéralogie., 111, 113-137.
Doukhan J.C. and Paterson M.S. (1986). 'Solubility of water in quartz. A revision.' Bull. Minéral., 109, 193-198.
Doukhan J.C. and Trépied L. (1985). 'Plastic deformation of quartz single crystals.' Bull. Minéral., 108, 97-123.
Gratier J.P. and Jenatton L. (1984). 'Deformation by solution - deposition and reequilibration of fluid inclusions in crystals depending on temperature, internal pressure and stress.' Journ. Struct. Geol., 5, 329-339.
Kats A. (1962). 'Hydrogen in alpha-quartz.' Philips Res. Rep., 17, 1-31, 133-279.
Kirby S.H. and McCormick J.W. (1979). 'Creep of hydrolytically weakened synthetic quartz crystals oriented to promote {2110} <0001> slip: a brief summary of work to date.' Bull. Mineral.,102, 124-137.
Kronenberg A.K., Kirby S.H., Aines R.D. and Rossman G.R. (1986). 'Solubility and diffusional uptake of hydrogen in quartz at high water pressures: implications for hydrolytic weakening.' Journ. Geoph. Res., 91, B12, 12723-12744.
Lemmlein G.C. and Kliya M.O. (1954). 'Changes in fluid inclusions under the effect of temporary heating up of a crystal'. Akad. Nauk. SSSR Doklady, 94, 233-236 (in russian).

352

Leroy J. (1979). 'Contribution à l'étalonnage de la pression interne des inclusions fluides lors de leur décrépitation.' Bull. Minéral., 102, 584-593.

McLaren A.C. A.C., Cook R.F., Hyde S.T. and Tobin R.C. (1983). 'The mechanisms of the formation and growth of water bubbles and associated dislocation loops in synthetic quartz.' Phys. Chem. Minerals, 9, 79-94.

Michot G., Weil B. and George A. (1984). 'In situ observation by synchrotron X-Ray topography of the evolution with temperature of fluid inclusions in synthetic quartz.' Journal Crystal Growth, 69, 627-630.

Pêcher A. (1979). 'Les inclusions fluides des quartz d'exsudation de la zone du M.C.T. himalayen du Népal central'. Bull. Minéral.,.102, 537-554.

Pêcher A. (1981). 'Experimental decrepitation and reequilibration of fluid inclusions in synthetic quartz.' Tectonophysics, 78 , 567-584.

Pêcher A. (1984). 'Chronologie et rééquilibrage des inclusions fluides: ;quelques limites à leur utilisation en microthermométrie.' Thermométrie et barométrie géologiques, éd. M.Lagache, Soc. Fr. Minéral. et de Cristall., 463-485.

Pêcher A. et Boullier A.M. (1984). 'Evolution à pression et température élevées d'inclusions fluides dans un quartz synthétique.' Bull. Minéral., 107, 139-153.

Poty B., Leroy J. et Jachimowicz L. (1976). 'Un nouvel appareil pour la mesure des températures sous le microscope: l'installation de microthermométrie Chaix Méca'. Bull. Soc. fr. Minéral. Cristallogr., 99, 182-186.

Régrény A. (1973). 'Recristallisation hydrothermale du quartz.' Thèse Doct. Ing. Paris VI, 133p.

Sabouraud C. (1981). 'Décrépitation expérimentale d'inclusions sous pression. Application au cas d'inclusions primaires de fluorine.' C. R. Acad. Sc. Paris, 292, 729-732.

Sauniac S. and Touret J. (1983). 'Petrology and fluid inclusions of a quartz-kyanite segregation in the main thrust zone of the Himalayas.' Lithos, 16 , 35-45.

Smith D.L. and Evans B. (1984). 'Diffusional crack healing in quartz.' J. Geoph. Res., 89, 4125-4135.

Swanenberg H.E.C. (1980). 'Fluid inclusions in high-grade metamorphic rocks from SW Norway.' Thesis 25, Rijksuniversiteit Utrecht, 147p.

Wilkins R.W.T. and McLaren A.C. (1981). 'The formation of primary inclusions from etch pits in crystals.' N. Jb. Miner. Mh., 5, 220-224.I

Zang Y.G. and Frantz J.D. (1987). 'Determination of the homogenization temperatures and densities of supercritical fluids in the system $NaCl-KCl-CaCl_2-H_2O$ using synthetic fluid incusions.' Chemical Geology, 64 in press.

TABLE 1 - EXPERIMENTS.

Sample	t	T	Pc	Thi
2A7	69h 15'	448°C	200MPa	246.5°C
2A1	70h 15'	447°C	200MPa	246.5°C
2A9	153h 40'	448°C	200MPa	246.5°C
2A2	162h 35'	447°C	200MPa	246.3°C
2A14a	325h 15'	446°C	200MPa	246.0°C
2A10	351h	448°C	200MPa	244.6°C
2A4	409h	447°C	200MPa	246.3°C
2A3	1220h	448°C	200MPa	246.3°C
2A6	3064h	447°C	200MPa	247.6°C
2A12c	71h	447°C	350MPa	251.8°C
2A11c	192h 3'	447°C	350MPa	251.9°C
2A12b	484h 28'	447°C	350MPa	249.9°C
2A11b	1036h 52'	446°C	350MPa	250.1°C
2A8a ·	2086h 304	446°C	350MPa	249.7°C (large I.F. 246.5°C (small I.F.)
			Pc(H_2O)	
2A13b	288h 40'	447°C	200MPa	246.5°C
2A8c	281h 55'	446°C	350MPa	248.6°C

Table 1. Summary of the experiments with different duration (t) at the temperature T and under confining pressure P_c. Thi is the mean of the initial homogenezation temperature (before experiments).

354

Figure 1. Microphotograph of a thick section in the synthetic quartz crystal seed. Note the large fluid inclusions which are subparallel to the c axis. Compare with the X-ray topograph before experiment (figure 2).

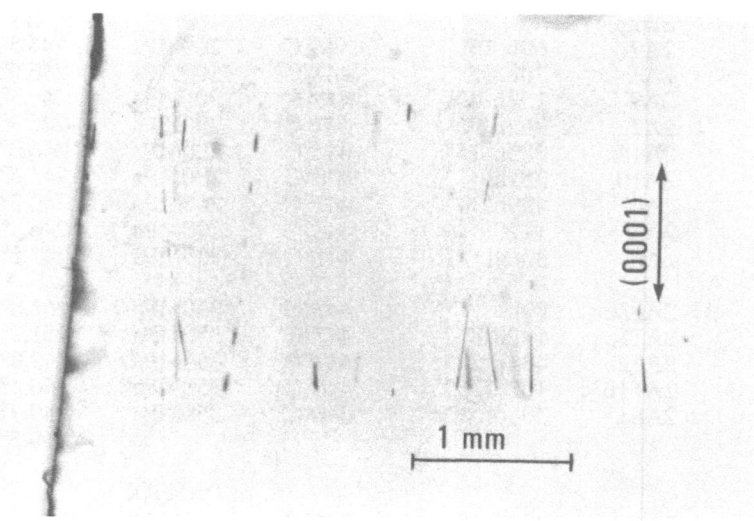

Figure 2. X-ray topograph of the 2A14a sample before experiment. Note the long and straight dislocations subparallel to the c axis and crossing the seed into the hydrothermally grown quartz crystal.

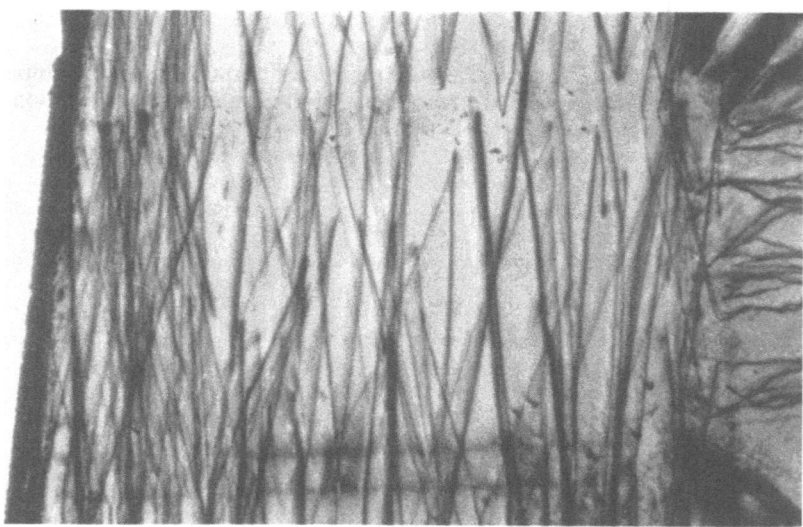

Figure 3. Homogenization temperature of fluid inclusions in the 2A10 sample before experiment. The distribution indicates a filling coefficient of 0.8 which is consistent with the growing conditions for the quartz crystal.

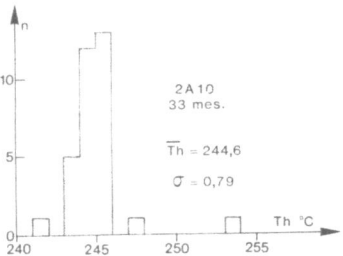

Figure 4. Pressure and temperature diagram showing the (P,T) experimental conditions and the isochore corresponding to the (P,T) hydrothermal growing conditions and to the initial Th of the fluid inclusions. The pressure difference (P_i-P_c) is deduced from the experimental conditions and the isochore position.

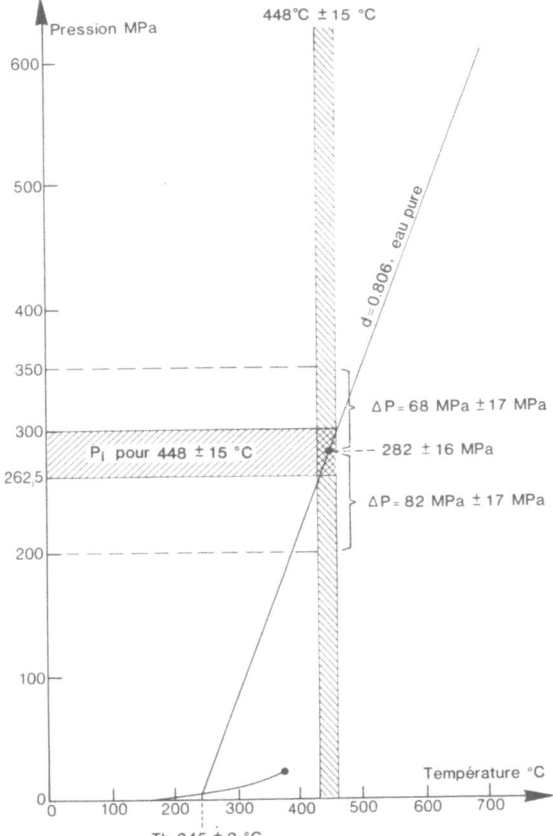

Figure 5. Evolution in time of the shape ratio of the fluid inclusions for both types of experiment. An equilibrium ratio of 2.7 (P_i>P_c, losanges) or 1.8 (P_i<P_c, squares) is reached after 350h of annealing. The filled symbols correspond to experiments under water confining pressure.

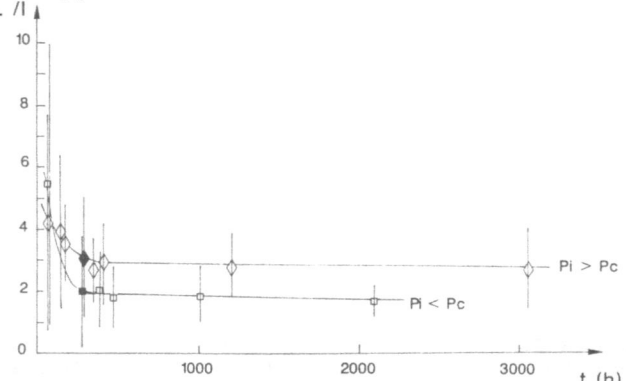

Figure 6. Experiments at T=448°C and P_C=200MPa. Histograms of homogenization temperatures of fluid inclusions after experiments. The hatched band represents initial Th; the star corresponds to experiment with water as the confining medium

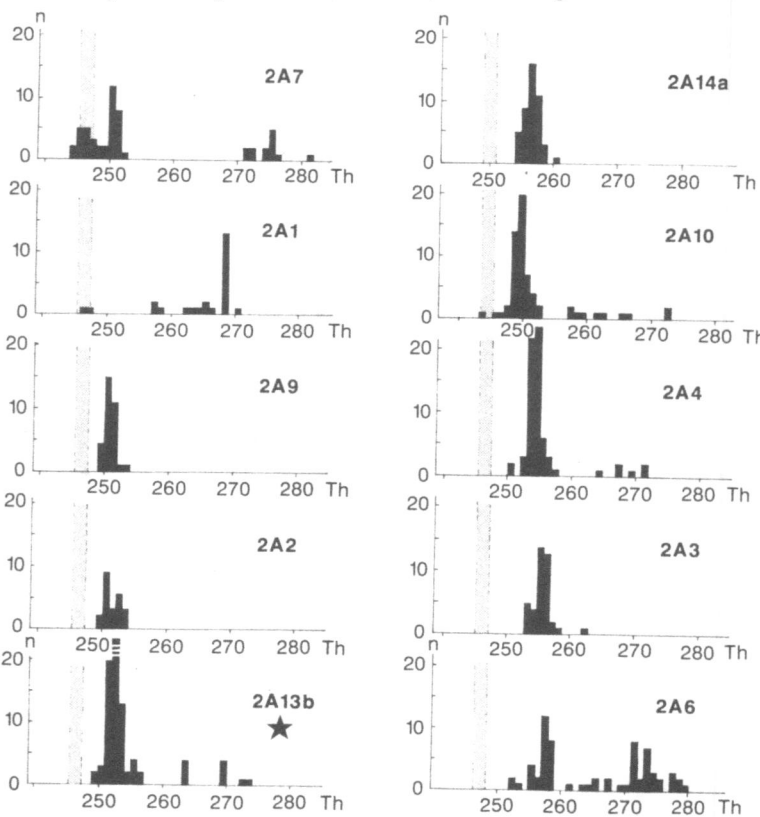

Figure 7. Th variations versus duration of experiments in the case where $P_i<P_c$.

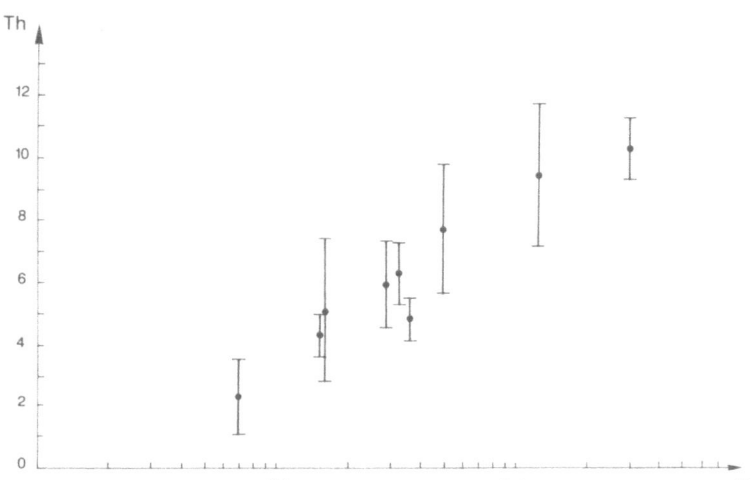

Figure 8. Experiments at T=448°C and P_c=350MPa. Histogramms of homogenization temperatures of inclusions after experiments. The hatched band represents initial Th; the star corresponds to experiment with water as the confining medium.

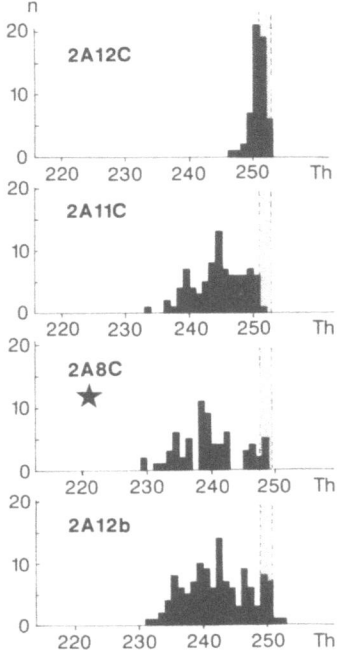

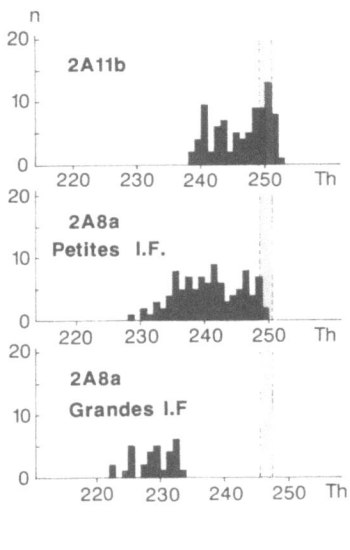

Figure 9. Variations of Th versus $L.l^2$ (proportional to the inclusion volume for initial cylindrical inclusions) for the 2A8c sample ($P_i < P_c$, t=281h 55mn). This diagram does not represent an equilibrium state but indicates that greater the fluid inclusion, larger (or faster) the Th variation.

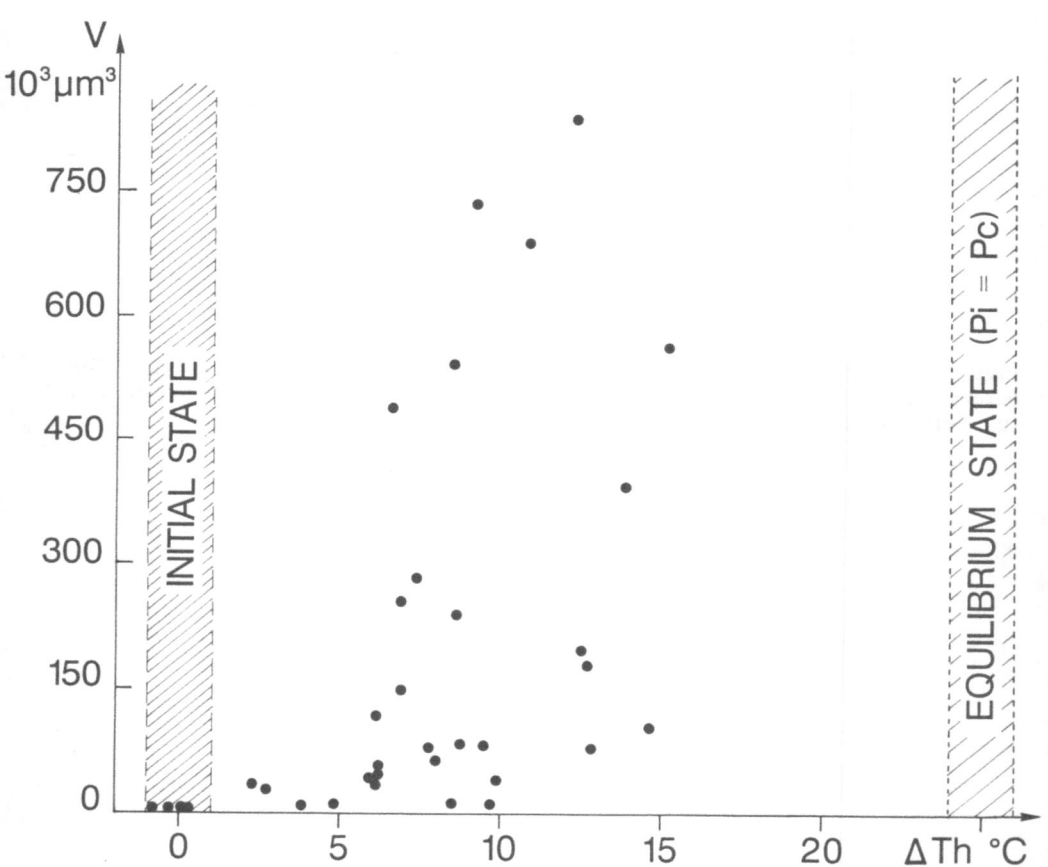

Figure 10. X-ray topographs in a Lang chamber for samples after experiments. a) 2A14a sample, (10$\bar{1}$1) reflexion (P$_C$=200MPa, T=448°C). Compare this topograph with the figure 2 (same sample before experiment). b) 2A8a sample, (10$\bar{1}$$\bar{1}$) reflexion (P$_C$=350MPa, T=448°C).

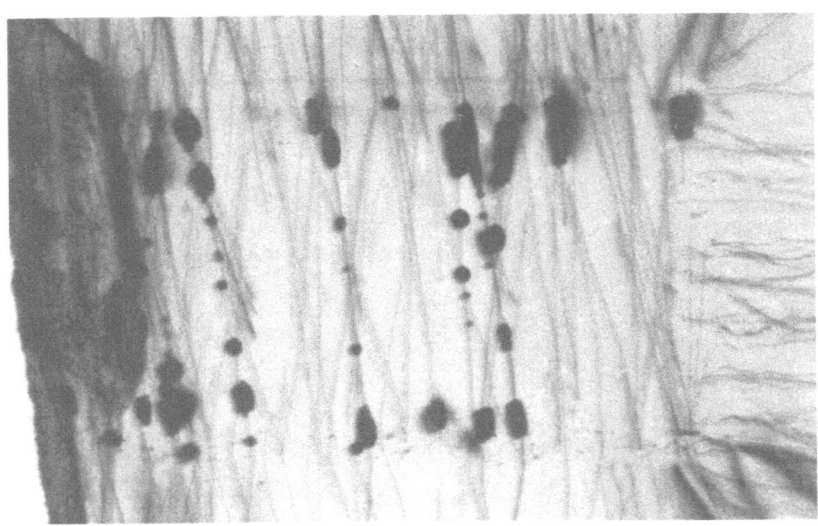

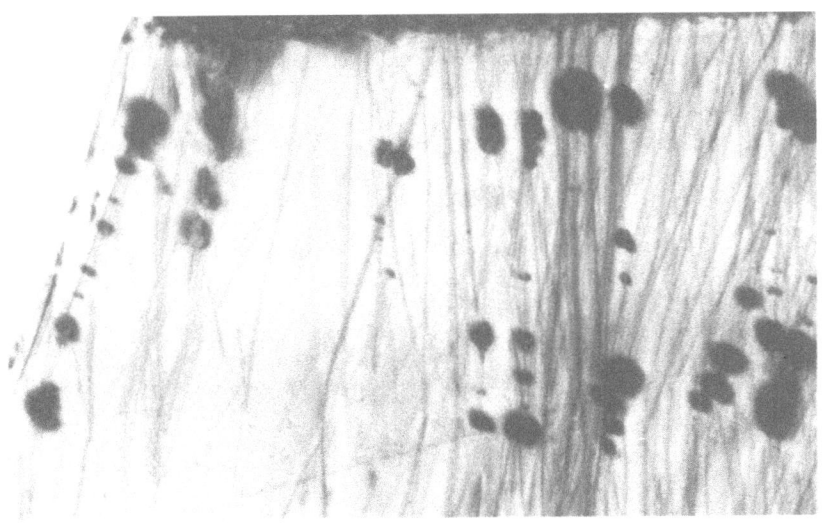

360

Figure 11. Microphotograph on transmission electron microscope showing dislocations near a fluid inclusion in the 2A4a sample after experiment. The dislocation density is about $10^7 cm^{-2}$.

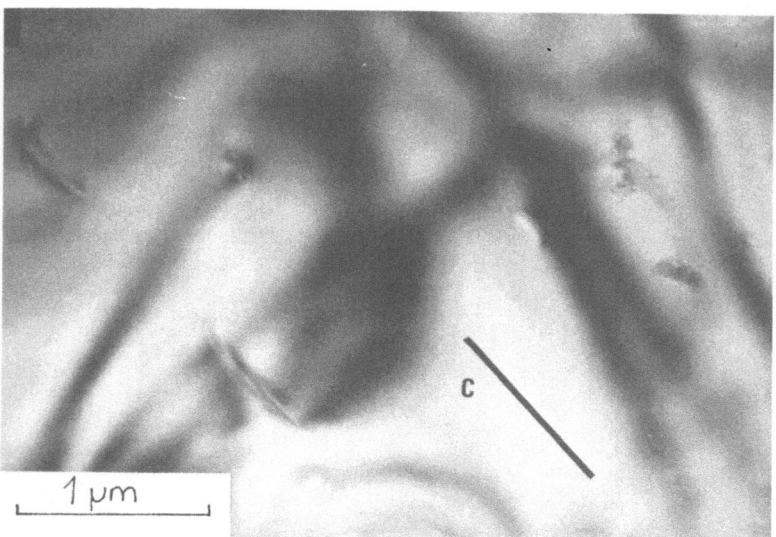

Figure 12. Infra-red spectra in samples before and after experiments. A sketch represents a schematic section of the seed with the two families of fluid inclusions. The beam was focalized near a fluid inclusion. Spectrum a: sample 2A11 before experiment, in the center of the seed. Spectrum b: sample 2A11b after experiment in the center of the seed. Spectrum c: sample 2A11 before experiment in the extremity of the seed with large fluid inclusions. Spectrum d: sample 2A8a after experiment in the extremity of the sample with large fluid inclusions. Scale is identical for the four spectra but a and b have been shifted.

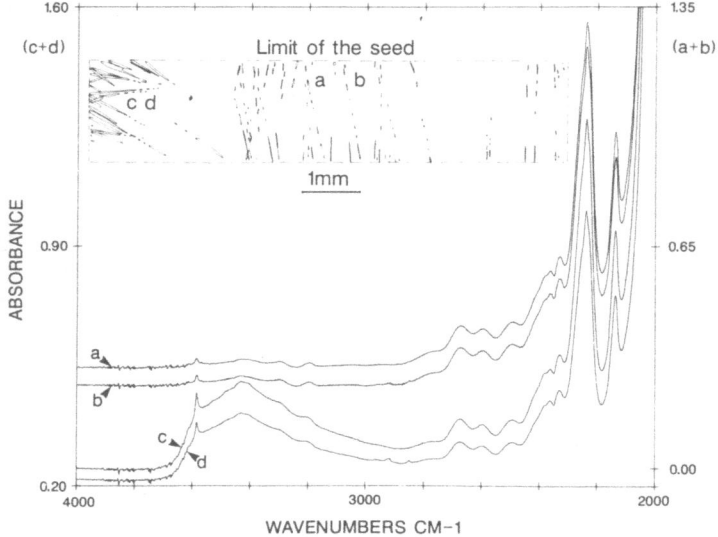

Permeability, bulk modulus and complex resistivity in crystalline rocks

Huenges, E. & G. Will
Mineralogical Institute of the University
Poppelsdorfer Schloß, D 5300 Bonn, West Germany

Abstract:

The pressure dependence of the permeability of dry rock and of the complex resistivity of saturated rocks have been determined on gneisses from the Urach drilling (Schwaebische Alp, West Germany) and the Kunklerwald drilling (Schwarzwald, West Germany). Permeability varied between 10^{-16} and 10^{-22} m^2 depending on the rock and pressure. Decreased permeability coincided with increased electrical bulk resistivity and increased dielectric permittivity. The pressure dependence parameter of the permeability, the permeability B_o at 0 bar and the permeability decrease pressure P_o, at which the permeability reaches a value of 1/8 B_o, were determined. There is a relationship between P_o and the measured bulk modulus K_o, so that it is possible to determine the permeability of a material at depth if K_o is available. Anisotropic stress fields up to 3.5 kbar produced increased permeability in a coarse- and decreased permeability in a fine-grained gneiss. Experiments on saw cut rocks show that lithostatic pressure up to 1 kbar are not sufficient to close the cracks to fluids.

1. Introduction

There is a need to derive physical properties of rocks in the earth's crust from geophysical deep soundings experiments. In order to have a meaningfulinterpretation of such field data laboratory measurements are needed in a wide varity of parameters. This requires especially defined conditions such as pressure, temperature, humidity and salinity, which can be controlled quite well in the laboratory.

The goal of this study is to explain low electrical resistivity layers, which have been recorded for example by Berktold et al. (1986) in the Schwarzwald area (West-Germany) at about 12 km depth. Several authors such as Brace (1971) or Haak & Hutton (1986) assume a highly conducting pore fluid to be responsible for such a behaviour. This requires however that sufficient pore space is available at these depths for this effect. According to Fyfe et al. (1978), interconnected pore space of up to 8 % shall be available at depth down to 10 km. The

361

D. Bridgwater (ed.), Fluid Movements – Element Transport and the Composition of the Deep Crust, 361–375.
© *1989 by Kluwer Academic Publishers.*

question whether a rock is able to increase its pore space under these pressure conditions can only be answered in the laboratory. Permeability-pressure dependencies of rocks have been studied on granites by Brace et al. (1968) and by Zoback & Byerlee (1975). In order to explain resistivity anomalies associated with such highly conducting fluids permeability must be compared with resistivity measured on the same rocks (Brace et al. 1968 and Trimmer et al. 1980).

The frequency dependence of the electrical resistivity of rocks has been investigated by Olhoeft (1981), (1985) and by Nover et al. (1985). The result of a frequency dependence measurement of the resistivity is a dispersion which Nover et al. (1985) interpreted as the effects of relaxation processes. Because different polarisation processes occurs, it is necessary to relate the permeability to the frequency dependent resistivity. This paper presents data of laboratory measurements of permeability, complex resistivity, and bulk modulus on different samples from the same crystalline rocks.

2. Experimental

The permeability measurements were done in an appartus built and developped in this laboratory (figure 1). The equipment allows measurements under hydrostatic conditions up to 2 kbar. The cylindrical sample (black) is placed in a glued shrink tube between pistons. This pistons apply load and contain a drilling to bring the pore fluid into the sample. Uniaxial pressure up to 3.5 kbar can be achieved with a hydraulic ram at confining pressure up to 2 kbar.

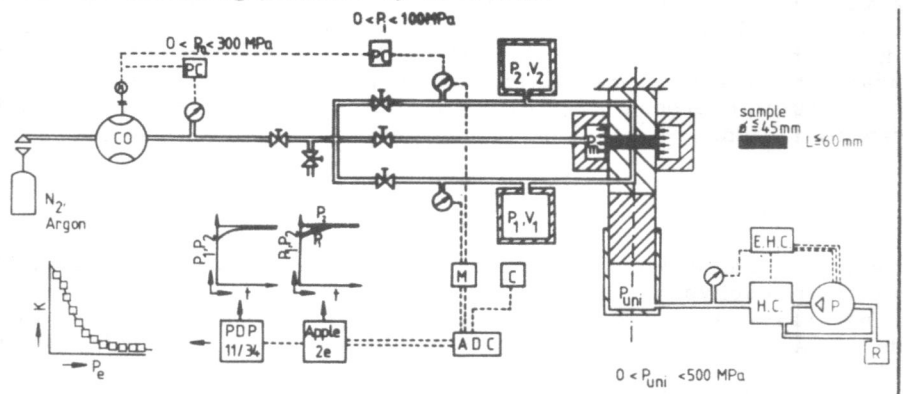

Figure 1: Experimental arrangement to measure the permeability as a function of uniaxial and confining pressure. PC Pressure Controller, CO Compressor, M Multiplexer, C Clock, ADC Analog-Digital-Converter, EHC Electronic Hydraulic Controller, HC Hydraulic Controller, P Pump, R Reservoir.

Permeability is determined with the pressure transient method (see Brace, 1968), where the pressure on one side of the sample is kept constant and the pressure on the other side is set to a lower pressure at the start and then the increase of this pressure is recorded as a function of time by an Apple IIe-computer. A LSQ-fit of equation (1) to the measured data points, in a PDP-11/34-computer, yields the permeability. Equation (1) describes the pressure (P) increase with time (t) (see Zoback & Byerlee, 1975).

$$P_2(t) = P_1 + (P_2(0) - P_1)\, e^{-\frac{BA}{L\eta\beta V_2}\, t} \tag{1}$$

where V is the volume. The subscribt 1 refers to the volume before and subscribt 2 to the volume behind the sample. L is the length of the sample, n is the fluid viscosity, ß the fluid compressibility, and B the permeability of the sample.

To perform the measurement P_1 was kept constant at 100 bar and the $P_2(t)$ started at $P_2(0)= 80$ bar. To calculate the permeability with (1) the viscosity of the streaming gas was taken from tables (Weast, 1974) and for the compressibility of the gas the reciprocal value of the pore pressure was put in. The size of the cylindrical samples was 30 mm in diameter and 10 mm in height.

To measure the bulk modulus the height of cylindrical samples (before: 16 mm diameter and 16 mm height) in a piston cylinder apparatus were measured as a function of pressure up to 20 kbar. The bulk modulus K_0 was determined by fitting equation (2) to the data, which is similar to the Murnaghan equation (Murnaghan, 1951).

$$P = \frac{K_0}{4}\left(\left(\frac{V_0}{V}\right)^4 - 1\right) \tag{2}$$

with V_0 original volume and V volume at pressure P.

The electrical resistivity was measured in the frequency range between 10^{-4} and 10^5 Hz using the AC-impedance system made by EG&G. Between 1 and 10^5 Hz a lock-in-analyser and at lower frequencies than 10 Hz a potentiostat with a fast fourier transform technique is used to determine the resistivity.

The resistivity of wet rocks as a function of pressure was measured in the pressure cell described by Nover et al. (1984) in which the confining pressure can be controlled up to 2 kbar at different pore fluid pressures on a cylindrical sample with 30 mm diameter and 10 mm height. In this cell a 2-electrode arrangement was used for the measurements. To separate the electrode polarisation 4-electrode measurements in a cell similar to a cell of Olhoeft (1981) without pressure was added.

3. Sample characterisation and results

For the investigations samples were taken from the deep drilling holes in Urach (Schwaebische Alp, West Germany), in Kunklerwald, and in Geschahse (both Schwarzwald, West Germany). Stenger (1982) described the petrology and geochemistry of the rocks in the Urach borehole. The lithologic profiles of the drillings in the Schwarzwald were documented

364

by Jenckner and Schädel (1986). Additional measurements were done on an
eclogite sample from Saualpe (Austria). Table 1 summarizes the lithologic describtion of the samples.

Table 1. Condition and lithology of the core
samples. Condition fr= fresh, hydr.= hydrothermal
altered, f= fine, c= coarse grained

Borehole, depth/m	condition		lithology
Urach			
1635	fr,	c	Plagioclase–biotit
1853	hydr.,	c	(–hornblende)
1950	hydr.,	f	
2125	hydr.,	c	Plagioclase–biotit
2308	fr,	f	(cordierite–) gneiss
2416	fr,	c	
2798	hydr.,	f	
2934	hydr.,	f	
2937	fr,	f	
3053	fr,	f	Diatectic rocks
3257	fr,	c	
Kunklerwald			
13.7		c	paragneiss, anatectic
20.		f	
26.3		c	granulitic paragneis
100.6		f	
153.7		c	
164.7		f	
280.		c	
Geschahse			
84.2		c	syenite
94.65		c	metablastic orthogneiss
106.2		c	metablastic flasergneiss
123.7		f	granite

The results of the measurements of the bulk moduli are shown in figure
2, where the volume of the gneisses of Urach and Kunklerwald were plotted as a function of pressure. The compression is between 2 and 4% at 20
kbar.

The bulk moduli listed in table 2, lie between 400 and 800 kbar. The
fine textured Urach gneisses from 2308 m and from 1950 m have anisotropic bulk moduli. The bulk moduli are lower parallel than perpendicular
to the bedding. The sample from 3053 m depth however shows no anisotropy
in its bulk modulus. This may be due to its lack of texture (Stenger,
1982) with respect to the more anisotropic rocks in table 2.

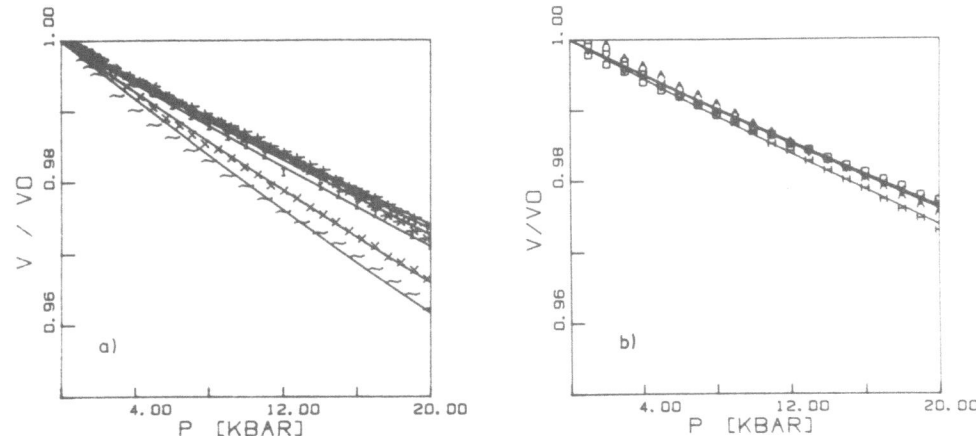

Figure 2: Compression of dry core samples. a) Urach at 1635m(*),
1950m(+), 2125m(-), 2308m(o), 2934m(x), 3053m(I) and b) Kunklerwald at
20mS (A), 100.7mS(V), 164mS(o) and 280mS(H). S= indicates compression
measured perpendicular to the borehole axis, all other are due to measurements parallel to the borehole axis. Solid lines= fit of Eq. (2) to
the data.

Table 2: Bulk modulus of dry core samples, parallel and perpendicular to borehole axis. Stan-
darddeviation in parenthesis.

depth m	bulk modulus K_o kbar parallel	perpendicular
Urach (parallel = perpendicular to the texture)		
1635	682(3)	
1853		586(3)
1950	721(2)	557(3)
2125	473(3)	
2308	705(3)	626(7)
2934	539(2)	
3053	646(3)	647(3)
Kunklerwald (parallel = diagonal to the texture)		
20	785(9)	
100.7	788(3)	
164	796(7)	
280	714(5)	

The complex electrical resistivity was measured in the frequency range
between 10^{-4} and 10^5 Hz. It was determined as a function of pressure
up to 500 bar for the Urach gneiss from 2125 m and for the eclogite .
All samples were saturated with distilled water. Figures 3 and 4 show
the measured dispersion of the resistivity from both samples under
pressure plotted as magnitude and phase of the complex resistivity
versus frequency. In all measurements the resistivity decreases with

increasing frequency. The amount of this decrease depends on pressure.

URACH 2125 M 100 % WET

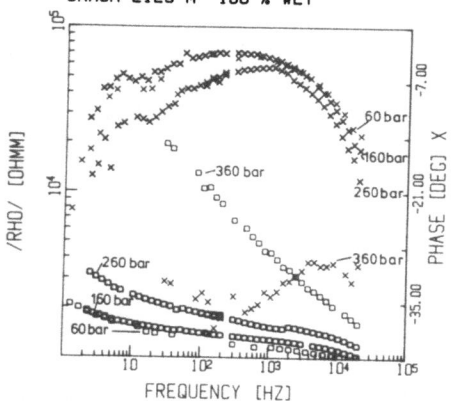

EKLOGIT (SAUALPE) 100 % WET

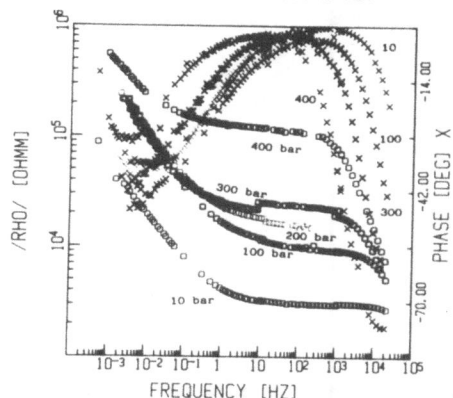

Figure 3 Magnitude and phase of the complex resistivity of saturated gneiss Urach 2125 m with a 2-pole arrangement at different pressures.

Figure 4 Magnitude and phase of the complex resistivity of saturated eclogite with a 2-pole arrangementatdifferent pressures.

Characteristic of figures 3 and 4 is a maximum in the phase between 10 and 1000 Hz with the exact frequency depending on sample and pressure. Figures 4 shows a plateau in the magnitude at about this frequency also. Following this plateau the magnitude of the complex resistivity increases more rapidly with decreasing frequency. Pressure influences both the frequency at which phase maximum well as the magnitude of the resistivity.

These dispersions are due to relaxation processes during the measurements. A simple description of an electric relaxation process is an arrangement of a capacitance and a resistance. The dispersions in figures 3 and 4 can be fitted with an equation of an equivalent circuit of 2 arangements of resistance and capacitance with the restriction that the capacitance is itself frequency dependent. Equation 3 calculates a relaxation process of a complex resistivity with 3 parameters, resistivity, relaxation time and exponent n.

$$\rho^*_{calc} = \left(\frac{1}{\rho_i} + \frac{\tau_i}{\rho_i} \cot\left(\frac{n_i \pi}{2}\right) \omega^{n_i} + i \frac{\tau_i}{\rho_i} \omega^{n_i} \right)^{-1} \quad (3)$$

Semicircle are observed for each relaxation process if the imaginary part is plotted versus the real part of the complex resistivity. Figure 5 and figure 6 show such a Cole-Cole-plot of the eclogite data and of the data of the Urach gneiss from 2125 m as a function of pressure together with a measurement of the complex resistivity in the 4-electrode arrangement at atmospheric pressure.

Characteristic for the 2-pole-dispersions in figure 5 and figure 6 is a minimum which separates two incomplete semicircles. The second semicircles in figure 5 and figure 6 at the low frequency side of the dispersion are missing in the 4-pole measurement. One can assume therefore, that only the high frequency relaxation process is due to the rock and can be called bulk effect. The bulk effects of the Urach gneiss from 2125 m differ from the bulk effects of the eclogite overlaying only a quarter circle in the Cole-Cole-plot. This shows that there are different frequency dependencies of the resistivity in these two rocks.

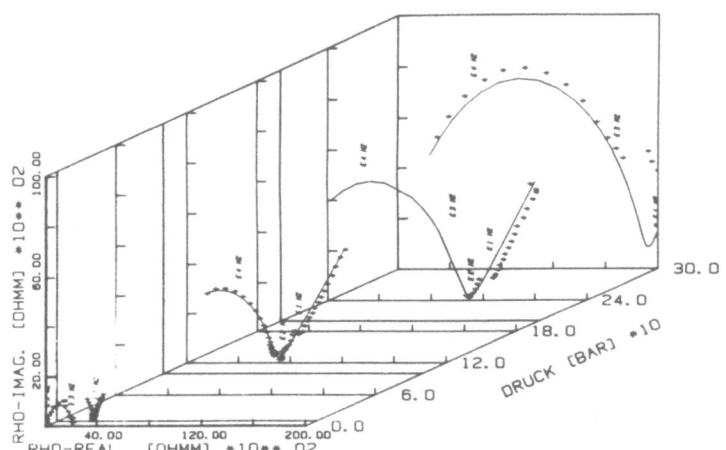

Figure 5 Cole-Cole-plots of the complex resistivity of saturated eclogite as a function of pressure measured with a 4-electrode arrangement at 0 bar and a 2-electrode arrangement at higher pressures. Solid line= fit of Eq. (3).

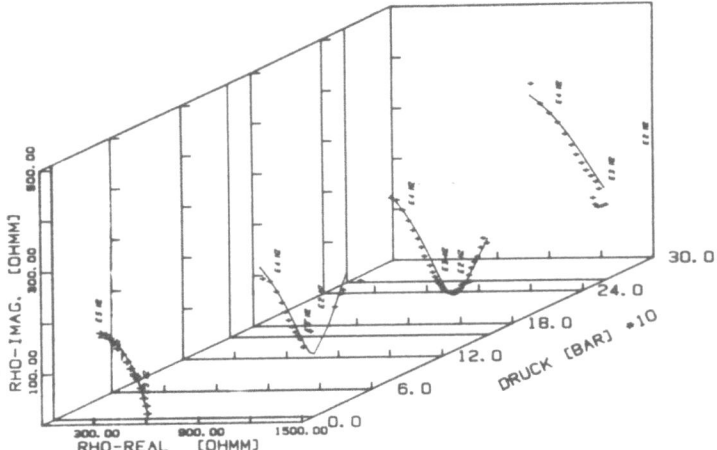

Figure 6 Cole-Cole-plots of the complex resistivity of the saturated gneiss Urach 2125 m as a function of pressure measured with a 4-electrode arrangement at 0 bar and a 2-electrode arrangement at higher pressures. Solid line= fit of Eq. (3).

Table 3 lists the parameters of the bulk effect of the measured data of several samples. The complex resisitivity at a given frequency can be recalculated with equation (3). For example the measured complex resistivity P* of the Urach gneiss from 2125 m, using the 4-pole measurement technique at 10 Hz yields P*= (536.2,0.25) Ohmm with P_B= 550 Ohmm, T_B= 1.0E-4 and n_B= 0.68 .

Table 3 Bulk-resistivity, bulk-relaxationtime, and bulk-exponent of saturated rocks at several pressures. Apparatus: (A)= 2 pole-pressure cell (NOVER et al., 1985); (4)= 4 pole-cell and (2)= 2 pole-cell (without pressure).

Sample	appa-ratus	at pressure bar	frequency decades log(v/Hz)	resis-tivity Ohm m	relaxa-tion time sec	exponent
Urach (parallel to the texture)						
2125	A	60	0...5	1200	3.2E-4	.60
		160	0...5	1300	3.1E-4	.60
		260	0...5	1600	4.7E-4	.58
		360	0...5	3100	1.2E-3	.60
	4	1	1...5	550	1.0E-4	.68
Kunklerwald (diagonal to the texture)						
13.7	2	1	-3...5	1.6E4	3.2E-4	.77
20.6	2	1	-3...5	5.7E4	8.2E4	.80
26.3	2	1	-3...5	670	4.0E-5	.72
100.6	2	1	1...5	1.4E5	2.0E-3	.77
153.7	2	1	1...5	3400	1.8E-3	.54
164.7	4	1	1...5	1.3E4	5.7E-4	.75
280	4	1	1...5	1.1E4	7.8E-4	.71
eclogite	A	10	-4...5	2800	7.1E-6	.93
		100	-4...5	9900	3.6E-5	.91
		200	-3...5	1.7E4	8.0E-5	.80
		300	-3...5	2.0E4	3.0E-4	.76
		400	-3...5	1.0E5	5.6E-4	.80
	4	1	1...5	2000	4.3E-5	.92

The permeability of 22 samples from Urach and Kunklerwald was determined as a function of effective confining pressure (= confining pressure minus pore pressure). Figure 9, 10 and 11 show the results of the cores of the drillings. Equation (4) was fitted to the data yielding 2 parameters, the permeability B_0 at 0 bar and the permeability-decrease-pressure P_0.

$$B = \frac{B_0}{(1 + \frac{P}{P_0})^3} \qquad (4)$$

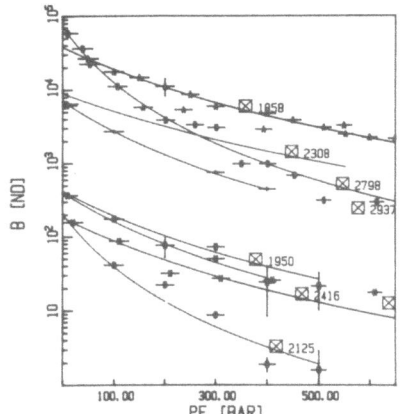

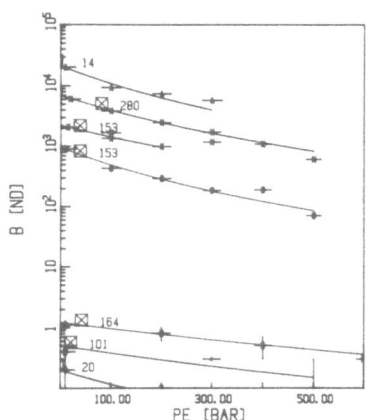

Figure 7 Permeability of Urach core samples as a function of pressure. The symbol X marks the pressure at depth from which the core was obtained using pressure measured in the borehole (Haenel, 1982). Solid lines= fit of eq. (4).

Figure 8 Permeability of Kunklerwald core samples as a function of pressure. The symbol X marks the pressure at depth from which the core was obtained assuming P= pgh and a p of 2.7 g/cm^3. Solid lines= fit of eq. (4).

In figure 7 the permeability scale comprises 5 orders of magnitude. One can clearly see a decrease of permeability with increasing pressure. Figure 7 shows that there are higher permeable rocks under dense horizons for example in 2798 m and 2416 m. Higher permeability appears in hydrothermally altered rocks (1853 m, 2308 m, 2798m), but not all hydrothermally altered rock have a higher permeability (2125 m). Also the pressure dependence is different. The permeability of the gneiss 2798 m is decreased at 500 bar 2 orders of magnitude, but the gneiss 2308 m only 1 order.

Figure 8 shows that the Kunklerwald gneisses comprise 6 orders of magnitude in permeability. Comparing these results with table 1 shows higher permeability for coarse grained rocks than for fine grained rocks.

Hydrostatic pressure is not the only possible stress on a rock in depth. In figure 9 are the results of permeability measurements on two rocks at 500 bar confining pressure with increasing uniaxial pressure plotted.

Figure 9 shows two different behavior of rocks in an anisotropic stress regime. The permeability of the coarse grained gneiss from 2416 m increased at 2800 bar about one order of magnitude while the permeability of the fine grained gneiss from 2937 m decreased more than one order of magnitude.

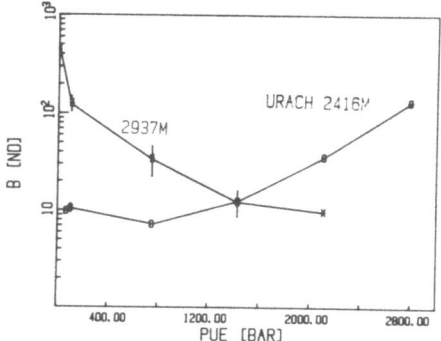

Figure 9   Dependence of the permeability on uniaxial pressure P_{uni}
500 bar confining pressure P_C ($P_{UE}= P_{uni}-P_C-P_i$) for Urach gneisses fr
2416 m and 2937 m.

To investigate the permeability of cut samples under pressure,
samples were taken from the same fine-grained granitic rock specim
with a low anisotropy of grain alignment and crack orientation.

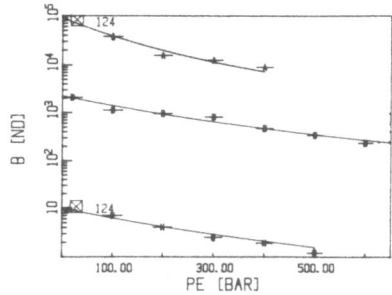

Figure 10   Permeability as a function of effective confining pressu
from 3 samples of the Geschahse granites from 123.7 m. N= intact,
axial cut and D= diagonal cut sample reset again. Solid lines= Fit
eq. (4).

The lowest curve in figure 10 shows the micropermeability of t
intact sample. The sample sawed parallel to the axis and through t
middle showed 4 orders higher permeability than the intact sampl
Pressures higher than 700 bar are necessary to obtain the micropermeab
lity for such a sample. The permeability of the diagonal cut sample li
between the above two data sets.
How much inelastic deformation remaines in the pore space of a ro
after a pressure cycle ? Figure 11 shows the permeability of t
Geschahse syenite from 84.2 m measured during 4 cycles of increasing a
decreasing pressure. After the 4th cycle there was no more hysteres
observed. The permeability of the unconfined sample decreased from i
original value of 97 ud to a value of 23 ud after the 4th cycle. Furth

it was observed that the permeability becomes less pressure dependent. This is documented too by the parameter P_o of equation (4), which increases from 170 (20) to 540 (30) at the last cycle.

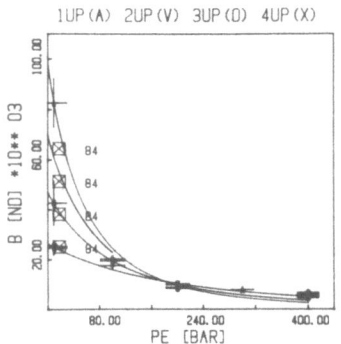

Figure 11   Permeability of the Geschahse syenite from 84.2 m depth during cyclic loading.

4. Pressure dependent permeability, bulk modulus, and complex resistivity

There is a correlation between the permeability-decrease-pressure P_o of eq. (4) and the bulk modulus K_o. In figure 12, K_o for the gneisses from Urach and Kunklerwald is plotted versus the results of fitting P_o with eq. (4) to the permeability data of the same samples.

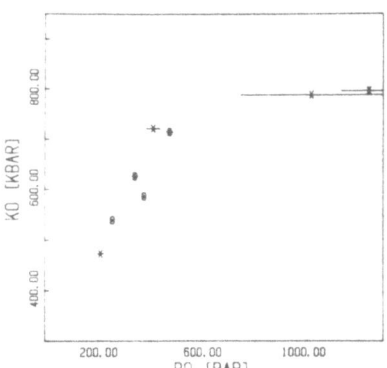

Figure 12   Bulk modulus K_o versus permeability-decrease-pressure P_o of the gneisses from Urach and Kunklerwald (o)= B_o higher than 1 ud; (x)= B_o smaller than 1 ud.

Figure 12 shows a distinct increase of K_O with increasing P_O independent of the permeability B_O at O bar. The relationship between P_O and K_O exists, because both are determined on rocks in which cracks and pores are closed as a function of pressure. High values of those parameters means a weak pressure dependence of the permeability or volume decrease of the rock, while low values of P_O and K_O indicate a strong pressure dependence.

Since the bulk modulus can be determined from seismic compression the practical meaning of figure 12 is that the P_O-values can be determined from bulk modulus as a function of depth . With eq. (4) we can calculate how the cracks and pores in rocks can be opened or closed as a function of pressure.

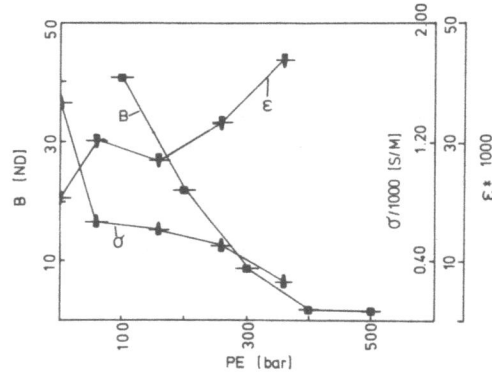

Figure 13   Pressure dependence of the permability B(=●) of a Urach gneiss (dry) from 2125 m. From the saturated sample: bulk conductivity (=▲) and dielectric permittivity   (=▼).

To see the relationship between the permeability and the complex resistivity results we can compare the measured pressure dependence of the permeability and complex resisitivity of the Urach gneiss from 2125 m. From the three resistivity bulk parameters, only a pressure dependence of bulk resistivity and bulk relaxation time was observed (see table 3). The dielectric permittivity, the electrical conductivity, and the permeability of this rock is plotted as a function of pressure in figure 13. For clarity in the figure, conductivity (the reciprocal resistivity) is used.

Figure 13 shows that conductivity decreases and dielectric permittivity increases with decreasing permeability. The decrease in conductivity is the result of reduced pore size. The increased dielectric permittivity means an increase in capacitance since the polarisation responsible for the capacitance requires a longer time with smaller pore channels.

In figure 14 the resistivity of Urach gneisses taken from the borehole log (Haenel, 1982) and of Kunklerwald gneiss (Jenckner, 1986) is plotted versus the measured permeability of the same rock. In the same figure the resistivity measured for saturated rocks with aqua dest.

373

(see table 3) is plotted versus the permeability (see figure 8).

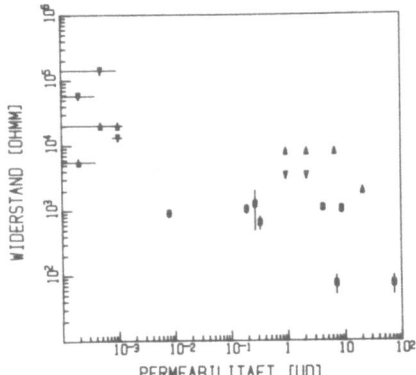

Figure 14   Resistivity versus permeability of gneisses. Resistivity from
borehole logs from Urach (=●) (Haenel, 1982) and from Kunklerwald (=▲)
(Jenckner, 1986). Resistivity from aqua dest. saturated gneisses (=▼).

Figure 14 shows a decrease of resistivity with increasing permeability. The spread of the data is due to uncertainities in the resistivity
of the electrolyte. For the same permeability, the Urach gneisses have a
lower resistivity than the Kunklerwald gneisses. This effect is consistent with observations of Nover et al. (1984), who found NaCl-crystals
at grain boundaries of the Urach gneisses, and of Behr et al. (1986),
who found less saline pore fluids in the Kunklerwald gneisses. Because
of these complications, these results give only a crude impression of
the relationship between permeability and resistivity. Thus before a
general relationship between permeability and resistivity can be derived, measurements should be made on several rock types at defined
condition such as pressure and pore fluid salinity. Particular important
is the latter because of the problems noted here for gneisses from
various depth in Urach and Kunklerwald boreholes.

References

Archie, G. 1942. 'The electrical resistivity log as an aid in determing
    some reservoir characteristics'. Trans. Amer. Inst. Mineral. Met. New
    York. 146 m, 54-62.
Behr, H.J., Gerler, J. (in prep.). 'Inclusions of sedimentary brines in
    post-Variscan mineralisations in the Federal Republic of Germany. A
    study by neutron activation analysis'. Chemical Geology.
Berktold, A., Musmann, G., Tezkan, B., Wohlenberg, J. 1985. 'Electrical
    conductivity studies, Schwarzwald, 2nd International Symposium on
    Observation of Continental Crust through Drilling'. Seeheim, Abstr.-
    70. (Alfred Wegener-Stiftung, Bonn)
Brace, W.F., J.B. Walsh, W.T. Frangos, 1968. 'Permeability of Granite
    under High Pressure' J. Geophys. Res., 73, 2225-2236.

374

Brace, W.F. 1971. 'Resistivity of satured at Crustal Rocks to 40 km Based on Laboratory Measurements; The structure and Physical Properties of the Earth's Crust' Geophys. Monogr. Ser., 14, AGU, Washington D.C., 265-273.

Cole, K.S., Cole, R. 1941. 'Dispersion and adsorption in dielectries I, alternating current characteristics'. J. Chem. Phys. V9, 341-351.

Duba, A.G. and Shankland, T.J. 1982. 'Free carbon and electrical conductivity in the earth's mantle'. Geophysical Research Letters, Vol. 9, No. 11, 1271-1274.

Fyfe, W.S.; Price, N.J., Thompson, A.B. 1986. 'Fluids in the Earth's Crust'. Developments in Geochemistry 1, Amsterdam.

Haak, V. & Hutton, R. 1986. 'Electrical resisitivity in continental lower crust'. In: Dawson, J.B., Carswell, D.A., Hall, J. & Wedepohl, K.H. (eds.) The Nature of the Lower Continental Crust, Geological Society Special Publication No. 24, 35-49.

Haenel, R. 1982. The Urach Geothermal Project. Schweizerbart'sche Verlagsbuchhandlung, Stuttgart.

Jenckner, B., Schädel, K. 1986. 'Geothermievorbohrungen im Mittleren Schwarzwald für das kontinentale Tiefbohrprogramm der Bundesrepublik, Zwischenbericht 1984/85'. Geologisches Landesamt Baden-Würtenberg, Freiburg i.Br..

Knapp, R.B., Knight, J.E. 1977. 'Differential Thermal Expansion of Pore Fluids: Fracture Propagation and Microearthquake Production in Hot Pluton Environments.' J. Geophys. Res., 82, No.17, 2515-2522.

Murnaghan, F.D. 1951. Finite Deformation of an Elastic Solid, Wiley & Sons, London.

Nover, G., E. Hinze and G. Will 1984. 'Elektrische Leitfähigkeitsmessungen an Gesteinen in Abhängigkeit von Druck, Temperatur und Gesteinsstatus' BMFT Forschungsbericht T 84-279, Fachinformatiohszentrum Karlsruhe.

Nover, G., E. Huenges and G. Will 1985. 'Petrophysical Properties and Electrical Conductivity of Core Samples from the Research Borehole Konzen, Hohes Venn (West Germany)'. N. Jb. Geol. Paläont. Abh., 171 (1-3), 169-181.

Olhoeft, G.R. 1981. 'Electrical properties of granite with implications for the lower crust'. J. Geophys. Res., 86, 931-936.

Olhoeft, G.R. 1985. 'Low-frequency electrical properties'. Geophysics, Vol. 50, No. 12, 2492-2503.

Shankland T.J. and Waff, H.S. 1974. 'Conductivity in Fluid-Bearing Rocks'. J. of Geophys. Res. 79, No. 32, 4863-4868.

Shankland, T.J. and WAFF, H.S. 1977. 'Partial Melting and Electrical Conductivity Anomalies in the Upper Mantle'. J. of Geophys. Res. 82, No. 33, 5409-5417.

Schön, J. 1983. Petrophysik, Enke Verlag, Stuttgart, 78ff.

Trimmer, L., Bonner, B., Heard, H.C., Duba, A. 1980. 'Effect of Pressure and Stress on Water Transport in Intact and Fractured Gabbro and Granite'. J. Geophys. Res. 85, 7059-7071.

Waff, H.F. 1974. 'Theoretical Considerations of Electrical Conductivity in a Partially Molten Mantle and Implications for Geothermometry, J. Geophys. Res. 79, No. 26, 4003-4010.

Weast, R.C. 1974. Handbook of Chemistry and Physics, 55th Edition, CRC Press Cleveland, Ohio, F-47ff und F-16ff.

Will, G., E. Hinze and G. Nover 1983. 'Porosity, Electrical Conductivity and Permeability of Rocks from the Deep Drilling Urach 3 and the Hot Dry Rock Project of Falkenberg (West Germany)'. J. Geomagn. and Geoelectr., 35, 787-804.

Zoback, M.D. and J.D. Byerlee 1975. 'The Effect of Microcrack Dilatancy on the Permeability of Westerly Granite.' J. Geophys. Res., 80 (5), 752-755.

Acknowledgement

This work was supported by the Bundesministerium für Forschung und Technologie (BMFT), Bonn, West Germany, under contract 03E-6187-A, which is gratefully acknowledged. We also wish to thank G.Nover, Bonn and Dr. A. Duba, Livermore, for many helpful discussions and critical reading of the manuscript.

Rock-fluid interaction: interpretation by   zeta potential   and   complex
resistivity measurements

G. Nover and G. Will
Mineralogisches Institut der Universität Bonn, Lehrstuhl für
Kristallographie, Poppelsdorfer Schloß, D-5300 Bonn 1,
West Germany

Abstract: The   petrophysical   parameters   porosity,   permeability,
compressibility, thermal expansion and   zeta potential were   determined
for selected core samples from the Falkenberg granite massif in northeastern Bavaria, FRG. This rock status   determination forms the   basis
for   the   interpretation   of   the   complex   electrical   resistivity
measurements as a function of frequency   (.0001 Hz up to 100 kHz).   The
measurements were   performed   on   cylindrical   core   samples.   In situ
conditions were obtained in an autoclave, with pressures from 10 bar up
to 150   bar   at   room   temperature. Solutions   of   NaCl   and   KCl   with
different molarity, and distilled water were used as pore fillings. The
observed shift   in   the relaxation   time   was attributed   to   different
polarization processes   whose existence   depends on   saturant type   and
concentration.

1.0 Introduction

The electrical   properties   of crustal   rocks   strongly depend   on   the
chemistry of the natural formation waters in the pores, the pore volume
available, and   the temperature   and   pressure conditions   (Keller   and
Frischknecht 1966; Schön   1983; Duba   et al   1988). In   rocks with   low
porosity (< 5   vol. %) two   processes contribute to   the total   current
transport (σ_{rock}),   conduction current (σ_{el}) and   dissipation   current
$(\sigma_{surface})$.

$$\sigma_{rock}=\sigma_{el}+\sigma_{surface}$$

Both, the conduction current, due to the transport of charge   carriers,
e.g. cations   and anions,   and the   dissipation current   are   frequency
dependent as   a result   of drifting   of the   ions, liquid   viscosities,
local forces and the   interaction between the dipoles|of the   formation
water (e.g. H_2O) and   the charges   on the   inner surface   of the   pore
structure (Hasted 1973; Olhoeft, G.R., 1986). The displacement   current
is associated with time changes   in the distribution of bound   charges,
that are not free to drift in the electrolyte (Anorgan and Madden 1977).
An orientation   of the   dipoles (Fuller and   Ward 1970,   Cammann   1973;

377

D. Bridgwater (ed.), Fluid Movements – Element Transport and the Composition of the Deep Crust, 377–397.
© *1989 by Kluwer Academic Publishers.*

Schultze 1986; Washburn 1982) can be observed in this electrochemical double layer (DL) (Schmickler, 1986). Such polarization effects are found for example at the boundary layers electrolyte-mineral grain (Fig. 1).

As result of the rock-fluid interactions, the resistivity of the core samples must be described as a complex quantity. If an ac field is applied to such a system, and the complex resistivity is measured over a wide frequency range, then different polarization effects can be recognized and determined (Armstrong and Firman, 1973; Mansfeld, 1981; Brauer and Piroth 1986; Will and Nover 1986; Duba et al 1988).

In general the phase shift in the high frequency (0.01 - 100 kHz) region is due to the bulk polarization of the sample, and in the low frequency (5 x 10^{-5} - 10Hz) region mainly polarization effects with longer relaxation times which are due to surface polarization effects can be seen. This electrical response of rocks can be modelled by an electric relaxation circuit consisting of a resistor and a capacitor in parallel (Jonscher 1978).

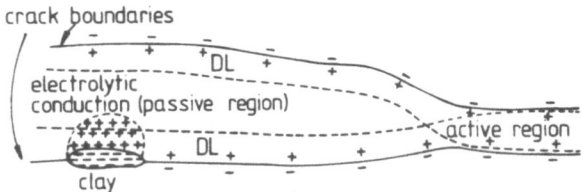

Fig. 1: Schematic drawing of a flat crack in a crystalline rock. Shown are the regions of electrolytic conduction (passive region), and the DL on each side of the crack, where the conduction process is due to a dissipation current (active region, surface conductivity).

For the interpretation of the electrical resistivity data of core samples the knowledge of the petrophysical rock parameters porosity, permeability, density and inner surface is necessary. The zeta potential of the rock forming minerals must be measured as a function of the electrolyte type and concentration too, in order to get the thickness of the DL (Davies and Rideal 1961; Delahay 1965; Ahmed and Maksimov 1969; Hills 1970; Ney 1973) and an independently determined value for the capacity of the DL. We therefore have measured beside the petrophysical rock parameters both, zeta potential (ZP) and complex electrical resistivity (CR) of several core samples from the Falkenberg granite as a function of electrolyte type and concentration and the hydrostatic pressure conditions.

2.0 Experimental investigations and results

2.1 Geological situation and petrography
The core samples are taken from the hot-dry-rock test site in the Falkenberg granite massif (West Germany), part of the pre-Permian

crystalline basement in northeastern Bavaria. The basement consists of a series of metamorphic rocks differing in composition and degree of metamorphism. The last metamorphism occurred about 300 Ma ago. The Falkenberg granite is the oldest of these granites and it forms the southern part of the Fichtelgebirge (Rummel, 1979).
This granite is coarse-grained with alkali feldspar crystals up to 8 cm in length composing about 40 vol.% of the rock. The matrix is formed by quartz (30 vol. %), plagioclase (15 vol. %), biotite (10 vol.%) and muscovite (4 vol. %), with a grain size of 1 to 10 mm. Accessories include apatite, zirkon, rutile, ilmenite and magnetite. The quartz crystals show irregular grain boundaries and a system of cracks with two different orientations. Inclusions in quartz include fluids, muscovite and rutile. All plagioclase crystals are nearly idiomorphic but strongly sericitized. Most of the biotite is altered into chlorite with inclusions of apatite, rutile, ilmenite and magnetite (Nover, et al., 1984).

2.2 Porosity

The porosity of the core samples was measured with a mercury intrusion method on samples 30 mm in diameter and 35 mm in height. For details see (Becker et al 1986). The measurement gives the gradual increase of the total pore volume with pressure, the radial distribution ($\Delta r_i = \delta 2 \cos \Phi / P_i$) and the inner surface area of the sample $s_m = \Sigma s_i = \Sigma (2 \Delta V / r_i)$. These equations are valid for cylindrical capillaries going through the rock sample, with δ knwon surface tension of mercury (0.48 Nm^{-1}) and Φ (140°) the contact angle for mercury on glass we arrive directly at the radial distribution of the pore diameter $2 \Delta r_i$, as a function of pressure p_i. Sm is the sum over all surfaces for each radii class (s_i), this is approximately equal to the sum over all volume elements (ΔV_i) for the correlated mean radii (r_i). Results are included in Table 1.

sample	depth (m)	condition	porosity (vol.%)	inner surface (m^2/g)	density (g/cm^3)
HB4a	25,40	fresh	1,61	1,03	2,674
HB4a	50,00	fresh	1,53	0,76	2,674
HB4	70,40	hydrothermal	1,20	0,73	2,648
HB4	71,00	altered	1,59	1,04	2,644
HB4a	279,55	fresh	1,33	1,01	2,712
HB4a	286,80	fresh	1,46	0,94	2,691

Table 1: Porosity, inner surface area and density for Falkenberg granite core samples from various depth (Nover et al. 1984).

2.3 Permeability

The permeability measurements were performed on cylindrical core samples of 30 mm in diameter and 15 mm in height in an apparatus that is described in Huenges, 1988. The permeability was determined with a pressure transient method. The pressure (P_1) on one side of the sample is kept constant and a lower pressure P_2 is on the other side of the sample. The pressure increase of P_2 was recorded as a function of time up to equilibrium conditions. The permeability can be calculated from these data using the equation:

$$P_2(t)= P_1 + (P_2(0)-P_1)\ e^{-(B/A)/.\ln\beta V)t}$$

where V is the volume and the subscripts 1 and 2 refer to the volumes before and behind the sample, l is the length of the sample, B is the permeability, n the gas viscosity and ß the gas compressibility. All measurements were done with Ar as gas.

The hydrostatic pressure conditions were realized by the controlled variation of the uniaxial pressure P_{uni} and the mantle pressure P_m. The following conditions were controlled: $P_{uni} = P_m$ and $P_{hydrostatic} = $

$P_{confining} = P_{uni} - P_i$, where P_i is the internal pore pressure, which

was kept constant at a level of 49-52 bar.

sample depth	permeability µD at a pressure of (bar)			
	60	100	200	400
HB4a 25,40	-	116,0(3)	54,9(8)	-
HB4a 50,00	5,21(3)	4,93(2)	2,68(2)	1,81(2)
HB4 70,40	94,0(12)	38,0(5)	[110,0(2)	(cracked)]
HB4 71,00	-	44,0(5)	22,0(1)	-
HB4a 279,55	1,85(1)	1,65(2)	1,90(4)	-
HB4a 286,80	-	13,0(1)	2,9(7)	-

Table 2: Permeability data for the Falkenberg granite samples at confining pressures of 60, 100, 200 and 400 bars.

2.4 Zeta potential

The surface potential Φ_o (the wall potential) of the rock forming minerals (Fig. 1) is one of the electrochemical parameters that are responsible for the occurance of polarization processes in the interface solid/liquid. The structure of such liquid-solid interfaces have been investigated by Helmholtz 1879; Gouy 1910; Chapman 1913; Stern 1924; McCafferty and Zettelmoyer 1977). Different layers within the electrochemical double-layer (DL) can be distinguished. Of special interest are the Stern-layer (potential determining layer) in a rock-water system, and the diffuse part of the double layer. The thickness is approximately 80 nm for distilled water and 2 nm for a 2 molar electrolyte. Within the diffuse part of the DL the potential of the "Outer Sternplane" is the zeta potential (ZP), which can be determined by electrophoretic measurements. The electrophoretic method gives a proportionality between the surface charge of a particle and its velocity in an applied dc field when dispersed in an electrolyte.

This method is based on the fact that a mineral particle with a specific surface charge obtains a characteristic velocity in an applied dc field when disperged in an electrolyte. The velocity of the particle is proportional to its surface charge and allows the calculation of the zeta potential (ZP):

$$ZP= 4\pi n(t)/\varepsilon(t) * (L*l)/(t*U)$$

where n is the viscosity of the electrolyte at temperature t, ε is its dielectric permitivity at temperature t, L/t is the velocity of the particle, l is the cell constant of the measuring equipment and U is the strength of the applied field. Debye and Hückel (Debeye, et al., 1923) give a formula to calculate the thickness of the double layer (1/x):

$$1/x=((1000 \ k \ T)/(8\pi e^2 NI))^{1/2}$$

where k is the Boltzmann constant, T is the absolute temperature, ε is the dielectric permitivity, e_2 is the elementary charge, N is the Lodschmid constant, $I= 1/2\Sigma c_i z_i^2$, c_i is the concentration and z_i the charge of the ion.

Increasing electrolyte concentration reduces the zeta potential. The strongest decrease of the ZP of more than 58 % (from - 59.3 to -25 mV) for an increase in concentration of NaCl from 0 g/l up to 0.03 M could be observed for quartz (Table 3, Fig. 2). The decrease of the ZP for the feldspar crystals in NaCl solutions is about 52 % and for mica we have observed almost no change of the value for the ZP in NaCl solutions. A similar behaviour could be detected for the decrease of the ZP for quartz in KCl solutions of different concentrations (Table 3, Fig.3). In this case the decrease of the ZP is about 52 % for quartz in the concentration range 0 g/l up to 0.03 M, 49 % for feldspar and for mica we get a decrease of 39 %.

mineral	dist. water	0,006	0,017	0,03	(M NaCl)
Quartz	59,3	45,9	27,0	25,0	
feldspar ·	49,3	45,9	37,6	23,9	
mica	32,6	32,3	33,8	28,7	
Quartz	59,3	44,1	32,9	25,6	(M KCl)
feldspar	49,3	38,7	36,8	30,3	
mica	32,6	39,5	31,3	20,1	
1/x (nm)	80,0	6,3	3,7	2,7	

Table 3 : Zeta potential for the major minerals comprising the Falkenberg granite. Measurements were performed in distilled water and solutions of different molarity of NaCl and KCl. Thickness of the DL (1/x) according to Deby and Hückel 1923.

A different behaviour is found for solutions of $MgCl_2$, (Fig. 4, Table 4). In this case the ionic charge must be considered. A higher concentration on ions with higher charges in the potential determining layer leads to a rapid decrease of the ZP even at low electrolyte concentrations. At a concentration of 1.0 g/l we measure only a ZP of 17.2 mV for quartz, 14.4 mV for feldspar and 10.5 mV for mica. This indicates that the point of zero charge of the minerals above will be found at low electrolyte concentrations. A measurement of this point was not possible because the current in the ZP measuring cell increased very rapidly with increasing electrolyte concentration. This behaviour holds also for the higher electrolyte concentrations in NaCl and KCl if the salt concentration is above 2.0 g/l.

c (g/l)$MgCl_2$	,005	,01	,1	,2	,35	,1
quartz	40,5	39,5	28,8	21,3	24,5	17,2
feldspar	32,9	31,4	28,9	20,1	22,2	14,4
mica	35,1	30,4	23,8	14,4	14,9	10,5
1/x (A)	450,0	310,0	100,0	72,0	54,0	32,0

Table 4: Zeta potential in (-mV) for the major minerals comprising the Falkenberg granite in solutions of $MgCl_2$ in water.

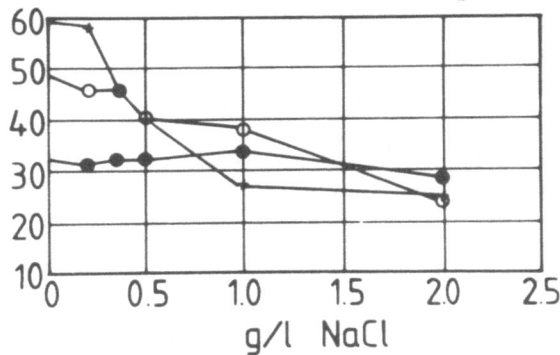

Fig. 2 Zeta potential for the major comprising minerals of the Falkenberg granite (+ quartz, . mica, o fieldspar) measured in solutions of NaCl of different molarities.

c(g/l)	,1 NaCl + ,1 KCl	,25 NaCl + ,25 KCl
quartz	39,0	36,7
feldspar	45,1	42,1
mica	32,4	30,0

Table 5: Zeta potential in (-mV) of the main mineralogical components of the Falkenberg granite in solutions of equal amounts of NaCl and KCl.

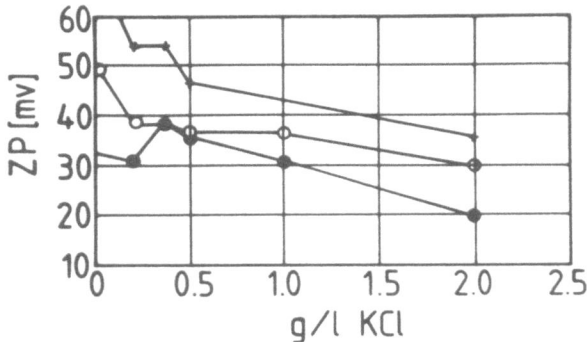

Fig. 3 Results of the zeta potential measurements in KCl solutions
(same symbols for the minerals as in Fig. 2).

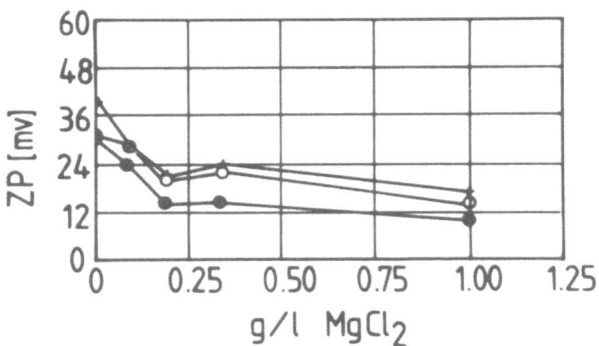

Fig. 4 Results of the zeta potential measurements in $MgCl_2$.

3.0 Complex electrical resistivity
The frequency dependent complex electrical resistivity measurements
were performed on different core samples with dimensions of 16 and 30
mm in diameter and 3 and 10 mm in height. An electrochemical cell (EG +
G, K47) was used for the determination of the transfer resistance (R_b)
and the capacity of the electrochemical double layer (C_{dl}) for
different types and concentrations of electrolytes at 1 bar confining
pressure. The measurements under higher pressure conditions (up to 500
bar) were performed in a pressure vessel (Nover et al 1984; 1987).
These samples had a diameter of 30 mm and a thickness of approximately
10 mm.
The frequency dependent complex resistivity measurements were done in
the frequency range between 10^{-5} Hz upto 10^5 Hz using an AC impedance
system (EG + G). Between 5 Hz and 10^5 Hz a lock-in analyzer (LIA) was
used, while at frequencies below 10 Hz a potentiostat with a Fast
Fourier Transform (FFT) technique was used to measure the impedance
(Fig. 5). M 178 is an electrometer. M 173 a potentiostat operating in
the frequency range 10^{-5} up to 10 Hz using a Fast Fourier Technique,
and M5206 is a lock in analyzer (LIA). M276 is an IEE 488 Interface,

384

controlling via the Apple computer both, the Lock in Analyzer and the potentiostat. The connections between M173 and M5206 are for the signal inputs and the frequency oscillator.

The complex input signal experiences a phase shift due to the electrical characteristics of the rock sample (Table 6). If we describe the electrical phenomena in the frequency domain, then the input voltage is E(x,w) where the vector x denotes the space dependence, and w= 2πf. The electric field input E(x, w) is separated into two compounts, the real part (conduction σ(w) of current $I_c(x,w)$) and the imaginary part (dissipation I_d(iε K(x, w))). In this way the sample is described to be linear, homogeneous, isotropic and time invariant, but each of the electrical parameters is still permitted to be a complex function of frequency. The physical meaning of σ*(w) is the representation of transport of free charge, but current density and electric field are not necessarily in phase. This means that the "free charges" do not follow instantaneously the variations of the applied electric field.

The physical meaning of ε*(w) is the description of various electric polarizations, which are complex values. Their changes in time

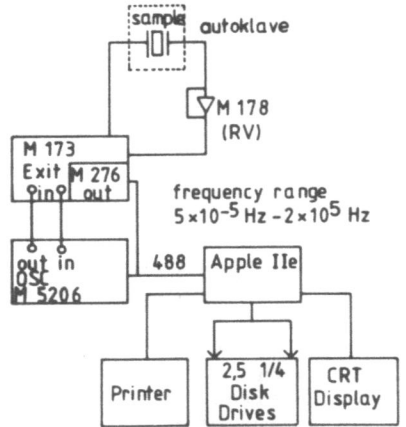

Fig. 5: Principle of the equipment for the measurement of the frequency dependent complex electrical resistivity of core samples

contribute to the total observed current density. The complex admittance Y* of the sample can therefore be described by the sum of the real and imaginary part of the admittance:

$$1/Z* = Y* = Y' + iY''$$

where Z*= the impedance and Y' and Y'' are the real and imaginary part of the admittance. For the interpretation of the complex response of the sample simple equivalent circuit models are used (Fig. 6c), they allow the determination of the values of the resistor and the capacitor

frequency (Hz)	physical/chemical effect	measuring techniques	
		field	laboratory
10^{-5}	ion conductivity	SP	DC
10^{-3}	ion exchange	MT	FFT
10^{-1}	interfacial phenomena	CR	FFT
	oxidation-reduction	IP	FFT
10^{1}	diffusion/migration	AMT	LIA
10^{3}		VLF	LIA
	ice relaxation	AAMT	LIA
10^{5}	H_2O in layers	Radar	

Table 6: Frequency range over which various physical/chemical
processes may be observed and techniques which can be used to
measure them in the field. Explanation of abbreviations:SP=
streaming potential, MT= magneto tellurics, VLF= very low
frequency electromagnetics, IP= induced polarization AAMT=
active audio magneto tellurics, DC= direct current, FFT= fast
fourier transform, CR= complex resistivity, LIA= lock in
analyser

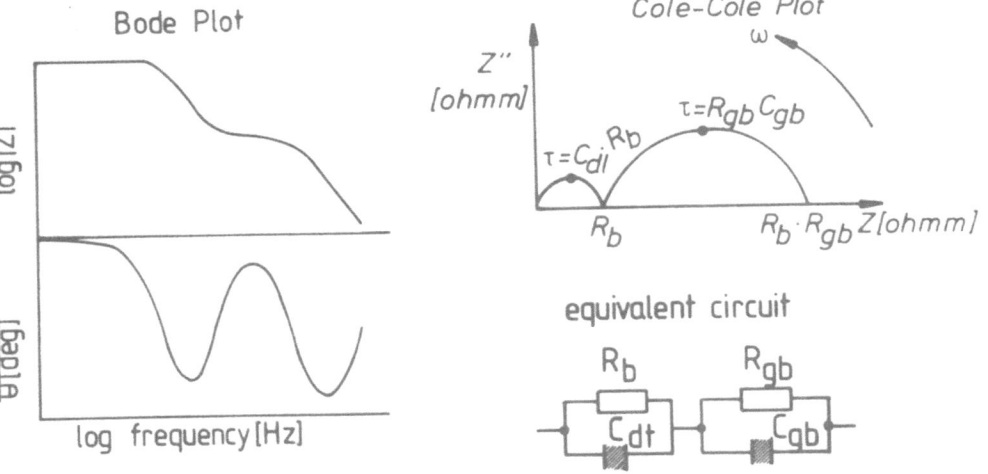

Fig. 6 a-c: Complex electrical resistivity data shown in a) a Bode
plot where the absolute value of the impedance and the
phase angle are plotted as a function of frequency, b)
Cole-Cole plot where the real part is plotted versus the
imaginary part of the impedance, and c) in terms of an
equivalent circuit model consisting of an array of RC
elements in series. The indices b= bulk, gb= grain
boundary and dl= double layer

(Grahame 1952; Mund 1986), for one relaxation process. If the relaxation times τi of the different polarization effects differ by about an order of magnitude, then we can expect well separated semicircles. In such an equivalent circuit model represent the parameters Zi, Ci and τi (i= 1, 2,...) resistors, capacitors, and relaxation times for the different processes (Cole and Cole 1941; Mason et al 1974; Sheppard and Grant 1974; Jonscher 1978). The resistivity values can be read at the intersection of the semicircles with the real axis, while the relaxation times τi can be read at the top of the semicircle where $w_{max} = 1/C*R$ or $w * \tau = 1$.

$$Z*_{calc} = \Sigma \ (1/Zi + \tau i/Zi \cot (n_i \pi/2) + w^n + i \tau_i/Z_i^n)^{-1}$$

3.1 Results

The chemical and mineralogical homogeneous Falkenberg granite samples are marked by different porosities, pore radii distributions, inner surface areas and permeabilities. Of special interest was the measurement of the electrical data, using the same type of pore saturand, in order to detect deviations of the electrical response. These deviations should then be compared, and if possible, be correlated with the parameters of the rock status. In general the volume resistivity increases with decreasing porosity Table 7. The phase angle increases with increasing inner surface area. The time constant τ_{gb} decreases for the relaxation process with decreasing surface area.

sample	25.40	50.00	70.40	71.00	279.55	286.80
R_{Archie} ($\times 10^4$)	7.1	7.7	25	14	9.5	8.3
R_{bulk} [$\times 10^4$]	0.22	0.76	0.96	0.32	0.3	0.23
τ_b (sec) ($\times 10^{-5}$)	7.9	4.2	6.3	0.32	–	18

sample	25.40	50.00	70.40	71.00	279.55	286.80
R0.5 M NaCl	306	127	–	217	–	–
	245	265	390	250	–	–
R2M NaCl	50	25	39	56	–	–
(ohm*m)	51	55	81	52	–	–

Table 7: Bulk resistivity data, (R_b) bulk relaxation times (τ_b) of the Falkenberg samples. Pore filling were distilled water 0.5 M and 2.0 M NaCl. Parameters of the Archie equation a= 1.5, m=

1.58, $_{el}$= 1 x 10^5, 0.24, 0.05 ohm-m, resp. porosity data are from Table 1. The time constant τ_b for the volume relaxation process was calculated from the data with distilled water as a pore fluid.

Results of the resistivity measurements in distilled water with a resistivity of 1 x 10^5 ohmm on the Falkenberg core samples from 25.4, 50.00, 70.40, 71.00 m in depth are compiled in Fig. 7. The phase angle in frequency range 1 x 10^4 Hz up to 1 x 10^5 Hz shows two polarization processes. In the high frequency region decreases the phase with increasing porosity, and the time constant of this volume polarization process is decreased. In the low frequency region dominates the contribution of surface conductivity, due to the decrease in the crack width and the interaction of the DL on each side of the crack. Figure 8 shows the results of the measurement on sample 70.4 m using distilled water, .5 M NaCl and 2 M NaCl as a pore electrolyte. In this case the pore geometry is fixed, but the thickness of the DL is decreased with increasing electrolyte concentration. The rate of interaction between the DL's on each side of the crack is thereby reduced.

NaCl	τ_b (sec)*10^{-5}	τ_{gb} (sec)*10^{-2}	R_b (ohm m)*10^3	R_{gb} (ohm m)*10^2
dist. water	4.1	2.9	7.6	–
.00005	4.1	2.8	5.4	3.9
.0001	1.5	2.7	3.8	1.9
.001	0.78	2.7	1.9	2.2
.01	0.44	0.74	0.82	3.5
.1	0.29	0.42	0.36	2.8
.5	0.22	0.10	0.13	1.4
KCl				
0.001	0.71	0.014	1.5	17.2
0.01	0.40	0.11	0.95	17.2
0.1	0.16	0.32	0.79	0.8

Table 8: Parameters of the equivalent circuit for Falkenberg sample HB 50.00 in NaCl and KCl solutions of different molarities.

The decrease of the impedance as a function of the electrolyte conductivity covers more than 3 orders of magnitude and cannot be explained by the Archie equation (Table 7). The phase angle decreases in the high frequency part of the spectrum, and increases in the low frequency region with increasing electrolyte concentration. The thickness of the electrochemical DL is thereby decreased (Table 3) and the surface area involved in the interfacial polarizations increased. The relaxation time for the bulk effect of the sample is thereby decreased in the case of 2 M NaCl electrolyte pure ohmic conduction and no dissipation current is from approximately 100 kHz down to 0.5 Hz.

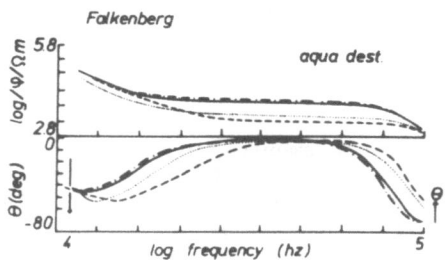

Fig. 7: Bode plots of the electrical data of the Falkenberg granite samples HB 25.4 (dashed line), 50.00 (dash and point), 70.4 (solid) and 71.00 (dotted line) measured with distilled water as a pore fluid

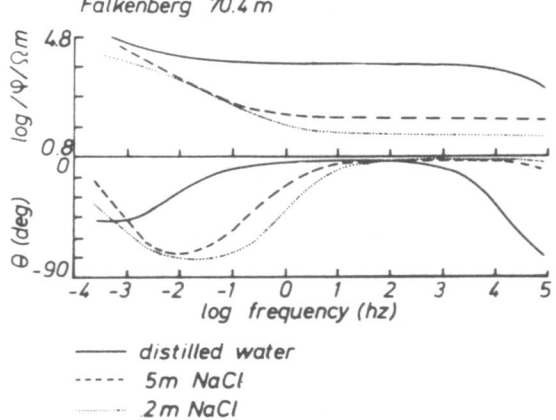

Fig. 8: Bode plots of the Falkenberg sample HB 70.4 with electrolytes of different molarity in the pore space of the sample

An increase of the confining pressure on core samples reduces the permeability due to a closing of the cracks. The electrochemical result is an overlapping and an interaction of the DL's on each side of a crack. Now two effects
a) surface conductivity and b) decreasing pore volume
are operating contrary. The closing of the cracks increases the rock resistivity, but the overlapping DL's increase the contribution of the surface conductivity with the result, that the impedance of the sample (HB 70.40 in Fig. 9) is decreased up to a pressure of 150 bar. In this way increasing load produces an exchange reaction between the two double layers on each side of the crack. The electrical effect is a shift of the time constant of the relaxation process to shorter times, when the double layers begin to overlap. A further increase of the confining pressure leads to a rapid increase of the impedance caused by a decrease of the conduction paths. An increase of the pressure decreases the phase angle in the high frequency region, thus decreasing the relaxation time for the samples bulk polarization process. In the low frequency region one can observe an increase of the phase angle

under increasing pressure conditions, too. The relaxation times due to surface effects are thereby increased. A further increase of the confining pressure leads to a rapid increase of the impedance caused by a closing of the conduction pathes in the pore system.

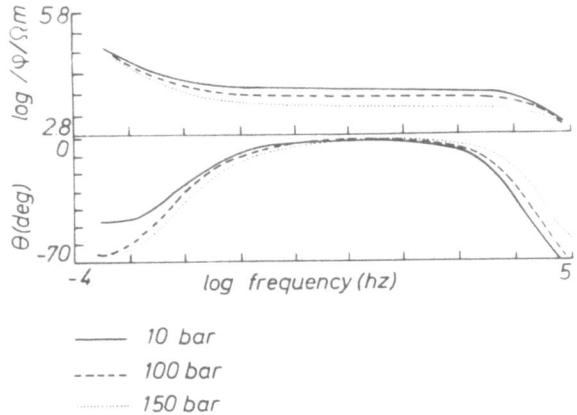

Fig. 9   Bode plot of the Falkenberg sample 70.40 with distilled water as a pore saturand, measured under different confining pressure conditions

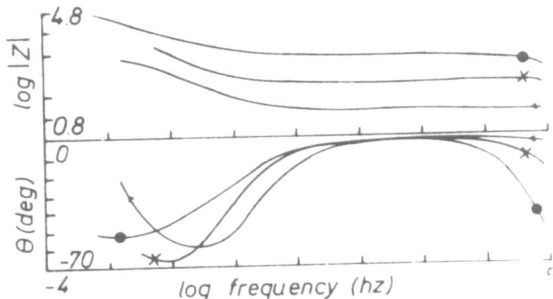

Fig. 10 Bode plot of the Falkenberg sample 25.40 m (· distilled water, x 0.5 m NaCl, < 2 m NaCl) at a fixed pressure of 10 bar

4.0 Discussion

The conduction process in rocks with a narrow crack system is influenced by the degree of interaction of the double layers on each side of a crack. The time constant of the volume and grain boundary polarization is varied if either the characteristics of the electrolyte are changed, or if the geometry of the pore space is decreased by an external pressure. To separate the bulk polarization effect from the surface polarization effects, both, the "thickness" of the electrochemical double-layer (DL), and the crack geometry were varied in our experiments.

Increasing the electrolyte concentration means to increase the contribution of the ohmic electrolytic conduction process. In this way the contribution of the surface conductivity is reduced, due to the decreasing thickness of the double layer. The bulk resistivity data are then comparable to the calculated "Archie" resistivities (Table 7) of the core sample.

The low frequency relaxation is shifted towards lower frequencies with shorter time constants for τ_{gb} with increasing salt concentration in the pore fluid (Table 7, Fig. 10).

An increase of an axial load perpendicular to the crack orientation leads to a closing of cracks. The time constant for polarizations at the inner surface of a rock sample depend on the surface area that is involved in the water/rock interfacial polarization. If we consider an electrolyte of low salinity, then the contribution of surface conductivity can't be neglected, because the thickness of the double layer is strongly increased in low electrolyte concentrations. The surface area involved for that particular polarization is thereby decreased, if the thickness of the DL exceeds the crack width. If the thickness of the immobile part of double layer is in the dimensions of narrow cracks in a rock, then we will get different viscosities if two neighboured double layers overlap (active and passive region in Fig.1).

molarity M KCl	capacity F(ZP)	capacity F(CR)
0.0011	-	8.1×10^{-8}
0.003	9×10^{-13}	-
0.006	1×10^{-12}	-
0.01	-	6.4×10^{-7}
0.017	2×10^{-12}	
0.034	3×10^{-12}	
0.1	-	3.8×10^{-6}

Table 9 Capacities for a plate capacitor (ZP) as calculated from the ZP measurements and complex resistivity data (CR). The factor of geometry for equal surface areas is not regarded.

The mobilities of ions in these regions are different, and this can be measured either by frequency dependent complex resistivity measurements, due to the fact that the relaxation time of polarization processes are different, or by direct measurement of the mobilities of particles. In each of the two experiments such an effect is correlated with a shift of the capacity of the double layer. So one should compare the change of the capacitors from both experiments zeta-potential and complex resistivity (Table 9).

We can see that in both experiments the capacity decreases with increasing electrolyte concentration due to the fact of the decreasing thickness of the double layer.

The results discussed so far have shown that it is possible to distinguish by frequency dependent complex electrical resistivity measurements between the bulk polarization of the sample and the polarizations on the inner surface. Fluid flow phenomena are strongly influenced by the area of the solid/liquid interface in a pore system, and thereby different polarizations should influence the fluid flow characteristics.

Acknowledgement

This work was supported by the Bundesministerium für Forschung und Technologie (BMFT) Bonn, West Germany, under grant O3E - 6187 - A, which is gratefully acknowledged. We also wish to thank Prof. Dr. Al Duba (Lawrence Livermore Laboratories) for helpful discussions.

References

Ahmed S, and Maksimov D (1969) Studies of the double layer on cassiterite and rutile. J Coll Interf Sci 29, 1: 97-104
Anorgan Y and Madden TR (1977) Induced polarization: a preliminary study of its chemical basis. Geophysics 42, 4: 788-803
Armstrong RD and Firman RE (1973) Impedance plane display of the adatom model for metal dissolution/deposition. Electronanal Chem Interf Electrochem 45: 257-266
Becker R, Lentz H, Hinze E, Nover G and Will G (1986) Ein Quecksilberporosimeter zur Charakterisierung mineralischer Stoffe. Berichte der Bunsenges Phys Chem 90: 833-838.
Brauer E und Piroth J (1986) Impedanzspektroskopie in der Elektrochemie. GIT Fachz Lab 6: 533-543
Cammann K (1973) Das Arbeiten mit ionenselektiven Elektroden. Springer, Berlin Heidelberg New York
Chapman DL (1913) Philas Mag 25 6: 475
Cole KS and Cole R (1941) Dispersion and absorption in dielectrics. J Chem Phys V9: 341-351
Davies JT and Rideal EK (1961) Interfacial phenomena. Academic Press New York London 56-107
Debeye P and Hückel E (1923) Phys Z 24: 185
Delahay P (1965) Evolution of ideas on the electrical double layer. Double layer and electrokinetics. John Wiley & Sons, New York
Duba Al, Huenges E, Nover G and Will G (1988) Impedance of Black Shale from Münsterland 1 Borehole: An Anomalously Good Conductor? Geophys J 94 413-419
Fuller BE and Ward SH (1970) Linear System Description of the Electrical Parameters of Rocks. IEEE Transact on Geosci Elect GE-8 1: 7-13
Gouy G (1910) J Phys 9 9: 457
Grahame DC (1952) Mathematical theory of the faradaic admittance. J electrochem Soc 99: 370-385

Hasted JB (1973) Water- a comprehensive treatise. Plenum New
  York vol. 1: pp. 255-458; vol 2: pp 405-458.
Helmholtz H (1879) Wied Ann 7: 337
Hills GJ (ed) (1970) Electrochemistry. Chem Soc, Burlington
  House London, pp 117-167
Huenges E (1988) Messung der Permeabilität von niedrigpermeablen
  Gesteinsproben unter Drücken bis 4 kbar und ihre Beziehung zu
  Kompressibilität porosität und elektrischem Widerstand. PhD Thesis
  Univ Bonn
Jonscher AK (1978) Analysis of the alternating current
  properties of ionic conductors. J Math Sci 13: 553-562
Keller GV and Frischknecht FC (1966) Electrical methods in
  geophysical prospecting. Pergamon, New York
Mansfeld F (1981) Recording and analysis of AC impedance data
  for corrosion studies. Corrosion 36, 5: 301-308
Mason PR, Hasted JB and Moore L (1974) The use of statistical theorey
  in fitting equations to dielectric dispersion data.
  Advanc molec relax proc 6: 217-232
Mc Cafferty E and Zettelmoyer AC (1971) Adsorption of Water Vapor on
  α-Fe$_2$O$_3$. Disc Faraday Soc 52: 239
Mund K (1986) Untersuchung poröser Elektrodenstrukturen mit
  Impedanzmessungen. Dechema Monographien 102 Grundlagen von
  Elektrodenreaktionen VCH Verlagsgesellschaft
Ney P (1973) Zeta Potentiale. Springer Berlin Heidelberg New
  York
Nover G, Hinze E and Will G (1984) Elektrische Leitfähigkeitsmessungen
  an Gesteinen in Abhängigkeit von Druck, Temperatur und
  Gesteinsstatus. Forschungsbericht T84-279,
  Fachinformationszentrum, Karlsruhe
Nover G, Huenges E and Will G (1987) Messung der Frequenzabhängigkeit
  elektrischer Gesteinswiderstände unter in situ Bedingungen.
  Abschlußbericht zum Forschungsvorhaben 03E-6187-A, BMFT
Olhoeft GR (1986) Electrical properties of rocks and minerals.
  Short Course Notes, Golden Colorado
Rummel F (1979) Experimente an einem künstlich erzeugten Riß im
  flachen, niedrig permeablen Untergrund als Grundlage zur
  großtechnischen Gewinnung terrestrischer Wärme. RUB BMFT-ET
  Vorhaben 4150/CEC-Projekt E 8(D)
Schmickler W (1986) Die Doppelschicht in wässriger und nicht
  wässriger Lösung. In: Dechema Monographien 102 Grundlagen
  von Elektrodenreaktionen, VCH Verlagsgesellschaft
Schön J (1983) Petrophysik. Enke, Stuttgart
Schultze JW (ed.) (1980) Grundlagen von Elektrodenreaktionen. Vol 102
  Verlag Chemie Weinheim
Sheppard R J and Grant E H (1974) Alternative interpretations of
  dielectric measurements with particular reference to polar
  liquids. Advances in Molecular Relaxation Processes 6 61-67
Stern O (1924) Zur Theorie der elektrolytischen Doppelschicht.
  Z Elektrochemie 30 508

Washburn J C (1982) Parameterization of spectral induced polarization data and Laboratory and in situ spectral induced polarization measurements: West Shasta copper-zinc district, Shasta, CA. Thesis, Colorado School of Mines, Golden Colerado

Will G and Nover G (1986) Measurment of the frequency dependence of the electrical conductivity and some other petro-physical parameters of core samples from the Konzen (West Germany) drill hole. Annales geophysicae 4 B2 173-182

SUBJECT INDEX

Abitibi greenstone belt 52 53
Acadian granulite facies metamorphism 117-119
Achankovil shear belt, southern Kerala 34
aCO2 (see also under Carbon, CO2)
192
addition of K and Rb to gneisses adjacent to pegmatites 179 281
admittance 384
advective transport of fluids 46
agmatitic complex 238
aH2O (see also under water) 57 97 106 192
airborn radiometric surveys for U 289
Al as immobile element 220
Al enrichment of ultramafic rocks due to fluid movements 187
alkali mobility associated with deformation 151 222 226
allanite   39 42 46 47 161 288
allanite as control of REE distribution 288
allanite mantled by epidote 161
alpine peridotites 121-122
Ameralik (fjord, southern West Greenland) 23 319 320 322 334
Ameralik dykes 321
Ameralik fjord chronology, regional framework 334
Amîtsoq gneisses 319 320 321 323 303 333
   Amîtsoq gneiss average Pb isotope composition 333
   Amîtsoq gneisses - Pb isotopic composition 289 303 333
Ammassalik -see Angmagssalik
ammonium content of biotite and muscovite (see also Nitrogen) 115
amphibolite - charnockite transition zones (see also granulite
   amphibolite transition) 29-38
anatectic granites (see also partial melts) 30 61 105
   anatectic removal of granitic components 61
anatexis as cause for water loss from granulites 106
   anatexis as suggested method for anorthosite formation 90
   anatexis in Scourian complex 98
Angmagssalik "charnockite" complex. 71-94
   Angmagssalik complex, P-T estimates 79
   Angmagssalik complex, fluid inclusions 80 81 82 83 84
   Angmagssalik complex, mineral compositions 78
   Angmagssalik complex, thermal history 87
anhydrous rocks as protoliths to granulite facies gneisses 2
ankerite 253 255
anorthosite 6 9 12 54 55 72 75 76 80 90 254
   anorthosite veins 72 75 76
   anorthosite P-wave velocity 55
   anorthosite formed by anatexis, discussion 90